Rapid & Automated Methods in Microbiology & Immunology: a bibliography 1976-1980

Compiled & edited by Wendy J.Palmer

Advisor: Dr. E.S.Krudy

Published by Information Retrieval Ltd.
London and Washington DC

Information Retrieval Ltd.,
1 Falconberg Court,
London W1V 5FG, England

Information Retrieval Inc.,
1911 Jefferson Davis Highway,
Arlington, VA 22202 USA.

ISBN 0 904147 19 3
Printed in England by Information Printing Ltd.

FOREWORD

This bibliography contains approximately 3300 references from a 5-year period (1976-1980) and reflects the increasing amount of automation that is being introduced into microbiology and immunology. It is the second bibliography published by IRL on the subject, the first being produced in 1976 containing more than 2500 references for a period 1967-1975.

The lay-out of this volume has been simplified and the contents are arranged mainly by techniques and methods. Citations have been allocated to the most relevant section and have been cross-referred only when considered of value, e.g. a citation on purification of a product by chromatography can be found in the 'Chromatography' section, whereas a citation on general purification technique that is not covered by any other section can be found in the 'Purification' section. Citations which are on more than one technique will be found in the most important section and cross-referred to sections on the other mentioned techniques Thus, citations that appear in the more general sections (e.g. diagnostic techniques, purification and concentration, general biochemical and immunological techniques) are those which cannot readily be allocated to a more specific section.

Citations do not include foreign-language titles, but translated English titles appear in square brackets. Key words are used where necessary to enrich the title.

The subject index entries are based upon techniques, taxonomic terms, names of chemicals and other meaningful entries, such as micromethods, purification, media, culture, identification etc.

Diseases caused by infectious agents are indexed under the aetiological agent and not the disease, e.g. syphilis papers can be found entered under 'Treponema pallidum'.

CONTENTS:

CONTENTS: (continued)

LIST OF SOURCE JOURNALS

Acta Biologica et Medica Germanica
Acta Hepato-Gastroenterologica
Acta Medica Okayama
Acta Microbiologica Hellenica
Acta Microbiologica Polonica
Acta Pathologica et Microbiologica Scandinavica, Series B
Acta Virologica (English Edition)
Agricultural and Biological Chemistry
Aktuelle Rheumatologie
American Journal of Clinical Pathology
American Journal of Tropical Medicine and Hygiene
Americal Journal of Veterinary Research
Analytical Biochemistry
Annales Medicales de Nancy
Annales de Microbiologie
Annales de Phytopathologie
Annales de Technologie Agricole
Annali Sclavo Rivista di Microbiologie e di Immunologia
Annals of Applied Biology
Annals of Clinical Laboratory Science
Antibiotiki
Antimicrobial Agents and Chemotherapy
Applied and Environmental Microbiology
APPS Newsletter
Aquaculture
Archiv fur Experimentelle Veterinarmedizin
Archives of Disease in Childhood
Archives of Pathology and Laboratory Medicine
Archives Roumaines de Pathologie Experimentale et de Microbiologie
Archives of Virology
Archivum Immunologiae et Therapiae Experimentalis
Arquivos da Escola de Veterinaria
Australian Journal of Dairy Technology
Australian Veterinary Journal
Avian Pathology

Biochemistry (Washington)
Biology of the Actinomycetes and Related Organisms
Biotechnology and Bioengineering
Bollettino dell'Istituto Sieroterapico Milanese
Bordeaux Medical
Brewer's Digest
British Journal of Haematology
British Journal of Venereal Diseases
British Medical Journal
British Phycological Journal
British Veterinary Journal
Bulletin de l'Academie Veterinaire de France
Bulletin de l'Institut Pasteur
Bulletin de la Societe Royale de Botanique de Belgique
Bulletin of the World Health Organization

Canadian Journal of Comparative Medicine
Canadian Journal of Medical Technology
Canadian Journal of Microbiology
Cancer Research
Casopis Lekaru Ceskych
Chinese Journal of Microbiology
Chinese Medical Journal
Clinica Chimica Acta
Clinical Allergy
Clinical Chemistry
Clinical and Experimental Immunology
Communications in Soil Science and Plant Analysis
Comptes Rendus des Seances de la Societe Biologie et de ses Filiales
Current Medical Practice

Deutsche Gesundheitswesen
Deutsche Medizinische Wochenschrift
Developments in Biological Standardisation
Developments in Industrial Microbiology

Enzyme and Microbial Technology
European Journal of Applied Microbiology
European Journal of Biochemistry
European Journal of Clinical Pharmacology
Experientia
Experimental Cell Biology
FEBS Letters
Florida Scientist
Folia Microbiologica

(continued)

Gastroenterology

Hautarzt
Holzforschung
Hormone Research
Hospital Practice

Immunochemistry
Immunitat und Infektion
Indian Journal of Animal Sciences
Indian Journal of Experimental Biology
Indian Journal of Medical Research
Indian Veterinary Journal
Infection and Immunity
International Archives of Allergy and Applied Immunology
International Biodeterioration Bulletin
International Journal of Applied Radiation and Isotopes
International Journal of Bio-Medical Computing
International Journal of Cancer
International Journal of Systematic Bacteriology
Intervirology
Investigative Ophthalmology and Visual Science

Japanese Journal of Medical Science and Biology
Journal of Animal Science
Journal of Antibiotics
Journal of Applied Bacteriology
Journal of the Association of Official Analytical Chemists
Journal of Bacteriology
Journal of Biological Chemistry
Journal of the Canadian Institute of Food and Science Technology
Journal of Chromatography
Journal of Clinical Endocrinology and Metabolism
Journal of Clinical Microbiology
Journal of Clinical Pathology
Journal of Experimental Botany
Journal of Fermentation Technology
Journal of Food Protection
Journal of Food Science
Journal of General Microbiology
Journal of General Virology
Journal of Histochemistry and Cytochemistry
Journal of Hygiene, Epidemiology, Microbiology and Immunology
Journal of Immunological Methods
Journal of Immunology
Journal of Infectious Diseases
Journal of Invertebrate Pathology
Journal of Laryngology and Otology
Journal of Medical Microbiology
Journal of Medical Virology
Journal of Microscopy (Oxford)
Journal of Neurology
Journal of Nuclear Medicine
Journal of Parasitology
Journal of Pathology

Klinische Wochenschrift
Kolloidnij Zhurnal
Kvansny Prumysi

Laboratory Practice
Lait
Lancet
Leber Magendarm
Lichenologist
Limnology and Oceanography
Magyar Allatorvosok Lapja
Marine Biology
Marine Science Communications
Medecine Tropicale
Medical Journal of Australia
Medical Laboratory Sciences
Medical Microbiology and Immunology
Meditsinskaya Parazitologiya i Parazitarnye Bolezni
Medycna Weterynaryjna
Microchemical Journal
Microscopica Acta
Monatsschrift fur Kinderheilkunde
Mutation Research
Mycologia

Nature
Nederlands Tijdschrift voor Geneeskunde
Netherlands Journal of Plant Pathology
New Phytologist
New Zealand Medical Journal
New Zealand Veterinary Journal
Nouvelle Presse Medicale

(continued)

Nucleic Acids Research

Oceanography and Marine Biology Annual Review
Oral Surgery, Oral Medicine and Oral Pathology

Parasitology
Philippine Journal of Veterinary Medicine
Phytopathology
Plant and Cell Physiology
Plant Disease Reporter
Postepy Higieny i Medycyny Doswiadczalnej
Poultry Science
Prikladnaya Biokhimiya i Mikrobiologiya
Proceedings of the National Academy of Sciences of the United States of America
Proceedings of the Society for Experimental Biology and Medicine
Proceedings of the University of Otago Medical School
Przemysi Fermentacyjny i Rolny

Recueil de Medecine Veterinaire, Ecole d' Alfort, Paris
Research Communications in Chemical Pathology and Pharmacology
Research in Veterinary Science
Revista do Instituto Adolfo Lutz
Revista de Saude Publica, Sao Paulo
Revue Francaise de Transfusion et Immuno-Haematologie
Revue Roumaine de Medicine, Serie Virologie

Sabouraudia
Scandinavian Journal of Immunology
Scandinavian Journal of Infectious Diseases
Schwizerische Medizinische Wochenschrift
Singapore Medical Journal
Soil Biology and Biochemistry
South African Journal of Medical Science
Steroids
Surgical Forum

Texas Reports on Biology and Medicine
Thrombosis Research
Tohoku Journal of Experimental Medicine
Transactions of the American Microscopical Society
Transactions of the British Mycological Society
Transactions of the Royal Society of Tropical Medicine and Hygiene
Transactions of the St. John's Hospital Dermatology Society
Transfusion
Transplantation

Umschau
Union Medicale du Canada

Vestnik Dermatologii i Venerologii
Veterinary Microbiology
Veterinary Parasitology
Veterinary Record
Virology
Voprosy Meditsinskoi Khimi
Voprosy Virusologii
Vox Sanguinis

Water and Sewage Works
Wiener Klinische Wochenschrift

Zentralblatt fur Bakteriologie, Parasitenkunde, Infektionskrankheiten und Hygiene, I (Abt. A)
Zentralblatt fur Veterinarmedizin, B
Zhurnal Mikrobiologii, Epidemiologii i Immunobiologii

Culture, isolation and identification techniques

1-U2 **Start-up of chemostat: application of fed-batch culture.** Yamane,T.; Sada,E.; Takamatsu,T. (Dep. Chem. Eng., Kyoto Univ., Kyoto 606, Japan) *Biotechnol. Bioeng., 21(1), 111-129 (1979)* En;en.

2-U2 **A procedure for culture of cells from mouse tail biopsies.** Lander,M.R.; Moll,B.; *Rowe,W.P. (Lab. Viral Dis., Natl. Inst. Allergy and Infect. Dis., Natl. Inst. Health, Public Health Serv., US Dep. Health, Educ., and Welfare, Bethesda, MD 20014, USA) *J. Natl. Cancer Inst., 60(2), 477-478 (1978)* En;en.

3-U2 **A simple optimization technique for fed-batch culture.** Yamane,T.; Kume,T.; Sada,E.; Takamatsu,T. (Dep. Chem. Eng., Kyoto Univ., Kyoto 606, Japan) *J. Ferment. Technol., 55(6), 587-598 (1977)* En;en.

4-U2 **Indirect methods of automatic substrate dosing for aerobic cultivation of microorganisms.** Stros,F.; Adamek,L.; Rut,M. (Vyzkumny Ustav Krmivarskeho Prum. a Sluzeb, Pecky, Odbor Mikrob. Vyrob, Praha, Czechoslovakia) *Kvasny Prum., 24(6), 128-131 (1978)* Cs;cs,de,en,ru.

5-U2 **Large scale production of human interferon in monolayer cell cultures.** Edy,V.G. (Dep. Hum. Biol., Div. Microbiol., Rega Inst., Univ. Leuven, B 3000 Leuven, Belgium) *Tex. Rep. Biol. Med., 35, 132-137 (1977)* En.

6-U2 **Semi-automatic apparatus for determination of the consistency of microbiological culture media and gel-forming substances.** Costin,I.D. (Pharma-Qualitatskontrolle der Fa. E. Merck, D-6100 Darmstadt 2, Postfach 4119, GFR) *Zentralbl. Bakteriol. Parasitenkd. Infektionskr. Hyg., I Abt. A, 234(4), 536-553 (1976)* De;de,en.

7-U2 **Simple method for identification of plasmid-coded proteins.** Sancar,A.; Hack,A.M.; *Rupp,W.D. (Yale Univ. Sch. Med., Dep. Biochem., New Haven, CT 06510, USA) *J. Bacteriol., 137(1), 692-693 (1979)* En,en.

8-U2 **Recent advances in the laboratory diagnosis of enteric virus infections.** Flewett,T.H. (Reg. Virus Lab., Birmingham, B9 5ST, UK) *In:* New perspectives in clinical microbiology. Brumfitt,W. (ed.) *Publ. by:* Kluwer Publishing Ltd., Forge House, 20 Market Place, Brentford, Middx. TW8 8EQ, UK. 1978 p. 97-108 ISBN 09-033-9335-2 En;en.

9-U2 **A rapid microscale technique for isolation of recombinant plasmid DNA suitable for restriction enzyme analysis.** Klein,R.D.; Selsing,E.; Wells,R.D. (Dep. Biochem., Coll. Agric. and Life Sci., Univ. Wisconsin, Madison, WI 53706, USA) *Plasmid, 3(1), 88-91 (1980)* En;en.

10-U2 **A rapid method for the identification of plasmid deoxyribonucleic acid in bacteria.** Eckhardt,T. (Dep. Microbiol., New York Univ. Sch. Med., New York, NY 10016, USA) *Plasmid, 1(4), 584-588 (1978)* En;en.

11-U2 **Isolation and characterisation of plasmids in *Pseudomonas aeruginosa*.** Stanisich,V.A. (Dep. Bacteriol., Med. Scb., Bristol Univ., Bristol BS8 1TD, UK) *Bull. Inst. Pasteur, 74(3), 285-294 (1976)* En;en.

12-U2 **A technique for the analysis of non-replicating and replicating forms of R6K plasmid DNA in *Escherichia coli*: relationship to the folded chromosome.** Sheehy,R.J.; Clark,C.W.; Archibold,E.R. (Dep. Biol., Morehouse Coll., Atlanta, GA 30314, USA) *In:* Plasmids. Medical and theoretical aspects. Mitsuhashi,S.; Rosival,L.; Kremery,V. (eds.) *Publ. by:* Springer-Verlag KG, D-1000 Berlin 33, Heidelberger Platz 3, GFR 1977 p. 265-275 ISBN 3-540-07946-7 En.

13-U2 **A simple method for phage or plasmid DNA's isolation suitable as a 'screening test' after molecular cloning.** Kozlov,Y.I.; Gening,L.V.; Strongin,A.Y.; Debabov,V.G. (Inst. Genet. and Selection Ind. Microorganisms, PO Box 825, 113545, Moscow, USSR) *Anal. Biochem., 86(1), 316-319 (1978)* En.

14-U2 **Micro-assay systems for infectious laryngotracheitis virus.** Robertson,G.M.; Egerton,J.R. (Dep. Vet. Med., Univ. Sydney, Private Bag, Camden, NSW, Australia 2570) *Avian Dis., 21(1), 133-135 (1977)* En.

15-U2 **Optimal method for recovery of cytomegalovirus from urine of renal transplant patients.** Lee,M.S.; *Balfour,H.H.,Jr. (Box 437, Mayo Mem. Build., Univ. Minnesota Health Sci. Cent., Minneapolis, MN 55455, USA) *Transplantation, 24(3), 228-230 (1977)* En.

16-U2 **Isolation of dengue viruses in mosquito cell cultures under field conditions [Dipt., Culicidae].** Race,M.W.; Fortune,R.A.J.; Agostini,C.; Varma,M.G.R. (Pan American Health Organ., Caribbean Epidemiol. Cent., PO Box 164, Port of Spain, Trinidad) *Lancet, 1(8084), 48-49 (1978)* En.

17-U2 **A new and simple method for recuperation of enteroviruses from water.** Sarrette,B.A.; Danglot,C.D.; Vilagines,R. (Lab. Virol., Serv. Controle des Eaux de la Ville de Paris, 144 Ave. P.V. Couturier, 75014 Paris, France) *Water Res., 11(4), 355-358 (1977)* En;en.

18-U2 **A greatly simplified method of establishing B-lymphoblastoid cell lines.** Tohda,H.; Oikawa,A.; Kudo,T.; Tachibana,T. (Dep. Pharmacol., Res. Inst. Tuberculosis and Cancer, Tohoku Univ., Sendai 980, Japan) *Cancer Res., 38(10), 3560-3562 (1978)* En;en.

19-U2 **Virus isolation test for feline leukemia virus.** de Noronha,F.; Poco,A.; Post,J.E.; Rickard,C.G. (Oncol. Lab., Pathol. Dep., New York State Coll. Vet. Med., Cornell Univ., Ithaca, NY 14853, USA) *J. Natl. Cancer Inst., 57(1), 129-131 (1976)* En;en.

20-U2 **One year experience on the Lindholm B medium used in large-scale FMD virus production on BHK cells in suspension.** Jensen,M.H.; Soerensen,F.O. (State Vet. Inst. Virus Res., Lindholm, DK 4771, Kalvehave, Denmark) *Dev. Biol. Stand., 35, 45-53 (1977)* En;en,fr.

21-U2 Certain aspects of foot-and-mouth disease. Virus production in growing BHK suspended cell cultures. Barteling,S.J. (Cent. Vet. Inst., Virol. Dep., Houtribweg 39, Lelystad, Netherlands) *Dev. Biol. Stand., 35, 55-60 (1977)* En;en,fr.

22-U2 [Multiplication and evaluation of IFFA 3 cells]. Mougeot,H.; Preaud,J.M.; Rouchouse,J.; Favre,H.; Dubouclard, C. (IFFA-Merieux, 254 rue Marcel Merieux, 69007 Lyon, France) *Dev. Biol. Stand., 35, 33-44 (1977)* Fr;en,fr.

23-U2 An air sampling technique for hepatitis B surface antigen. Petersen,N.J.; Bond,W.W.; Marshall,J.H.; Favero,M.S.; Raij,L. (Phoenix Lab. Div., Bur. Epidemiol., Cent. Dis. Control, Public Health Serv., Q Dep. Health, Educ. and Welfare, 4402 North Seventh St., Phoenix, AZ 85014, UA) *Health Lab. Sci., 13(4), 233-237 (1976)* En;en.

24-U2 Isolation of herpes simplex virus clones and drug resistant mutants in microcultures. Klein,R.J. (Dep. Microbiol., New York Univ. Sch. Med., 550 First Ave., New York, NY 10016, USA) *Arch. Virol., 49(1), 73-80 (1975)* En;en.

25-U2 Plaque morphology of herpes simplex virus in various cells under liquid overlay as a marker for its type differentiation. Shinkai,K. (Cent. Res. Lab., Sankyo Co., Ltd., 1-2-58 Hiromachi, Shinagawa-ku, Tokyo 140, Japan) *Jap. J. Microbiol., 19(6), 459-462 (1975)* En.

26-U2 A new method for producing virus plaques in cell cultures. Whitcutt,J.M. (Virus Cancer Unit, Natl. Inst. Virol., Johannesburg, South Africa) *S. Afr. Med. J., 55(9), 340-341 (1979)* En;en.

27-U2 Study of cell surface antigens induced by herpes simplex virus. Note 1. Identification of the antigens of means of protein A-containing staphylococci. Ghyka,Gr.; Mutiu,A.; Gajal,N. ('Stefan S. Nicolau' Inst. Virol., 285, Sos. Mihai Bravu, 74339, Bucharest, Romania) *Rev. Roum. Med., Ser. Virol., 30(2), 95-101 (1979)* En;en,fr.

28-U2 Rapid identification and subtyping of herpes simplex virus by complement-dependent cytotoxicity. Olofsson,S.; Jeansson,S.; *Lycke,E. (Dep. Virol., Inst. Med. Microbiol., Univ. Goteborg, Guldhedsgatan 10 B, S-413 46 Goteborg, Sweden) *Intervirology, 10(1), 40-43 (1978)* En;en.

29-U2 The use of temperature sensitivity and selective cell culture systems for differentiation of herpes simplex virus types 1 and 2 in a clinical laboratory. Nordlund,J.J.; Anderson,C.; *Hsiung,G.D.; Tenser,R.B. (Virol. Lab., VA Hosp., West Haven, CT 06516, USA) *Proc. Soc. Exp. Biol. Med., 155(1), 118-123 (1977)* En;en.

30-U2 Differentiation of herpes simplex virus serotypes 1 and 2 by DNA-DNA-hybridization Schulte-Holthausen,H; Schneweis,K.E.; (Inst. Klin. Virol., Univ. Erlangen-Nurnberg,D-8520 Erlangen, Loschgestr. 7, GFR) *Med. Microbiol. Immunol., 161(4), 279-285 (1975)* En.

31-U2 Skim passage: a rapid method of adapting influenza viruses to new host tissues. de St. Groth,S.F.; Tees,R. (Basel Inst. for Immunol., Grenzacherstr. 487, CH-4058 Basel, Switzerland) *Intervirology, 5(6), 335-341 (1975)* En;en.

32-U2 Growth of some attenuated influenza viruses in hamster tracheal organ cultures. Reeve,P.; Gerendas,B.; Walzl,H. (Sandoz Forschungsinst., Wien, Austria) *Med. Microbiol. Immunol., 166(1-4), 141-150 (1978)* En;en.

33-U2 Further studies on the Cowan strain of *Staphylococcus aureus* as an aid for the diagnosis of influenza. Zalan,E.; Wilson,C. (Virus Lab., Ont. Minist. Health, Toronto, Ont., Canada) *Arch. Virol., 56(1-2), 177-180 (1978)* En;en.

34-U2 Propagation of UB-1 strain of Marek's disease virus in embryonating hen eggs. Raghavan,R. (Dep. Vet. Microbiol., Vet. Coll., Univ. Agric. Sci., Bangalore 560024, India) *Indian Vet. J., 54(9), 685-687 (1977)* En;en.

35-U2 Multiplication of turkey herpes virus and Marek's disease virus in chick embryo skin cell cultures. Prasad,L.B.M.; Spradbrow,P.B. (Commonw. Serum Lab., 45 Poplar Rd., Parkville, Vic. 3052, Australia) *J. Comp. Pathol., 87(4), 515-520 (1977)* En;en.

36-U2 The laboratory diagnosis of Nairobi sheep disease. Davies,F.G.; Mungai,J.N.; Taylor,M. (Vet. Res. Lab., PO Kabete, Kenya) *Trop. Anim. Health Prod., 9(2), 75-80 (1977)* En;en,es,fr..

37-U2 Methods for the quantitative assessment of nuclear-polyhedrosis virus in soil. Evans,H.F.; Bishop,J.M.; Page,E.A. (NERC, Unit Invertebr. Virol., 5 South Parks Road, Oxford OX1 3YB, UK) *J. Invertebr. Pathol., 35(1), 1-8 (1980)* En;en.

38-U2 Methods for monitoring bacteriophage in cheese factories. Hull,R.R. (Dairy Res. Lab., Div. Food Res., CSIO, Highett, Vic., Australia) *Aust. J. Dairy Technol., 32(2), 63-64 (1977)* En.

39-U2 The red plaque test: a rapid method for identification of excision defective variants of bacteriophage lambda. Enquist,L.W.; Weisberg,R.A. (Lab. Mol. Genet., Natl. Inst. Child Health, and Hum. Dev., Bethesda, MD 20014, USA) *Virology, 72(1), 147-153 (1976)* En;en.

40-U2 Simple tests for the diagnosis of picornavirus epidemic conjunctivitis (acute haemorrhagic conjunctivitis). Yin-Murphy,M. (Dep. Microbiol., Fac. Med., Univ. Singapore, Singapore) *Bull. WHO, 54(6), 675-679 (1976)* En;en,fr.

41-U2 Intratypic serodifferentiation of poliomyelitis virus strains by strain-specific antisera. van Wezel,A.L.; Hazendonk,A.G. (Rijks Inst. Volksgezondheid, Antonie van Leeuwenhoeklaan 9, POB 1, Bilthoven, Netherlands) *Intervirology, 11(1), 2-8 (1979)* En;en.

42-U2 New serological method for detection of potato virus diseases. 1. Detection of potato virus Y. Hinostroza de

Lekeu,A.M. (Lab. Pathol. Veg., Fac. Sci. Agron. Etat, B-5800 Gembloux, Belgium) *Phytopathol, Z., 95(4) 342-345 (1979)* En;en.

43-U2 Rabies virus study by the plaque forming system. A simplified technique. Salaun,J.-J. (Lab. Biol. Med., Hop. Militaire H.-Larrey, 4, place Saint-Pierre, 31066 Toulouse Cedex, France) *Ann. Biol. Clin., 36(1), 23-25 (1978)* Fr;en,fr.

44-U2 Rotavirus isolation and cultivation in the presence of trypsin. Babiuk,L.A.; Mohammed,K.; Spence,L.; Fauvel,M.; Petro,R. (Dep. Vet. Microbiol., Western Coll. Vet. Med., Univ. Saskatchewan, Saskatoon, S7N 0W0, Canada) *J. Clin. Microbiol., 6(6), 610-617 (1977)* En;en.

45-U2 Replication of murine paramyxoviruses in hamster tracheal organ culture and comparison with standard tissue culture methods. Schiff,L.J. (Life Sci. Div., IIT Res. Inst., Chicago, IL 60616, USA) *J. Clin. Microbiol., 4(3), 248-252 (1976)* En;en.

46-U2 A simple and high-yielding method for vaccinia virus multiplication in cell cultures. Dowjat,K.; Plachcinska,J. (00-725 Warszawa, Chelmska 30, Cent. Lab. Surowic i Szezepionek, Poland) *Med. Dosw. Mikrobiol., 29(1), 43-49 (1977)* Pl;en,pl,ru.

47-U2 [Technical note on arbovirus isolation by means of baby mice inoculation. Preparation of ground mosquitoes]. Cornet,M.; Dejardin,J.; Jan,C.; Coz,J.; Adam,C.; Valade,M. (Serv. Sante des Armees, Cent. ORSTOM de Dakar-Hann, BP 1386, Dakar, Senegal) *Bull. Soc. Pathol. Exot., 70(2), 137-143 (1977)* Fr;en,fr.

48-U2 The use of human foreskin cell cultures for isolation of herpesvirus group in the diagnostic laboratory. Sharon,N. (Dep. Pathol. and Lab. Med., Evanston Hosp., Evanston, IL 60201, USA) *Am. J. Clin. Pathol., 66(1), 65-72 (1976)* En;en.

49-U2 Large-scale production of virus. Nicklin,P.M.; House,W. (Imp. Cancer Res. Fund Lab., London WC2, UK) *Biotechnol. Bioeng., 18(5), 723-727 (1976)* En.

50-U2 Interferon production by individual cells in culture. Kronenberg,L.H. (Dep. Pediatr., Sch. Med., Univ. Califoria, San Diego, La Jolla, CA 92093, USA) *Virology, 76(2), 634-642 (1977)* En;en.

51-U2 Human interferon: mass production in a newly established cell line, MG-63. Billiau,A.; Edy,V.G.; Heremans,H.; van Damme,J.; Desmyter,J.; Georgiades,J.A.; De Somer,P. (Univ. Leuven, Dep. Human Biol., Rega Inst. Med. Res., B-3000 Leuven, Belgium) *Antimicrob. Agents Chemother., 12(1), 11-15 (1977)* En;en.

52-U2 An apparatus for rapid sampling of growing cell cultures. Dalbow,D.G.; Young,R. (Dep. Pathol., Colorado State Univ., Ft. Collins, CO 80521, USA) *Anal. Biochem., 71(2), 540-543 (1976)* En.

53-U2 Nitrite agar method for the isolation and enumeration of denitrifying bacteria. Mycielski,R.; Gucwa,D. (Dep. Environ. Microbiol., Inst. Microbiol., Warsaw Univ., Karowa 18, 00-324 Warsaw, Poland) *Acta Microbiol. Pol., 26(3), 317-318 (1977)* En.

54-U2 An adaptation of a rapid sensitivity method combined with a rapid identification scheme for urinary pathogens. Wells,J.M. (Public Health Lab., Gloucestershire R. Hosp., Southgate St., Gloucester GL1 1UD, UK) *Med. Lab. Technol., 32(3), 215-218 (1975)* En;en.

55-U2 Rapid method for identification and enumeration of oral *Actinomyces*. Marucha,P.T.; Keyes,P.H.; Wittenberger,C.I.; *London,J. (Microbiol. Sect., Lab. Microbiol. and Immunol., Natl. Inst. Dent. Res., Bethesda, MD 20014, USA) *Infect. Immun., 21(3), 786-791 (1978)* En;en.

56-U2 Rapid identification of Actinomycetaceae and related bacteria. Kilian,M. (Dep. Microbiol., R. Dent. Coll., DK-8000 Aarhus C, Denmark) *J. Clin. Microbiol., 8(2), 127-133 (1978)* En;en.

57-U2 Isolation of antibiotic-producing actinomycetes from soil. Pinney,R.; Kalil,M. (Dep. Biol., Univ. Wisconsin Cent.-Marshfield-Wood, Marshfield, WI 54449, USA) *Am. Biol. Teach., 37(7), 436-937 (1975)* En.

58-U2 Rapid method to enumerate and isolate soil actinomycetes antagonistic towards rhizobia. Panthier,J.J.; Diem,H.G.; Dommergues,Y. (CNRS/ORSTOM, Lab. Microb., BP 1386, Dakar, Senegal) *Soil Biol. Biochem., 11(4), 443-445 (1979)* En.

59-U2 Medium for the presumptive identification of *Aeromonas hydrophila* and Enterobacteriaceae. Kaper,J; Seidler,R.J.; Lockman,H.; *Colwell,R.R. (Dep. Microbiol., Univ. Maryland, College Park, MD 20742, USA) *Appl. Environ. Microbiol., 38(5), 1023-1026 (1979)* En;en.

60-U2 Modification of the pH-auxostat culture method for the mass cultivation of bacteria. Oltmann,L.F.; Schoenmaker,G.S.; Reijnders,W.N.M.; Stouthamer,A.H. (Biol. Lab., Dep. Microbiol., Free Univ. de Boelelaan, 1087 Amsterdam, Netherlands) *Biotechnol. Bioeng., 20(6), 921-925 (1978)* En.

61-U2 Rapid techniques for identification of microorganisms based on DNA hybridization. Antonov,A.S.; Belousova,A.A.; Lysenko,A.M.; Turova,I.P. (Moscow State Univ., Moscow, USSR) *Mikrobiologiya, 47(6), 1049-1054 (1978)* Ru;en,ru.

62-U2 A recombination test to classify mutants of *Bacillus subtilis* of identical phenotype. Galizzi,A.; Siccardi,A.G.; Mazza,G.; Canosi,U.; Poisinelli,M. (Ist. Genet., Univ. Pavia, Via S. Epifanio 14, 27100 Pavia, Italy) *Genet. Res., 27(1), 47-58 (1976)* En;en.

63-U2 Isolation and identification of *Bacillus thuringiensis* from environmental objects. Daburov,K.N.; Korolik,V.V.; Deriglazov,A.D.; Golidonova,N.D. (II Moscow Med. Inst., Moscow, USSR) *Gig. Sanit., No. 11, 71-75 (1978)* Ru.

64-U2 **A simplified method for the isolation of *Bacteroides nodusus* from ovine foot-rot and studies on its colony morphology and serology.** Thorley,C.M. (Dep. Bacteriol., Wellcome Res. Lab., Langley Court, Beckenham, Kent BR3 3BS, UK) *J. Appl. Bacteriol., 40(3), 301-309 (1976)* En;en.

65-U2 **Identification and antimicrobial susceptibility of 250 *Bacteroides fragilis* subspecies tested by broth microdilution methods.** Jones,R.N.; Fuchs,P.C. (Clin. Microbiol. Div., Kaiser Found. Hosp. Lab., Portland, OR 97217, USA) *Antimicrob. Agents Chemother., 9(4), 719-721 (1976)* En;en.

66-U2 **Valine, malic, and pyruvic dehydrogenase tests in the differentiation of *Bacteroides*.** Funderburk,N.R.; Kester,A.S. (Dep. Pathol., Univ. Texas-Health Sci. Cent., San Antonio, TX 78274, USA) *Can. J. Microbiol., 21(9), 1369-1371 (1975)* En;en,fr.

67-U2 **Rapid identification of *Bacteroides fragilis* with bile and antibiotic disks.** Draper,D.L.; *Barry,A.L. (Clin. Microbiol. Lab., Med. Cent., Sacramento, CA 95817, USA) *J. Clin. Microbiol., 5(4), 439-443 (1977)* En;en.

68-U2 **Identification of clinical isolates of selected species of *Bacteroides*: production of phenylacetic acid.** Mayrand,D. (Dep. Biochim., Fac. Sci. et Genie, Univ. Laval, Que. G1K 7P4, Canada) *Can. J. Microbiol., 25(8), 927-928 (1979)* En;en,fr.

69-U2 **Rapid enrichment of *Beggiatoa* from soil.** Joshi,M.M.; Hollis,J.P. (Dep. Plant Pathol., Louisiana State Univ., Baton Rouge, LA 70803, USA) *J. Appl. Bacteriol., 40(2), 223-224 (1976)* En.

70-U2 **[Preparation of *Brucella canis* antigens for agglutinin research using the rapid plate test.]** Godoy,A.M.; Peres,J.N.; Barg,L. (Univ. Fed. Minas Gerais, Inst. Cienc. Biol., Dep. Microbiol., Bacteriol. Zoonoses Lab., Belo Horizonte, MG, Brazil) *Arq. Esc. Vet., 30(2), 165-169 (1978)* Pt;en,pt.

71-U2 **A simple technique for mass cultivation of *Campylobacter fetus*.** Simon,P.C. (Anim. Pathol. Lab., Health of Anim. Branch, Agric. Canada, Pacific Area Lab., 3802 West 4th Ave., Vancouver, BC V6R 1P5, Canada) *Can. J. Comp. Med., 40(3), 318-319 (1976)* Er en,fr.

72-U2 **New simplified culture technique for *Chlamydia trachomatis*.** Ripa,K.T.; Mardh,P.-A. (Inst. Med. Microbiol., Univ. Lund, Solvegatan 23, Sweden) *In:* Nongonococcal urethritis and relation infections. Hobson,D.; Holmes,K.K. (eds.) *Publ. by:* American Society for Microbiology, 1913 I St., N.W., Washington, DC 20006, USA 1 Aug 1977 p. 323-327 ISBN 0-914826-11-5 En.

73-U2 **Susceptibility of chlamydiae to chemotherapeutic agents.** Treharne,J.D.; Day,J.; Yeo,C.K.; Jones,B.R.; Squires,S. (Inst. Ophthalmol., Univ. London, London WC1H 9QS, UK) *In:* Nongonococcal urethritis and related infections. Hobson,D.; Holmes,K.K. (eds) *Publ. by:* American Society for Microbiology, 1913 I St., N.W., Washington, DC 20006, USA 1 Aug 1977 p. 214-222 ISBN 0-914826-11-5 En.

74-U2 **Isolation of *Chlamydia* from ocular infections.** Darougar,S.; Woodland,R.M.; Forsey,T.; Cubitt,S.; Allami,J.; Jones,B.R. (Inst. Ophthalmol., Univ. London, London WC1H 9QS, UK) *In:* Nongonococcal urethritis and related infections. Hobson,D.; Holmes,K.K. (eds.). *Publ. by:* American Society for Microbiology, 1913 I St., N.W., Washington, DC 20006, USA 1 Aug 1977 p. 295-298 ISBN 0-914826-11-5 En.

75-U2 **Laboratory procedures for the isolation of *Chlamydia trachomatis* foom the human genital tract.** Reeve,P.; Owen,J.; Oriel,J.D. (Dep. Genito-Urinary Med., University Coll. Hosp., Gower St., London WC1E 6AU, UK) *J. Clin. Pathol., 28(11), 910-914 (1975)* En;en.

76-U2 **Quantitative aspects of the growth of *Chlamydia trachomatis* in diagnostic tissue culture procedures.** Johnson,F.W.A.; Hobson,D.; Rees,E.; Tait,I.A. (Dep. Med. Microbiol., New Med. Sch., Univ. Liverpool, Liverpool L69 3BX, UK) *In:* Nongonococcal urethritis and related infections. Hobson,D.; Holmes,K.K. (eds.) *Publ. by:* American Society for Microbiology; 1913 I St., N.W. Washington, DC 20006 (USA). 1 Aug 1977 p. 309-313 ISBN 0-914826 11 5. En.

77-U2 **An improved method for demonstrating the growth of chlamydiae in tissue culture.** Johnson,F.W.A.; Chancerelle,L.Y.J.; Hobson,D. (Dep. Med. Microbiol., Univ. Liverpool, Liverpool L69 3BX, UK) *Med. Lab. Sci. 35(1), 67-74 (1978)* En;en.

78-U2 **A simple method for the isolation and determination of *Clostridium perfringens*.** Debevere,J.M. (Lab. Gen. and Ind. Microbiol., Fac. Agric. Sci., Univ. Gent, Coupure, 533, B-9000 Gent, Belgium) *Eur. J. Appl. Microbiol. Biotechnol., 6(4), 409-414 (1979)* En;en.

79-U2 **An improved cooked meat medium for the detection of *Clostridium botulinum*.** Quagliaro,D.A. (FDA, 850 Third Ave., Brooklyn, NY 11232, USA) *J. Assoc. Off. Anal. Chem., 60(3), 563-569 (1977)* En;en.

80-U2 **Use of ganglioside affinity filters to identify toxigenic strains of *Clostridium botulinum* types C and D.** Hayes,S.(Dep. Mol. Biol. and Biochem., Univ. California, Irvine, CA 92717, USA) *Infect. Immun., 26(1), 150-156 (1979)* En;en.

81-U2 **Amino acid utilization patterns in clostridial taxonomy.** Elsden,S.R.; Hilton,M.G. (Sch. Biol. Sci., Univ. East Anglia, Norwich NR4 7RJ, UK) *Arch. Microbiol., 123(2), 137-141 (1979)* En;en.

82-U2 **Coliform enumeration in butter.** Loane,P.; Tommerup,J.; Stokoe,J.; Newton,E.; Forbes,L. (Otto Madsen Dairy Res. Lab., Dep. Primary Ind., Hamilton, Queensl., Australia) *Aust. J. Dairy Technol., 32(2), 72-74 (1977)* En;en.

83-U2 **Identification and quantification of fecal coliforms using violet red bile agar at elevated temperature.** Klein,H.; Fung,D.Y.G. (Dep. Microbiol., Pennsylvania State Univ., University Park, PA 16802, USA) *J. Milk Food Technol., 39(11), 768-770 (1976)* En;en.

84-U2 Development of a rapid coliform test for pasteurised milk. Sudarsanam,T.S.;Nambudripad,V.K.N. (Natl. Dairy Res. Inst., Karnal-132001, India) *Indian J. Dairy Sci., 31(3), 282-284 (1978)* En;en.

85-U2 A new method for the detection of coliform bacteria with a reduced incubation time. Rapp,M.; Ihle,P. (Staatl. Milchwirtschaft. Lehr- und Forschungsanst., Dr. Oskar Farny Inst., Wangen in Allgau, GFR) *Milchwissenshaft, 34(8), 471-474 (1979)* De;de,en.

86-U2 Rapid bacteriological analyses of washouts from various objects in the environment. Kartsev,V.V.; Kapaseba,P.I.; Ozepets,A.V.; Tsdkapdshvdld,T.A.; Mebvebeva,I.N. (Address not stated) *Gig. Sanit.,No. 6, 74-75 (1979)* Ru;ru.

87-U2 Standard micromethod for identification of the coryneform bacteria. Perrier,J.; Gounot,A.M. (Univ. Claude-Bernard, Lyon 1, Dep. Biol. Veg., Serv. Microbiol., 43 blvd. 11-Novembre-1918, 69621 Villeurbanne, France) *Rev. Inst. Pasteur Lyon, 8(2), 133-154 (1975)* Fr;en,fr.

88-U2 Computer assisted identification of coryneform bacteria. Hill,L.R.; Lapage,S.P.; Bowiew,I.S. (deceased). (Natl. Collect. Type Cult., Cent. Public Health Lab., Colindale, London NW9, UK) *In*Coryneform bacteria (Special publications of the Society for General Microbiology.1.) Bousfield,J.J.; Callely,A.G. (eds.). *Publ. by:* Academic Press (London) Ltd., 24-28 Oval Road, London NW1 7DX, UK. Nov 1978. p. 181-215 ISBN 0-12-119550-X En.

89-U2 Computer assisted identification of coryneform bacteria. Hill,L.R.; Lapage,S.P.; Bowiew,I.S. (deceased) (Natl. Collect. Type Cult. Cent. Public Health Lab., Colindale, London NW9, UK) *In:* Coryneform bacteria. Boursfield,I,J; Callely,A.G. (eds.). *Publ. by:* Academic press (London) Ltd. 24-28 Oval Road, London NW1 TX, UK. Nov 1978 p. 181-215 ISBN 0-12-119550-X En.

90-U2 Four hour-tests for the identification of Enterobacteriaceae. Nord,C.-E.; Lindberg,A.A.; Dahlback,A. (Dep. Bacteriol., Natl. Bacteriol. Lab., S-10521 Stockholm, Sweden) *Med. Microbiol. Immunol., 161(4), 231-238, (1975)* En;en.

91-U2 Evaluation of the Enterotube system for rapid identification of Enterobacteriaceae from the genital system and from fetuses of horses. Sonnenschein,B.; Weiss,R. (3 Hannover, Bischoisholer Damm 15, GFR) *Dtsch. Tierarztl., 83(4), 146-148 (1976)* De;de,en.

92-U2 Rapid test for acetyl-methyl-carbinol formation by Enterobacteriaceae. Hussain Qadri,S.M.; Nichols,C.W.; Qadri,S.G.M.; Villarreal,A. (Univ. Oklahoma Health Sci. Cent., Oklahoma City, OK 73125, USA) *J. Clin. Microbiol., 8(4), 463-464 (1978)* En;en.

93-U2 [A rapid test for acetoin in the microbiological diagnosis of Enterobacteriaceae]. Tiecco,G.; Piccininno,G.; Vignolo Del Bon,L. (Address not stated) *Riv. Zootec. Vet., 6, 517-518 (1976)* It;en,fr,it.

94-U2 Comparison of amino acid decarboxylase and dihydrolase results by Moeller, rapid, and replicator plate methods. Jones,R.N.; Fuchs,P.C.; Sniderman,S. (Dep. Pathol., Kaiser Found. Hosp. and Lab., Portland, OR 97217, USA) *J. Clin. Microbiol., 3(1), 75-76 (1976)* En;en.

95-U2 Simultaneous detection of salmonellae, *Escherichia coli* type I *E. coli* and other Enterobacteriaceae in dehydrated foods for infants and children. Mulindwa,D.K.; Kaferstein,F.K. (Inst. Veterinarmed., Bundesgesundheitsamtes, Postfach D-1000 Berlin 33, GDR) *Z. Bakteriol. Parasitenkd. Infektionskr. Hyg., I Abt. B, 166(1), 72-80 (1978)* En;de,en.

96-U2 Identification of Enterobacteriaceae by the API 20E system. Holmes,B.; Willcox,W.R.; Lapage,S.P. (Natl. Collection of Type Cultures, Cent. Public Health Lab., Colindale, London NW9 5HT, UK) *J. Clin. Pathol., 31(1), 22-30 (1978)* En;en.

97-U2 A practical system for the speciation of Enterobacteriaceae from clinical material. Murphy,R. (St. Paul's Hosp., Vancouver, BC, Canada) *Can. J. Med. Technol., 38(1), 8-13 (1976)* En.

98-U2 Use of Enterotube test system for identification of enterobacteria isolated from patients. Menshikov,D.D.; Yanisker,G.Ya. (Moscow Res. Inst., Moscow, USSR) *Zh. Mikrobiol. Epidemiol. Immunobiol., 56(2), 73-77 (1979)* Ru;en,ru.

99-U2 Simplified 48-hour IMVic test: an agar plate method. Powers,E.M.; Latt,T.G. (US Army Natick Res. and Dev. Command, Natick, MA 01760, USA) *Appl. Environ. Microbiol., 34(3), 274-279 (1977)* En;en.

100-U2 Factors affecting the value of a simple biochemical scheme for identifying Enterobacteriaceae: the reproducible recognition of biotypes. Barr,J.G.; Mahood,R.J.; Curry,K.P.W. (Dep. Clin. Bacteriol., R. Victoria Hosp., Belfast, UK) *J. Clin. Pathol., 30(6), 495-504 (1977)* En;en.

101-U2 Evaluation of the efficacy of composite media for the rapid identification of members of the Enterobacteriaceae. Ahuja,S.; Mago,M.L.; Saxena,S.N. (Natl. Salmonella and Escherichia Cent., Central Res. Inst., Kasauli, India) *Indian J. Pathol. Microbiol., 22(3), 249-253 (1979)* En;en.

102-U2 An evaluation of the PathoTec system for identification of Enterobacteriaceae. Gregersen,T.; Halgaard,C. (Bulowsvej 34 A III, DK-1870 Kobenhavn V, Denmark) *Nord. Vet. Med., 28(6), 316-321 (1976)* Da;da,en.

103-U2 Clinical evaluation of the Minitek differential system for identification of Enterobacteriaceae. Finklea,P.J.; Cole,M.S.; Sodeman,T.M. (Dep. Pathol., Microbiol. Sect., University Hosp., Ann Arbor, MI 48109, USA) *J. Clin. Microbiol., 4(5), 400-404 (1976)* En;en.

104-U2 Enterobacteriaceae identification compared by MORLOC (a new system) and API 20E. Branson,D. (Grant Hosp., 309 E. State St., Columbus, OH 43215, USA) *Am. J. Med. Technol., 42(8), 267-276 (1976)* En;en.

105-U2 Comparative investigations on the biochemical differentiation of Enterobacteriaceae using Patho-Tec test strips. Reinhold,A. (Sekt. Tierprod. und Veterinarmed., Fachgruppe Veterinar.-Mikrobiol. und Tierseuchenlehre, Karl-Marx-Univ., 701 Leipzig, Margarete-Blankstr. 8, GDR) *Wiss. Z. Karl-Marx-Univ. Leipzig Mathematisch-Naturwiss., 24(3), 303-310 (1975)* De;de.

106-U2 Bioconversion from DL-homoserine to L-threonine. II. Application of the Simplex method of optimization. Dumenil,G.; Cremieux,A.; Phan-Tan-Luu,R.; Aune,J.P. (Lab. Microbiol., Fac. Pharm., F-13385 Marseille Cedex 4, France) *Eur. J. Appl. Microbiol., 1(3), 221-231 (1975)* En;en.

107-U2 Differentiation of *Pasteurella* and *Actinobacillus* from Enterobacteriaceae by use of the O/129 vibriostatic agent. Chatelain,R.; Bercovier,H.; Guiyoule,A.; Richard,C.; Mollaret,H.H. (Unite Ecol. Bact., Inst. Pasteur, 75724 Paris Cedex 15, France) *Ann. Microbiol., 130A(4), 449-454 (1979)* Fr;en,fr.

108-U2 Comparison of enteric identification systems. /|presented at the Laboratory Section of the 104th Annual Meeting of the American Public Health Association, held in Miami Beach, Florida, on 20 October, 1976]. Borchardt,K.A.; Gibson,J. (Clin. Microbiol., USPHS Hosp., 15th Ave. and Lake St., San Francisco, CA 94118, USA) *Health Lab. Sci., 14(1), 5-10 (1977)* En;en.

109-U2 Reproducibility of three microdilution systems for identification of Enterobacteriaceae, compared with API 20E and micro-ID test systems. Barry,A.L.; Badal,R.E.; Effinger,L.J. (Dep. Pathol., Microbiol. Lab., Univ. California, Davis, Med. Cent., Sacramento, CA 95817, USA) *Curr. Microbiol., 3(1), 21-25 (1979)* En;en.

110-U2 An evaluation of the modified R/B Enteric Differental System for the identification of the Enterobacteriaceae. Hayek,L.J.; Willis,G.W. (Dep. Pathol., Torbay Hos , Torquay, Devon, UK) *J. Clin. Pathol., 30(2), 154-156 (1977)* En;en.

111-U2 Comparison of miniaturized multitest systems with conventional methodology for identification of Enterobacteriaceae from foods. Guthertz,L.S.; Okoluk,R.L. (Letterman Army Inst. Res., Dep. Nutr., Food Hyg. Div., Presidio of San Francisco, CA 94129, USA) *Appl. Environ. Microbiol., 35(1), 109-112 (1978)* En;en.

112-U2 Evaluation of the Repliscan system for identification of Enterobacteriaceae. Brown,S.D.; *Washington,J.A.,II. (Mayo Clin. and Mayo Found., Rochester, MN 55901, USA) *J. Clin. Microbiol., 8(6), 695-699 (1978)* En;en.

113-U2 Identification of Enterobacteriaceae in frozen microdilution trays prepared by micro-media systems. Barry,A.L.; Badal,R.E.; Effinger,L.J. (Microbiol. Lab.-Univ. California, Davis, CA 95616, USA) *J. Clin. Microbiol., 10(4), 492-496 (1979)* En;en.

114-U2 Rapid identification of Enterobacteriaceae from blood cultures with the Micro-ID systems. Edberg,S.C.; Clare,D.; Moore,M.H.; Singer,J.M. (Dep. Pathol., Montefiore Hosp. and Med. Cent. Bronx, NY 10467, USA) *J. Clin. Microbiol., 10(5), 693-697 (1979)* En;en.

115-U2 Use of the API 20E system for rapid identification of Enterobacteriaceae (six hours). Pinon,G.; Loulergue,J.; Quentin,R.; Darchis,J.P.; Prieur,D. Vargues,R.; (Lab. Bacteriol., CHU Bretonneau, 37044 Tours Cedex, France) *Ann. Biol. Clin., 37(4), 221-223 (1979)* Fr;en,fr.

116-U2 Clinical evaluation of the MICRO-ID, API 20E, and conventional media systems for identification of Enterobacteriaceae. Edberg,S.C.; Atkinson,B.; Chambers,C.; Moore,M.H.; Palumbo,L.;Zorzon,C.F.; Singer,J.M. (Div. Microbiol., Dep. Pathol., Montefiore Hosp. and Med. Cent., Albert Einstein Coll. Med., New York, NY 10467, USA) *J. Clin. Microbiol., 10(2), 161-167 (1979)* En;en.

117-U2 Rapid identification of Enterobacteriaceae by using noncommercial micro-tests in conjunction with API 20E profile data. George,S.; *Davis,G.H.G. (Dep. Microbiol., Univ. Queensland, Brisbane, Australia) *J. Clin. Microbiol., 10(4), 399-403 (1979)* En;en.

118-U2 Preparation of synchronous cultures of *Escherichia coli* by continuous-flow size selection. Evans,J.B. (Dep. Microbiol., University Coll., Cardiff CF2 1TA,UK) *J. Gen. Microbiol., 91(1), 188-190 (1975)* En.

119-U2 Ethanol formed from arabinose: a rapid method for detecting *Escherichia coli*. Coloe,P.J. (Dep. Microbiol., Monash Univ. Med. Sch., Prahran, Vic. 3181, Australia) *J. Clin. Pathol., 31(4), 361-364 (1978)* En;en.

120-U2 A rapid technique for isolation of *Escherichia coli* type I in food. Kaferstein,F.K.; Rassai,A.; Schulz,H. (Inst. Veterinarmed. (Robert van Ostertag-Inst.), Bundesgesundheitsamtes, D-1000 Berlin 33, Postfach, GFR) *Zentralbl. Bakteriol. Parasitenkd. Infektionskr. Hyg., I Abt. B, 161(5-6), 540-544 (1976)* De;de,en.

121-U2 Use of a Coulter Counter to detect discrete changes in cell numbers and volume during growth of *Escherichia coli*. Smither,R. (Lab. Gov. Chem., Dep. Ind., Cornwall House, Stamford St., London SE1 9NQ, UK) *J. Appl. Bacteriol., 39(2), 157-165 (1975)* En;en.

122-U2 Rapid identification of Enterobacteriaceae. II. Use of a β-glucuronidase detecting agar medium (PGUA agar) for the identification of *E. coli* in primary cultures of urine samples. Kilian,M.; Bulow,P. (Dep. Microbiol., R. Dent. Coll., Vennelyst Blvd., DK-8000 Arhus C, Denmark) *Acta Pathol. Microbiol. Scand., Ser. B, 87(5), 271-276 (1979)* En;en.

123-U2 Effect of potential water pollutants and enzyme inhibitors on an automated rapid test for *Escherichia coli*. Moran,J.W.; Smith,T.L.; *Witter,L.D. (Dep. Food Sci., Univ. Ilhnois at Urbana-Champaign, Urbana, IL 61801, USA) *Appl. Environ. Microbiol., 32(4), 645-646 (1976)* En;en.

124-U2 Rapid differentiation of *E. coli* colonies among coliform colonies. Jacquet,J.; Coiffier,O. (Lab. Microbiol.,

UER Sci. Vie et Comportement, 14032 Caen Cedex, France) *Lait, 58(573-574), 111-117 (1978)* Fr;en.

125-U2 **Use of the Minitek system for biotyping *Haemophilus* species.** Back,A.E.; *Oberhofer,T.R. (Microbiol. Sect., Dep. Pathol., Madigan Army Med. Cent., Tacoma, WA 98431, USA) *J. Clin. Microbiol., 7(3), 312-313 (1978)* En;en.

126-U2 **Evaluation of a rapid β-lactamase test for detecting ampicillin-resistant strains of *Haemophilus influenzae* type b.** Scheifele,D.W.; Syriopoulou,V.P.; Harding,A.L.; Emerson,B.B.; Smith,A.L. (Div. Infect. Dis., Child. Hosp. Med. Cent., 300 Longwood Ave., Boston, MA 02115, USA) *Pediatrics (Springfield), 58(3), 382-387 (1976)* En;en.

127-U2 **Rapid biochemical characterization of *Haemophilus* species by using the micro-ID.** Edberg,S.C.; Melton,E.; Singer,J.M. (Div. Microb. and Immunol., Dep. Pathol. Montefiore Hosp. and Med. Cent., The Albert Einstein Coll. Med., New York, NY 10467, USA) *J. Clin. Microbiol., 11(1), 22-26 (1980)* En;en.

128-U2 **Rapid penicillinase paper strip test for detection of beta-lactamase-producing *Haemophilus influenzae* and *Neisseria gonorrhoeae*.** Jorgensen,J.H.; Lee,J.C.; Alexander,G.A. (Dep. Pathol., Univ. Texas Health Sci. Cent., San Antonio, TX 78284, USA) *Antimicrob. Agents Chemother., 11(6), 1087-1088 (1977)* En;en.

129-U2 ***Haemophilus vaginalis (Corynebacterium vaginale)*: method for isolation and rapid biochemical identification.** Greenwood,J.R.; Pickett,M.J.; Martin,W.J.; Mack,E.G. (Dep. Bacteriol., Univ. California, Los Angeles, CA 90024, USA) *Health Lab. Sci., 14(2), 102-106 (1977)* En;en.

130-U2 **Rapid speciation of *Haemophilus* with the porphyrin production test versus the satellite test for X.** Lund,M.E.; Blazevic,D.J. (Dep. Lab. Med. and Pathol., Univ. Minnesota, Minneapolis, MN 55455, USA) *J. Clin. Microbiol., 5(2), 142-144 (1977)* En;en.

131-U2 **An easy method for the determination of the optical types of lactic acid produced by lactic acid bacteria.** Okada,S.; Toyoda,T.; Kozaki,M. (Tokyo Univ. Agric., 1-1 Sakuragaoka 1-chome, Setagaya-ku, Tokyo, Japan) *Agric. Biol. Chem., 42(9), 1781-1783 (1978)* En.

132-U2 **Application of swelling test for the identification of *Leptospira* groups.** Anon. (Inst. Hyg. and Microbiol., Zhejiang People's Hyg. Exp. Acad., Hangzhou, China) *Acta Microbiol. Sin., 15(3), 243-245 (1975)* Ch;ch,en.

133-U2 **The computer in fermentation process research.** Unden,A.; Rindone,W.P.; Heden,C.-G. (AB Fermenta, Strangnas, Sweden) *Process Biochem., 14(3), 8, 9-10, 12 (1979)* En;en.

134-U2 **A replica-plating method for the identification of Micrococcaceae.** Bibel,D.J.; Smiljanic,R.J.; LeBrun,J.R. (Dep. Dermatol. Res., Letterman Army Inst. Res., San Francisco, CA 94129, USA) *Can. J. Microbiol., 21(11), 1676-1680 (1975)* En;en,fr.

135-U2 **A simple transporting and maintenance medium for *Mycoplasma synoviae*.** Talburt,D.E.; Stinson,R.S. (Dep. Bot. and Bacteriol., Univ. Arkansas, Fayetteville, AR 72701, USA) *Poult. Sci., 55(5), 1996-1997 (1976)* En;en.

136-U2 **Computer-assisted system for a mycobacteria tuberculosis laboratory using MUMPS.** Noguchi,H.; Ogushi,Y. (Osaka Prefect. Habikino Hosp., Habikino, Osaka, Japan) *Med. Inform., 3(4), 305-315 (1978)* En;en,fr.

137-U2 **Three simple tests as an adjunct to the niacin test for the small mycobacteriology laboratory.** Gruft,H. (Div. Lab. and Res., New York State Dep. Health, Albany, NY 12201, USA) *Health Lab. Sci., 13(3), 179-183 (1976)* En;en.

138-U2 **Simple procedure for detection of *Mycobacterium gordonae* in water causing false-positive acid-fast smears.** Dizon,D.; Mihailescu,C.; *Bae,H.C. (Bacteriol. Lab., Dep. Lab., Long Island Coll. Hosp., Brooklyn, NY 11201, USA) *J. Clin. Microbiol., 3(2), 211 (1976)* En;en.

139-U2 **Rapid urease test for mycobacteria: preliminary observations.** Cox,F.R.; Cox,M.E.; Martin,J.R. (Dep. Pathol., Schumpert Med. Cent., Shreveport, LA 71101, USA) *J. Clin. Microbiol., 5(6), 656-657 (1977)* En;en.

140-U2 **A simplified tetrazolium reduction test for *Mycoplasma pneumoniae*.** Johnson,J.E.; Smith,T.F. (Sect. Clin. Microbiol., Mayo Clin. and Mayo Found., Rochester, MN 55901, USA) *Med. Lab. Sci., 33(3), 235-236 (1976)* En.

141-U2 **Differentiation of rapidly growing scotochromogenic mycobacteria employing the ZYM API system as substratum.** Sabater,J.F.G.; Amador,A.; Perales,J. (Serv. Med. Prev. y Anal. Clin., Secc. Bacteriol., Hosp. Mil., Valencia, Spain) *Acta Microbiol Acad. Sci. Hung., 26(4), 345-349 (1979)* En;en.

142-U2 **Reliable urease test for identification of mycobacteria.** Steadham,J.E. (Bur. Lab., Texas Dep. Health, Austin, TX 78756, USA) *J. Clin. Microbiol., 10(2), 134-137 (1979)* En;en.

143-U2 **Comparative study of some methods of isolation and identification of mycoplasmas in cell cultures.** Cracea,E.; Botez,D.; Ioanid,L.; Constantinescu,S.; Ionescu,M.D.; Petrovici,A. (Inst. Dr. I. Cantacuzino, Spl. Independentei 103, Bucuresti 35, Romania) *Arch. Roum. Pathol. Exp. Microbiol., 36(1), 5-11 (1977)* En;en,fr,ru.

144-U2 **An evaluation of the applicability of DAPI for detecting *Mycoplasma* in cell cultures.** Jagielski,M.; Zaleska,M.; Kaluwzewski,S.; Polna,I. (00-791 Warszawa, ul. Chocimska 24, Zaklad Bakteriol., PZH, Poland) *Med. Dosw. Mikrobiol., 28(2), 161-173 (1976)* Pl;en,pl,ru.

145-U2 **An easy method for the isolation of fruiting myxobacterial from various substrates.** Singh,N.B.; Yadava,J.N.S. (Cent. Drug Res. Inst., Lucknow 226001, India) *Curr. Sci., 47(15), 541-542 (1978)* En.

146-U2 Simplified method for cultural diagnosis of gonorrhoea. Willcox,R.R.; John,J. (Praed St. Clin., St. Mary's Hosp., London W2, UK) *Br. J. Vener. Dis., 52(4), 256 (1976)* En.

147-U2 Improved transport and culture system for the rapid diagnosis of gonorrhoea. Jephcott,A.E.; Bhattacharyya,M.N.; Jackson,D.H. (Public Health Lab., Northern Gen. Hospital, Herries Road, Sheffield S5 7AU, UK) *Br. J. Vener. Dis., 52(4), 250-252 (1976)* En;en.

148-U2 Stability of working reagents for the Modified Rapid Fermentation Test (MRFT). Brown,W.J. (Cent. Dis. Control, Public Health Serv., USDHEW, Atlanta, GA 30333, USA) *Health Lab. Sci., 14(3), 172-176 (1977)* En;en.

149-U2 Evaluation of methods for the rapid identification of *Neisseria gonorrhoeae* in a routine clinical laboratory. Pollock,H.M. (Dep. Lab. Med. and Microbiol., Univ. Washington, Seattle, WA 98104, USA) *J. Clin. Microbiol., 4(1), 19-21 (1976)* En;en.

150-U2 Detection of *N. gonorrhoeae* via Jembec system. Author(s) not stated (State Hyg. Lab., Univ. Iowa Med. Lab. Build., Iowa City, IA 52241, USA) *J. Iowa Med. Soc., 66(11), 444-445 (1976)* En.

151-U2 A comparison of three fermentation methods for the confirmation of *Neisseria gonorrhoeae*. Brown,W.J. (Cent. Dis. Control, Public Health Serv., US Dep. Health, Educ., and Welfare, Atlanta, GA 30333, USA) *Health Lab. Sci., 13(1), 54-58 (1976)* En;en.

152-U2 Rapid micro-carbohydrate test for confirmation of *Neisseria gonorrhoeae*. Yong,D.C.T.; Prytula,A. (Windsor Reg. Public Health Lab., Lab. Serv. Branch, Ontario Minist. Health, Windsor, Ont. N9A 6S2, Canada) *J. Clin. Microbiol., 8(6), 643-647 (1978)* En;en.

153-U2 Growth on Congo red agar: possible means of identifying penicillin-resistant non-penicillinase-producing gonococci. Payne,S.M.; *Finkelstein,R.A. (Dep. Microbiol., Univ. Texas Southwestern Med. Sch. and Grad. Sch. Biomed. Sci., Dallas, TX 75235, USA) *J. Clin. Microbiol., 6(5), 534-535 (1977)* En;en.

154-U2 Rapid identification of *Neisseria gonorrhoeae* and *Neisseria meningitidis* by using enzymatic profiles. D'Amato,R.F.; Eriquez,L.A.; Tomfohrde,K.M.; Singerman,E. (Analytab Products, Plainview, NY 11803, USA) *J. Clin. Microbiol., 7(1), 77-81 (1978)* En;en.

155-U2 The detection of *Neisseria gonorrhoeae* in liquid medium using a modification of the oxidase test. Geary,I.; Mellersh,A. (Dep. Med. Microbiol., Univ. Sheffield Med. Sch., Beech Hill Road, Sheffield S10 2RX, UK) *Med. Lab. Sci., 35(2), 195 (1978)* En.

156-U2 Simple disk-plate method for the biochemical confirmation of pathogenic *Neisseria*. /[presented in part at the 74th Annual Meeting of the American Society for Microbiology, held in Chicago, Illinois, 12-17 May 1974]. Valu,J.A. (Microbiol. Res. Lab., Akro-Medic Eng., Inc., Denville, NJ 07834, USA) *J. Clin. Microbiol., 3(2), 172-174 (1976)* En;en.

157-U2 Evaluation of rapid carbohydrate degradation tests for identification of pathogenic *Neisseria*. Pizzuto,D.J.; *Washington,J.A.,II (Sect. Clin. Microbiol., Mayo Clin. and Mayo Found., Rochester, MN 55901, USA) *J. Clin. Microbiol., 11(4), 394-397 (1980)* En;en.

158-U2 Rapid carbohydrate fermentation test for confirmation of the pathogenic *Neisseria* using a $Ba(OH)_2$ indicator. Slifkin,M.; Pouchet,G.R. (Microbiol. Sect., Singer Res. Inst., Allegheny Gen. Hosp., Pittsburgh, PA 15212, USA) *J. Clin. Microbiol., 5(1), 15-19 (1977)* En;en.

159-U2 Culturing of *Neisseria gonorrhoeae* - development of easy method. Itani,Z.S. (Univ.-Hautklin. Dusseldorf, Moorenstr. 5, D-4000 Dusseldorf, GFR) *Dermatologica, 154(5), 273-276 (1977)* De;de,en.

160-U2 Evaluation of the microcult system for isolating and identifying *Neisseria gonorrhoeae*. Williams,R.J.; Ratnatunga,C.S.; Hamilton-Miller,J.M.T.; Brumfitt,W. (Dep. Med. Microbiol., R. Free Hosp., London NW3, UK) *J. Clin. Pathol., 31(3), 209-212 (1978)* En;en.

161-U2 A simple carbohydrate fermentation test for identification of the pathogenic *Neisseria*. Reddick,A. (Bur. Lab., South Carolina Dep. Health and Environ. Control, Columbia, SC 29201, USA) *J. Clin. Microbiol., 2(1), 72-73 (1975)* En;en.

162-U2 Evaluation of the Microcult-GC kit as a screening method for the detection of *Neisseria gonorrhoeae*. Sachs,G.; Hofherr,L. (Dep. Lab. Med. and Pathol., Univ. Minnesota, Minneapolis, MN 55455, USA) *Am. J. Med. Technol., 44(10), 937-940 (1978)* En;en.

163-U2 [A simple procedure for the incubation of gonococci cultures]. Tiller,F.W.; Rott,G. (Bezirkshygieneinst., Abt. Mikrobiol., Burgstr. 2, DDR-65 Gera, GDR) *Dermatol. Monatsschr., 165(2), 139-140 (1979)* De.

164-U2 *Neisseria* confirmation by an enriched, bicarbonate-containing carbohydrate medium. Graves,J.O.; Magee,L.A. (Mississippi State Board of Health, Jackson, MS 39205, USA) *J. Clin. Microbiol., 8(5), 525-528 (1978)* En;en.

165-U2 Penicillinase-producing *Neisseria gonorrhoeae*. Phillips,C.W.; Aller,R.D.; Cohen,S.N. (Dep. Lab. Med., Univ.California, San Francisco, CA 94143, USA) *Lancet, 2(7992), 960 (1976)* En.

166-U2 A simple manganous chloride and Congo red disc method for differentiating *Neisseria gonorrhoeae* from *Neisseria meningitidis*. Odugbemi,T.O.; *McEntegart,M.G.; Hafiz,S. (Dep. Med. Microbiol., Univ. Sheffield Med. Sch., Beech Hill Road, Sheffield S10 2RX, UK) *J. Clin. Pathol., 31(10), 936-938 (1978)* En;en.

167-U2 The use of Microcult GC in women. Clay,J.C.; Willcox,R.R. (Praed Street Clin., St. Mary's Hosp., London W2, UK) *Br. J. Clin. Pract., 31(10), 149-151 (1977)* En;en.

168-U2 Identification of *Neisseria gonorrhoeae* by carbohydrate disc reactions on a modified fermentation medium. Odugbemi,T.O.; Hafiz,S. (Dep. Med. Microbiol., Univ. Sheffield Med. Sch., Sheffield, S. Yorks., UK) *J. Trop. Med. Hyg., 81(6), 106-109 (1978)* En;en.

169-U2 Cultural diagnosis of gonorrhoea with modified New York City (MNYC) medium. Young,H. (Dep. Bacteriol., University Med.Sch., Edinburgh, UK) *Br.J. Vener. Dis., 54(1), 36-40 (1978)* En;en.

170-U2 Use of bacteriophages as an adjunct in the identification of *Pasteurella multocida*. Gadberry,J.L.; Miller,N.G. (Dep. Med. Microbiol., Univ. Nebraska Coll. Med., Omaha, NB 68105, USA) *Am. J. Vet. Res., 38(1), 129-130 (1977)* En;en.

171-U2 Evaluation of the sodium polyanethol sulfonate disk test for the identification of *Peptostreptococcus anaerobius*. Wideman,P.A.; Vargo,V.L.; Citronbaum,D.; Finegold,S.M. (Med. and Res. Serv., Wadsworth Hosp. Cent., Veterans Adm., Los Angeles, CA 90073, USA) *J. Clin. Microbiol., 4(4), 330-333 (1976)* En;en.

172-U2 A method for the isolation of phage mutants altered in their response to lysogenic induction. Mount,D.W. (Dep. Microbiol., Coll. Med., Univ. Arizona, Tucson, AZ 85724, USA) *Mol. Gen. Genet., 145(2), 165-167 (1976)* En;en.

173-U2 The identification of pseudomonads and related bacteria in a clinical laboratory. King,A.; Phillips,I. (Dep. Microbiol., St. Thomas's Hosp. Med. Sch., London SE1 7EH, UK) *J. Med. Microbiol., 11(2), 165-176 (1978)* En;en.

174-U2 Improved medium for recovery and enumeration of *Pseudomonas aeruginosa* from water using membrane filters. Brodsky,M.H.; Ciebin,B.W. (Ontario Minist. Health, Lab. Serv. Branch, Environ. Bacteriol., Toronto, Ont. M5W 1R5, Canada) *Appl. Environ. Microbiol., 36(1), 36-42 (1978)* En;en.

175-U2 Rapid selection of microorganisms belonging to the genus *Pseudomonas* and the order Actinomycetales and producing exocellular lecithinases. Kolesnikova,I.G.; Kuimova,T.F. (Inst. Microbiol., Acad. Sci. USSR, Moscow, USSR) *Mikrobiologiya, 45(6), 987-990 (1976)* Ru;en,ru.

176-U2 Differentiation of fluorescent pseudomonads by their effect on milk agar. Reynolds,M.T.; Falkiner,F.R.; Hardy,R.; Keane,C.T. (Dep. Clin. Microbiol., Trinity Coll., Adelaide Hosp., Peter St., Dublin 8, Eire) *J. Med. Microbiol., 2(3), 379-382 (1979)* En;en.

177-U2 Identification of *Pseudomonas aeruginosa* with the API-20E system. Amato,S.; Vanik,J.; *Kocka,F.E. (Clin. Microbiol. Lab., Veterans Adm. Med. Cent., North Chicago, IL 60064, USA) *Can. J. Microbiol., 26(4), 554-555 (1980)* En;en,fr.

178-U2 Identification of *Pseudomonas pseudomallei* in the clinical laboratory. Ashdown,L.R. (Microbiol. Dep., Australian Dep. Health, Pathol. Lab., MSO Box 5278, Townsville, Queensld. 4810, Australia) *J. Clin. Pathol., 32(5), 500-504 (1979)* En;en.

179-U2 An improved indicator plant method for the detection of *Pseudomonas solanacearum* race 3 in soil. Graham,J.; Lloyd,A.B. (Dep. Microbiol. and Genet. Univ., New England, Armidale, NSW 2351, Australia) *Plant Dis. Rep., 62(1), 35-37 (1978)* En;en.

180-U2 A selective medium for the rapid isolation of pseudomonads associated with poultry meat spoilage. Mead,G.C.; Adams,B.W. (Agric. Res. Council, Food Res. Inst., Colney Lane, Norwich NR4 7UA, UK) *Br. Poult. Sci., 18(6), 661-670 (1977)* En;en.

181-U2 The methylene blue reduction disc test (M.R.D.-test): a simplified method for the identification of *Pseudomonas* species. Schubert,R.H.W.; Esanu,J.G. (Zent. Hyg., Klin. Johann Wolfgang Goethe-Univ. Frankfurt a. M., Abt. Allg. und Umwelthyg., Frankfurt a. M., GFR) *Zentralbl. Bakteriol. Parasitenkd. Infektionskr. Hyg., I Abt. A, 239(4), 504-509 (1977)* De;de,en.

182-U2 A further simplified procedure for the detection of *Pseudomonas aeruginosa* in contaminated aqueous substrata. Mossel,D.A.A.; de Vor,H.; Eelderink,I. (Lab. Microbiol., Dep. Sci. Food of Anim. Orig., Fac. Vet. Med., Univ. Utrecht, Utrecht, Netherlands) *J. Appl. Bacteriol., 41(2), 307-309 (1976)* En;en.

183-U2 A simple and inexpensive culture tube method for assaying acetylene reduction by facultative and anaerobic nitrogen-fixing bacteria. Tu,C.M. (Res. Inst. Agric. Canada, London, Ont. N6A 5B7, Canada) *Commun. Soil Sci. Plant Anal., 9(3), 243-247 (1978)* En;en.

184-U2 A semi-solid enrichment medium in the isolation of *Salmonella* from minced meat. de Blaauw,L.H.; de Pijper,F.W. (Keurmeester-Lab., Vleeskeuringsdienst, Amsterdam, Netherlands) *Tijschr. Diergeneeskd., 103(17), 889-893 (1978)* Nl;en,nl.

185-U2 A simplified biochemical system to screen *Salmonella* isolates from poultry for stereotyping. Cox,N.A.; Williams,J.E. (USDA, ARS, Anim. Prod. Res. Lab., Richard B. Russell Agric. Res. Cent., Athens, GA 30604, USA) *Poult. Sci., 55(5), 1968-1971 (1976)* En;en.

186-U2 Shortened identification of *Salmonella* by the use of four-tube test. Hoszowski,A.; Truszczynski,M. (Al. Partyzantow 53, 24100 Pulawy, Poland) *Med. Weter., 33(12), 738-740 (1977)* Pl;en,pl,ru.

187-U2 An economic and rapid diagnostic procedure for the detection of *Salmonella/Shigella* using the polyvalent *Salmonella* phage O-1. Fey,H.; Burgi,E.; Margadant,A.; Boller,E. (Vet. Bacteriol. Inst., Univ. Bern, Langgass. Str. 122, CH-3012 Bern, Switzerland) *Zentralbl. Bakteriol. Parasitenkd. Infektionskr. Hyg., I Abt. A, 240(1), 7-15 (1978)* En;de,en.

188-U2 Rapid confirmation of suspect *Salmonella* colonies by use of the Minitek system in conjunction with serological tests. Cox,N.A.; Mercuri,A.J. (Anim. Prod. Res. Lab.,

Richard B. Russell Agric. Res. Cent., USDA, ARS, PO Box 5677, Athens, GA 30604, USA) *J. Appl. Bacteriol., 41(3), 389-394 (1976)* En;en.

189-U2 Isolation of salmonella using a standardised inoculum and a rotary plating technique. Lynton-Moll,C.A.; Johnson,T. (Public Health Lab., East Birmingham Hosp., Bordesley Green East, Birmingham B9 5ST, UK) *J. Clin. Pathol., 31(9), 904-906 (1978)* En.

190-U2 Timed-release capsule method for the detection of salmonellae in foods and feeds. Sveum,W.H.; *Hartman,P.A. (Dep. Bacteriol., Iowa State Univ., Ames, IA 50011, USA) *Appl. Environ. Microbiol., 33(3), 630-634 (1977)* En;en.

191-U2 Laboratory and clinical investigation of recovery of *Salmonella typhi* from blood. Watson,K.C. (Cent. Microbiol. Lab., Western Gen. Hosp., Edinburgh, UK) *J. Clin. Microbiol., 7(2), 122-126 (1978)* En;en.

192-U2 A method of detecting *Salmonella* in water samples with a high pathogen concentration. Kruger,W.; Kraft-Kugler,K. (Bezirks-Hyg. Inst. Berlin, DDR-1055 Berlin, Schneeglockchenstr. 26, GDR) *Z. Gesamte Hyg. Grenzgeb., 22(10), 771-773 (1976)* De;de,en,ru.

193-U2 [A rapid and precise kit for the demonstration of salmonellae infecting food]. Cirilli,G. (Associazione Ital. Anal.Chim., Bologna, Italy) *Ind. Aliment., 17(12), 938-939 (1978)* It.

194-U2 Application of the rapid lysine decarboxylase test for early isolation and detection of salmonellae in sewage and other wastewaters. Phirke,P.M. (Bacteriol. Cell, Life Sci. Div., Natl. Environ. Eng. Res. Inst., Nehru Marg, Nagpur 440 020, India) *Appl. Environ. Microbiol., 34(4), 453-455 (1977)* En;en.

195-U2 Duodenal isolation of *Salmonella typhi* by string capsule in acute typhoid fever. Gilman,R.H.; Hornick,R.B. (Univ. Maryland Sch. Med., Baltimore, MD 21202, USA) *J. Clin. Microbiol., 3(4), 456-457 (1976)* En;en.

196-U2 An economic and rapid diagnostic procedure for the detection of *Salmonella/Shigella* using the polyvalent *Salmonella* phage O-1. Fey,H.; Burgi,[illegible]; Margadant,A.; Boller,E. (Vet. Bacteriol. Inst., Univ. Bern, Langgass. Str. 122, CH-3012 Bern, Switzerland) *Zentralbl. Bakteriol. Parasitenkd. Infektionskr. Hyg., 1 Abt. A, 240(1), 7-15 (1978)* En;de,en.

197-U2 [One-step method for identification of *Salmonella* and *Shigella* by biochemical tests. I. Evaluation and selection of the differentiating tests]. Tyc,Z.; Macierewicz,M.; Zaleska,H. (00-791 Warszawa, ul. Chocimska 24, Zaklad Bakteriol. PZH, Poland) *Med. Dosw. Mikrobiol., 28(2), 121-130 (1976)* Pl;en,ru..

198-U2 A method of cultivation of typhoid bacilli under conditions of automatic regulation of glucose supply. Savranskaya,S.Ya.; Gorbachev,I.D.; Zhdanova,L.G. (Inst. Vaccines and Sera, Moscow, USSR) *Zh. Mikrobiol. Epidemiol. Immunobiol., 55(5), 43-46 (1978)* Ru;en,ru.

199-U2 A simple procedure for screening of salmonellae using a semi-solid enrichment and a semi-solid indicator medium. Chau,P.Y.; Huang,C.T. (Dep. Microbiol., Univ. Hong Kong, Hong Kong) *J. Appl. Bacteriol., 41(2), 283-294 (1976)* En;en.

200-U2 Iron bacteria of the genus *Siderocapsa* in mineral waters. Scorcova,L. (Konevova 20, Karlovy Vary, Czechoslovakia) *Z. Allg. Mikrobiol. Morphol. Physiol. Okol. Mikroorg., 15(7), 553-557 (1975)* En;en.

201-U2 Micromethod for biochemical identification of coagulase-negative staphylococci. Brun,Y.; *Fleurette,J.; Forey,F. (Lab. Bacteriol., Fac. Med. Alexis Carrel, 69372 Lyon Cedex 2, France) *J. Clin. Microbiol., 8(5), 503-508 (1978)* En;en.

202-U2 A rapid, simple agar-overlay method for the detection of penicillinase-producing *Staphylococcus aureus* in the clinical bacteriology laboratory. Wong,K.W.; Soo-Hoo,T.S. (Dep. Med. Microbiol., Fac. Med., Univ. Malaya, Kuala Lumpur, Malaysia) *Jap. J. Microbiol., 20(2), 143-154 (1976)* En.

203-U2 Rapid identification of *Staphylococcus aureus* by using lysostaphin sensitivity. Severance,P.J.; *Kauffman,C.A.; Sheagren,J.N. (Infect. Dis. Sect., Ann Arbor Veterans Adm. Med. Cent. and Univ. Michigan Med. Sch., Ann Arbor, MI 48105, USA) *J. Clin. Microbiol., 11(6), 724-727 (1980)* En;en.

204-U2 A simple test for the assessment of the suitability of pork plasma for incorporation into a Baird-Parker base medium for the enumeration of *Staph. aureus* in foods and other specimens. Mossel,D.A.A.; Eelderink,I. (Dep. Food. Microbiol., Fac. Vet. Med., Univ. Utrecht, Bilstraat 172, Utrecht, Netherlands) *Lab. Pract., 28(6), 623 (1979)* En.

205-U2 Evaluation of three rapid tests for identification of *Staphylococcus aureus* isolated in bovine milk. Poutrel,B.; Ducelliez,M. (INRA Stn. Pathol. Reprod., 37380 Nouzilly, France) *Ann. Rech. Vet., 10(1), 125-129 (1979)* En;en,fr.

206-U2 Accelerated procedure for the enumeration and identification of food-borne *Staphylococcus aureus*. Lachica,R.V. (Unified Food Control Lab., Inst. Nutr. Cent. America and Panama, Guatemala City, Guatemala) *Appl. Environ. Microbiol., 39(1), 17-19 (1980)* En;en.

207-U2 Simplified thermonuclease test for rapid identification of *Staphylococcus aureus* recovered on agar media. Lachica,R.V.F. (Unified Food Control Lab., Inst. Nutr. Central America and Panama, Guatemala City, Guatemala) *Appl. Environ. Microbiol., 32(4), 633-634 (1976)* En;en.

208-U2 Fecal contamination - the water analyst's responsibility. Part II. Galvani,M.M. (St. Paul Water Dep., St. Paul, MN, USA) *Water and Sewage Works, 122(1), 68-70 (1975)* En.
[Faecal streptococci, media].

209-U2 Methods to improve detection of pneumococci in respiratory secretions. Dilworth,J.A.; Stewart,P.; Gwaltney,J.M.,Jr.; Hendley,O.; *Sande,M.A. (Dep. Intern. Med., Univ. Virginia Sch. Med., Charlottesville, VA 22901, USA) *J. Clin. Microbiol., 2(5), 453-455 (1975)* En;en.

210-U2 A rapid test for the identification of enterococci. Hahn,G.; Tolle,A. (Inst. Hyg., Bundesanstalt Milchforsch., Kiel, GFR) *Milchwissenschaft., 30(10), 585-597 (1975)* De;de,en.

211-U2 Multiple inocula (replicator) CAMP test for presumptive identification of group B streptococci. Fuchs,P.C.; Christy,C.; Jones,R.N. (Microbiol. Sect., Dep. Pathol., St. Vincent Hosp. and Med. Cent., Portland, OR 97225, USA) *J. Clin. Microbiol., 7(2), 232-233 (1978)* En;en.

212-U2 Rapid sodium chloride tolerance test for presumptive identification of enterococci. Qadri,S.M.H.; Nichols,C.W.; Qadri,S.G.M. (Dep. Pathol. and Lab. Med., Univ. Texas Med. Sch., Houston, TX 77030, USA) *J. Clin. Microbiol., 7(2), 238 (1978)* En;en.

213-U2 Modified Christie-Atkins-Munch-Petersen (CAMP) test for direct identification of hemolytic and nonhemolytic group B streptococci on primary plating. Bae,B.H.C.; Bottone,E.J. (Microbiol. Lab., Dep. Lab., Long Island Coll. Hosp., Brooklyn, NY 11201, USA) *Can. J. Microbiol., 26(4), 539-542 (1980)* En;en,fr.

214-U2 Presumptive identification of enterococci from other D streptococci by a rapid sodium chloride tolerance test. Qadri,S.M.H.; deSilva,M.J.; Qadri,S.G.M.; Villarreal,A. (University Hosp. and Clin., Oklahoma City, OK 73125, USA) *Med. Microbiol. Immunol., 167(3), 197-203 (1979)* En;en.

215-U2 CAMP-disk test for presumptive identification of group B streptococci. Wilkinson,H.W. (Cent. Dis. Control, Atlanta, GA 30333, USA) *J. Clin. Microbiol., 6(1), 42-45 (1977)* En;en.

216-U2 A shortened scheme for the identification of indifferent streptococci. Waitkins,S.; Ball,L.C.; Fraser,C.A.M.; (Reg. Public Health Lab., Fazakerley Hosp., Lower Lane, Liverpool L9 7AL, UK) *J. Clin. Pathol., 33(1), 47-52 (1980)* En;en.

217-U2 Use of the API-ZYM system in rapid identification of *α* and non-haemolytic streptococci. Waitkins,S.A.; Ball,L.C.; Fraser,C.A.M. (Reg. Public Health Lab., Fazakerley Hosp., Lower Lane, Liverpool L9 7AL, UK) *J. Clin. Pathol., 33(1), 53-57 (1980)* En;en.

218-U2 Identification of viridans streptococci on the Minitek Miniaturised differentiation system Holloway,Y.; Schaareman,M.; *Dankert,J. (Lab. Med. Microbiol., Univ. Hosp. Groningen. Oostersingel 59, Groningen, Netherlands) *J. Clin. Pathol., 32(11), 1168-1173 (1979)* En;en.

219-U2 Serological identification of group A streptococci from throat scrapings before culture. El Kholy,A.; Facklam,R.; Sabri,G.; Rotta,J. (Biomed. Res. Cent. Infect. Dis., Cairo, Egypt) *J. Clin. Microbiol., 8(6), 725-728 (1978)* En;en.

220-U2 Rapid recognition of group-B streptococci. Islam,A.K.M.S. (Dep. Pathol., St. Peter's Hosp. and Inst. Virol., London WC2H 9AE, UK) *Lancet, 1(8005), 256-257 (1977)* En.

221-U2 A modified method for the selection of mutants of *Actinomyces fradiae* having higher antibiotic activity. Kibarska,T.T.; Markov,K.I.; Naidenova,M.M. (Inst. Microbiol., Bulgarian Acad. Sci., Sofia, Bulgaria) *Dokl. Bolg. Akad. Nauk., 30(9), 1329-1330 (1977)* En.

222-U2 Rapid screening of *Veillonella* by ultraviolet fluorescence. Chow,A.W.; Patten,V.; Guze,L.B. (Dep. Med., Harbor Gen. Hosp., Torrance, CA 90509, USA) *J. Clin. Microbiol., 2(6), 546-548 (1975)* En;en.

223-U2 Rapid presumptive identification of vibrios by immobilization in distilled water. Chester,B.; Poulos,E.G. (Lab. Serv., Veterans Adm. Hosp., Miami, FL 33125, USA) *J. Clin. Microbiol., 11(5), 537-539 (1980)* En;en.

224-U2 A simple laboratory method for the diagnosis of *V. cholerae*. Huq,M.I. (Cholera Res. Lab., GPO Box 128, Dacca-2, Bangladesh) *Trans. R. Soc. Trop. Med. Hyg., 73(5), 553-536 (1979)* En;en.

225-U2 [Modification of Pessoa and Da Silva medium for a rapid identification of bacteria of the genera *Yersinia* and *Pasteurella*]. Louzis,C.; Dubois-Darnaudpeys,A. (Minist. Agric. Dir. Qual., Serv. Vet. Lab. Cent. Rech. Vet., 22 rue Pierre-Curie, 94700 Maisons-Alfort, France) *Recl. Med. Vet., Ec. Alfort, Paris, 153(6), 435-438 (1977)* Fr;en,es,fr.

226-U2 PLP test as a rapid method for preliminary identification of *Yersinia enterocolitica* and *Yersinia pseudotuberculosis*. Zaremba,M. (15-952 Bialystok, ul. Mickiewicza 2C, Zaklad Mikrobiol. Inst. Biostrukt. AM, Poland) *Med. Dosw. Mikrobiol., 30(4), 243-246 (1978)* Pl;en,pl,ru.

227-U2 Alkali method for rapid recovery of *Yersinia enterocolitica* and *Yersinia pseudotuberculosis* from foods. Aulisio,C.C.G.; Mehlman,I.J.; Sanders,A.C. (Div. Microbiol., Food and Drug Adm., Washington, DC 20204, USA) *Appl. Environ. Microbiol., 39(1), 135-140 (1980)* En;en.

228-U2 Rapid enumeration of psychrotrophic bacteria in raw and pasteurized milk. Oliveria,J.S.; Parmelee,C.E. (FTA-UNICAMP, CP 1170, 13100 Campinas, Sao Paulo, Brazil) *J. Milk Food Technol., 39(4), 269-272 (1976)* En;en.

229-U2 Use of a rapid fermentation test for identification of anaerobic bacteria. Lindquist,B.L.; Kjellander,J. (Dep. Clin. Bacteriol., Cent. Cty. Hosp., S-701 85 Orebro, Sweden) *Med. Microbiol. Immunol., 165(1), 67-72 (1978)* En;en.

230-U2 Evaluation of two rapid methods for identification of commonly encountered nonfermenting or oxidase-positive, gram-negative rods. Dowda,H. (Div. Diagn. Microbiol., South Carolina Dep. Health and Environ. Control, Columbia,

SC 29201, USA) *J. Clin. Microbiol., 6(6), 605-609 (1977)* En;en.

231-U2 **Simplified method for bacteriological urine culture.** Toni,M.; Menozzi,M.G.; Allevato,F.; Schito,G.C. (Univ. Parma, Ist. Microbiol., Parma, Italy) *Ann. Sclavo Riv. Microbiol. Immunol., 19(6), 1177-1188 (1977)* It;en.

232-U2 **Simple equipment for the growth of photo-synthetic bacteria.** Stevens,S.E.,Jr.; *Fox,J.L. (Dep. Zool., Univ. Texas at Austin, Austin, TX 78712, USA) *J. Appl. Bacteriol., 42(2), 275-278 (1977)* En;en.

233-U2 **A microperfusion chamber for studying the growth of bacterial cells.** Duxbury,T. (Dep. Microbiol., Univ. Sydney, Sydney 2006, NSW, Australia) *J. Appl. Bacteriol., 43(2), 247-251 (1977)* En;en.

234-U2 **Improved chamber for the isolation of anaerobic microorganisms.** Cox,M.E.; Mangels,J.I. (Int. Shellfish Enterprises, Moss Landing, CA 95039, USA) *J. Clin. Microbiol., 4(1), 40-45 (1976)* En;en.

235-U2 **A simple dynamic method of K_La determination in laboratory fermenter.** Mukhopadhyay,S.N.; Ghose,T.K. (Biochem. Eng. Div., Sch. Bioeng. and Biosci. Stud., IIT Hauz Khas, New Delhi-110029, India) *J. Ferment. Technol., 54(6), 406-419 (1976)* En;en.

236-U2 **Semi-batch culture of methanol-assimilating bacteria with exponentially increased methanol feed.** Yamane,T.; Kishimoto,M.; Yoshida,F. (Chem. Eng. Dep., Kyoto Univ., Kyoto, Japan) *J. Ferment. Technol., 54(4), 229-240 (1976)* En;en.

237-U2 **Characterisation of bacteria by Curie-point-pyrolysis, gas chromatography and computerisation of the pyrograms.** Wasserfallen,K.; Rinderknecht,F. (Inst. Lebensmittelchem., Univ. Bern. Postfach, CH-3000 Bern 9, Switzerland) *Chromatographia, 11(3), 128-136 (1978)* De;de,en.

238-U2 **A method for the production of pre-reduced anaerobically sterilized culture media.** Gosden,P.E.; Ware,G.C. (Dep. Bacteriol., Univ. Bristol., Bristol BS8 1TD, UK) *J. Appl. Bacteriol., 42(1), 77-79 (1977)* En;en.

239-U2 **Rapid method for identification of gram-negative, nonfermentative bacilli.** Otto,L.A.; *Pickett,M.J. (Dep. Bacteriol., Univ. California, Los Angeles, CA 90024, USA) *J. Clin. Microbiol., 3(6), 566-575 (1976)* En;en.

240-U2 **Fecal monitoring of caged birds.** Dolphin,R.E.; Olsen,D.E. (Cat and Bird Hosp., 11002 Pacific Ave., Tacoma, WA 98444, USA) *Vet. Med. Small Anim. Clin., 72(6), 1081-1085 (1977)* En.

241-U2 **A rapid glutamic decarboxylase test for identification of bacteria.** Freier,P.A.; Graves,M.H.; Kocka,F.E. (Clin. Microbiol. Lab., Univ. Chicago, Chicago, IL 60637, USA) *Ann. Clin. Lab. Sci., 6(6), 537-539 (1976)* En;en.

242-U2 **A simplified method for the growth and identification of strictly anaerobic bacteria.** Giglio,M.M.; Grossman,C.G. (Microbiol. Lab., San Juan de Dios Hosp., Santiago, Chile) *Rev. Med. Chile, 106(6), 453-458 (1978)* Es;en.

243-U2 **Evaluation of the API 20E system for identification of nonfermentative gram-negative bacteria.** Shayegani,M.; Maupin,P.S.; McGlynn,D.M. (Div. Lab. and Res., New York State Dep. Health, Albany, NY 12201, USA) *J. Clin. Microbiol., 7(6), 539-545 (1978)* En;en.

244-U2 **Computer-assisted identification of anaerobic bacteria.** Kelley,R.W.; Kellogg,S.T. (Dep. Microbiol., Univ. Hawaii, Honolulu, HI 96822, USA) *Appl. Environ. Microbiol., 35(3), 507-511 (1978)* En;en.

245-U2 **Comparison of two methods for assessing the removal of total organisms and pathogens from the skin.** Ayliffe,G.A.J.; Babb,J.R.; Bridges,K.; Lilly,H.A.; Lowbury,E.J.L.; Varney,J.; Wilkins,M.D. (MRC Burns Unit, Birmingham Accident Hosp., Birmingham, UK) *J. Hyg., 75(2), 259-274 (1975)* En;en.

246-U2 **Comparative investigation on recovering bacteria from the artificially contaminated hand.** Manner,F.; Rotter,M.; Mittermayer,H. (Salzstadlplatz 7, D-845 Amberg, GFR) *Zentralbl. Bakteriol. Parasitenkd. Infektionskr. Hyg., I Abt. B, 160(4-5), 412-431 (1975)* De;de,en.

247-U2 **Performance of converted pressure cookers and two conventional jars for anaerobic bacterial culture.** Gargan,R.A.; Phillips,I. (Dep. Microbiol., St. Thomas's Hosp. Med. Sch., London SE1 7EH, UK) *J. Clin. Pathol., 31(5), 426-429 (1978)* En;en.

248-U2 **A simple device for cultivation of anaerobic bacteria.** Melinder,P.C. (Veterinarfak., Sveriges Jordbruksuniv., S-751 23 Uppsala, Sweden) *Nord. Vet. Med., 30(10), 430-433 (1978)* Sv;en,sv.

249-U2 **Rapid detection of selected gram-negative bacteria by aminopeptidase profiles.** Peterson,E.H.; Hsu,F.J. (DHFW/PHS Food and Drug Adm., NCMI, 240 Hennepin Ave., Minneapolis, MN 55401, USA) *J. Food Sci., 43(6), 1853-1856 (1978)* En;en.

250-U2 **API computer profiles: correlation of API 20E with API 10S.** Phillips,S.B.; *Amsterdam,D. (Dep. Microbiol., Isaac Albert Res. Inst. Kingsbrook Jewish Med. Cent., Brooklyn, NY 11203, USA) *J. Clin. Microbiol., 6(6), 645-646 (1977)* En;en.

251-U2 **Computer-assisted bacterial identification utilizing antimicrobial susceptibility profiles generated by Autobac 1.** Sielaff,B.H.; Johnson,E.A.; Matsen,J.M. (Pfizer, Inc., Groton, CT 06340, USA) *J. Clin. Microbiol., 3(2), 105-109 (1976)* En;en.

252-U2 **Evaluation of simplified dichotomous schemata for the identification of anaerobic bacteria from clinical material.** Porschen,R.K.; Stalons,D.R. (Microbiol. Sect., Veterans Adm. Hosp., Long Beach, CA 90801, USA) *J. Clin.*

Microbiol., 3(2), 161-171 (1976) En;en.

253-U2 Rapid fermentation testing of anaerobic bacteria. Schreckenberger,P.C.; Blazevic,D.J. (Curriculum in Med. Lab. Sci., Univ. Illinois Med. Cent., Chicago, IL 60612, USA) *J. Clin. Microbiol., 3(3), 313-317 (1976)* En;en.

254-U2 Evaluation of the rapid decarboxylase and dihydrolase test for the differentiation of nonfermentative bacteria. Oberhofer,T.R.; Rowen,J.W.; Higbee,J.W.; Johns,R.W. (Microbiol. Sect., Lab. Activity, Brooke Army Med. Cent., Fort Sam, Houston, TX 78234, USA) *J. Clin. Microbiol., 3(2), 137-142 (1976)* En;en.

255-U2 Automated methods for identification of bacteria from clinical specimens. Bascomb,S.; Spencer,R.C. (St. Mary's Hosp. Med. Sch., Univ. London, Dep. Bacteriol., Wright-Fleming Inst., London W21PG, UK) *J. Clin. Pathol., 33(1), 36-46 (1980)* En;en.

256-U2 Rapid miniaturized tests for bacteriuria: microstix and bacturcult urine tests. Winter,C.C. (Div. Urol., Ohio State Univ. Med. Cent., Columbus, OH 43210, USA) *J. Urol., 114(5), 755-757 (1975)* En;en.

257-U2 Investigations on microorganisms in thickening agents. II. Number of organisms of anaerobic sporulating species. Souw,P.; Rehm,H.J. (Inst. Mikrobiol., Univ. Munster, Munster, GFR) *Chem. Mikrobiol. Technol. Lebensm., 4(3), 71-74 (1975)* De;de,en,fr.

258-U2 Rapid detection of bacteremia by an early subculture technic. Todd,J.K.; Roe,M.H. (Children's Hosp. Denver, 1056 East Nineteenth Ave., Denver, CO 80218, USA) *Am. J. Clin. Pathol., 64(5), 694-699 (1975)* En;en.

259-U2 Design of a microculture chamber to observe cell division of bacterial L-forms in liquid medium. Nagy,S.S.; *Gilpin,R.W. (Dep. Microbiol., Med. Coll. Pennsylvania, Philadelphia, PA 19129, USA) *Appl. Environ. Microbiol., 31(3), 444-445 (1976)* En;en.

260-U2 The automicrobic system for urines. Nicholson,D.P.; Koepke,J.A. (Dep. Pathol., Univ. Iowa, Iowa City, IA 52242, USA) *J. Clin. Microbiol., 10(6), 823-833 (1979)* En;en.

261-U2 Semiquantitative catalase test as an aid in identification of oxidative and nonsaccharolytic gram-negative bacteria. Chester,B. (Veterans Adm. Hosp., Miami, FL 33125, USA) *J. Clin. Microbiol., 10(4), 525-528 (1979)* En;en.

262-U2 Sensitivity of a nitrite indicator strip method in detecting bacteriuria in preschool girls. Kunin,C.M.; DeGroot,J.E. (Veterans Adm. Hosp., 2500 Overlook Terrace, Madison, WI 53705, USA) *Pediatrics (Springfield), 60(2), 244-245 (1977)* En.

263-U2 Use of a single infusion bottle to culture and prepare potentially hazardous microorganisms for biochemical analysis. Weijman,A.C.M. (Centraalbur. Schimmelcultures, Baarn, Netherlands) *Can. J. Microbiol., 24(9), 1097-1098 (1978)* En;en,fr.

264-U2 A new anaerobic jar. Burt,R.; Philips,K.D. (Public Health Lab., Luton and Dunstable Hosp., Lewsey Road, Luton LU4 0DZ, UK) *J. Clin. Pathol., 30(11), 1082-1084 (1977)* En.

265-U2 Experiences with the API 20 A System in routine species identification of anaerobes. Essers,L.; Haralambie.E. (Inst. Med. Mikrobiol., Univ.-klin., Hufelandstr. 55, D-4300 Essen, GFR) *Zentralbl. Bakteriol. Parasitenkd. Infektionskr. Hyg., I Abt. A, 238(3), 394-401 (1977)* De;de.

266-U2 Cultivation of microorganisms spoiling beer. Savel,J.; Prokopova,M. (Jihoceske pivovary, n.p., Ceske Budejovice, Czechoslovakia) *Kvasny Prum., 24(11), 246-249 (1978)* Cs;cs;de,en,ru.

267-U2 Acceleration of tetrazolium reduction by bacteria. Bartlett,R.C.; *Mazens,M.; Greenfield,B. (Dep. Pathol., Hartford Hosp., Hartford, CT 06115, USA) *J. Clin. Microbiol., 3(3), 327-329 (1976)* En;en.

268-U2 Modification of the Minitek miniaturized differentiation system for characterization of anaerobic bacteria. Stargel,M.D.; Thompson,F.S.; Phillips,S.E.; Lombard,G.L.; Dowell,V.R.,Jr. (Cent. Dis. Control, Atlanta, GA 30333, USA) *J. Clin. Microbiol., 3(3), 291-301 (1976)* En;en.

269-U2 The API ZYM system in the identification of gram-negative anaerobes. Tharagonnet,D.; Sisson,P.R.; Roxby,C.M.; *Ingham,H.R.; Selkon,J.B. (Dep. Microbiol., Newcastle Gen, Hosp., Westgate Road, Newcastle upon Tyne, UK) *J. Clin. Pathol., 30(6), 505-509 (1977)* En;en.

270-U2 Simple method for carbon determination in microbiological experiments. Novak,B. (Drnovska 507, Praha 6, Ruzyne, Czechoslovakia) *Zentralbl. Bakteriol. Parasitenkd. Infektionskr. Hyg., II, 131(7), 588-591 (1976)* En;de,en.

271-U2 Urine culture: a new approach. Freedel,D.; Amortegui,A.J.; Feinberg,S. (Dep. Microbiol., Magee-Womens Hosp., Univ. Pittsburgh Sch. Med., Pittsburgh, PA 15213, USA) *Am. J. Med. Technol., 41(12), 454-456 (1975)* En;en.

272-U2 A comparison of the dip innoculum transport medium (Dipspoon) and Uricult methods for detecting bacteriuria. McKenna,H.; Johnston,D. (Pathol. Dep., R. Women's Hosp., Brisbane, Queensl. 4029, Australia) *Aust. J. Med. Technol., 3(3), 89-90 (1972)* En;en.

273-U2 Early detection and preliminary susceptibility testing of positive pediatric blood cultures with the steers replicator. Paisley,J.W.; Todd,J.K.; Roe,M.H. (Dep. Clin. Microbiol., Mayo Clin., Rochester, MN 55901, USA) *J. Clin. Microbiol., 6(4), 367-372 (1977)* En;en.

274-U2 Rapid identification and quantitation of small numbers of microorganisms by a chemiluminescent immunoreaction. Haimann,M.; Velan,B.; Sery,T. (Israel Inst.

Biol. Res., Ness-Ziona, Israel) *Appl. Environ. Microbiol., 34(5), 473-477 (1977)* En;en.

275-U2 Comparative study between the standard method and the improved technic with SMCA, for the detection and count of proteolytic bacteria. Espejo,S.J.; Perez,F.D. (Dep. Tecnol. y Bioquim. de los Alimentos, Fac. Vet., Univ. Cordoba, Cordoba, Spain) *Arch. Zootec., 27(107), 291-299 (1978)* Es;en,es.

276-U2 Laboratory evaluation of a multitest system for identification of gram-negative organisms. Rosenthal,S.L.; Freundlich,L.F.; Washington,W. (Room 6N22, Jacobi Hosp., Bronx Munic. Hosp. Cent., Bronx, NY 10461, USA) *Am. J. Clin. Pathol., 70(6), 914-917 (1978)* En;en.

277-U2 A modified tube method for the cultivation and enumeration of anaerobic bacteria. Ogg,J.E.; Lee,S.Y.; Ogg,B.J. (Colorado State Univ., Fort Collins, CO 80523, USA) *Can. J. Microbiol., 25(9), 987-990 (1979)* En;en,fr.

278-U2 Computer-aided numerical identification of gram-negative fermentative rods on a desk-top computer. Schindler,J.; Duben,J.; Lysenko,O. (Dep. Med. Microbiol., Charles Univ., Studnickova 7, 128 00 Prague 2, Czechoslovakia) *J. Appl. Bacteriol., 47(1), 45-51 (1979)* En;en.

279-U2 Alternative replica plating technique. Lindstrom,E.B. (Dep. Microbiol., Univ. Umea, S-901 87 Umea, Sweden) *Appl. Environ. Microbiol., 34(2), 225-227 (1977)* En;en.

280-U2 Rapid identification and antibiotic sensitivity testing of bacteria isolated from clinical infections. Lindberg,A.A.; *Nord,C.-E.; Dahlback,A. (Dep. Bacteriol., Natl. Bacteriol. Lab., Statens Bakteriol. Lab., S-105 21 Stockholm, Sweden) *Med. Microbiol. Immunol., 163(1), 13-24 (1977)* En;en.

281-U2 Testing reliability of the biochemical tests of the PathoTec system for identification of bacteria. Khomenko,N.A. (Inst. Vaccines and Sera, Moscow, USSR) *Zh. Mikrobiol. Epidemiol. Immunobiol., 54(4), 42-46 (1977)* Ru;en,ru.

282-U2 Development and use of single 'polytrophic' diagnostic tubes for the approximate taxonomic grouping of bacteria isolated from foods, water and medicinal preparations. Mossel,D.A.A.; Eelderink,I.; Sutherland,J.P. (Lab. Microbiol., Dep. Sci. Food Anim. Origin, Fac. Vet. Med., Univ. Utrecht, Bilstraat 172, Utrecht, Netherlands) *Zentralbl. Bakteriol. Parasitenkd. Infektionskr. Hyg., I Abt. A, 238(1), 66-79 (1977)* En;de,en.

283-U2 A five-hour system for identification of bacteria. Lorian,V.; Waluschka,A. (Div. Microbiol. and Epidemiol., Dep. Pathol., Bronx-Lebanon Hosp. Cent., Bronx, NY 10456, USA) *Am. J. Med. Technol., 45(7), 618-627 (1979)* En;en.

284-U2 A reliable test for differentiation and presumptive identification of certain clinically significant anaerobes. Hansen,S.L.; Stewart,B.J. (Lab. Serv., Veterans Adm. Hosp., 3900 Loch Raven Blvd., Baltimore, MD 21218, USA) *Am. J. Clin. Pathol., 69(1), 36-40 (1978)* En;en.

285-U2 API ZYM: a simple rapid system for the detection of bacterial enzymes. Humble,M.W.; King,A.; *Phillips,I. (Dep. Microbiol., P. Thomas's Hosp. Med. Sch., London SE1 7EH, UK) *J. Clin. Pathol., 30(3), 275-277 (1977)* En;en.

286-U2 An improved cooked meat medium for the growth of anaerobic bacteria. Wren,M.W.D. (Microbiol. Dep., North Middlesex Hosp., Edmonton, London N18 1QX, UK) *Med. Lab. Sci., 36(2), 197-199 (1979)* En.

287-U2 An anaerobic glove box for the isolation and cultivation of methanogenic bacteria. Cox,D.J.; Herbert,S.D. (Dep. Biol.,Univ. York, York YO1 5DD, UK) *J. Appl. Bacteriol., 45(3), 411-415 (1978)* En;en.

288-U2 Comparison of a commercial identification kit and conventional biochemical tests used for the identification of enteric gram-negative rods. McCarthy,L.R.; Mayo,J.B.; Bell,G.; *Armstrong,D. (Diagnostic Microbiol. Lab., Mem. Sloan-Kettering Cancer Cent., 1275 York Ave., New York, NY 10021, USA) *Am. J. Clin. Pathol., 69(2), 161-164 (1978)* En;en.

289-U2 Evaluation of the Oxi/Ferm tube system with selected gram-negative bacteria. Oberhofer,T.R.; Rowen,J.W.; Cunningham,G.F.; *Higbee,J.W. (Dep. Pathol. and Area Lab. Serv., Brooke Army Med. Cent., Fort Sam Houston, TX 78234, USA) *J. Clin. Microbiol., 6(6), 559-566 (1977)* En;en.

290-U2 Productivity of the roll-streak method to perform anaerobic bacteriology in the routine clinical laboratory. Neblett,T.R. (Henry Ford Hosp., 2799 West Grand Blvd., Detroit, MI 48202, USA) *Med. J., Henry Ford Hosp., 24(2), 103-118 (1976)* En;en.

291-U2 [What advantages do quadratic plastic petri dishes offer in a routine bacteriological laboratory?]. Seidler,M. (Staatl. Vet.-Untersuchungsamt. Hannover, Hannover, GFR) *Fleischwirtschaft, 56(4), 517-518 (1976)* De;de.

292-U2 Multichannel electrochemical microbial detection unit. Wilkins,J.R.; Young,R.N.; Boykin,E.H. (Natl. Aeronautics and Space Admin., Langley Res. Cent., Hampton, VA 23665, USA) *Appl. Environ. Microbiol., 35(1), 214-215 (1978)* En;en.

293-U2 Micromethod for identification of anaerobic bacteria: design and operation of apparatus. Wilkins,T.D.; Walker,C.B.; Moore,W.E.C. (Anaerobe Lab, Virginia Polytech. Inst. and State Univ., Blacksburg, VA 24061, USA) *Appl. Microbiol., 30(5), 831-837 (1975)* En;en.

294-U2 Modification of the Powell culture chamber for the observation and micromanipulation of growing bacteria. Isaac,L.; Leonard,P.G.; Ware,G.C. (Dep. Bacteriol., Med. Sch., Univ. Bristol, University Walk, Bristol BS8 1TD, UK) *Lab. Pract., 24(10), 667-669 (1975)* En.

295-U2 Microstix - a reagent strip for urine culture. Bailey,R.R.; Pearson,S. (Dep. Renal Med., Christhurch

Hosp., Christchurch, New Zealand) *N.Z. Med. J., 82(552), 331-333 (1975)* En;en.

296-U2 Some aspects concerning the intensification of production of bacterial preparations. Baskanyan,I.A.; Zaporozhtsev,L.N. (Inst. Vaccines and Sera, Mechnikova, Moscow, USSR) *Zh. Mikrobiol. Epidemiol. Immunobiol., 53(11), 30-34 (1976)* Ru;en,ru.

297-U2 Identification of anaerobes on the Minitek system, compared to a conventional system. Holloway,Y.; *Dankert,J. (Lab. Med. Microbiol., Dep. Hosp. Epidemiol., University Hosp. Groningen, Oostersingel 59, Groningen, Netherlands) *Zentralbl. Bakteriol. Parasitenkd. Infektionskr. Hyg., I Abt. A. 245(3), 324-331 (1979)* En;de,en.

298-U2 Nonfermentative bacilli: evaluation of three systems for identification. Otto,L.A.; Blachman,U. (FDA, Los Angeles, CA 90015, USA) *J. Clin. Microbiol., 10(2), 147-154 (1979)* En;en.

299-U2 Rapid method for distinction of gram-negative from gram-positive bacteria. Gregersen,T. (Vet. Fac. FAO Fellows,R. Vet. and Agric. Univ., 13 Bulowsvej, DK-1870 Copenhagen V, Denmark) *Eur. J. Appl. Microbiol. Biotechnol., 5(2), 123-127 (1978)* En;en.

300-U2 Micromethod for rapid identification of gram-negative, nonfermentative bacteria. Gibson,J.B.; Crull,S.L.; *Borchardt,K.A. (Clin. Microbiol. Lab., US Public Health Serv. Hosp., 15th and Lake St., San Francisco, CA 94118, USA) *Health Lab. Sci., 15(1), 9-14 (1978)* En;en.

301-U2 Evaluation of the Mini-Tek®, API-20E®, Enterotube®, and Oxi/Ferm® bacterial identification systems with industrial isolates. Bessicks,W.E. (Haskell Lab. Toxicol. and Ind. Med. E.I. du Pont de Nemours and Co. Newark, DE 19711, USA) *Dev. Ind. Microbiol., 20, 563-570 (1979)* En;en.

302-U2 [On the culture of anaerobes (glove box method)]. Bernhardt,H.; Mannchen,E.; Knoke,M.; Hadicke,K. (Med. Klin. Ber. Med. Ernst-Moritz-Arndt-Univ., Friedrich-Loeffler-Str. 23a, DDR-22 Greifswald, GDR) *Urologe, Sect. B, 34(22), 1046-1049 (1979)* De;de,en,ru.

303-U2 Rapid identification of gram-negative mastitis pathogens by means of the 'Urotube' Roche. A field trial. Beglinger,F. (Florastr. 67, CH-8610 Uster, Switzerland) *Schweiz. Arch. Tierheilkd., 122(1), 45-48 (1980)* De;de,en,fr,it.

304-U2 New approaches to the identification of microorganisms. Heden,C.-G.; Illeni,T. (eds.) *Publ. by:* John Wiley and Sons Ltd., Baffins Lane, Chichester, Sussex, England, March, 1975 ISBN: 0-471-36746-X £15-85. at S26.90. En.

305-U2 Rapid spot test for the determination of esculin hydrolysis. Edberg,S.C.; Gam,K.; Bottenbley,C.J.; Singer,J.M. (Div. Microbiol. and Immunol., Dep. Pathol., Montefiore Hosp. and Med. Cent., New York, NY 10467, USA) *J. Clin. Microbiol., 4(2), 180-184 (1976)* En;en.

306-U2 Multi-laboratory evaluation of an automated microbial detection/identification system. Smith,P.B.; Gavan,T.L.; Isenberg,H.D.; Sonnenwirth,A.; Taylor,W.L.; Washington,J.A.H.; Balows,A. (Cent. Dis. Control, Public Health Serv., Dep. Health, Educ., and Welfare, Atlanta, GA 30333, USA) *J. Clin. Microbiol., 8(6), 657-666 (1978)* En;en.

307-U2 Automatization of bacteriological water analysis: description of a new method for measuring faecal pollution. Trinel,P.A.; Leclerc,H. (INSERM U-146, Microbiol. Aquat. et Med., Institut Pasteur, Domaine du CERTIA, 59650 Villeneuve d'Ascq, France) *Ann. Microbiol., 128A(4), 419-432 (1977)* Fr;en,fr.

308-U2 Evaluation of the Oxi/Ferm tube system for identification of nonfermentative gram-negative bacilli. Shayegani,M.; Lee,A.M.; McGlynn,D.M. (Div. Lab. and Res., New York State Dep. Health, Albany, NY 12201, USA) *J. Clin. Microbiol., 7(6), 533-538 (1978)* En;en.

309-U2 Rapid detection of bacterial growth in blood samples by a continuous-monitoring electrical impedance apparatus. Specter,S.; Throm,R.; Strauss,R.; *Friedman,H. (Dep. Microbiol. and Immunol., Albert Einstein Med. Cent., Philadelphia, PA 19141, USA) *J. Clin. Microbiol., 6(5), 489-493 (1977)* En;en.

310-U2 Preparation of prereduced anaerobically sterilized media and their use in cultivation of anaerobic bacteria. Chan,E.C.S.; DeVries,J.; Harvey,R.F. (Dep. Microbiol. and Immunol., McGill Univ., Montreal, Que., Canada) *J. Clin. Microbiol., 8(2), 123-126 (1978)* En;en.

311-U2 Anaerobic bacteria: their recognition and significance in the clinical laboratory. Sutter,V.L.; Finegold,S.M. (Res. Serv., Wadsworth Veterans Adm. Hosp. Cent., Los Angeles, CA 90073, USA) *Med. Rev. Mex., 56(1198), 236-253 (1975)* Es;en.

312-U2 A rapid and simple method for the identification of enteropathogenic bacteria. Engelbrecht,E. (Bacteriol.-Serol. Dep., Stichting Samenwerking Delftse Ziekenhuizen, Delft, Netherlands) *Pathol. Microbiol., 42(4), 237 (1975)* En.

313-U2 Differentiation of pathogenic, gram-negative, bacterial genera in the presence of β-2-thienylalanine. Brown,K.J.; Lines,D.R. (Sect. Child Health Res., Dep. Paediatr., Univ. Auckland Sch. Med., Auckland, New Zealand) *Can. J. Microbiol., 22(12), 1673-1679 (1976)* En;en,fr.

314-U2 [Experiences with a simple culture tool for the semiquantitative determination of urine microorganisms]. Knorr,M.; Koch,W.; Neukirchner,J. (Hyg.-Inst. Dessau, Mikrobiol. Abt., 45 Dessau, Am Georgengarten 22, GDR) *Dtsch. Gesundheitswes., 31(11), 519-522 (1976)* De;de,en,ru.

315-U2 A method for the regulation of microbial population density during continuous culture at high growth rates. Martin,G.A.; *Hempfling,W.P. (Dep. Biol., Univ. Rochester, River Station, Rochester, NY 14627, USA) *Arch. Microbiol., 107(1), 41-47 (1976)* En;en.

316-U2 An all-metal block for replica plating of microbial colonies. Skodova,H.; Weisgerber,J.; Skoda,J. (Res. Inst. for Biofactors and Vet. Drugs, Pohori-Chotoun, Czechoslovakia) *Folia Microbiol., 23(11), 27-29 (1978)* En;en.

317-U2 Automated microbiological detection/identification system. Aldridge,C.; *Jones,P.W.; Gibson,S.; Lanham,J.; Meyer,M.; Vannest,R.; Charles,R. (McDonnell Douglas Astronautics Co.-East, St. Louis, MO 63166, USA) *J. Clin. Microbiol., 6(4), 406-413 (1977)* En;en.

318-U2 Use of a single infusion bottle to culture and prepare potentially hazardous microorganisms for biochemical analysis. Weijman,A.C.M. (Centraalbur. Schimmelcultures, Baarn, Netherlands) *Can. J. Microbiol., 24(9), 1097-1098 (1978)* En;en,fr.

319-U2 A method for cultivation of anaerobic microorganisms of manure samples. Pokorna-Kozova,J. (Res. Inst. Plant Prod., Inst. Plant Nutr., Drnovska 507, 16106 Praha-Ruzyne, Czechoslovakia) *Zentralbl. Bakteriol. Parasitenkd. Infektionskr. Hyg., II, 130(6), 560-562 (1975)* De;de,en.

320-U2 Clinical laboratory evaluation of an automated microbial detection and identification system (AMS™). Kayser,F.H.;Kolar,I. (Inst. Med. Mikrobiol., Univ. Zurich, Postfach, CH-8028 Zurich, Switzerland) *Zentralbl. Bakteriol. Parasitenkd. Infektionskr. Hyg., I Abt. A, 245(1/2), 222-228 (1979)* De;de,en.

321-U2 A rapid method for measuring specific growth rate of microorganisms. Boyle,D.T. (Biol. Sci. Branch, New Technol. Div., BP Res. Cent., Chertsey Road, Sunbury-on-Thames, Middlesex TW16 7LN, UK) *Biotechnol. Bioeng., 19(2), 297-300 (1977)* En.

322-U2 Assessing the efficiency of various defoamers. Adamek,L.; Stros,F. (Vyzkumny ustav Krmivarsheho Prumvslu a Sluzeb v Peckach, Odbor Mikrobialnich Vyrob, Prague, Czechoslovakia) *Kvasny Prum., 21(9), 205-208 (1975)* Cs;cs,de,en,ru.

323-U2 Methods of microbiological testing of fruit juices and beverages with pH less than 3.7, sold in trade packings. Przybylowicz,S. (PZPOW, Nowy Sacz, Poland) *Przem. Ferment. Rolny, 22(4), 12-16 (1978)* Pl;en,pl,ru.

324-U2 Simple equipment for establishing continuous synchronous cultures of microorganisms. Kjaergaard,L.; Joergensen,B.B. (Dep. Appl. Biochem., Tech. Univ. Denmark, Block 223, DK 2800 Lyngby, Denmark) *Biotechnol. Bioeng., 21(1), 147-151 (1979)* En.

325-U2 Computer-assisted fermentation developments. Humphrey,A.E. (Univ. Pennsylvana, Philadelphia, PA 19174, USA) *Dev. Ind. Microbiol., 18, 58-70 (1977)* En;en.

326-U2 Use of a simple duodenal capsule to study upper intestinal microflora. Gracey,M.; Suharjono; Sunoto (Gastroenterol. Res. Unit, Princess Margaret Hosp. Child., GPO Box D 184, Subiaco, Perth, WA 6001, Australia) *Arch. Dis. Child., 52(1), 74-76 (1977)* En;en.

327-U2 Diagnostic preparations - stable paper indicator systems for the rapid identification of microorganisms. Blokhina,I.N.; Lavrovskaya,V.M.; Altman,R.Sh. (Sci.-Res. Inst. Epidemiol. and Microbiol., Gorky, USSR) *Zh. Mikrobiol. Epidemiol. Immunobiol., 55(7), 112-115 (1978)* Ru;en,ru.

328-U2 Simple equipment for establishing continuous synchronous cultures of microorganisms. Kjaergaard,L.; Joergensen,B.B. (Dep. Appl. Biochem., Tech. Univ. Denmark, Block 223, DK-2800 Lyngby, Denmark) *Biotechnol. Bioeng., 21(1), 147-151 (1979)* En.

329-U2 Time- and media-saving testing and identification of microorganisms by multipoint inoculation on undivided agar plates. Burman,L.G.; Ostensson,R. (Dep. Clin. Bacteriol., Univ. Umea, S-901 85 Umea, Sweden) *J. Clin. Microbiol., 8(2), 219-227 (1978)* En;en.

330-U2 Identification methods for microbiologists, 2nd. Edition. /[The Society for Applied Bacteriology Technical Series No. 14]. Skinner,F.A.; Lovelock,D.W. (eds.) *In:* Skinner,F.A.; Lovelock,D.W. (eds.) *Publ.by:* Academic Press Inc. (London) Ltd., 24-28 Oval Road, London NW1 7DX, UK. 1980. 315 pp. ISBN 0-12-647750-7. at £14.80 or US $34.50. En.

331-U2 Collecting aquatic microorganisms: the glass slide method. Brown,H.D. (Clayton Junior Coll., Morrow, GA 30260, USA) *Am. Biol. Teach., 37(5), 302-303 (1975)* En.

332-U2 Detection of microorganisms that produce extracellular maltases. Wang,L.-H.; McWethy,S.J.; Hartman,P.A. (Taiwan Sugar Res. Inst., Tainan, Taiwan 700) *Anal. Biochem., 76(1), 380-381 (1976)* En.

333-U2 A column fermenter for the continuous cultivation of micro-organisms attached to surfaces. Ashby,R.E.; Bull,A.T. (Biol. Lab., Univ. Kent at Canterbury, Canterbury CT2 7NJ, UK) *Lab. Pract., 26(4), 327-329 (1977)* En.

334-U2 Chamber for culturing microorganisms. Coleman,D.E. (Dep. Anim. Sci., Univ. Hawaii, Honolulu, HI 96822, USA) *Prog. Fish. Cult., 41(2), 110 (1979)* En.

335-U2 [A simple method for the isolation of soil microorganisms which degrade gallic acid]. Percuoco,G.; di Cristo,C.; Percuoco,S. (Fac. Sci. Agrar., Univ. Napoli, Portici, Italy) *Annali, 11, 42-45 (1977)* It;en,it.

336-U2 Recent advances in the rapid detection of brewery microorganisms and development of a microcolony method. Harrison,J.; Webb,T.J.B. (Allied Brew. (Prod.) Ltd., 107 Station St., Burton-Upon-Trent, UK) *J. Inst. Brew., 85(4), 231-234 (1979)* En;en.

337-U2 The miniloop: a small-scale air-lift microbial culture vessel and photobiological reactor. Pirt,S.J.; Panikov,N.; Lee,Y.K.(Microbiol. Dep., Queen Elizabeth Coll. (Univ. London), Campden Hill Road, London W8 7AH, UK) *J. Chem. Technol. Biotechnol., 29(7), 437-441 (1979)* En;en.

338-U2 **Chamber for culturing microorganisms.** Coleman,D.E. (Dep. Anim. Sci., Univ. Hawaii, Honolulu, HI 96822, USA) *Prog. Fish Cult., 41(2), 110 (1979)* En.

339-U2 **Clinical laboratory evaluation of automated microbial detection/identification system in analysis of clinical urine specimens.** Isenberg,H.D.; Gavin,T.L.; Sonnenwirth,A.; Taylor,W.I.; Washington,J.A.II (Long Island Jewish-Hillside Med. Cent., New Hyde Park, NY 11042, USA) *J. Clin. Microbiol., 10(2), 226-230 (1979)* En;en.

340-U2 **Preprototype of an automated microbial detection and identification system: a developmental investigation.** Sonnenwirth,A.C. (Dep. Pathol. and Lab. Med., Jewish Hosp. St. Louis, St. Louis, MO 63110, USA) *J. Clin. Microbiol., 6(4), 400-405 (1977)* En;en.

341-U2 **Autobac 1® and Kirby-Bauer: a comparative study in a clinical microbiology laboratory.** Mathie,W.G. (Westman Lab., Brandon, Manit., Canada) *Can. J. Med. Technol., 40(6), 177-180 (1978)* En;en,fr.

342-U2 **The fistulated diffusion chamber and its use for the study of microbial associations in vivo.** Mikhno,I.L.; Barshtein,Yu.A.; Kondratenko,V.N. (Kiev Res. Inst. Epidemiol., Microbiol., and Parasitol., Kiev, USSR) *Byull. Eksp. Biol. Med., 85(2), 247-249 (1978)* Ru;en,ru.

343-U2 **Electrical impedance measurements: rapid method for detecting and monitoring microorganisms.** Cady,P.; Dufour,S.W.; Shaw,J.; Kraeger,S.J. (Bactomatic, Inc., Palo Alto, CA 94303, USA) *J. Clin. Microbiol., 7(3), 265-272 (1978)* En;en.

344-U2 **The photobioluminometer, an instrument for the study of ecological factors affecting photosynthesis.** Tchan,Y.T.; Chiou,A.C.M.; New,P.B.; Funnell,G.R. (Dep. Microbiol., Univ. Sydney, NSW 2006, Australia) *Microb. Ecol., 3(4), 327-332 (1977)* En;en.

345-U2 **An adaptable system for timed aseptic sampling and storage of microbial cultures.** Newman,P.B. (Agric. Res. Counc. Meat Res. Inst., Langford, Bristol BS18 7DY, UK) *J. Appl. Bacteriol., 41(3), 497-504 (1976)* En;en.

346-U2 **Computer applications to fermentation operations** Dobry,D.D.; Jost,J.L. (Upjohn Co. Kalamazoo, MI, USA) *In:* Annual reports on fermentation processes. Vol. 1, Perlman,D. (ed.) *Publ. by:* Academic Press Inc. (London) Ltd., 24-28, Oval Rd, London NW1 7DX, UK, 1977, p. 95-114, ISBN: 0-12-040301-3. En.

347-U2 **A simple culture tube closure method for prevention of contamination by airborne fungi and mites.** Richardson,L.T. (Agric. Canada, Res. Inst., Univ. Sub Post Office, London, Ont. N6A 5B7, Canada) *Phytopathology, 65(7), 833-834 (1975)* En;en.

348-U2 **A method for the production of *Allomyces macrogynus* mitospores.** Sandstedt,V.; Aronson,J.M. (Dep. Bot. and Microbiol., Arizona State Univ., Tempe, AZ 85281, USA) *Exp. Mycol., 2(1), 114-117 (1978)* En;en.

349-U2 **A new apparatus for automatic growth estimation of mold cultured on solid media.** Okazaki,N.; Sugama,S. (Natl. Res. Inst. Brewing, 2-6 Takinogawa, Kita-ku, Tokyo 114, Japan) *J. Ferment. Technol., 57(5), 413-417 (1979)* En;en.

350-U2 **A rapid method for the isolation of tetrads in higher Basidiomycetes.** Epp,B.D. (Lehrstuhl Allgemeine Bot., Ruhr-Univ. Bochum, Bochum, GFR) *Mycologia, 69(1), 210-211 (1977)* En.

351-U2 **Differentiation between white- and brown-rot fungi by cultivation on cellophane.** Danninger,E.; Messner,K.; Rohr,M.; Stachelberger,H. (Inst. Biochem. Technol. und Mikrobiol., Getreidemarkt 9, 1060 Wien, Austria) *Mater. Org., 14(4), 287-299 (1979)* De;de,en,es,fr.

352-U2 **Simple method for maintaining cultures of *Blastocladiella emersonii*.** Webster,J.; Davey,R.A. (Dep. Biol. Sci., Univ. Exeter, Exeter, UK) *Trans. Br. Mycol. Soc., 67(3), 543-544 (1976)* En.

353-U2 **A simple medium for isolation and identification of *Candida albicans* directly from clinical specimens.** Gunasekaran,M.; Hughes,W.F. (Infect. Dis. Serv., St. Jude Child Res. Hosp., Memphis, TN 38101, USA) *Mycopathologia, 61(3), 151-157 (1977)* En;en.

354-U2 **Rapid methods for identification of yeasts.** Huppert,M.; Harper,G.; Sun,S.H.; Delanerolle,V. (Mycol. Res. Dep., Vet. Adm. Hosp., Long Beach, CA 90801, USA) *J. Clin. Microbiol., 2(1), 21-34 (1975)* En;en.

355-U2 **Evaluation of a dipstick for *Candida*.** Weissberg,S.M. (6201 S.W. 70 St., South Miami, FL 33143, USA) *Obstet. Gynecol., 52(4), 506-509 (1978)* En;en.

356-U2 **Rapid method for the determination of wild yeast contamination in bakers' yeast by lipid analysis.** Biacs,P.A. (Dep. Agric. Chem. Technol., Tech. Univ. Budapest, H-1111 Budapest, Gellert ter 4, Hungary) *Acta Aliment., 8(1), 57-67 (1979)* En;en.

357-U2 **[Rapid method of identification of *Candida* of medical interest].** Tuttobello,L. (Ist. Super. Sanita, Serv. Biol., Roma, Italy) *Boll. Soc. Ital. Biol. Sper., 55(15), 1474-1480 (1979)* It;en.

358-U2 **Use of Autobac 1 for rapid assimilation testing of *Candida* and *Torulopsis* species.** Ngui Yen,J.H.; *Smith,J.A. (Div. Microbiol., Vancouver Gen. Hosp. Vancouver, BC V5Z 1M9, Canada) *J. Clin. Microbiol, 7(2), 118-121 (1978)* En;en.

359-U2 **A strip test for detecting *Candida* in the oral cavity.** Berdicevsky,I.; Ben-Arych,H.; Glick,D.; Gutman,D. (Dep. Microbiol., Technion Sch. Med., PO Box 9649, Haifa, Israel) *Oral Surg. Oral Med. Oral Pathol., 44(2), 206-209 (1977)* En;en.

360-U2 **The nascent culture and its use.** Feo,M. (Secc. Micol. Med., Inst. Med. Trop., Apartado 8250, Caracas, Venezuela) *Mycopathologia, 63(3), 185-186 (1978)* En.

361-U2 Rapid in vitro conversion and identification of *Coccidioides immitis*. Sun,S.H.; *Huppert,M.; Vukovich,K.R. (Mycol. Res. Lab., VA Hosp., Long Beach, CA 90801, USA) *J. Clin. Microbiol., 3(2), 186-190 (1976)* En;en.

362-U2 Immunological procedure for the rapid and specific identification of *Coccidioides immitis* cultures. Standard,P.G.; Kaufman,L. (Lab. Training and Consultation Div., Cent. Dis. Control, Atlanta, GA 30333, USA) *J. Clin. Microbiol., 5(2), 149-153 (1977)* En;en.

363-U2 Improved version of the exoantigen test for identification of *Coccidioides immitis* and *Histoplasma capsulatum* cultures. Kaufman,L.; Standard,P. (Mycol. Div., Cent. Dis. Control, Atlanta, GA 30333, USA) *J. Clin. Microbiol., 8(1), 42-45 (1978)* En;en.

364-U2 A rapid pigmentation test for identification of *Cryptococcus neoformans*. Paliwal,D.K.; Randhawa,H.S. (Dep. Med. Mycol., Vallabhbhai Patel Chest Inst., Univ. Delhi, Delhi-110007, India) *Antonie van Leeuwenhoek, 44(2), 243-246 (1978)* En;en.

365-U2 Evaluation of a caffeic acid-ferric citrate test for rapid identification of *Cryptococcus neoformans*. Wang,H.S.; Zeimis,R.T.; *Roberts,G.D. (Mayo Clin. and Mayo Found., Rochester, MN 55901, USA) *J. Clin. Microbiol., 6(5), 445-449 (1977)* En;en.

366-U2 Rapid selective urease test for presumptive identification of *Cryptococcus neoformans*. Zimmer,B.L.; Roberts,G.D. (Sect. Clin. Microbiol., Dep. Lab. Med., Mayo Clin. and Mayo Found., Rochester, MN 55901, USA) *J. Clin. Microbiol., 10(3), 380-381 (1979)* En;en.

367-U2 Further simplification of the *Guizotia abyssinica* seed medium for identification of *Cryptococcus neoformans* and *Cryptococcus bacillispora*. Saliin,I.F. (Lab. Mycol., Div. Lab. and Res., New York State Dep. Health, Albany, NY 12201, UDSA) *Can. J. Microbiol., 25(9), 1116-1118 (1979)* En;en,fr.

368-U2 The production and germination of resting spores of *Entomophthora virulenta* (Entomophthorales; Entomophthoraceae). Matanmi,B.A.; Libby,J.L. (Dep. Plant Sci., Univ. Ife, Ile-Ife, Nigeria) *J. Inver. ebr. Pathol., 27(3), 279-285 (1976)* En;en.

369-U2 An automatic spore trap for collecting pycnidiospores of *Leptosphaeria nodorum* and other fungi from the air during rain and maintaining them in a viable condition. Faulkner,M.J.; Colhoun,J. (Cryptogram, Bot. Lab., Univ. Manchester, Manchester M13 9PL, UK) *Phytopathol. Z., 89(1), 50-59 (1977)* En;de,en.

370-U2 A simple technique for inducing sporulation in cultures of Myxomycetes. Venkataramani,R.; Kalyanasundaram,I.; Kalyanasundaram,R. (Univ. Bot. Lab., Madras-600 005, India) *Trans. Br. Mycol. Soc., 69(2), 320-322 (1977)* En.

371-U2 A rapid method of examining honey bees individually for *Nosema* spores. Clinch,P.G. (Wallacevile Anim. Res. Cent., Res. Div., Min. Agric. and Fish., PB, Upper Hutt, New Zealand) *N.Z. J. Exp. Agric., 4(1), 125-126 (1976)* En;en.

372-U2 An improved selective medium for isolation of *Phaeolus schweinitzii*. Barrett,D.K. (Dep. Forest., Univ. Oxford, Oxford OX1 3BD, UK) *Trans. Br. Mycol. Soc., 71(3), 507-508 (1978)* En.

373-U2 Rapid production of *Phytophthora parasitica* var. *nicotianae* zoospores in vitro from infested cloth. Taylor,G.S. (Valley Lab., Connecticut Agric. Exp. Stn., PO Box 248, Windsor, CT 06095, USA) *Plant Dis. Rep., 62(3), 281-282 (1978)* En;en.

374-U2 Rapid determination of the viability of conidia of *Phytophthora infestans* (Mont.) d By. Karaseva,E.V.; Minaeva,L.A.; Filippov,A.V.; Mochalkin,A.I. (Address not stated) *Mikol. Fitopatol., 10(4), 336-338 (1976)* Ru.

375-U2 A simplified method for sporangial production by *Phytophthora cinnamomi*. Hwang,S.C.; Ko,W.H.; Aragaki,M. (Dep. Plant Pathol., Univ. Hawaii, Beaumont Agric. Res. Cent., Hilo, HI 96720, USA) *Mycologia, 67(6), 1233-1234 (1975)* En.

376-U2 Rapid identification of *Prototheca* species by the API 20C system. Padhye,A.A.; Baker,J.G.; D'Amato,R.F. (Mycol. Div.,Cent. Dis. Control, Public Health Serv., USDHEW, Atlanta, GA 30333, USA) *J. Clin. Microbiol., 10(4), 579-582 (1979)* En;en.

377-U2 A shake culture method for Pythiaceae applicable to rapid small-scale assay of vegetative physiology. Rawn,C.D.; Van Etten,J.L. (Dep. Plant Pathol., Univ. Nebraska, Lincoln, NE 68583, USA) *Phytopathology, 68(9), 1384-1388 (1978)* En;en.

378-U2 The rotorofermentor. II. Application to ethanol fermentation. Margaritis,A.; Wilke,C.R. (Dep. Chem. and Biochem. Eng., Univ. Western Ontario, London, Ont., Canada) *Biotechnol. Bioeng., 20(5), 727-753 (1978)* En;en.

379-U2 A rapid method for the characterization of individual yeast colonies. Nashed,N. (Lab. Mutagenitats-forsch., Exp. Pathol. Toxikol., C. H. Boehring Sohn, 650 Ingelheim/Rhein, GFR) *Microb. Genet. Bull., 39, 1 (1975)* En.

380-U2 Dissolved oxygen as a parameter of automatic control of the nutrient medium inflow in the yeast cultivation process. Part I. Principles and characteristics of an automatic nutrient medium inflow control. Miskiewicz,T.; Lesniak,W. (Inst. Technol. Przemyslu Chem. i. Spozywczego, Wroclaw, Poland) *Przem. Ferment. Rolny, 19(10), 15-17 (1975)* Pl;pl,ru.en.

381-U2 Selective spore survival during replicating of fission yeast. Egel,R. (Inst. Biol. III, Freiburg i. Br., Schanzlestr. 9-11, D-7800 Freiburg i. Br., GFR) *Arch. Microbiol., 112(1), 109-110 (1977)* En;en.

382-U2 A simple technique for direct enumeration of viable *Sclerotium rolfsii* sclerotia in soil. Toribio,J.A. (Stn. Pathol. Veg., Cent. Rech. Agron. Antilles et Guyane INRA, Domaine Duclos, 97170 Petit-Bourg, Guadeloupe) *Ann. Phytopathol., 9(2), 177-182 (1977)* Fr;en,fr.

383-U2 Wet-sieving floatation technique for isolation of sclerotia of *Sclerotium cepivorum* from muck soil. Utkhede,R.S.; Rahe,J.E. (Pestology Cent., Dep. Biol. Sci., Simon raser Univ., Burnaby, B.C. V5A 1S6, Canada) *Phytopathology, 69(3), 295-297 (1979)* En;en.

384-U2 A quantitative method to examine large numbers of soil samples for *Synchytrium endobioticum*, the cause of potato wart disease. Hampson,M.C.; Thompson,P.R. (Res. Branch, Agric. Can., POB 7098, St. John's West, Canada) *Plant and Soil, 46(3), 659-664 (1977)* En;en.

385-U2 Early detection and identification of *Trichophyton verrucosum*. Kane,J.; Smitka,C. (Lab. Serv. Branch, Ontario Minist. Health, Toronto, Ont. M5W 1R5, Canada) *J. Clin. Microbiol., 8(6), 740-747 (1978)* En;en.

386-U2 Influence of temperature and humidity on the efficiency and cleerness of the Bauch test applied in *Ustilago scitaminea* Syd. Toffano,W.B. (Inst. Biol. Sao Paulo, CP 7119, 01000 Sao Paulo, SP Brazil) *Arq. Inst. Biol., Sao Paulo, 43(1-2), 43-48 (1976)* Pt;en,pt.

387-U2 Induction of synchronous growth in the yeast phase of *Wangiella dermatitidis*. Roberts,R.L.; Lo,R.J.; *Szaniszlo,P.J. (Dep. Microbiol., Univ. Texas at Austin, Austin, TX 78712, USA) *J. Bacteriol., 141(2), 981-984 (1980)* En;en.

388-U2 [Simplified fermentation and assimilation tests for the identification of yeasts]. Quast,R. (Herminenstr. 17f, D-3062 Buckeburg, GFR) *Mykosen, 21(3), 81-86 (1978)* De;de,en.

389-U2 Selective enrichment technique for isolation of methanol-utilizing yeasts. Mehta,R.J. (Dep. Microbiol., Smith Kline and French Lab., 1500 Spring Garden Street, Philadelphia, PA 19101, USA) *Experientia, 35(12), 1576 (1979)* En;en.

390-U2 Evaluation of a modified dye pour-plate auxanographic method for the rapid identification of clinically significant yeasts. Comparison with two commercial systems. Weymann,L.H.; Stager,C.E.; *Qadri,S.G.M.; Villarreal,A.; Qadri,S.M.H. (Dep. Pathol., Univ. Hosp. and Clin., Oklahoma City, OK 73125, USA) *Med. Microbiol. Immunol., 167(1), 11-20 (1979)* En;en.

391-U2 New methods for the detection of viable microorganisms. Portno,A.D.; Molzahn,S.W. (Bass Production Ltd., Burton-on-Trent, UK) *Brew. Dig., 52(3), 44-50 (1977)* En;en.

392-U2 Evaluation of a commercial multitest system for identification of yeasts. Qadri,S.M.H.; Nichols,C.W. (Dep. Pathol. and Lab. Med., Univ. Texas Med. Sch. at Houston, Houston, TX 77030, USA) *Am. J. Med. Technol., 44(5), 368-372 (1978)* En;en.

393-U2 Method and apparatus for automatic drawing of growth curve. Minami,K.; Yamamura,M.; Shimizu,S. (Res. Dev. Cent., Maruzen Oil Co. Ltd., PO Box 1, Satte, Saitama 340-01, Japan) *J. Ferment. Technol., 56(3), 229-236 (1978)* En;en.

394-U2 Evaluation of the new API 20C strip for yeast identification against a conventional method. Land,G.A.; Harrison,B.A.; Hulme,K.L.; Cooper,B.H.; Byrd,J.C. (Div. Lab. Med., Univ. Cincinnati Med. Cent., Cincinnati, OH 45267, USA) *J. Clin. Microbiol., 10(3), 357-364 (1979)* En;en.

395-U2 Microstix®-*Candida*. A rapid and simple method of demonstration of yeasts in the vagina. Skoven,I.; Bugge,M.; Pedersen,G.T. (Dermato-venerol. Afd., Odense sygehys, DK-5000 Odense C, Denmark) *Ugeskr. Laeg., 141(25), 1691-1693 (1979)* Da;da,en.

396-U2 Evaluation of a multitest microtechnique for yeast identification. Miller,R.E.,Jr.; Lu,L.-P. (Dep. Biochem., Jewish Hosp. of St. Louis, St. Louis, MO 63110, USA) *Am. J. Med. Technol., 42(7), 238-242 (1976)* En;en.

397-U2 Evaluation of the micromethod API 20 C for yeast identification. Gille,Y.; Guinet,R. (Lab. Microbiol., Hop. Antiquaille, 1, rue d l'Antiquaille, 69325 Lyon Cedex 1, France) *Rev. Inst. Pasteur Lyon, 11(1), 55-63 (1978)* Fr;en,fr..

398-U2 Rapid method for detection of urea hydrolysis by yeasts. Paliwal,D.K.; *Randhawa,H.S. (Dep. Med. Mycol., Vallabhbhai Patel Chest Inst., Univ. Delhi, Delhi 110007, India) *Appl. Environ. Microbiol., 33(2), 219-220 (1977)* En;en.

399-U2 An enrichment method for temperature-sensitive and auxotrophic mutants of yeast. Walton,E.F.; Carter,B.L.A.; *Pringle,J.R. (Div. Biol. Sci., Univ. Michigan, Ann Arbor, MI 48109, USA) *Mol. Gen. Genet., 17(1), 111-114 (1979)* En;en.

400-U2 Isolation and rapid identification of yeasts from compromised hosts. Land,G.A.; Dorn,G.L.; Fleming,W.H.,III; Beadles,T.A.; Foxworth,J.J.(Dep. Microbiol., Wadley Inst. Mol. Med., Dallas, TX 75235, USA) *Mycopathologia, 65(1-3), 123-131 (1978)* En;en.

401-U2 Tube carbohydrate assimilation method for the rapid identification of clinically significant yeasts. Hussan Qadri,S.M.; Nichols,C.W. (Dep. Pathol., Univ. Texas Med. Sch. at Houston, PO Box 20708, Houston, TX 77025, USA) *Med. Microbiol. Immunol., 165(1), 19-27 (1978)* En;en.

402-U2 Evaluation of a new system for rapid identification of clinically important yeasts. Segal,E.; Ajello,L. (Cent. Dis. Cont., Public Health Serv., US Dep. Health, Educ. and Welf., Atlanta, GA 30333, USA) *J. Clin. Microbiol., 4(2), 157-159 (1976)* En;en.

403-U2 Evaluation of commercial systems for the identification of clinical yeast isolates. Bowman,P.L. Ahearn,D.G. (Dep. Biol., Georgia State Univ., Atlanta, GA 30303, USA) *J. Clin. Microbiol., 4(1), 49-53 (1976)* En;en.

404-U2 Improved auxanographic method for yeast assimilations: a comparison with other approaches. Land,G.A.; Vinton,E.C.; Adcock,G.B.; Hopkins,J.M. (Dep. Clin. Microbiol., Wadley Inst. Mol. Med., Dallas, TX 75235, USA) *J. Clin. Microbiol., 2(3), 206-217 (1975)* En;en.

405-U2 Experience in the use of culture fluorescence for monitoring fermentations. Ristroph,D.L.; Watteeuw,C.M.; Armiger,W.B.; Humphrey,A.E. (Chem. and Biochem. Eng. Dep., Univ. Pennsylvania, Philadelphia, PA 19104, USA) *J. Ferment. Technol., 55(6), 599-608 (1977)* En;en.

406-U2 A simple device for stationary cultivation of microorganisms. Kybal,J.; Vicek,V. (Res. Inst. Pharm. and Biochem., Prague, Czechoslovakia) *Biotechnol. Bioeng., 18(12), 1713-1718 (1976)* En;en.

407-U2 A new method for the rapid identification of pathogenic fungi on the skin. Gip,L. (Dep. Dermatol., Sundsvalls Sjukhus, S-851 86 Sundsvall, Sweden) *Curr. Ther. Res., Clin. Exp., 22(1), 57-64 (1977)* En;en.

408-U2 A simple method of isolation and purification of cultures of wood-rotting fungi. Ginterova,A.; Janotkova,O. (Res. Inst. Distill. and Canning Ind., 885 30 Bratislava, Czechoslovakia) *Folia Microbiol., 20(6), 519-520 (1975)* En;en.

409-U2 A new technique for the preparation of permanent reference cultures of fungi. Hofherr,L. (Dep. Lab. Med. and Pathol., Univ. Minnesota, Minneapolis, MN 55420, USA) *Can. J. Microbiol., 24(10), 1275-1277 (1978)* En;en,fr.

410-U2 A convenient method for preparing inocula of homogenized mycelia. Zweck,S.; Huttermann,A.; Chet,I. (Forstbot. Inst., Univ. Gottingen, 3400 Gottingen-Weende, GFR) *Exp. Mycol., 2(4), 377-378 (1978)* En;en.

411-U2 A new lid closure for fungal culture vessels giving complete protection against mite infestation and microbiological contamination. Smith,R.S. (Fish. and Environ. Canada, For. Directorate, W stern For. Prod. Lab., Vancouver, BC V6T 1X2, Canada) *Mycologia, 70(3), 499-508 (1978)* En;en.

412-U2 Simple method for extraction of predacious fungi from the soil profile. Shagalin,S.F. (Address not stated) *Izv. Akad. Nauk Turmensk. SSR, Ser. Biol., No. 1, 81-83 (1978)* Ru;en.

413-U2 Rapid agar-stick breaking-radius test to determine the ability of fungi to degrade wood. Safo-Sampah,S.; Graham,R.D.(Dep. For. Prod., Sch. For., Oregon State Univ., Corvallis, OR 97331, USA) *Wood Sci., 9(2), 65-69 (1976)* En;en.

414-U2 Chitin as a measure of fungal growth in stored corn and soybean seed. Donald,W.W.; Mirocha,C.J. (Dep. Plant Pathol., Univ. Minnesota, St. Paul, MN 55108, USA) *Cereal Chem., 54(3), 466-474 (1977)* En;en.

415-U2 A new method for isolating fungi spores to obtain pure cultures. Kirilenko,T.S. (Inst. Mikrobiol. and Virol., Acad. Sci. Ukr. SSR, U, Zabolotnogo 59, Kiev, USSR) *Mikrobiol. Zh., 39(2), 233-235 (1977)* Uk;en.

416-U2 An improved approach for the accurate determination of fungal pathogens in diseased plants. Toppan,A.; Esquerre-Tugaye,M.T.; Touze,A. (Cent. Physiol. Veg., LA 241 CNRS, Univ. Paul Sabatier, 118 route de Narbonne, 31077 Toulouse Cedex, France) *Physiol. Plant Pathol., 9(3), 241-251 (1976)* En;en.

417-U2 A continuous temperature gradient incubator as a tool in studies of the temperature characteristics of fungi. Eriksen,J. (Dep. Microbiol. and Plant Physiol., Univ. Brgen, Allegt. 70, N-5014 Bergen Univ., Norway) *Lab. Pract., 28(5), 510-511 (1979)* En.

418-U2 A chemical method for the estimation of mould in tomato products. Jarvis,B. (Leatherhead Food R.A., Randalls Rd., Leatherhead, Surrey KT22 7RY, UK) *J. Food. Technol., 12(6), 581-591 (1977)* En;en.

419-U2 A permanent mounting medium for fungi. Omar,M.B.; Bolland,L.; Heather,W.A. (Address not stated) *Bull. Br. Mycol. Soc., 13(1), 31-32 (1979)* En.

420-U2 A variable-medium culture chamber for fungi. Padgett,D.E.; Lundeen,C.V. (Dep. Biol., Univ. North Carolina at Wilmington, NC 20481, USA) *Mycologia, 69(4), 835-837 (1977)* En.

421-U2 Apparatus for rapid inspection of corn for aflatoxin contamination. Barabolak,R.; Colburn,C.R. (deceased); Just,D.E.; *Kurtz,F.A.; Schleichert,E.A. (CPC International, Inc., Moffett Tech. Cent., PO Box 345, Agro, IL 60501, USA) *Cereal Chem., 55(6), 1065-1067 (1978)* En.

422-U2 Computer control and optimization in the cultivation of yeast. Blachere,H.; Peringer,P.; Lane,A.G.; Bourdaud,D.; Foulard (Stn. Genie Microbiol., INRA, Dijon, France) *In:* Global impacts of applied microbiology, IVth International Conference. J.S. Furtado. *Publ. by:* Sociedade Brasileira de Microbiologia, Revista de Microbiologia, Sao Paulo, Brazil. Part 2, 823-830. En.

423-U2 Small-scale chemostat for the growth of mesophilic and thermophilic microorganisms. Gilbert,P.; Stuart,A. (Dep. Pharm., Univ. Aston, Birmingham, B4 7ET, UK) *Lab. Pract., 26(8), 627-628 (1977)* En.

424-U2 Computer simulation of the biomass production rate of cyclic fed batch continuous culture. Keller,R.; Dunn,I.J. (Tech. Chem. Lab., Swiss Fed. Inst. Technol. (ET, 8092 Zurich, Switzerland) *J. Appl. Chem. Biotechnol., 28(11), 784-790 (1978)* En;en.

425-U2 An improved method for the detection of diatoms. Udermann,H.; Schuhmann,G. (Inst. Gerichtl. Med. Univ., A-8020 Graz, Universitatsplatz 4, Austria) *Z. Rechtsmedizin,*

76(2), 119-122 (1975) De;de,en.

426-U2 A photographic method for accurately measuring the growth of crustose and foliose saxicolous lichens. Hooker,T.N.; Brown,D.H. (Dep. Bot., Univ. Bristol, Woodland Rd., Bristol BS8 1UG, UK) *Lichenologist, 9(1), 65-75 (1977)* En;en.

427-U2 A continuous synchronous culture and sampling system for algae. Ricketts,T.R. (Cell Biol. Unit, Dep. Bot., Sch. Biol. Sci., Univ. Nottingham, Nottingham NG7 2RD, UK) *J. Exp. Bot., 28(103), 416-424 (1977)* En;en.

428-U2 Discontinuous gradients of light and temperature for culturing of microalgae in liquid cultures. Norland,S.; 'Heldal,M.; Dagestad,D. (Inst. Gen. Microbiol., Univ. Bergen, Allegt, 70, N 5014 Bergen Univ., Norway) *Lab. Pract., 26(11), 863-864 (1977)* En.

429-U2 The use of cage cultures in studies of the biochemistry and ecology of marine phytoplankton. Sakshaug,E.; Jensen,A. (Biol. Stn., Univ. Trondheim, Trondheim, Norway) Barnes.H. (ed.) *Publ.by:* Aberdeen University Press; Aberdeen (UK) 1978 p. 81-106 *Oceanogr. Mar. Biol. Annu. Rev., v. 16* En.

430-U2 [Description of a system of temperature and light conditioned chambers for vegetative ecophysiological studies: first results obtained using the rhodophyte *Acrochaetium* sp]. /[presented at a meeting of the French Phycological Society, held on Jan 19, 1975]. Meheut,G.; Ducher,M.; Michaux,O. (Lab. Phytotron, C.N.R.S.-91190, Gif-sur-Yvette, France) *Bull. Soc. Phycol. Fr., 21, 12-17 (1976)* Fr;en,fr.

431-U2 A rapid procedure for selective enrichment of photosynthetic electron transport mutants. Schmidt,G.W.; Matlin,K.S.; Chua,N.-H. (Rockefeller Univ., New York, NY 10021, USA) *Proc. Natl. Acad. Sci. USA, 74(2), 610-614 (1977)* En;en.

432-U2 Large-scale mass culture of the marine blue-green alga, *Gomphosphaeria aponina*. Eng-Wilmot,D.L.; Martin,D.F. (Dep. Chem., Univ. South Florida, Tampa, FL 33620, USA) *Fla. Sci., 40(2), 193-197 (1977)* En;en.

433-U2 A chemostat device adapted to planktonic *Oscillatoria* cultivation. Feuillade,J.; Feuillade,M. (Stn. Hydrobiol., Lacustre, INRA, F-74203 Thonon, France) *Limnol. Oceanogr., 24(3), 562-564 (1979)* En;en.

434-U2 A chemostat device adapted to planktonic *Oscillatoria* cultivation. Feuillade,J.; Feuillade,M. (Stn. Hydrobiol., Lacustre, INRA, F-74203 Thonon, France) *Limnol. Oceanogr., 24(3), 562-564 (1979)* En;en.

435-U2 Long-term maintenance of fertile algal clones: experience with *Pandorina* (Chlorophyceae). Coleman,A.W. (Div. Biol. and Med. Sci., Brown Univ., Providence, RI 02912, USA) *J. Phycol., 11(3), 282-286 (1975)* En;en.

436-U2 An improved culture tube for axenic cultures of microalgae. Groeneweg,J.; Soeder,C.J. (GSF, Abt. Algenforschung und Algentechnol., Bunsen-Kirchhoff-Str. 13, 4600 Dortmund, GFR) *Br. Phycol. J., 13(4), 337-340 (1978)* En;en.

437-U2 An unusual method for culture of unicellular marine algae. Ukeles,R.; Bishop,J. (Natl. Mar. Fish. Serv., Middle Atlantic Coast. Fish. Cent., Milford Lab., Milford, CT 06460, USA) *J. Phycol., 12(3), 332-335 (1976)* En;en.

438-U2 A Winkler procedure for making precise measurements of oxygen concentration for productivity and related studies. Bryan,J.R.; Riley,J.P.; Williams,P.J.LeB. (Dep. Oceanogr., Southampton, Univ., Southampton, UK) *J. Exp. Mar. Biol. Ecol., 21(3), 191-197 (1976)* En;en.

439-U2 An improved apparatus for the culture of algae under varying regimes of temperature and light intensity. Yarish,C.; Lee,K.W.; Edwards,P. (Biol. Dep., Univ. Connecticut, Scofieldtown Road, Stamford, CT 06903, USA) *Bot. Mar., 22(6), 395-397 (1979)* En.

440-U2 A method for cultivation. of benthic algae by artificial floatation Gibor,A. (Dep. Biol. Sci., Univ. California, Santa Barbara, CA 93106, USA) *Bot. Mar., 19(6), 397-399 (1976)* En;en.

441-U2 An incubator method for estimating the actual daily planktonalgae primary production. Gargas,E.; Nielsen,C.S.; Lonholdt,J. (Water Quality Inst., 11 Agern Alle, DK 2970 Horsholm, Denmark) *Water Res., 10(10), 853-860 (1976)* En;en.

442-U2 The miniloop: a small-scale air-lift microbial culture vessel and photobiological reactor. Pirt,S.J.; Panikov,N.; Lee,Y.K. (Microbiol. Dep., Queen Elizabeth Coll. (Univ. London), Campden Hill Road, London W8 7AH, UK) *J. Chem. Technol. Biotechnol., 29(7), 437-441 (1979)* En;en.

443-U2 Improved method for axenizing *Blepharisma* by means of antibiotics. Marti,E.J.; Slavin,R.J.; Hirshfield,H.I. (Middlesex Community Coll., 100 Training Hill Rd., Middletown, CT 06457, USA) *J. Protozool., 16(1), 133-134 (1979)* En;en.

444-U2 A simple technique for preparing clone cultures of *Entamoeba histolytica*. Farri,T.A. (Dep. Med. Protozool., London Sch. Hyg. and Trop. Med., Keppel St., London WC1E 7HT, UK) *Trans. R. Soc. Trop. Med. Hyg., 72(2), 205-206 (1978)* En.

445-U2 Haemoflagellates: commercially available liquid media for rapid cultivation. Hendricks,L.D.; Wood,D.E.; Hajduk,M.E. (Dep. Biol., Div. Med. Chem., Walter Reed Army Inst. Res., Walter Reed Army Med. Cent., Washington, DC 20012, USA) *Parasitology, 76(3), 309-316 (1978)* En;en.

446-U2 Isolation of the intracellular stages of *Leishmania mexicana amazonensis* using cellulose column. Brazil,R.P. (Dep. Parasitol., Liverpool Sch. Trop. Med., Pembroke Place, Liverpool L3 5QA, UK) *Ann. Trop. Med. Parasitol., 72(6), 579-580 (1978)* En.

447-U2 *Plasmodium falciparum*: continuous cultivation in a semiautomated apparatus. Jensen,J.B.; Trager,W.; Doherty,J. (Rockefeller Univ., New York, NY 10021, USA) *Exp. Parasitol., 48(1), 36-41 (1979)* En;en.

448-U2 *Plasmodium falciparum* in culture: improved continuous flow method. Trager,W. (Lab. Parasitol., Rockefeller Univ., New York, NY 10021, USA) *J. Protozool., 26(1), 125-129 (1979)* En;en.

449-U2 Rapid, large-scale isolation of *Plasmodium berghei* sporozoites from infected mosquitoes. Pacheco,N.D.; Strome,C.P.A.; Mitchell,F.; Bawden,M.P.; Beaudoin,R.L. (Immunoparasitol., Dep. Naval Med. Res. Inst., Bethesda, MD 20014, USA) *J. Parasitol., 65(3), 414-417 (1979)* En;en.

450-U2 A rapid method for isolating purified *Toxoplasma* tachyzoites from peritoneal exudate and cell culture. Tyron,J.C.; Weider,E.; Larson,A.D.; Hart,L.T. (Dep. Zool. and Physiol., Louisiana State Univ., Baton Rouge, LA 70803, USA) *J. Parasitol., 64(6), 1127-1129 (1978)* En.

451-U2 A comparison of five methods for the detection of *Trichomonas vaginalis* in clinical specimens. Levett,P.N. (Dep. Microbiol., Univ. Surrey, Guildford, Surrey GU2 5XH, UK) *Med. Lab. Sci., 37(1), 85-88 (1980)* En;en.

452-U2 Disposable bacteriological loops and vaginal discharge. Sparks,R.A.; Davies,A.J. (St. David's Hosp., Cardiff, UK) *Br. Med. J., 1(6016), 1017-1018 (1976)* En.

453-U2 A simple cultivation method for field diagnosis of avian trypanosomes. Kucera,J. (Inst. Parasitol., Czechoslovak Acad. Sci., Prague, Czechoslovakia) *Folia Parasitol., 26(4), 289-293 (1979)* En;en,ru.

454-U2 Comparison of different methods for the detection of intestinal protozoa and helminths in stool. Akhtarazzaman,K.M.; *Bienzle,U.; Rosenkaimer,F.; Guggenmoos,R.; Dietrich,M. (Klin. Abt. Bernhard-Nocht-Inst., Bernard Nocht Str. 74, D 2000 Hamburg 4, GFR) *Tropenmed. Parasitol., 29(4), 427-431 (1978)* En;de,en.

455-U2 A simple system for continuous culture of rumen ciliates. Michalowski,T. (Dep. Anim. Physiol., Zool. Inst., Univ. Warsaw, Zwirkii Wigury 93, 02-089 Warsaw, Poland) *Bull. Acad. Pol. Sci., Ser. Sci. Biol., 27(7), 581-583 (1979)* En;en,ru.

456-U2 A method for mass cultivation of sessile peritrich protozoa. Rose,C.R.; Finley,H.E. (deceased) (Dep. Zool., Howard Univ., Washington, DC 20001, USA) *Trans. Am. Microsc. Soc., 95(4), 541-544 (1976)* En;en.

457-U2 A new, simple method for gently collecting planktonic protozoa. Graham,L.B.; Colburn,A.D.; Barke,J.C. (Woods Hole Oceanogr. Inst., Woods Hole, MA 02543, USA) *Limnol. Oceanogr., 21(2), 336-341 (1976)* En;en.

458-U2 Comparison of blotters and guaiacol agar for detection of *Helminthosporium oryzae* and *Trichoconis padwickii* in rice seeds. Kulik,M.M. (Seed Quality Lab., AMRI, ARS, USDA, Beltsville Agric. Res. Cent.-West, Beltsville, MD 20705, USA) *Phytopathology, 65(11), 1325-1326 (1975)* En;en.

459-U2 A micromethod for evaluating turkey lymphocyte responses to phytomitogens. Maheswaran,S.K.; Thies,E.S.; Greimann,C. (Dep. Vet. Biol., Coll. Vet. Med., St. Paul, MN 55101, USA) *Am. J. Vet. Res., 36(9), 1397-1400 (1975)* En;en.

460-U2 A simple method for obtaining peritoneal macrophages from chickens. Sabet,T.; Hsia,W.-C.; Stanisz,M.; el-Domeiri,A.; van Alten,P. (Dep. Histol., Univ. Illinois, Med. Cent., Chicago, IL 60612, USA) *J. Immunol. Methods, 14(2), 103-110 (1977)* En;en.

461-U2 A miniaturized in vitro diffusion culture system. Eipert,E.F.; Adorini,L.; Coulderc,J. (Dep. Anat., Sch. Med., Univ. California, Los Angeles, CA 90024, USA) *J. Immunol. Methods, 22(3-4), 283-292 (1978)* En;en.

462-U2 A rapid method of obtaining cultures of chick embryo cells, not producing leukosis virus or chicken group-specific antigens. Shmel'kova,V.I.; Kuznetsov,O.K.; Batrakova,V.P.; Fedorova,S.M.; Sokolova,A.N. (Lab. Biochem., Order of the Red Banner Res. Inst., Mz, SSSR) *Vopr. Onkol., 25(6), 125-126 (1979)* Ru.

463-U2 A microculture method for the generation of primary immune responses in vitro. Pike,B.L. (Walter and Eliza Hall Inst. Med. Res., PO R. Melbourne Hosp., Victoria 3050, Australia) *J. Immunol. Methods, 9(1), 85-104 (1975)* En;en.

464-U2 In vitro culture of lymphocyte colonies in agar capillary tubes after PHA-stimulation. Maurer,H.R.; Maschler,R.; Dietrich,R.; Goebel,B. (Pharm. Inst., Freien Univ. Berlin, Konigin-Luise-Str. 2 + 4, D-1000 Berlin 33, GFR) *J. Immunol. Methods, 18(3-4), 353-364 (1974)* En;en.

465-U2 A simple reliable system for studying antigen-specific murine T cell proliferation. Lee,K.C.; Singh,B.; Barton,M.A.; Procyshyn,A.; Wong,M. (Dep. Immunol., Univ. Alberta, 845E Med. Sci. Build., Edmonton, Alta T6G 2H7, Canada) *J. Immunol. Methods, 25(2), 159-170 (1979)* En;en.

466-U2 Automated scanning of bone marrow cell colonies growing in agar-containing glass capillaries. Maurer,H.R.; Henry,R. (Pharm. Inst. Freien Univ. Berlin, D-1000 Berlin 33, GFR) *Exp. Cell Res., 103(2), 271-277 (1976)* En;en.

467-U2 Rapid, quantitative human lymphocyte separation and purification in a closed system. Ludefer,A.A.; Zine,A.R.; Hess,D.M.; Henyan,J.N.; Odstrchel,G. (Corning Glass Works, Sullivan Sci. Park, Res. and Dev. Lab., Corning, NY 14830, USA) *Mol. Immunol., 16(8), 621-624 (1979)* En;en.

468-U2 Development of a methodology for the production of foot-and-mouth disease virus from BHK 21 C13 monolayer cells grown in a 100 l (20 m^2) glass sphere propagator. Whiteside,J.P.; Whiting,B.R.; Spier,R.E. (Anim. Virus Res. Inst., Pirbright, Surrey GU24 ONF, UK) *Dev. Biol. Stand., 42, 113-119 (1979)* En;en.

469-U2 A simple technique for harvesting lymphocytes cultured in Terasaki plates. O'Brien,J.; Knight,S.; Quick,N.A.; Moore,E.H.; Platt,A.S. (Clin. Res. Cent., Div. Surg. Sci., Watford Road, Harrow, Middx. HA1 3UJ, UK) *J. Immunol. Methods, 27(3), 219-223 (1979)* En;en.

470-U2 One-stage stimulation of human T-lymphocyte colony-forming units (TL-CFU) in a micro agar culture in glass capillaries. Ulmer,A.J.; Flad,H.-D. (Lab. Immunol., Dep. Microbiol., Univ. Ulm, 7900 Ulm, GFR) *Immunology, 38(2), 393-399 (1979)* En;en.

471-U2 Simultaneous identification of T and B cells in blood smears using antibody coated bacteria. /[presented in part at the 12th Annual Meeting of the Reticuloendothelial Society, held in Miami, Florida, USA, on 4-8 Dec 1975]. Teodorescu,M.; Mayer,E.P.; Dray,S. (Dep. Microbiol., Univ. Illinois at the Med. Cent., Chicago, IL 60612, USA) *Cell. Immunol., 24(1), 90-96 (1976)* En;en.

472-U2 A greatly simplified method of establishing B-lymphoblastoid cell lines. Tohda,H.; Oikawa,A.; Kudo,T.; Tachibana,T. (Dep. Pharmacol., Res. Inst. Tuberculosis and Cancer, Tohoku Univ., Sendai 980, Japan) *Cancer Res., 38(10), 3560-3562 (1978)* En;en.

473-U2 Rapid mixed lymphocyte culture (MLC) - an aid for selection of recipients of cadaveric renal allografts. Butt,K.M.H.; Glass,N.R.; Parsa,I.; Rao,T.K.S.; Adamsons,R.J.; Kountz,S.L. (Assoc. Professor Surg., Dir. Transplant., Downstate Med. Cent., 450 Clarkson Ave., Brooklyn, NY 11203, USA) *Transplant Proc., 11(1), 763-766 (1979)* En.

474-U2 Canine vaginal flora: a technique for sampling and clinical observations. Stockner,P.K.; Brudvik,A.M.; Baker,D. (7134 Bentley Road East, Puyallup, WA 98371, USA) *Canine Pract., 6(1), 18-19 + 22 (1979)* En.

See: 476, 1142, 1163, 1164, 1294, 1298, 1302, 1347, 1680, 1735, 1777, 1785, 1786, 1843, 2044, 2045, 2093, 2294, 3020, 3053, 3069, 3091, 30 3104, 3146, 3151, 3160, 3285, 3295, 1936, 2072, 3280

Growth chambers

475-U2 Procedures for rapid detection of virus and viruslike diseases of grapevine. Mink,G.I.; Parsons,J.L. (Dep. Plant Pathol., Washington State Univ., Irrigated Agric. Res. and Extension Cent., Prosser, WA 93350, USA) *Plant Dis. Rep., 61(7), 567-571 (1977)* En;en.

476-U2 A simple humidity chamber for germinating spores. Omar,M.B.; Heather,W.A. (Address not stated) *Bull. Br. Mycol. Soc., 12(2), 128-129 (1978)* En.

Inoculation techniques

477-U2 Simplified method for efficient intravascular inoculation of chicken embryos. Kelling,C.L.; Schipper,I.A. (Dep. Vet. Sci., North Dakota State Univ., Fargo, ND 58102, USA) *J. Clin. Microbiol., 4(1), 104-105 (1976)* En;en.

478-U2 Efficient mechanical inoculation of turnip yellow mosaic virus using small volum.es of inoculum Fraser,L.; Matthews,R.E.F. (Dep. Cell Biol., Univ. Auckland, Private Bag, Auckland, New Zealand) *J. Gen. Virol., 44(2), 565-568 (1979)* En;en.

479-U2 A simple method of inoculating large woody roots in situ with *Phytophthora cinnamomi* Rands. Rockel,B.A. (Div. For. Res., CSIRO, Kalmscott, WA 6111, Australia) *Aust. For. Res., 7(4), 271-272 (1977)* En.

480-U2 A simple field technique for development of neck blast through artificial infection. Padhi,B. (Pathol. Div., Cent. Rice Res. Inst., Cuttack-6 (Orissa), India) *Indian Phytopathol., 27(1), 122-123 (1974)* En.

481-U2 Simple techniques of epicutaneous inoculation of guinea-pigs with dermatophytes. Weigl,E. (Dep. Biol., Med. Fac., Palacky Univ., Olomouc, Czechoslovakia) *Mycopathologia, 59(3), 149-150 (1976)* En;en.

Genetic techniques

482-U2 Molecular cloning of DNA. An introduction into techniques and problems. Vosberg,H.-P. (Max-Planck-Inst. Med. Forsch., Abt. Mol. Biol., Jahnstr. 29, D-6900 Heidelberg, GFR) *Hum. Genet., 40(1), 1-72 (1977)* En;en.

483-U2 Structural gene identification and mapping by DNA.mRNA hybrid-arrested cell-free translation. Paterson,B.M.; Roberts,B.E.; Kuff,E.L. (Lab. Biochem., Natl. Cancer Inst., Bethesda, MD 20014, USA) *Proc. Natl. Acad. Sci. USA, 74(10), 4370-4374 (1977)* En;en.

484-U2 A general method for inserting specific DNA sequences into cloning vehicles. Bahl,C.P.; Marians,K.J.; *Wu,R.; Stawinsky,J.; Narang,S.A. (Sect. Biochem., Mol. and Cell Biol., Cornell Univ., Ithaca, NY 14853, USA) *Gene, 1(1), 81-92 (1977)* En;en.

485-U2 A simple method for phage or plasmid DNA's isolation suitable as a 'screening test' after molecular cloning. Kozlov,Y.I.; Gening,L.V.; Strongin,A.Y.; Debabov,V.G. (Inst. Genet. and Selection Ind. Microorganisms, PO Box 825, 113545, Moscow, USSR) *Anal. Biochem., 86(1), 316-319 (1978)* En.

486-U2 A suitable method for construction and cloning hybrid plasmids containing EcoRI-fragments of *E. coli* genome. Kozlov,J.I.; Kalinina,N.A.; Gening,L.V.; Rebentish,B.A.; Strongin,A.Y.; Bogush,V.G.; Debabov,V.G. (Inst. Genet. and Sel. Ind. Microorg., PO Box 825, Moscow

113545, USSR) *Mol. Gen. Genet., 150(2), 211-219 (1977)* En;en.

487-U2 Qualitative complementation test for temperature-sensitive mutants of herpes simplex virus Chu,C.-T.; *Schaffer,P.A. (Dep. Virol. and Epidemiol., Baylor Coll. Med., Houston, TX 77025, USA) *J. Virol., 16(5), 1131-1136 (1975)* En,en.

488-U2 A simplified method for mapping deletion/substitution mutants of bacteriophage lambda. Fanning,T.G.; Schreier,P.H.; Buchel,D.E.; Davies,R.W. (Inst. Genet. Univ. Koln, Weyertal 121, 5 Koln 41, GFR) *Anal. Biochem., 81(1), 57-64 (1977)* En;en.

489-U2 A simple way to measure the parameter R (ζ) of half-chromatid chiasma. Shcherbakov,V.P.; Plugina,L.A.; Kudryashova,E.A. (Div. Inst. Chem. Phys., Acad. Sci. USSR, Moscow Region, USSR) *Genetika, 16(6), 967-974 (1980)* Ru;en,ru.

490-U2 Radiobiological method of examination of partly diploid area structure in phage T4. Matvienko,N.I.; Tikhomirova,L.P. (Inst. Biochem. and Physiol. Microorg., Acad. Sci. USSR, Moscow Region, USSR) *Genetika, 14(1), 129-137 (1978)* Ru;en,ru.

491-U2 A rapid method for detecting and mapping homology between heterologous DNAs. Evaluation of polyomavirus genomes. Howley,P.M.; Israel,M.A.; Law,M.F.; Martin,M.A. (Lab. Pathol., Natl. Cancer Inst., Natl. Inst. Allergy and Infect. Dis., NIH, Bethesda, MD 20205, USA) *J. Biol. Chem., 254(11), 4876-4883 (1979)* En;en.

492-U2 Method for rapidly screening revertants of reovirus temperature-sensitive mutants for extragenic suppression. Ramig,R.F.; Fields,B.N. (Dep. Microbiol. and Mol. Genet., Harvard Med. Sch., 25 Shattuck St., Boston, MA 02115, USA) *Virology, 81(1), 170-173 (1977)* En;en.

493-U2 A rapid test for the *rel A* mutation in *E. coli.* Uzan,M.; Danchin,A. (Inst. Biol. Phys.-Chim., rue P. et M. Curie, 75005 Paris, France) *Biochem. Biophys. Res. Commun., 69(3), 751-758 (1976)* En;en.

494-U2 Simplified method for interruption of conjugation in *Escherichia coli.* Zipkas,D.; *Riley,M. (Biochem. Dep., State Univ. New York, Stony Brook, NY 11794, USA) *J. Bacteriol., 126(1), 559-560 (1976)* En;en.

495-U2 Use of a simplified fluctuation test to detect low levels of mutagens. Green,M.H.L.; Muriel,W.J.; Bridges,B.A. (MRC Cell Mutat. Unit, Univ. Sussex, Falmer, Brighton BN1 9QG, UK) *Mutat. Res., 38(1), 33-42 (1976)* En;en.

496-U2 Mutagen testing using TRP$^+$ reversion in *Escherichia coli.* Green,M.H.L.; Muriel,W.J. (MRC Cell Mutat. Unit, Univ. Sussex, Falmer, Brighton BN1 9QG, UK) *Mutat. Res., 38(1), 3-32 (1976)* En;en.

497-U2 A suitable method for construction and cloning hybrid plasmids containing EcoRI-fragments of *E. coli* genome. Kozlov,J.I.; Kalinina,N.A.; Gening,L.V.; Rebentish,B.A.; Strongin,A.Y.; Bogush,V.G.; Debabov,V.G. (Inst. Genet. and Select. Ind. Microorg., PO Box 825, Moscow 113 545, USSR) *Mol. Gen. Genet., 150(2), 211-219 (1977)* En;en.

498-U2 The identification of mutants of *Escherichia coli* deficient in formate dehydrogenase and filtrate reductase activities using dye indicator plates. Begg,Y.A.; Whyte,J.N.; *Haddock,B.A. (Dep. Biochem., Med. Sci. Inst., Univ. Dundee, Dundee DD1 4HN, UK) *FEMS Microbiol. Lett., 2(1), 47-50 (1977)* En.

499-U2 Mutagen screening by a simplified bacterial fluctuation test. Use of microsomal preparations and whole liver cells for metabolic activation. Green,M.H.L.; Bridges,B.A.; Rogers,A.M.; Horspool,G.; Muriel,W.J.; Bridges,J.W.; Fry,J.R. (MRC Cell Mutat. Unit, Univ. Sussex, Falmer, Brighton BN1 9QG, UK) *Mutat. Res., 48(3-4), 287-294 (1977)* En;en.

500-U2 [Construction and molecular cloning of hybrid plasmids containing certain fragments of *Escherichia coli* DNA.] Kuzlov,Y.I.; Kalinina,N.A.; Strongin,A.Y.; Gening,L.V.; Rebentish,B.A.; Dolganov,G.M.; Shemiakin,M.F.; Bogush,V.G.; Savejiev,V.L.; Permogorov,V.I.; Debabov,V.G (Inst. Genet. and Sel. Ind. Miroorg., Moscow, USSR) *Mol. Biol., 12(1), 108-115 (1978)* Ru;en;ru.

501-U2 Detection of mutagenic derivatives of cyclophosphamide and a variety of other mutagens in a 'Microtitre'® fluctuation test, without microsomal activation. Gatehouse,D. (Pathol. Lab., Allen and Hanburys Res. Ltd., Priory Street, Ware, Herts., UK) *Mutat. Res., 53(3), 289-296 (1978)* En;en.

502-U2 Genetic transformation assays for identification of strans of *Moraxella urethralis.* Juni,E. (Dep. Microbiol., Univ. Michigan, Ann Arbor, MI 48109, USA) *J. Clin. Microbiol., 5(2), 227-235 (1977)* En;en.

503-U2 Genetic transformation as a tool for detection of Neisseria gonorrhoeae. Janik,A.; Juni,E.; Heym,G.A. (Ames Res. Lab., Ames Co., Div. Miles Lab. Inc., Elkhart, IN 46514, USA) *J. Clin. Microbiol., 4(1), 71-81 (1976)* En;en.

504-U2 A simplified method for the induction of 8-azaguanine resistance in *Salmonella typhimurium.* Bignami,M.; Crebelli,R. (Inst. Superiore Sanita, Viale Regina Elena 299, Roma, Italy) *Toxicol. Lett., 3(3), 169-175 (1979)* En;en.

505-U2 Rapid methods for generalized transduction of *Salmonella typhimurium* mutants. Rosenfeld,S.A.; *Brenchley,J. (Dep. Biol., Purdue Univ., West Lafayette, IN 47907, USA) *J. Bacteriol., 138(1), 261-263 (1977)* En;en.

506-U2 The computer assisted bacterial test for mutagenesis. Claxton,L.; Baxter,R. (Biochem. Branch, HERL, US Environ. Prot. Agency, Environ. Res. Cent. MD68, Research Triangle Park, NC 27711, USA) *Mutat. Res., 53(3), 345-350 (1978)* En.

507-U2 A combined bacterial and liver test system for detection and classification of carcinogens as mutagens. Ames,B.N. (Biochem. Dep., Univ. California, Berkeley, CA 94720, UA) *Genetics, 78(1), 91-95 (1974)* En.

508-U2 A quick method for testing recessive lethal damage with a diploid strain of *Aspergillus nidulans*. Morpurgo,G.; Puppo,S.; Gualandi,G.; Conti,L. (Ist. Orto Bot., Univ. Roma, Largo Cristina di Suezia, 24, 00165 Rome, Italy) *Mutat. Res., 54(2), 131-137 (1978)* En;en.

509-U2 A quick method of the selection of *Aspergillus niger* strains taking part in the citric acid synthesis process. Ilczuk,Z.; Zulikowski,W. (Inst. Mikrobiol. i Biochem. UMCS, Lublin Poland) *Przem. Ferment. Rolny, 20(11), 22-24 (1976)* Pl;en,pl,ru.

510-U2 A method of preparation and application of nitrous acid as a mutagen in *Claviceps purpurea*. Strnadova,K. (Res. Inst. Pharm. and Biochem., 130 00 Prague 3, Czechoslovakia) *Folia Microbiol., 21(6), 455-458 (1976)* En;en.

511-U2 Microtiter plate assay for sexual agglutination in the yeast *Hansenula wingei*. Crandall,M. (Thomas Hunt Morgan Sch. Biol. Sci., Univ. Kentucky, Lexington, KY 40506, USA) *J. Bacteriol., 137(2), 1051-1052 (1979)* En;en.

512-U2 A technique for the rapid selection of drug-sensitive and auxotrophic mutants of *Phycomyces blakesleeanus*. Brunke,K.J.; Mercurio,R.; Goodgal,S.H. (Dep. Microbiol., Sch. Med., Univ. Pennsylvania, Philadelphia, PA 19174, USA) *J. Gen. Microbiol., 102(2), 287-293 (1977)* En;en.

513-U2 Simple and sensitive procedure for screening yeast mutants that lyse at nonpermissive temperatures. Cabib,E.; Duran,A. (Natl. Inst. Arthritis, Metab., and Digest. Dis., NIH, Bethesda, MD 20014, USA) *J. Bacteriol., 124(3), 1604-1606 (1975)* En;en.

514-U2 A simple method for the isolation and characterization of thymidylate uptaking mutants in *Saccharomyces cerevisiae*. Brendel,M. (Arbeitsgruppe Mikrobengenet. im Fachbereich Biol., J.W. Goethe-Univ., D-6000 Frankfurt/Maine, GFR) *Mol. Gen. Genet., 147(2), 209-215 (1976)* En;en.

515-U2 Cloning of yeast genes coding for glycolytic enzymes. Holland,M.J.; Holland,J.P.; Jackson,K.A. (Dep. Biochem., Univ. Connecticut Health Cent., Farmington, CT 06032, USA) *Methods Enzymol., 68, 408-419 (1979)* En.

516-U2 A capillary technique for cloning *Amoeba* from single cells. Ord,M.J. (Dep. Biol., Univ. Southampton Med. and Biol. Sci. Build., Bassett Crescent East, Southampton SO9 3TU, UK) *Cytobios, 21(8), 57-69 (1978)* En;en.

517-U2 The determination of lymphoid cell chimerism using peripheral blood lymphocytes from murine bone marrow chimeras. Skidmore,B.J.; Miller,L.S. (Dep. Cell. and Dev. Immunol., Scripps Clin. and Res. Found., La Jolla, CA 92037, USA) *J. Immunol. Methods, 24(3-4), 337-343 (1978)* En;en.

See: 62, 431, 542, 627

Assays and immunochemical assays

518-U2 Improved technique for accurate and convenient assay of biological reactions liberating $^{14}CO_2$. Sissons,C.H. (Dep. Zool., Univ. Edinburgh, Edinburgh, EH9 3JT, Scotland, UK) *Anal. Biochem., 70(2), 454-462 (1976)* En;en.

519-U2 Rapid gentamicin assay; 3 years experience with urease method. Edmunds,P.N.; Heddle,A.C.; Fox,C.C. (Dep. Med. Microbiol., Fife Area Lab., Hayfield Rd., Kirkcaldy KY2 5AG, UK) *J. Antimicrob. Chemother., 3(6), 625-626 (1977)* En.

520-U2 Assay of gentamicin concentrations in serum speciments by microbiological and radioenzymatic methods. Dornbusch,K. (Statens Bakeriol. Lab., S 105 21 Stockholm, Sweden) *Scand. J. Infect. Dis., 9(3), 227-231 (1977)* En;en.

521-U2 Automated immunoassay with a silver sulphide ion-selective electrode. Solsky,R.L.; *Rechnitz,G.A. (Dep. Chem., Univ. Delaware, Newark, DE 19711, USA) *Anal. Chim. Acta, 99(2), 241-246 (1978)* En;en.

522-U2 A simple visual surface immunology test. Giaever,I. (Corporate Res. and Devel., General Electric Company, POB 8, Schenectady, NY 12301, USA) *J. Immunol. Methods, 24(1-2), 57-61 (1978)* En;en.

523-U2 A new method for fluorescence immunoassay using plane surface solid phases (FIAPS). Sedlacek,H.H.; Muck,K.-F.; Rehkopf,R.; Baudner,S.; Seiler,F.R. (Res. Lab. Behringwerke AG, 3550 Marburg/Lahn, GFR) *J. Immunol. Methods, 26(1), 11-24 (1979)* En;en.

524-U2 Luminescence immunoassay (LIA): a solid-phase immunoassay monitored by chemiluminescence. Olsson,T.; Brunius,G.; Carlsson,H.E.; Thore,A. (Natl. Defence Res. Inst., Dep. 4, Div. Appl. Microbiol., S-172 04 Sundyberg, Sweden) *J. Immunol. Methods, 25(2), 127-135 (1979)* En;en.

525-U2 A new immunoassay based on fluorescence excitation by internal reflection spectroscopy. Kronick,M.N.; Little,W.A. (Lab. Chem. Biodynamics, Lawrence Berkeley Lab., Univ. California, Berkeley, CA 94720, USA) *J. Immunol. Methods, 8(3), 235-240 (1975)* En;en.

526-U2 In situ immunoassays for gene translation products in phage plaques and bacterial colonies. Skalka,A.; Shapiro,L. (Dep. Cell Biol., Roche Inst. Mol. Biol., Nutley, NJ 07110, USA) *Gene, 1(1), 65-79 (1977)* En;en.

527-U2 Spin immunoassay for opiates in urine - results of screening military personnel. Cate,J.,IV; Clarkson,M.; Strickland,J.; D'Amato,N.A. (Dep. Pathol., Holston Valley Community Hosp., Kingsport, TN 37763, USA) *Clin. Toxicol., 9(2), 235-243 (1976)* En.

528-U2 Immunoassay of enzymes. Forman,D.T. (Div. Clin. Biochem., Evanston Hosp., Evanston, IL 60201, USA) *Ann.*

Clin. Lab. Sci., 7(4), 329-334 (1977) En;en.

529-U2 **Electroimmunoquantitation of serum proteins cross-linked by glutaraldehyde.** Lou,K.; Bowers,J. (Immunol. Res., ICL Sci., Fountain Valley, CA 92708, USA) *J. Immunol. Methods, 9(1), 69-79 (1975)* En;en.

530-U2 **Assay of adenylate cyclase by use of polyacrylamide-boronate gel columns.** Hageman,J.H.; Kuehn,G.D. (Dep. Chem., New Mexico State Univ., Las Cruces, NM 88003, USA) *Anal. Biochem., 80(2), 547-554 (1977)* En;en.

531-U2 **Rapid assay of interferon.** Green,J.A. (Lab. Viral Dis., Natl. Inst. Allergy and Infect. Dis., Natl. Inst. Health, Bethesda, MD 20014, USA) *Tex. Rep. Biol. Med., 35, 167-172 (1977)* En;en.

532-U2 **Conventional assay systems.** Oie,H.K. (NCI Veterans Adm. Med. Oncol. Branch, Veterans Adm. Hosp. and Natl. Cancer Inst., Washington, DC 20422, USA) *Tex. Rep. Biol. Med., 35, 154-160 (1977)* En;en.

533-U2 **New rapid bioassay of gentamicin based on luciferase assay of extracellular ATP in bacterial cultures.** Nilsson,L. (Dep. Clin. Bacteriol., Linkoping Univ., Regionsjukhuset, S-581 85 Linkoping, Sweden) *Antimicrob. Agents Chemother., 14(6), 812-816 (1978)* En;en.

534-U2 **A rapid in vitro assay for carcinogenicity of chemical substances in mammalian cells utilizing an attachment-independence endpoint.** Traul,K.A.; Kachevsky,V.; Wolff,J.S. (Pfizer Inc., Maywood, NJ 07607, USA) *Int. J. Cancer, 23(2), 193-196 (1979)* En;en,fr.

535-U2 **Spin membrane immunoassay: simplicity and specificity.** Chan,S.W.; Tan,C.-T.; *Hsia,J.C. (Dep. Pharmacol., Fac. Med., Univ. Toronto, Toronto, Ont. M5S 1A8, Canada) *J. Immunol. Methods, 21(1-2), 185-195 (1978)* En;en.

536-U2 **Platelet immuno-assay (PIA) as a simple and sensitive technique for the detection of antigens.** Schultz,J.; Disciullo,S.O.; *Abuelo,J.G. (Rhode Island Hosp., Providence, RI 02902, USA) *J. Immunol. Methods, 17(1,2), 47-55 (1977)* En;en.

537-U2 **Extraction and quantification of solutes in solidified agar culture media.** Buynitzky,S.J.; *Howe,H.B.,Jr.; Shellhorse,Y. (Dep. Microbiol., Univ. Georgia, Athens, GA 30602, USA) *Appl. Environ. Microbiol., 37(2), 202-207 (1979)* En;en.

538-U2 **A simple spot technique for thin layer immunoassays (TIA) on plastic surfaces.** Elwing,H.; Nilsson,L.A.; Ouchterlony,O. (Inst. Med. Microbiol., Univ. Goteborg, Goteborg, Sweden) *J. Immunol. Methods, 17(1-2), 131-145 (1977)* En;en.

539-U2 **A systems approach to fluorescent immunoassay: general principles and representative applications.** Curry,R.E.; Heitzman,H.; Riege,D.H.; Sweet,R.V.; Simonsen,M.G. (Bio-Rad Lab., 2200 Wright Ave., Richmond, CA 98404, USA) *Clin. Chem., 25(9), 1591-1595 (1979)* En;en.

540-U2 **Microbiological assay utilizing an automatic zone scanner.** Geigert,J.; Hansen,D.; McDowell,C.; Merrill,R.; Ward,C. (Cetus Corp., 600 Bancroft Way, Berkeley, CA 94710, USA) *Dev. Ind. Microbiol., 17, 153-156 (1976)* En;en.

541-U2 **Plasmid detection and sizing in single colony lysates.** Barnes,W.M. (Mol. Biol., Hills Road, Cambridge CB2 2QM, UK) *Science (Wash.), 195(4276), 393-394 (1977)* En;en.

542-U2 **In situ immunoassays for gene translation products in phage plaques and bacterial colonies.** Skalka,A.; Shapiro,L. (Dep. Cell Biol., Roche Inst. Mol. Biol., Nutley, NJ 07110, USA) *Gene, 1(1), 65-79 (1977)* En;en.

543-U2 **Rapid procedure for the detection of plasmids in *Staphylococcus epidermidis*.** Wilson,C.R.; Totten,P.A.; *Baldwin,J.N. (Dep. Microbiol., Univ. Georgia, Athens, GA 30602, USA) *Appl. Environ. Microbiol., 36(2), 368-374 (1978)* En;en.

544-U2 **Plastic multiwell plates to assay avian infectious bronchitis virus in organ cultures of chicken embryo trachea.** Yachida,S.; Aoyama,S.; Takahashi,N.; Iritani,Y.; Katagiri,K. (Aburahi Lab., Shionogi and Co. Ltd., Koka-cho, Shiga, 520-34, Japan) *J. Clin. Microbiol., 8(4), 380-387 (1978)* En;en.

545-U2 **Observations on the growth and plaque assay of BK virus in cultured human and monkey cells.** Seehafer,J.; Carpenter,P.; Downer,D.N.; Colter,J.S. (Dep. Biochem., Univ. Alberta, Edmonton, Alta. T6G 2H7, Canada) *J. Gen. Virol., 38(2), 383-387 (1978)* En;en.

546-U2 **Direct syncytial assay for the quantitation of bovine leukemia virus.** Benton,C.V.; Soria,A.E.; Gilden,R.V. (Viral Oncol. Program, Natl. Cancer Inst. Frederick Cancer Res. Cent., Frederick, MD 21701, USA) *Infect. Immun., 20(1), 307-309 (1978)* En;en.

547-U2 **Development of an in vitro infectivity assay for the C-type bovine leukemia virus.** Ferrer,J.F.; Diglio,C.A. (Sect. Viral Oncol., Comp. Leukemia Studies Unit, Sch. Vet. Med., Univ. Pennsylvania, New Bolton Center, Kennett Sq., PA 19348, USA) *Cancer Res., 36(3), 1068-1073 (1976)* En;en.

548-U2 **A reverse transcriptase assay for detection of the bovine leukemia virus.** Graves,D.C.; Diglio,C.A.; Ferrer,J.F. (Comp. leukemia Studies Unit, New Bolton Cent., Univ. Pennsylvania, Kennett Sq, PA 19348, USA) *Am. J. Vet. Res., 38(11), 1739-1744 (1977)* En;en.

549-U2 **A new plaque system for canine distemper: characteristics of the green strain of canine distemper virus.** Ho,C.K.; Babiuk,L.A. (Dep. Vet. Microbiol., Western Coll. Vet. Med., Univ. Saskatchewan, Saskatoon, Sask. S7N 0W0, Canada) *Can. J. Microbiol., 25(6), 680-685 (1979)* En;en,fr.

550-U2 **Comparative studies of two alphaviruses by macroplaque assays and by a fluorescent focus method in vero cell lines.** Tignor,G.H.; Digoutte,J.P. (Inst. Pasteur de la

Guyane Francaise BP 304, 97305 Cayenne, French Guyana) *Ann. Microbiol., 127B(3), 439-444 (1976)* Fr;en,fr.

551-U2 **End-point dilution and plaque assay methods for titration of cricket paralysis virus in cultured *Drosophila* cells.** Scotti,P.D. (Entomol. Div., Dep. Sci. and Industrial Res., Auckland, New Zealand) *J. Gen. Virol., 35(2), 393-396 (1977)* En;en.

552-U2 **Rapid in vitro assay for thymotropic, leukemogenic murine C-type RNA viruses.** Lieberman,M.; Decleve,A.; Kaplan,H.S. (Cancer Biol. Res. Lab., Dep. Radiol., Stanford Univ. Sch. Med., Stanford, CA 94305, USA) *Virology, 90(2), 274-278 (1978)* En;en.

553-U2 **Titration of a cytoplasmic polyhedrosis virus by a tissue microculture assay: some applications.** Belloncik,S.; Chagnon,A. (Cent. Rech. Virol., Inst. Armand-Frappier, Univ. Quebec, Ville de Laval, CP 100, Que. H7V 1B7, Canada) *Intervirology, 13(1), 28-32 (1980)* En;en.

554-U2 **Detection of small amounts of viruses (enteroviruses) in drinking water.** Beytout,D.; Charrier,F.; Laveran,H.; Monghal,M. (Lab. Bacteriol. Virol., Fac. Med., 63001 Clermont-Ferrand Cedex, France) *Ann. Microbiol., 128A(2), 255-262 (1977)* Fr;en,fr.

555-U2 **An improved specific Igm antibody assay for Epstein-Barr virus.** Goetz,O.; Peller,P. (Univ. Child. Hosp., D-8 Munich 2, Lindwurmstr. 4, GFR) *Infection, 3(4),229-230 (1975)* En;de,en.

556-U2 **An evaluation of some methods of assay of foot-and-mouth disease antigen for vaccines.** Garland,A.J.M.; Mowat,G.N.; Fletton,B. (Anim. Virus Res. Inst., Pirbright, Woking, Surrey GU24 0NF, UK) *Dev. Biol. Stand., 35, 323-332 (1977)* En;en,fr.

557-U2 **Quantitation of Friend spleen focus-forming virus by a nine-day ^{59}Fe assay.** Menna,J.H.; Hankins,W.D.; Krantz,S.B. (Univ. Arkansas for Med. Sci., Little Rock, AR 72201, USA) *J. Clin. Microbiol., 4(6), 486-491 (1976)* En;en.

558-U2 **Reverse plaque formation by hog cholera virus of the GPE$^-$ strain inducing heterologous interference.** Fukusho,A.; Ogawa,N.; Yamamoto,H.; Sawada,M.; Sazawa,H. (First Assay Div., Natl. Vet. Assay Lab., Kokubunji, Tokyo, 185, Japan) *Infect. Immun., 14(2), 332-336 (1976)* En;en.

559-U2 **[^{125}I]deoxycytidine used in a rapid, sensitive, and specific assay for herpes simplex virus type 1 thymidine kinase.** Summers,W.C.; Summers,W.P. (Dep. Ther. Radiol., Yale Univ. Sch. Med., New Haven, CT 06510, USA) *J. Virol., 24(1), 314-318 (1977)* En;en.

560-U2 **A direct plaque assay for hog cholera virus.** Laude,H. (Stn. Rech. Virol. et Immunol. (INRA), Lab. Pathol. Porcine, 78850 Thiverval Grignon, France) *J. Gen. Virol., 40(1), 225-228 (1978)* En;en.

561-U2 **The assay of influenza antineuraminidase activity by an elution inhibition technique.** Appleyard,G.; Oram,J.D. (Microbiol. Res. Establ., Porton, Nr. Salisbury, Wiltshire, UK) *J. Gen. Virol., 34(1), 137-144 (1977)* En;en.

562-U2 **Solid phase antibody assay by means of enzyme conjugated to anti-immunoglobulin.** Leinikki,P.; Passila,S. (Virus Lab., Municip. Bacteriol. Lab., Aurora Hosp., Helsinki, Finland) *J. Clin. Pathol., 29(12), 1116-1120 (1976)* En;en.

563-U2 **Standardization of a rapid modified microneuraminidase-inhibition test (Essen-NIT) for influenza virus-neuraminidase antibody assay and comparison with the W.H.O. method.** Thraenhart,O.; Kuwert,E.K. (Inst. Med. Virol. and Immunol., Klinik. Univ. Essen-Gesamthochschule, Hufelandstr. 55, 4300 Essen 1, GFR) *J. Biol. Stand., 4(3), 224-241 (1976)* En;en.

564-U2 **Usefulness of methyl cellulose-overlay for the determination of plaque-forming units in Marek's disease vaccine.** Cakala,A.; Samorek-Salamonovicz,E. (Vet. Res. Inst., Pulawy, Dep. Poult. Dis., PL-24100 Pulawy, Poland) *Bull. Vet. Inst. Pulawy, 22(3-4), 43-48 (1978)* En;en.

565-U2 **Bioassay of local Marek's disease agent in cell culture.** Waseem,M.; Pathak,R.C.; Singh,D.P. (Dep. Bacteriol., UP Coll. Vet. Sci. and Anim. Husbandry, Mathura, India) *Indian J. Exp. Biol., 14(3), 333-336 (1976)* En;en.

566-U2 **A plaque assay for Mount Elgon bat virus based on intrinsic interference.** Patel,J.R. (Dep. Microbiol. and Parasitol., R. Vet. Coll., Royal College St., London NW1 0TU, UK) *J. Gen. Virol., 40(3), 659-664 (1978)* En;en.

567-U2 **Mouse hepatitis virus (MHV-2): plaque assay and propagation in mouse cell line DBT cells.** Hirano,N.; *Fujiwara,K.; Matumoto,M. (Dep. Anim. Pathol., Inst. Med. Sci., Univ. Tokyo, 4-6-1 Shirogane-dai, Minato-ku, Tokyo 108, Japan) *Jap. J. Microbiol., 20(3), 219-225 (1976)* En;en.

568-U2 **In vitro infectivity assay for mouse mammary tumor virus.** Vacquier,J.P.; *Cardiff,R.D. (Dep. Pathol., Sch. Med., Univ. California, Davis, CA 95616, USA) *Proc. Natl. Acad. Sci. USA, 76(8), 4117-4121 (1979)* En;en.

569-U2 **A short-term quantitative XC assay for murine leukemia virus.** Gautsch,J.W.; Meier,H. (Jackson Lab., Bar Harbor, ME 04609, USA) *Virology, 72(2), 509-513 (1976)* En;en.

570-U2 **Plaque assay of neonatal calf diarrhea virus and the neutralizing antibody in human sera.** Matsuno,S.; Inouye,S.; Kono,R. (Cent. Virus Diagn. Lab., Natl. Inst. Health, Musashimurayama, Tokyo, 190-12, Japan) *J. Clin. Microbiol., 5(1), 1-4 (1977)* En;en.

571-U2 **Peroral bioassay of technical-grade preparations of the Douglas-fir tussock moth nucleopolyhedrosis virus (Baculovirus).** Martignoni,M.E.; Iwai,P.J. (Pac. Northwest For. and Range Exp. Stn., POB 3141, Portland, OR 97208. USA) *Res. Pap., USDA For. Ser., Pac. Northwest For. and Range Exp. Stn., 1977 12 pp. PNW-222* En;en.

572-U2 **An improved technique for plaque assay of *Autographa californica* nuclear polyhedrosis virus on TN-368 cells.** Hink,W.F.; Strauss,F.M. (Dep. Entomol., Ohio State Univ., 1735 Neil Avenue, Columbus, OH 43210, USA) *J. Invertebr. Pathol., 29(3), 390-391 (1977)* En.

573-U2 **An agar overlay plaque assay method for *Autographa californica* nuclear polyhedrosis virus.** Wood,H.A. (Boyce Thompson Inst., 1086 North Broadway, Yonkers, NY 10701, USA) *J. Invertebr. Pathol., 29(3), 304-307 (1977)* En;en.

574-U2 **A plaque assay for nuclear polyhedrosis viruses using a solid overlay.** Brown,M.; Faulkner,P. (Dep. Microbiol. and Immunol., Queen's Univ., Kingston, Ontario, Canada K7L 3N6) *J. Gen. Virol., 36(2), 361-364 (1977)* En;en.

575-U2 **A very sensitive biochemical assay for detecting and quantitating avian oncornaviruses.** Tereba,A.; Murti,K.G. (Lab. Virol., St. Jude Child. Res. Hosp., 332 North Lauderdale, PO Box 318, Memphis, TN 38101, USA) *Virology, 80(1), 166-176 (1977)* En;en.

576-U2 **The molecular mechanism of virus induction. I. A procedure for the biochemical assay of prophage induction.** Smith,C.L.; Oishi,M. (Public Health Res. Inst., City of New York, Inc., 455 First Ave., New York, NY 10016, USA) *Mol. Gen. Genet., 148(2), 131-138 (1976)* En;en.

577-U2 **Technique for determining total bacterial virus counts in complex aqueous systems.** Ewert,D.L.; *Paynter,M.J.B. (Dep. Microbiol., Clemson Univ., Clemson, SC 29631, USA) *Appl. Environ. Microbiol., 39(1), 253-260 (1980)* En;en.

578-U2 **Prophage λ induction in *Escherichia coli* K12 *envA uvrB*: a highly sensitive test for potential carcinogens.** Moreau,P.; Bailone,A.; Devoret,R. (Sect. Radiobiol. Cell., Lab. Enzymol., CNRS, 91190 Gif-sur-Yvette, France) *Proc. Natl. Acad. Sci. USA, 73(10), 3700-3704 (1976)* En;en.

579-U2 **Restriction assay for integrative recombination of bacteriophage λ DNA in vitro: requirement for closed circular DNA substrate.** Mizuuchi,K.; Nash,H.A. (Lab. Mol. Biol., Natl. Inst. Arthritis, Metab. and Dig. Dis., Natl. Inst. Mental Health, Bethesda, MD 20014, USA) *Proc. Natl. Acad. Sci. USA, 73(10), 3524-3528 (1976)* En;en.

580-U2 **DNA replication with bacteriophage T4 proteins. Purification of the proteins encoded by T4 genes 41, 45, 44, and 62 using a complementation assay.** Nossal,N.G. (Lab. Biochem. Pharmacol., Natl. Inst. Arthritis, Metab., and Dig. Dis., NIH, Bethesda, MD 20205, USA) *J. Biol. Chem., 254(13), 6026-6031 (1979)* En;en.

581-U2 **The molecular mechanism of virus induction. I. A procedure for the biochemical assay of prophage induction.** Smith,C.L.; Oishi,M. (Public Health Res. Inst. City New York, Inc., 455 First Ave., New York, NY 10016, USA) *Mol. Gen. Genet., 148(2), 131-138 (1976)* En;en.

582-U2 **A rapid microassay for detecting antibodies against poliovirus based on [^{14}C]thymidine uptake of treated cell cultures.** Hilfenhaus,J.; Damm,H.; Ziegelmaier,R.; Gruschkau,H. (Behringwerke AG, 3550 Marburg (Lahn), GFR) *J. Immunol. Methods, 17(1-2), 31-38 (1977)* En;en.

583-U2 **Isolation and assay of rabies serogroup viruses in CER cells.** Smith,A.L.; Tignor,G.H.; Mifune,K.; Motohashi,T. (Dep. Epidemiol. and Public Health, Yale Arbovirus Res. Unit, Yale Univ. Sch. Med., 60 College St., New Haven, CT 06510, USA) *Intervirology, 8(2), 92-99 (1977)* En;en.

584-U2 **A comparison of methods for the potency test of rubella vaccine.** Rapicetta,M.; Santoro,R.; Grandolfo,M.E. (Dep. Virol., Inst. Superiore di Sanita, Rome, Italy) *J. Biol. Stand., 5(3), 231-236 (1977)* En;en.

585-U2 **Simultaneous assay for mRNA(guanine-7-)-methyltransferase and mRNA(2′-O-nucleoside)methyltransferase of purified vaccinia (WR) virions.** Pugh,C.S.G.; Borchardt,R.T.; Stone,H.O. (Dep. Biochem. and Microbiol., Univ. Kansas, Lawrence, KS 66045, USA) *Anal. Biochem., 88(2), 504-512 (1978)* En;en.

586-U2 **A simple histological assay to detect virus-like particles in the green alga *Cylindrocapsa* (Chlorophyta).** Stanker,L.H.; Hoffman,L.R. (Dep. Bot., Univ. Illinois, Urbana, IL 61801, USA) *Can. J. Bot., 57(7), 838-842 (1979)* En;en,fr.

587-U2 **Vesicular stomatitis virus plaque production in monolayer cultures with liquid overlay medium: description and adaptation to a one-day, human interferon-plaque reduction assay.** Green,J.A.; Stanton,G.J.; Goode,J.; Baron,S. (Lab. Viral Dis., Natl. Inst. Allergy and Infect. Dis., Natl. Inst. Health, Bethesda, MD 20014, USA) *J. Clin. Microbiol., 4(6), 479-485 (1976)* En;en.

588-U2 **Assay of bovine interferons in cultures of the porcine cell line IB-RS-2.** Ahl,R.; Rump,A. (Bundesforschungsanst. Viruskr. Tiere, D-74 Tubingen, GFR) *Infect. Immun., 14(3), 603-606 (1976)* En;en.

589-U2 **A sensitive interferon assay for many species of cells: encephalomyocarditis virus haemagglutinin yield reduction.** Jameson,P.; Dixon,M.A.; Grossberg,S.E. (Dep. Microbiol., Med. Coll. Wisconsin, 561 North 15th St., Milwaukee, WI 53233, USA) *Proc. Soc. Exp. Biol. Med., 155(2), 173-178 (1977)* En;en.

590-U2 **Detection and occurrence of enteric viruses in shellfish: a review.** Gerba,C.P.; Goyal,S.M. (Dep. Virol. and Epidemiol., Baylor Coll. Med., Houston, TX 77030, USA) *J. Food Prot., 41(9), 743-754 (1978)* En;en.

591-U2 **Variables affecting viral plaque formation in microculture plaque assays using homologous antibody in a liquid overlay.** Randhawa,A.S.; Stanton,G.J.; Green,J.A.; Baron,S. (Dep. Microbiol., Univ. Texas Med. Branch, Galveston, TX 77550, USA) *J. Clin. Microbiol., 5(5), 535-542 (1977)* En;en.

592-U2 [Method of agar overlay for plaque assay of viruses]. Girin,V.N.; Pashkov,V.P. (Kiev Med. Inst., Kiev, USSR) *Mikrobiol. Zh., 41(6), 697-700 (1979)* Uk;en.

593-U2 A rapid method for interferon titration. Kadyrova,A.A.; Novokhatsky,A.S.; Ershov,F.I. (D. I. Ivanovsky Inst. Virol., Acad. Med. Sci. USSR, Moscow, USSR) *Vopr. Virusol., No.6, 745-748 (1976)* Ru;en.

594-U2 Rapid quantitation of interferon with chronically oncornavirus-producing cells. Aboud,M.; Weiss,O.; *Salzberg,S. (Dep. Life Sci., Bar-Ilan Univ., Ramat-Gan, Israel) *Infect. Immun., 13(6), 1626-1632 (1976)* En;en.

595-U2 Application of a truncated Poisson law for the enumeration of virus plaques by micromethod. Variations of the relative dimness. Tixier,G. (IMTSSA Lab. Biol. Mol., 1398 Marseille-Armees, France) *C. R. Seances Soc. Biol. Fil., 170(6), 120(1214 (1976)* Fr;en,fr.

596-U2 Simple procedure for large-scale assays for chick interferon. Viehhauser,G. (Inst. Physiol. Chem. der Univ. Wurzburg, 87 Wurzburg, GFR) *Appl. Environ. Microbiol., 33(3), 740-743 (1977)* En;en.

597-U2 A microplaque reduction assay for human and mouse interferon. Campbell,J.B.; Grunberger,T.; Kochman,M.A.; White,S.L. (Dep. Microbiol. and Parasitol., Fac. Med., Univ. Toronto, Toronto, Ont. M5S 1A1, Canada) *Can. J. Microbiol., 21(8), 1247-1253 (1975)* En;en,fr.

598-U2 Microbiological method for assay of antibiotics in combination. Sen,M.; Dastidar,S.G. (Div. Microbiol., Dep. Pharm., Jadavpur Univ., Calcutta 700032, India) *Indian J. Exp. Biol., 13(3), 286-288 (1975)* En;en.

599-U2 *Bacillus subtilis*; a sensitive bioassay for patulin. Reiss,J. (Mikrobiol. Lab., Grahamhaus Studt K.G., 655 Bad Kreuznach, GFR) *Bull. Environ. Contam. Toxicol., 13(6), 689-691 (1975)* En;en.

600-U2 A rapid agar-diffusion test for the detection of antibiotic residues in kidneys from slaughter-animals. Rosdahl,U.T.; Christensen,S.G.; Jacobsen,M. (Inst. Med. Microbiol., Juliane Mariesuy 22, DK 2100 Copenhagen, Denmark) *Acta Vet. Scand., 20(3), 466-468 (1979)* En.

601-U2 Microbiological determination of sucrose by means of heterogeneous cultures of *Bacillus stearothermophilus*. Pollach,G. (Osterreichlisches Zuckerforsch.-Inst. Fuchsenbigl, A-2286 Haringsee, Austria) *Zucker, 29(9), 495-501 (1976)* De;de,en,fr.

602-U2 A biological assay for quantitative determination of roquefortin. Kopp,B.; Rehm,H.-J. (Inst. Mikrobiol., Univ. Munster, Tibusstr. 7-15, D-4400 Munster, GFR) *Z. Lebensm.-Unters.-Forsch., 169(2), 90-91 (1979)* De;de,en.

603-U2 A microbiological method for the rapid determination of antibiotic concentrations during therapy. Wahlig,H. (Firma E. Merck, Med. Forsch., Abt. Chemother., Frankfurter St. 250, D-6100 Darmstadt, GFR) *Infection, 5(2), 117-118 (1977)* De;de,en.

604-U2 Microbiological determination of penicillin G, ampicillin, and cloxacillin residues in milk. Vilim,A.B.; Moore,S.D.; Larocque,L. (Health and Welfare Canada, Drug Res. Lab., Health Prot. Branch, Ottawa, Ont., K1A 0L2 Canada) *J. Assoc. Off. Anal. Chem., 62(6), 1247-1250 (1979)* En;en.

605-U2 A simple method for rapid isolation and identification of *Bacteroides fragilis*. Bittner,J. (Lab. Anaerobic Infect., Inst Dr. I. Cantacuzino, Bucuresti 35, Romania) *Arch. Roum. Pathol. Exp. Microbiol., 34(3), 231-238 (1975)* En;en,fr,ru.

606-U2 Rapid assay of BCG vaccine viability by measuring the ATP content. Gheorghiu,M.; Lagranderie,M. (Lab. BCG. Inst. Pasteur, 75724 Paris Cedex 15, France) *Ann. Microbiol., 130B(2), 147-156 (1979)* Fr;en,fr.

607-U2 Immobilization of bacterial luciferase and FMN reductase on glass rods. Jablonski,E.; DeLuca,M. (Dep. Chem., Univ. California, San Diego, La Jolla, CA 92093, USA) *Proc. Natl. Acad. Sci. USA, 73(11), 3848-3851 (1976)* En;en.

608-U2 Effective method for activity assay of lipase from *Chromobacterium viscosum*. Horiuti,Y.; Koga,H.; Gocho,S. (Res. Lab., Toyo Jozo Co., Ltd., Mifuku, Ohito-cho, Tagata-gun, Shizuoka 410-23, Japan) *J. Biochem., 80(2), 367-370 (1976)* En;en.

609-U2 New medium for rapid screening and enumeration of *Clostridium perfringens* in foods. Erickson,J.E.; Deibel,R.H. (W.M. Plank, Res. Coordinator, Dep. Health, Educ. and Welfare, Food and Drug Adm., Brooklyn, NY 11232, USA) *Appl. Environ. Microbiol., 36(4), 567-571 (1978)* En;en.

610-U2 Membrane filter enumeration method for *Clostridium perfringens*. Bisson,J.W.; *Cabelli,V.J. (Marine Field Stn., Health Effects Res. Lab. Clin., US Environ. Protection Agency, West Kingston, RI 02892, USA) *Appl. Environ. Microbiol., 37(1), 55-66 (1979)* En;en.

611-U2 The use of suckling mice in assaying *Clostridium perfringens* type A enterotoxin. Torres-Anjel,M.J.; Darland,G.; Dowell,V.R.; Riemann,H.P. (Fac. Med. Vet. Zootec., Univ. Nac. Columbia, Apartado Aereo 11951, Bogota, Columbia) *Rev. Latinoam. Microbiol., 17(4), 195-197 (1975)* En;en,es.

612-U2 A simplified assay for phospholipase C. Krug,E.L.; Truesdale,N.J.; Kent,C. (Dep. Biochem., Purdue Univ., West Lafayette, IN 47907, USA) *Anal. Biochem., 97(1), 43-47 (1979)* En;en.

613-U2 Rapid microbiological assay for chloramphenicol and tetracyclines. Louie,T.J.; Tally,F.P.; Bartlett,J.G.; *Gorbach,S.L. (Infect. Dis. Serv., Tufts-New England Med. Cent., Boston, MA 02111, USA) *Antimicrob. Agents Chemother., 9(6), 874-878 (1976)* En;en.

614-U2 Rapid, single-step most-probable-number method for enumerating fecal coliforms in effluents from sewage

treatment plants. Munoz,E.F.; *Silverman,M.P. (Extraterrest. Res. Div., NASA, Ames Res. Cent., Moffett Field, CA 94035, USA) *Appl. Environ. Microbiol., 37(3), 527-530 (1979)* En;en.

615-U2 **Modification of M FC medium by eliminating rosolic acid.** Presswood,W.G.; Strong,D.K. (Gelman Instrument Co., Ann Arbor, MI 48106, USA) *Appl. Environ. Microbiol., 36(1), 90-94 (1978)* En;en.

616-U2 **Use of two rapid A-1 methods for the recovery of fecal coliforms and *Echerichia coli* from selected food types.** Andrews,W.H.; Duran,A.P.; McClure,F.D.; Gentile,D.E. (Div. Microbiol., FDA, 200 'C' St. S.W., Washington, DC 20204, USA) *J. Food Sci., 44(1), 289-291, 293 (1979)* En;en.

617-U2 **Evaluation of a rapid method for the quantitaive estimation of coliforms in meat by impedimetric procedures.** Martins,S.B.; Selby,M.J. (777-58 San Antonio, Palo Alta, CA 94303, USA) *Appl. Environ. Microbiol., 39(3), 518-524 (1980)* En;en.

618-U2 **Rapid identification of gram-negative rods using a three-tube method combined with a dichotomic key.** Lassen,J. (Natl. Inst. Public Health, Dep. Bacteriol., Oslo, Norway) *Acta Pathol. Microbiol. Scand., Ser. B, 83(6), 525-533 (1975)* En;en.

619-U2 **Standardization of a rapid microbiologic assay for aminoglycosides using *Enterobacter cloacae*. Its use in the presence of the newer cephalosporins.** Stevens,D.L.; Page,B.M.A.; *Adeniyi-Jones,C. (Dep. Lab. Med., St. Joseph's Hosp., 50 Charlton Ave. East, Hamilton, Ont. L8N 1Y4, Canada) *Am. J. Clin. Pathol., 70(5), 808-815 (1978)* En;en.

620-U2 **An automated rapid test for *Escherichia coli* in milk.** Moran,J.W.; Witter,L.D. (Dep. Food Sci., Univ. Illinois at Urbana-Champaign, Urbana, IL 61801, USA) *J. Food Sci., 41(1), 165-167 (1976)* En;en.

621-U2 **A continuous assay for an intracellular enzyme: the analysis of acetate kinase in *Escherichia coli*.** Koplove,H.M.; Cooney,C.L. (Dep. Nutr. and Food Sci., Massachusetts Inst. Technol., Cambridge, MA 02139, USA) *Anal. Biochem., 72(1/2), 297-304 (1976)* En;en.

622-U2 **A simple method for estimating the biological activity of irradiated *Escherichia coli* DNA by the use of the prophage.** Petranovic,D.; Petranovic,M.; Salaj-Smic,E.; Trgovcevic,Z. (Inst. 'Rudjer Boskovic', 41001 Zagreb, Croatia, Yugoslavia) *Int. J. Radiat. Biol., 29(2), 187-190 (1976)* En.

623-U2 **A rapid assay for *Escherichia coli* 4-thiouridine-tRNA sulfurtransferase.** Kayne,M.S.; LaBone,T. (Dep. Biol., Trenton State Coll., Trenton, NJ 08625, USA) *Anal. Biochem., 98(1), 146-153 (1979)* En;en.

624-U2 **A comparison of methods for the confirmation of *Escherichia coli* isolated from dairy products.** Cooke,B.C.; Hill,B.M.; Pitcher,P.A.; Thomson,A.E. (Dairy Div., Minist. Agric. and Fish., Mount Maunganui, New Zealand) *N. Z. J. Dairy Sci. Technol., 12(4), 270-271 (1977)* En;en.

625-U2 **Studies on a simple *Limulus* test, a slide method.** Goto,H.; Watanabe,M.; Nakamura,S. (Div. Intern.Med., Hosp. Inst. Med. Sci., Univ. Tokyo, Shirokanedai, Minato-ku, Tokyo 108, Japan) *Jap. J. Exp. Med., 47(6), 523-524 (1977)* En;en.

626-U2 **Microbiological assay of sulphonamide in blood.** Chattopadhyay,B.; Hassam,Z. (Public Health Lab., Whipps Cross Hosp., Leytonstone, London E11 1NR, UK) *Microbios Lett., 3(11-12), 179-181 (1976)* En;en.

627-U2 **Mutagenicity and DNA-modifying activity: a comparison of two microbial assays.** Rosenkranz,H.S.; Gutter,B.; Speck,W.T. (Dep. Microbiol., Coll. Physicians and Surgeons, Columbia Univ., New York, NY 10032, USA) *Mutat. Res., 41(1), 61-70 (1976)* En;en.

628-U2 **A rapid and sensitive determination of bacterial rRNA by means of hybridizatition-competion.** Oostra,B.A.; Zentinge,B.; van Goor,A.L.; van Ooyen,A.J.J.; Gruber,M. (Biochem. Lab., Rijksuniversiteit, Zernikelaan, Groningen, Netherlands) *Anal. Biochem., 74(2), 496-502 (1976)* En;en.

629-U2 **A rapid and direct plate method for enumerating *Escherichia coli* biotype 1 in food.** Anderson,J.M.; Baird-Parker,A.C. (Unilever Res. Lab., Colworth House, Sharnbrook, Bedford, UK) *J. Appl. Bacteriol., 39(2), 111-117 (1975)* En;en.

630-U2 **Rapid screening method for enterotoxigenic *Escherichia coli*.** Gurwith,M. (Dep. Med. Microbiol., Univ. Manitoba, Fac. Med., Winnipeg, Manit. R3E 0W3, Canada) *J. Clin. Microbiol., 6(3), 314-316 (1977)* En;en.

631-U2 **A simple whole-cell assay for *E. coli* anthranilate synthetase.** Bliss,R.D. (Dep. Biol., Univ. California, Riverside, CA 92521, USA) *Anal. Biochem., 96(1), 152-154 (1979)* En.

632-U2 **Fluorescent assay for estimating the binding of erythromycin derivatives to ribosomes.** Brandt-Rauf,P.; Vince,R.; LeMahieu,R.; *Pestka,S. (Roche Inst. Mol. Biol., Nutley, NJ 07110, USA) *Antimicrob. Agents Chemother., 14(1), 88-94 (1978)* En;en.

633-U2 **The purification from *Escherichia coli* of a protein relaxing superhelical DNA.** Burrington,M.G.; Morgan,A.R. (Dep. Biochem., Univ. Alberta, Edmonton, Alta. T6G 2H7, Canada) *Can. J. Biochem., 54(4), 301-306 (1976)* En;en,fr.

634-U2 **[Cylinder plate method for vitamin B_{12} assays in pharmaceutical preparations].** De Oliveira Dias,G. (Univ. Federal Rio de Janeiro, Rio de Janeiro, Brazil) *Rev. Latinoam. Microbiol., 18(1), 43-45, (1976)* Pt;en,es,pt.

635-U2 **A shortcut method for estimation of *Escherichia coli* in sewage and receiving water.** Grunnet,K.; Gundstrup,A.S.P. (Inst. Hyg., Universitetsparken, DK-8000 Aarhus C, Denmark) *Nord. Vet. Med., 28(9), 430-433 (1976)* En;da,en.

636-U2 Assay method for *Vibrio cholerae* and *Escherichia coli* enterotoxins by automated counting of floating Chinese hamster ovary cells in culture medium. Nozawa,R.T.; Yokota,T.; Kuwahara,S. (Dep. Bacteriol., Juntendo Univ. Sch. Med., Hongo, Tokyo 113, Japan) *J. Clin. Microbiol., 7(5), 479-485 (1978)* En;en.

637-U2 Serum gentamicin assays of 100 clinical serum samples by a rapid 40 C *Klebsiella* method compared with overnight plate diffusion and acetyltransferase assays. Shanson,D.C.; Hince,C. (Dep. Bacteriol., St. Stephen's Hosp., Chelsea, London SW10 9TH, UK) *J. Clin. Pathol., 30(6), 521-525 (1977)* En;en.

638-U2 Rapid microbiologic assay of tobramycin. /[presented at Symposium at the 9th International Congress of Chemotherapy, held in London, UK, on 13-18 July 1975]. Shanson,D.C.; Hince,C.J.; Daniels,J.V. (Dep. Med. Microbiol., London Hosp. Med. Coll., Turner St., London E1 2AD, UK) *J. Infect. Dis., 134(Suppl.), S104-S109 (1976)* En;en.

639-U2 Disc microbiological method for the determination of lysine or methionine in protein hydrolysates. Sotelo,A.; Sousa,V. (Inst. Mexicano Seguro Soc., Subjefatura Invest. Cient., Lab. Bromatol., Apartado Postal 73-032, Mexico 73 DF, Mexico) *Nutr. Rep. Int., 14(3), 337-344 (1976)* En;en.

640-U2 Antibiotics. Improved microbiological procedure for determining bacitracin in premixes and mixed feeds /[presented at the 89th Annual Meeting of the Association of Official Analytial Chemists, held in Washington, DC, on 13-16 Oct 1975]. Fassbender,C.A.; Katz,S.E. (Dep. Biochem. and Microbiol., Cook Coll., Rutgers Univ., New Brunswick, NJ 08903, USA) *J. Assoc. Off. Anal. Chem., 59(5), 1113-1117 (1976)* En;en.

641-U2 Comparison of analysis of variance and a quick method in the potency estimation of tuberculin. Zdidai,J.; Csizer,Z.; Joo,I. (Inst. Serobacteriol. Prod. and Res. 'HUMAN', 1475 Budapest PO Box 4, Hungary) *J. Biol. Stand., 4(1), 29-34 (1976)* En;en.

642-U2 A quantitative assay to study cell movement in the myxobacteria. Lonski,J.; Heromin,R.; Ingraham,D. (Dep. Biol., Bucknell Univ., Lewisburg, PA 17837, USA) *J. Cell Sci., 25, 173-178 (1977)* En;en.

643-U2 Plaque assay for measuring serum bactericidal activity against gonococci. Corbeil,L.B.; Wunderlich,A.C.; Ito,J.L.; McCutchan,J.A. (Dep. Infect. Dis., Coll. Vet. Med., Kansas State Univ., Manhattan, KS 66506, USA) *J. Clin. Microbiol., 8(5), 618-620 (1978)* En;en.

644-U2 Feasibility of screening for penicillinase-producing *Neisseria gonorrhoeae* from primary culture plates by using a rapid microacidometric test. Weissfeld,A.S.; Sanner,G.D.; Childress,J.R.; Dyckman,J.D.; Huber,T.W.; Williams,R.P. (Dep. Microbiol. and Immunol., Baylor Coll. Med., Houston, TX 77030, USA) *Antimicrob. Agents Chemother., 12(6), 703-706 (1977)* En;en.

645-U2 The limulus lysate test. A rapid test for diagnosis of *Pseudomonas* keratitis or endophthalmitis. Ellison,A.C. (Veterans Adm. Hosp., 7400 Merton Minter Blvd., San Antonio, TX 78284, USA) *Arch. Ophthalmol., 96(7), 1268-1271 (1978)* En;en.

646-U2 Evaluation of a most-probable -number technique for the enumeration of *Pseudomonas aeruginosa*. Highsmith,A.K.; Abshire,R.L. (CDC, Atlanta, GA 30333, USA) *Appl. Microbiol., 30(4), 596-601 (1975)* En;en.

647-U2 Rapid method for counting antibiotic-resistant rhizobia in soils. Cooper,J.E. (Agric. and Food Bacteriol., Dep. Queen's Univ. Belfast, Belfast, UK) *Soil Biol. Biochem., 11(4), 433-435 (1979)* En.

648-U2 A mutagen assay detecting forward mutations in an arabinose-sensitive strain of *Salmonella typhimurium*. Ruiz-Vazquez,R.; Pueyo,C.; Cerda-Olmedo,E. (Dep. Genet., Fac. Cienc., Univ. Sevilla, Sevilla, Spain) *Mutat. Res., 54(2), 121-129 (1978)* En;en.

649-U2 Metabolism of cigarette smoke condensates by human and rat homogenates to form mutagens detectable by *Salmonella typhimurium* TA 1538. Hutton,J.J.; Hackney,C. (Dep. Med., Veterans Adm. Hosp., Lexington, KY 40507, USA) *Cancer Res., 35(9), 2461-2468 (1975)* En;en.

650-U2 A simple technique for assaying certain microbial phytotoxins and its application to the study of toxins produced by *Spiroplasma citri*. Daniels,M.J. (John Innes Inst., Colney Lane, Norwich NR4 7UH, UK) *J. Gen. Microbiol., 114(2), 323-328 (1979)* En;en.

651-U2 Screening for staphylococcal enterotoxins in food. Tatini,S.R.; Cords,B.R.; Gramoli,J. (Dep. Food Sci. Nutr., Univ. Minnesota, St. Paul, MN 55108, USA) *Food Technol., 30(4), 64, 66, 70, 72-74 (1976)* En.

652-U2 The thermostable deoxyribonuclease (DNase) test as a rapid screening method for the detection of staphylococcal enterotoxin in milk and milk products. Batish,V.K.; Ghodekar,D.R.; Ranganathan,B. (Dairy Bacteriol. Div., Natl. Dairy Res., Inst., Karnal, India) *Microbiol. Immunol., 22(7), 437-441 (1978)* En.

653-U2 Detection and enumeration of coagulase positive staphylococci in dairy products. I. New method. Ibrahim,G.F. (Dairy Res. Cent., Dep. Agric., Richmond, NSW, Australia) *Aust.J. Dairy Technol., 31(2), 44-47 (1976)* En;en.

654-U2 Quantitative, radial diffusion slide assay for stapylocoagulase. Kohl,J.D.; *Johnson,M.G. (Dep. Food Sci. and Microbiol., Clemson Univ., Clemson, SC 29631, USA) *Appl. Environ. Microbiol., 29(2), 339-341 (1980)* En;en.

655-U2 Laser light scattering bioassay for 1-β-D-arabinofuranosylcytosine (ara-C, NSC-63878). Mellett,L.B.; Wyatt,P.J.; Woolley,C. (USV Pharmaceut. Corporation, 1 Scarsdale Rd., Tuckahoe, NY 10707, USA) *Res. Commun. Chem. Pathol. Pharmacol., 20(2), 379-398 (1978)* En;en.

656-U2 **A rapid semiquantitative assay that facilitates purification of endo-β-N-acetylglucosaminidase H from *Streptomyces plicatus*.** Tkacz,J.S. (Waksman Inst. Microbiol., Rutgers Univ., State Univ. New Jersey, PO Box 759, Piscataway, NJ 08854, USA) *Anal. Biochem., 84(1), 49-55 (1978)* En;en.

657-U2 **[Immunological diagnosis of streptococcal pharyngitis in infancy. Assessment of a rapid new polyvalent test for the detection of antistreptococcal antibodies].** Silenzi,M. (Clin. Mal. Infett., Univ. Firenze, Firenze, Italy) *Minerva Pediatr., 28(35), 2129-2136 (1976)* It;en,it.

658-U2 **A rapid bioassay method for gramicidin by measuring rubidium ion leakage from *Streptococcus faecalis*.** Miller,S.J.P. (Microbiol. Control, Roussel Lab., Kingfisher Drive, Covingham, Swindon, UK) *J. Appl. Bacteriol., 47(1), 161-165 (1979)* En;en.

659-U2 **A method of evaluation of choleragen activity on the tadpoles of *Rana temporaria*.** Ermolyeva,Z.V. (deceased); *Avstsyn,A.P.; Shakhlamov,V.A.; Khlystova,Z.S.; Andreev,S.V.; Trager,R.S.; Pasternak,N.A.; Kalinina,N.A.; Vedmina,E.A.; Balyn,I.R.; Polyakova,G.P.; Shenderovich,V.A. (Inst. Human Morphol., Acad. Med. Sci. USSR, Moscow, USSR) *Byull. Eksp. Biol. Med., 80(11), 57-60 (1975)* Ru;en,ru.

660-U2 **Affinity filters, a new approach to the isolation of *tox* mutants of *Vibrio cholerae*.** Mekalanos,J.J.; Collier,R.J.; Romig,W.R. (Dep. Bacteriol., Univ. California, Los Angeles, CA 90024, USA) *Proc. Natl. Acad. Sci. USA, 75(2), 941-945 (1978)* En;en.

661-U2 **Assay of aminoglycoside antibiotics in clinical specimens.** Giamarellou,H.; Zimelis,V.M.; Matulionis,D.O.; *Jackson,G.G. (Sect. Infect. Dis., Dep. Med., Abraham Lincoln Sch. Med., Box 6998, Chicago, IL 60680, USA) *J. Infect. Dis., 132(4), 407-414 (1975)* En;en.

662-U2 **Quantitation of imidazoles by agar-disk diffusion.** Grendahl,J.G.; Sung,J.P. (Res. Lab., Veterans Adm. Hosp., Fresno, CA 93703, USA) *Antimicrob. Agents Chemother., 14(3), 509-513 (1978)* En;en.

663-U2 **A new, fast, and very sensitive bioluminescence assay for phospholipases A and C.** Ulitzur,S.; Heller,M. (Dep. Food Eng. and Biotechnol., The Technion, Haifa, Israel) *Anal. Biochem., 91(2), 421-431 (1978)* En;en.

664-U2 **A simple and safe volumetric alternative to the method of Miles, Misra and Irwin for counting viable bacteria.** Slack,M.P.E.; Wheldon,D.B. (Dep. Bacteriol. and Reg. Public Health Lab., Radcliffe Infirmary, Oxford, UK) *J. Med. Microbiol., 11(4), 541-545 (1978)* En;en.

665-U2 **Iodometric assay method for beta-lactamase with various beta-lactam antibiotics as substrates.** Sawai,T.; Takahashi,I.; Yamagishi,S. (Fac. Pharm. Sci., Chiba Univ., Chiba, Japan) *Antimicrob. Agents Chemother., 13(6), 910-913 (1978)* En;en.

666-U2 **Limulus test, parenteral drugs and biological products: an approach.** Fumarola,D.; Jirillo,E. (Med. Microbiol.-Univ. Sch. Med., Bari, Italy) *G. Batteriol. Virol. Immunol. Ann. Osp. Maria Vittoria, Torino, Parte I Parte II, 69(1-6), 34-37 (1976)* En;en.

667-U2 **Collaborative study comparing the spiral plate and aerobic plate count methods.** Gilchrist,J.E.; Donnelly,C.B.; Peeler,J.T.; Campbell,J.E. (FDA, Div. Microbiol., 1090 Tusculum Ave., Cincinnati, OH 45226, USA) *J. Assoc. Off. Anal. Chem., 60(4), 807-812 (1977)* En;en.

668-U2 **A rapid semiautomated bioassay of gentamicin based on luciferase assay of bacterial adenosine triphosphate.** Nilsson,L.; Hojer,H.; Ansehn,S.; Thore,A. (Dep. Clin. Bacteriol., Linkoping Univ., Regionsjukhuset, S-581 85 Linkoping, Sweden) *Scand. J. Infect. Dis., 9(3), 232-236 (1977)* En;en.

669-U2 **The superoxide dismutase activity of various photosynthetic organisms measured by a new and rapid assay technique.** Henry,L.E.A.; Halliwell,B.; Hall,D.O. (Dep. Plant Sci., King's Coll., 68 Half Moon Lane, London SE24 9JF, UK) *FEBS Lett., 66(2), 303-306 (1976)* En.

670-U2 **Impedimetric screening for bacteriuria.** Cady,P.; Dufour,S.W.; Lawless,P.; Nunke,B.; Kraeger,S.J. (Bactomatic, Inc., Palo Alto, CA 94303, USA) *J. Clin. Microbiol., 7(3), 273-278 (1978)* En;en.

671-U2 **A quick method of determining the total aerobic count (viable organisms) on fresh meat by means of photometrically measurable extinction changes during the resazurin test.** Baumgart,J.; Portner,A.; Lassak,G. (Fachhochsch. Lippe, 4920 Lemgo 1, Liebigstr. 87, Lab. Lebensm.-Mikrobiol., Fachber, Lebensmitteltechnol., GFR) *Fleischwirtschaft, 55(7), 969-973 (1975)* De;de,en,fr.

672-U2 **Assessment of bacterial counts in the meat-processing industry. III. The stomacher method and the spiral plate method.** Gerats,G.E.; Snijders,J.M.A. (Afd. Hyg., Fac. Diergeneeskd., Rijksuniversiteit. Bilstr. 172, Utrecht, Netherlands) *Tijdschr. Diergeneeskd., 102(18), 1084-1092 (1977)* Nl;en,nl.

673-U2 **Assessment of bacterial counts in the meat-processing industry. II. The plate-loop method and the droplet method.** Gerats,G.E.; Snijders,J.M.A. (Vakgroep Voedingsmiddelen, Dier. Oorsprong, Afd. Hyg., Fac. Diergeneeskd., Rijksuniversiteit, Bilstr. 172, Utrecht, Netherlands) *Tijdschr. Diergeneeskd., 102(16), 975-982 (1977)* Nl;en,nl.

674-U2 **[Results of a rapid method for determination of microbial sensitivity to antibacterial drugs in a surgical clinic].** Lunacharskaya,T.V. (Inst. Surg., Acad. Med. Sci. USSR, Moscow, USSR) *Antibiotiki, 22(8), 753-756 (1977)* Ru;en,ru.

675-U2 **[A simple method for counting bacteria with an active electron transport system in water and sediment samples].** Iturriaga,R.; Rheinheimer,G. (Inst. Meereskd., Univ. Kiel, 23 Kiel, Dusternbrooker Weg 20, GFR) *Kiel. Meeresforsch., 31(2), 83-86 (1975)* De;de,en.

676-U2 A simple screening test for determining the β-lactamase activity of bacteria. Ullmann,U. (Dep. Microbiol., Hyg.-Inst., Eberhard-Karls-Univ. Tubingen, Silcherstr. 7, D-7400 Tubingen 1, GFR) *Microbios Lett., 3(9), 35-39 (1976)* En;en.

677-U2 Counts of aerobic bacteria using a droplet technique. Schoenmakers,M.J.G.; Bes,J. (Lab. van het Abattoir van Amsterdam, Amsterdam, Netherlands) *Tijdschr. Diergeneeskd., 101(5), 251-254 (1976)* Nl;en,nl.

678-U2 Urease assay and urease-producing species of anaerobes in the bovine rumen and human feces. Wozny,M.A.; *Bryant,M.P.; Holdeman,L.V.; Moore,W.E.C. (Dep. Dairy Sci., Univ. Illinois, Urbana, IL 61801, USA) *Appl. Environ. Microbiol., 33(5), 1097-1104 (1977)* En;en.

679-U2 A quantitative assay for bacterial RNA polymerases. Chamberlin,M.J.; Nierman,W.C.; Wiggs,J.; Neff,N. (Dep. Biochem., Univ. California, Berkeley, CA 94720, USA) *J. Biol. Chem., 254(20), 10061-10069 (1979)* En;en.

680-U2 Microbial assays for mutagenicity: a modified liquid culture method compared with the agar plate system for precision and sensitivity. de G. Mitchell,I. (Beecham Pharmaceuticals, Med. Res. Cent., Coldharbour Rd., The Pinnacles, Harlow, Essex, UK) *Mutat. Res., 54(1), 1-16 (1978)* En;en.

681-U2 A microtiter technique for assessing bacterial numbers in aquatic systems. Leffler,J.W.; Lambert,J.E. (Inst. Ecol., Univ. Georgia, Athens, GA 30602, USA) *Water Res., 13(2), 211-212 (1979)* En.

682-U2 New test for endotoxin potency based upon histamine sensitization in mice. Bergman,R.K.; Milner,K.C.; Munoz,J.J. (Natl. Inst. Allergy and Infect. Dis., Rocky Mountain Lab., Hamilton, MT 59840, USA) *Infect. Immun., 18(2), 352-355 (1977)* En;en.

683-U2 Demonstration of antibiotic residues from the milk by thin layer agar gel diffusion test. Horvath,I.; Szepes,G. (1389 Budapest, Pf. 110, Lehel ut 43/47, Hungary) *Magy. Allatorv. Lapja, 33(4), 273-275 (1978)* Hu;de,en,hu,ru.

684-U2 Enumeration of high numbers of bacteria using hydrophobic grid-membrane filters. Sharpe,A.N.; Michaud,G.L. (Bur. Microb. Hazards, Food Directorate, Health Prot. Branch, Health and Welfare, Tunney's Pasture, Ottawa, Ont., Canada) *Appl. Microbiol., 30(4), 519-524 (1975)* En;en.

685-U2 Rapid method for determining very high bacterial counts in foods. Model-investigations. Popken,A.M.; Bomar,M.T. (Bundesforschungsanst. Ernahr., Inst. Biol., Engesserstr. 20, D-7500 Karlsruhe 1, GFR) *Alimenta, 18(6), 163-172 (1979)* De;de,en.

686-U2 Rapid diagnosis of gram-negative bacterial meningitis by the *Limulus* endotoxin assay. Jorgensen,J.H.; Lee,J.C. (Dep. Pathol., Univ. Texas Health Sci. Cent. at San Antonio, San Antonio, TX 78284, USA) *J. Clin. Microbiol., 7(1), 12-17 (1978)* En;en.

687-U2 On the application of Limulus coagulation test to the hygienic bacteriological field. Kawasaki,H.; Kanoh,S. (Natl. Inst. Hyg. Sci., Osaka Branch, 6 Hoenzaka-machi, Higashi-ku, Osaka, Japan) *J. Food Hyg. Soc. Jap., 18(3), 266-272 (1977)* Ja;en,ja.

688-U2 [Microbial colony counting by opto-electronic counters (Part 1)]. Van Reusel,A.; Laloux,J. (Min. Agric., Cent. Rech. Agron, Etat, Stn. Laitiere, Chaussee de Namur, 24, B-5800 Gembloux, Belgium) *Rev. Agric., 32(4), 1019-1030 (1979)* Fr.

689-U2 Rapid determination of nicotinic acid by immobilized *Lactobacillus arabinosis*. Matsunaga,T.; *Karube,I.; Suzuki,S. (Res. Lab. Resourc. Utilization, Tokyo Inst. Technol., Nagatsuta-cho, Midori-ku, Yokohama 227, Japan) *Anal. Chim. Acta, 99(2), 233-239 (1978)* En;en.

690-U2 Limulus test in diagnosis of acute meningitis. Aristegui,J.; Juan,S.; Saitua,G.; Hernandez,M. (Dep. Pediatr., Fac. Med., Hosp. Civ. Basurto, Bilbao-13, Spain) *An. Esp. Pediatr., 10(11), 835-842 (1977)* Es;en,es.

691-U2 Rapid determination of deoxycycline based on luciferase assay of bacterial adenosine triphosphate. Hojer,H.; Nilsson,L. (Dep. Clin. Bacteriol., Linkoping Univ., Regionsjukhuset, S-581 85 Linkoping, Sweden) *J. Antimicrob. Chemother., 4(6), 503-508 (1978)* En;en.

692-U2 A new method for the quantitative determination of microorganisms on human skin. Staal,E.M.; Noordzij,A.C. (Intradal Res. Lab., Brabantsestr. 17, Amersfoort, Netherlands) *J. Cosmet. Chem., 29(10), 607-615 (1978)* En;en.

693-U2 A quantitative assay for bacterial RNA polymerases. Chamberlin,M.J.; Nierman,W.C.; Wiggs,J.; Neff,N. (Dep. Biochem., Univ. California, Berkeley, CA 94720, USA) *J. Biol. Chem., 254(20), 10061-10069 (1979)* En;en.

694-U2 A rapid microdilution method for total count of bacteria in foods. Zavanella,M.; Tagliabue,S.; Lodetti,E. (Ist. Zooprofilattico Sperim, Lombardia e Emilia, Brescia, Italy) *Ind. Aliment., 19(3), 226-229 (1980)* It;en,it.

695-U2 A possible method for enumerating bacteria in wood. Carey,J.K. (Build. Res. Establ., Princes Risborough Lab., Princes Risborough, Aylesbury, Bucks. HP17 9PX, UK) *Int. Biodeterior. Bull., 15(4), 119-123 (1979)* En;de,en,es,fr.

696-U2 Experience with the use of the drop method of inoculation in sanitary microbiologic studies of soils. Pertsovskaya,A.F.; Filimonova,E.V. (A.N. Sysin, Inst. Gen. and Munic. Hyg., Pogodinskaya ul. 10, Moscow 119 883, USSR) *Gig. Sanit., No. 8, 51-53 (1979)* Ru.

697-U2 Aerobic and anaerobic mixtures of human pathogens: a rapid 4-plate counting technique. Kelly,M.J. (Sect. Microbiol., Dep. Pathol., Cambridge Univ., Cambridge,

UK) *Br. J. Exp. Pathol., 58(5), 478-483 (1977)* En;en.

698-U2 **A simple micro agar diffusion method for the determination of antibiotic concentrations in blood and other body fluids.** Georgopoulos,A. (Sandoz Foschungsinst. GmbH, Brunner Str. 59, A-1235 Wien, Austria) *Zentralbl. Bakteriol. Parasitenkd. Infektionskr. Hyg., I Abt. A, 242(3), 387-393 (1978)* En;de,en.

699-U2 **Enhancement of the sensitivity of the *Limulus* assay for the detection of gram negative bacteria.** Coates,D.A. (Johnson and Johnson Ltd., Airebank, Gargrave, Nr. Skipton, Yorks, BD23 3RX, UK) *J. Appl. Bacteriol., 42(3), 445-449 (1977)* En;en.

700-U2 **A rapid technic for quantitating wound bacterial count.** Magee,C.; Haury,B.; Rodeheaver,G.; Fox,J.; Edgerton,M.T.; *Edlich,R.F. (Dep. Plastic Surg., Univ. Virginia Sch. Med., Charlotsville, VA 22901, USA) *Am. J. Surg., 133(6), 760-762 (1977)* En;en.

701-U2 **Improved filter paper method for semi-quantitative colony count in urine. Comparison with the counting plate.** Pilars de Pilar,C.E. (Kinderpoliklin., Univ. Pettenkoferstr. 8a, D-8000, Munchen 2, GFR) *Munch. Med. Wochenschr., 118(45), 1449-1452 (1976)* De;de,en.

702-U2 **[Results with a new test tape for the diagnosis of urinary tract infection].** Holzer,H.; Pogglitsch,H.; Katschnig,H. (Med. Univ.-Klin., Auenbruggerplatz 15, A-8036 Graz, Austria) *Wien. Med. Wochenschr., 126(48), 685-687 (1976)* De;de.

703-U2 **An automated loop method for determining the total count of bacteria in milk.** Fleming,M.G.; O'Connor,F.O. (An Foras Taluntais, Dairy Microbial. Dep., Moorepark Res. Cent., Fermoy, Co. Cork, Eire) *Ir. J. Agric. Res., 14(1), 27-32 (1975)* En;en.

704-U2 **[A new method for the evaluation of the bacteriology quality of refrigerated raw milk].** Cordellana,C.; Michel,J.P. (Address not stated) *Rev. Lait. Fr., No. 343, 369-375 (1976)* Fr.

705-U2 ***Limulus* test, parenteral drugs and biological products: an approach.** Fumarola,D.; J·illo,E. (Dep. Med. Microbiol., Sch. Med., Univ. Bari, 70124 Bari, Italy) *Dev. Biol. Stand., 34, 97-100 (1977)* En;en.

706-U2 **Limulus test as a detector of the occurrence of endotoxins.** Braito,A.; Ferrea,G.; Schiavoni,S. (I. Clin. Mal. Infett., Univ. Genova, Genova, Italy) *G. Mal. Infett. Parassit., 31(11), 890-894 (1979)* It;en,it.

707-U2 **A rapid method for counting bacteria in raw milk.** Cousins,C.M.; Pettipher,G.L.; McKinnon,C.H.; Mansell,R. (Address not stated) *Dairy Ind. Int., 44(4), 27-31 and 39(1979)* En.

708-U2 **A bacterial bioassay for measuring the copper-chelation capacity of seawater.** Gillespie,P.A.; Vaccaro,R.F. (Cawthron Inst., Box 175, Nelson, New Zealand) *Limnol. Oceanogr., 23(3), 543-548 (1978)* En;en.

709-U2 **Improved enzymatic assay of chloramphenicol.** Smith,A.L.; Smith,D.H. (Div. Infect. Dis., Child. Hosp. Med. Cent., 300 Longwood Ave., Boston, MA 02115, USA) *Clin. Chem., 24(9), 1452-1457 (1978)* En;en.

710-U2 **The limulus amebocyte lysate test: present and future application.** Pearson,F.C. (Dep. Biol., Rhode Island Coll., Providence, RI 02908, USA) *R.I. Med. J., 61(8), 315-319 (1978)* En.

711-U2 **[The inhibition of thrombocyte aggregation by LT enterotoxins - a rapid test for the study of clinical preparations].** Richter,B.; Tschape,H.; Kuhn,H. (Inst. Exp. Epidemiol., DDR-37 Wernigerode, Burgstr. 37, GDR) *Dtsch. Gesundheitswes., 33(23), 1094-1099 (1978)* De;de,en,ru.

712-U2 **Electrochemical microbioassay of vitamin B_1.** Matsunaga,T.; *Karube,I.; Suzuki,S. (Res. Lab. Resourc. Util., Tokyo Inst. Technol., Nagatsuta, Midori-ku, Yokohama 227, Japan) *Anal. Chim. Acta, 98(1), 25-30 (1978)* En;en.

713-U2 **Viable bacterial counts by agar-droplet technique.** Koller,W.; Jelinek,J.A. (Hyg. Inst., Univ. Wien, Kinderspitalgasse 15, A-1095 Wien, Austria) *Zentralbl. Bakteriol. Parasitenkd. Infektionskr. Hyg., I Abt. A, 235(4), 527-553 (1976)* De;de,en.

714-U2 **Rapid determination of bacteriological water quality by using *Limulus* lysate.** Evans,T.M.; *Schillinger,J.E.; Stuart,D.G. (Dep. Microbiol., Montana State Univ., Bozeman, MT 59717, USA) *Appl. Environ. Microbiol., 35(2), 376-382 (1978)* En;en.

715-U2 **A rapid microdilution technique for counting viable bacteria in food.** Kramer,J. (Food Hyg. Lab., Cent. Public Health Lab., Colindale Ave., London NW9 5HT, UK) *Lab. Pract., 26(9), 675-676 (1977)* En;en.

716-U2 **A miniaturized counting technique for anaerobic bacteria.** Sharpe,A.N.; Pettipher,G.L.; Lloyd,G.R. (Bur. Microb. Hazards, Food Directorate, Health Prot. Branch, Tunney's Pasture, Ottawa, Ont. K1A 0L2, Canada) *Can. J. Microbiol., 22(12), 1728-1733 (1976)* En;en,fr.

717-U2 **Bacterial counts using the spiral plate apparatus.** Ruosch,W. (Veterinar-Bakteriol. Inst., Univ. Zurich, Winterthurer Str. 270, CH-8057 Zurich, Switzerland) *Fleischwirtschaft, 56(8), 1147-1148 (1976)* De;de,en,fr.

718-U2 **A simple quantitative assay for bacterial motility.** Segel,L.A.; Chet,I.; Henis,Y. (Dep. Appl. Math., Weizmann Inst. Sci., Rehovot, Israel) *J. Gen. Microbiol., 98(2), 329-337 (1977)* En;en.

719-U2 **Chloramphenicol bioassay.** Bannatyne,R.M.; Cheung,R. (Dep. Bacteriol., Hosp. Sick Children, Toronto, Ont. M5G 1X8, Canada) *Antimicrob. Agents Chemother., 16(1), 43-45 (1979)* En;en.

720-U2 **Microbiological assay of lincomycin in swine and broiler feeds.** Stahl,G.L. (Upjohn Co., Kalamazoo, MI 49001, USA) *J. Assoc. Off. Anal. Chem., 61(1), 39-42 (1978)* En;en.

721-U2 Some methods for microbiological assay. Board,R.G.; Lovelock,D.W. (eds.) *Publ. by:* Academic Press Inc., (London) Ltd., 24-28 Oval Road, London NW1 7DX, UK. May 1975 ISBN: 0-12-108240-7 at £7.80 En.

722-U2 Microassay for amphotericin B. Bannatyne,R.M.; Cheung,R.; Devlin,H.R. (Dep. Bacteriol., Hosp. Sick Child., Toronto, Ont. M5G 1X8, Canada) *Antimicrob. Agents Chemother., 11(1), 44-46 (1977)* En;en.

723-U2 Which gentamicin assay method is the most practicable? Waterworth,P.M. (Dep. Microbiol., University Coll. Hosp., London WC1E 6AU, UK) *J. Antimicrob. Chemother., 3(1), 1-8 (1977)* En.

724-U2 Enumeration of microorganisms in food: a comparative study of five methods. Kramer,J.M.; Gilbert,R.J.(Food Hyg. Lab., Cent. Public Health Lab., Colindale Ave., London NW9 5HT, UK) *J. Hyg., 81(1), 151-159 (1978)* En;en.

725-U2 Assays and statistical analyses for antibiotic standards. Tarcza,E.; Garth,M.A. (Natl. Cent. Antibiot. Anal., Food and Drug Adm., Washington, DC 20204, USA) *J. Pharm. Sci., 67(8), 1050-1053 (1978)* En;en.

726-U2 Rapid screening, quantitation of microorganisms by bioluminescence. Baker,K.F.; McCormick,R.D. (Sci. and Process Div., DuPont Instruments, Wilmington, DE, USA) *Food Prod. Dev., 10(5), 93-94 (1976)* En.

727-U2 A comparative study of the microbiological assays currently available for nystatin raw material. Cosgrove,R.F.; Beezer,A.E.; Miles,R.J. (Squibb Inst. Med. Res., Reeds Lane, Moreton, Wirral, Merseyside L46 1QW, UK) *J. Pharm. Pharmacol., 31(3), 171-173 (1979)* En;en.

728-U2 [Mutagenicity assays as rapid tests for potential carcinogens.] Fahrig,R. (Zentrallab. Mutagenitatsprufung, Deutsch. Forschungsgemeinschaft, Freiburg i Br., GFR) *Umschau, 76(7), 224-225 (1976)* De;en.

729-U2 Assay for amatoxins in *Amanita phalloides* Fries (Basidiomycetes) with direct spectrometric measurements of chromatograms. Andary,C.; Enjalbert,F.; Privat,G.; Mandrou,B. (Lab. Bot. et Cryptogam., Fac. Pharm., 30460-Montpellier Cedex, France) *J. Chromatogr., 132(3), 525-532 (1977)* Fr;en.

730-U2 A rapid direct assay for the determination of the separate activities of the three arylsulphatastes of *Aspergillus oryzae*. Burns,G.R.J.; Wynn,C.H. (Dep. Biochem., Univ. Manchester, Manchester M13 9PL, UK) *Biochem. J., 166(3), 411-413 (1977)* En;en.

731-U2 A bioassay for appraising preservative protection of wood aboveground. Scheffer,T.C.; Gollob,L. (For. Res. Lab., Sch. For., Oregon State Univ., Corvallis, OR 97330, USA) *Holzforschung, 32(5), 157-161 (1978)* En;de,en.

732-U2 Open-vacuole method for measuring membrane potential and membrane resistance of Characeae cells. Tazawa,M.; Kikuyama,M.; Nakagawa,S. (Dep. Biol., Fac. Sci., Osaka Univ., Toyonaka, Osaka 560, Japan) *Plant Cell Physiol., 16(4), 611-622 (1975)* En;en.

733-U2 Assay for *Helminthosporium maydis* toxin-binding activity in plants. Ireland,C.R.; *Strobel,G.A. (Dep. Plant Pathol., Montana State Univ., Bozeman, MT 59715, USA) *Plant Physiol. 60(1), 26-29 (1977)* En;en.

734-U2 Enzyme assays using permeabilized cells of *Neurospora*. Basabe,J.R.; Lee,C.A.; *Weiss,R.L. (Dep. Chem., Univ. California, Los Angeles, CA 90024, USA) *Anal. Biochem., 92(2), 356-360 (1979)* En;en.

735-U2 Elutriation procedures for quantitative assay of soils for *Rhizoctonia solani*. Clark,C.A.; Sasser,J.N.; Barker,K.R. (Dep. Plant Pathol., Louisiana State Univ., Baton Rouge, LA 70803, USA) *Phytopathology, 68(8), 1234-1236 (1978)* En;en.

736-U2 A rapid objective method for measuring the yeast opsonisation activity of serum. Levinsky,R.J.; Harvey,B.A.M.; Paleja,S. (Immunol. Dep., Inst. Child Health, 30 Guildford St., London WC1, UK) *J. Immunol. Methods, 24(3-4), 251-256 (1978)* En;en.

737-U2 Rapid quantitative method for determining the sensitivity of yeast cells in the inactivation of nitrous acid. Devin,A.B. (I. V. Kurchatov Inst. Atomic Energy, Moscow, USSR) *Genetika, 14(4), 732-733 (1978)* Ru;en,ru.

738-U2 Bioassay of antifungal antibiotics by flow microcalorimetry. Beezer,A.E.; Chowdhry,B.Z.; Newell,R.D.; Tyrrell,H.J.V. (Chem. Dep., Chelsea Coll., Univ. London, Manresa Road, London SW3 6LX, UK) *Anal. Chem., 49(12), 1781-1784 (1977)* En;en.

739-U2 A novel assay for endonucleases acting at apurinic sites and its use in measuring AP endonuclease activity in repair-deficient mutants of *Saccharomyces cerevisiae*. Futcher,A.B.; Morgan,A.R. (Dep. Biochem., Univ. Alberta, Edmonton, Alta. T6G 2H7, Canada) *Can. J. Biochem., 57(6), 932-937 (1979)* En;en,fr.

740-U2 Test tube method of bioassay for *Thielaviopsis basicola* root rot of soybean. Maduewesi,J.N.C.; Lockwood,J.L. (Dep. Bot., Univ. Nigeria, Nsukka, Nigeria) *Phytopathology, 66(6), 811-814 (1976)* En;en.

741-U2 Automatic cellulase assay in computer coupled pilot fermentation. Leisola,M.; Virkkunen,J.; Karvonen,E.; Meskanen,A. (Dep. Chem. and Electrical Eng., Helsinki Univ. Technol., SF-02150 Espoo 15, Finland) *Enzyme Microb. Technol., 1(2), 117-121 (1979)* En;en.

742-U2 Rapid urea broth test for yeasts. Roberts,G.D.; Horstmeier,C.D.; *Land,G.A.; Foxworth,J.H. (Dep. Microbiol., Wadley Inst. Mol. Med., Dallas, TX 75235, USA) *J. Clin. Microbiol., 7(6), 584-588 (1978)* En;en.

743-U2 Quantitative microbiological determination of B-vitamins in beer and yeast. Part I. Assay procedures. Voss,H.; Piendl,A. (Inst. Brauereitechnol. und Mikrobiol., Technisch. Univ. Munchen, Weihenstephan, GFR) *Brew. Dig., 51(10),*

56-66, 80 (1976) En;en.

744-U2 Methods for assay of aflatoxins in coconut products. Samarajeewa,U.; Arseculeratne,S.N. (Dep. Bacteriol., Univ. Sri Lanka, Peradeniya, Sri Lanka) *J. Food Sci. Technol. (Mysore), 12(1), 27-31 (1975)* En;en.

745-U2 A rapid method for detection of aflatoxins from xylitol. Niskanen,A.; Lindroth,S.; Pensala,O. (Tech. Res. Cent., Finland, Food Res. Lab., Biologinkuja 1, SF-02150 Espoo 15, Finland) *Eur. J. Appl. Microbiol., 2(4), 307 (1976)* En.

746-U2 Automatic assay of cellulase activity during fermentation. Leisola,M.; Kauppinen,V. (Helsinki Univ. Technol., Dep. Chem., SF-02150 Espoo 15, Finland) *Biotechnol. Bioeng., 20(6), 837-846 (1978)* En;en.

747-U2 Methods for determination of aflatoxins. Kneist,S.; Koch,H.A. (Dep. Dermatol., Lab. Mycol., Med. Acad., Erfurt, GDR) *Mater. Med. Pol., 10(2), 116-119 (1978)* En;en.

748-U2 A new method for quantitation of aflatoxin in corn. Seitz,L.M.; Mohr,H.E. (US Grain Marketing Res. Cent., ARS, Manhattan, KS 66502, USA) *Cereal Chem., 54(1), 179-183 (1977)* En;en.

749-U2 The rapid determination of aflatoxin M_1 in dairy products. Stubblefield,R.D. (North. Reg. Res. Cent., Fed. Res., Sci. and Educ. Adm., USDA, Peoria, IL 61604, USA) *J. Am. Oil Chem. Soc., 56(9), 800-802 (1979)* En;en.

750-U2 Biological assay of fungicides against yeasts in vitro using a Coulter Counter. Brotherton,J. (Abt. Gynakol. Endokrinol., Klinik. Stelitz, Freien Univ. Berlin, Hindenburgdamm 30, D-1000 Berlin 45, GFR) *Mykosen, 19(10), 361-372 (1976)* En;de,en.

751-U2 Simple assay for the condensation component enzyme (β-ketoacyl synthetase) of fatty acid synthetase. Brown,O.R.; Stees,J.L. (John M. Dalton Res. Cent., Univ. Missouri, Columbia, MO 65201, USA) *Microbios, 17(67), 17-21 (1976)* En;en.

752-U2 A rapid microbiological assay for nystatin using Rb^+ enriched yeast cells. Cosgrove,R.F. (Int. Dev. Lab., Squibb Inst. Med. Res., E.R. Squibb and Sons Ltd., Reeds Lane, Moreton, Merseyside L46 1QW, UK) *J. Appl. Bacteriol., 44(2), 199-206 (1978)* En;en.

753-U2 An automated assay for the determination of nitrate reductase in marine phytoplankton. Stawyk,G.; Collos,Y. (Stn. Mar. Endoume, Cent. Univ. Luminy, Marseille, France) *Mar. Biol., 34(1), 23-26 (1976)* En;en.

754-U2 An automated assay for the determination of nitrate reductase in marine phytoplankton. Slawyk,G.; Collos,Y. (Lab. Oceanogr., Cent. Univ. de Luminy, Marseilles, France) *Mar. Biol., 34(1), 23-26 (1976)* En;en.

755-U2 A simple agar plate method, using microalgae, for herbicide bio-assay or detection. Wright,S.J.L. (Sch. Biol. Sci., Univ. Bath, Bath BA2 7AY, UK) *Bull. Environ. Contam. Toxicol., 14(1), 65-70 (1975)* En;en.

756-U2 Enumeration of natural *Microcystis* populations. Reynolds,C.S.; Jaworski,G.H.M. (Freshwater Biol. Assoc., Ferry House, Ambleside, Cumbria LA22 0LP, UK) *Br. Phycol. J., 13(3), 269-277 (1978)* En;en.

757-U2 [Automatic evaluation of an algae bioassay. Inhibition of the movement of a blue-green algae (*Phormidium* sp.) by the herbicide diquat]. Benecke,G. (Inst. Wasserforsch. GmbH, Dortmund D-5840, Schwerte 1, GFR) *Z. Wasser Abwasser Forsch., 10(6), 195-197 (1977)* De;de,en.

758-U2 A simplified method of phytoplankton counting. Willen,E. (Natl. Swedish Environ. Prot. Board, Limnol. Survey, S-75122 Uppsala, Sweden) *Br. Phycol. J., 11(3), 265-278 (1976)* En;en.

759-U2 Electronic particle counting in quantitative phytoplankton studies. Ilmavirta,V. (Dep. Bot., Univ. Helsinki, Unionink, 44, SF-00170 Helsinki 17, Finland) *Acta Bot. Fenn., 110, 87-90 (1979)* En;en.

760-U2 An improved method for rapid assaying of viability of cryopreserved unicellular algae. Gilboa,A.; Ben-Amotz,A. (Israel Oceanogr. and Limnol. Res., Ltd., PO Box 8030, Haifa, Israel) *Plant Sci. Lett., 14(4), 317-320 (1979)* En;en.

761-U2 Quantitation and cell size of *Naegleria fowleri* by electronic particle counting. Weik,R.R.; John,D.T. (Dep. Microbiol., Virginia Commonw. Univ., Richmond, VA 23298, USA) *J. Parasitol., 63(1), 150-151 (1977)* En.

762-U2 An improved method for the microbiological assay of available amino acids in proteins using *Tetrahymena pyriformis*. Shepherd,N.D. (deceased); Taylor,T.G.; Wilton,D.C. (Dep. Physiology and Biochem., Univ. Southampton, Southampton SO9 3TU, UK) *Br. J. Nutr., 38(2), 245-253 (1977)* En;en.

763-U2 A method for the assay of ablastin in the serum of rats infected with *Trypanosoma lewisi*. Ferrante,A.; Drew,P.A.; Jenkin,C.R. (Dep. Microbiol. and Immunol., Univ. Adelaide, Adelaide, 5000 Australia) *Aust. J. Exp. Biol. Med. Sci., 56(6), 741-745 (1978)* En;en.

764-U2 Comparison of *Tetrahymena pyriformis* W and rat bioassays for the determination of protein quality. Evancho,G.M.; Hurt,H.D.; Devlin,P.A.; Landers,R.E.; Ashton,D.H. (Campbell Inst. Food Res., Camben, NJ 08101, USA) *J. Food Sci., 42(2), 444-448 (1977)* En;en.

765-U2 A rapid assay for stimulation of human lymphocytes by tumor-associated antigens. Roth,J.A.; Grimm,E.A.; Morton,D.L. (Univ. California, Los Angeles, 54-140 Cent. Health Sci., Los Angeles, CA 90024, USA) *Cancer Res., 36(9), 3001-3010 (1976)* En;en.

766-U2 A simple and rapid method to distinguish between rosettes and non-specific aggregation of erythrocytes in the rosette assay for human T-lymphocytes. Kalland,T. (Inst. Anat., Univ. Bergen, Bergen, Norway) *J. Immunol. Methods,*

17(3-4), 279-283 (1977) En;en.

767-U2 **A simple method of cell-mediated cytotoxic assay by postlabelling of tritiated thymidine.** Toh,K.; Kikuchi,K. (Dep. Pathol., Sapporo Med. Coll., Sapporo 060, Japan) *Tohoku J. Exp. Med., 122(3), 259-265 (1977)* En;en.

768-U2 **Continuous-flow automation of the *Lactobacillus casei* serum folate assay.** Tennant,G.B. (Dep. Haematol., Welsh Natl. Sch. Med., Heath Park, Cardiff CF4 4XN, UK) *J. Clin. Pathol., 30(12), 1168-1174 (1977)* En;en.

769-U2 **Automated analysis of histamine released from leukocytes of allergic donors.** Siraganian,R.P. (Lab. Microbiol. and Immunol., Natl. Inst. Dent. Res., Bethesda, MD 20014, USA) *In:* Allergy and clinical immunology. Mathov,E.; Sindo,T.; Naranjo,P. (eds.) *Publ.by:* Excerpta Medica, Amsterdam, Netherlands 1977 p.404-410 ISBN: 90-219-0338-5 En.

770-U2 **Quantitation of leukotaxis in agarose by three different methods.** Orr,W.; Ward,P.A. (Dep. Pathol., Univ. Connecticut Health Cent., Farmington, CT 06032, USA) *J. Immunol. Methods, 20, 95-107 (1978)* En;en.

771-U2 **A rapid and quantitative assay of phagocytosis-connected oxygen consumption by leukocytes in whole blood.** Nakamura,M.; Nakamura,M.A.; Okamura,J.; Kobayashi,Y. (Dep. Biochem., Kyushu Cancer Cent., Minami-ku, Fukuoka 815, Japan) *J. Lab. Clin. Med., 91(4), 568-575 (1978)* En;en.

772-U2 **Improved assay for monocyte chemotaxis using frozen stored responder cells.** Dean,D.A.; Strong,D.M. (Dep. Clin. and Exp. Immunol., Naval Med. Res. Inst., Bethesda, MD 20014, USA) *J. Immunol. Methods, 14(1), 65-72 (1977)* En;en.

773-U2 **Automated histamine analysis for in vitro allergy testing. II. Correlation of skin test results with in vitro whole blood histamine release in 82 patients.** Siraganian,R.P. (Lab. Microbiol., Natl. Inst. Dent. Res., Natl. Inst. Health, Bethesda, MD 20014, USA) *J. Allergy Clin. Immunol., 59(3), 214-222 (1977)* En;en.

774-U2 **A direct binding assay for rheumatoid factor serum antiglobulins using fluorescein-labelled Fc fragment of human immunoglobulin-G.** Singh,I.; *Francis,G.E. (Dep. Biochem. and Chem., Med. Coll. St. Bartholomew's Hosp., Charterhouse Squ., London EC1M 6BQ, UK) *J. Clin. Pathol., 31(10), 963-973 (1978)* En;en.

775-U2 **A new method for measuring simultaneously the phagocytic and bactericidal capacity of human leukocytes.** Grange,M.J.; Eche,F.; Dresch,C.; Najean,Y. (Lab. Cent. Med. Nucl., Cent. Rech. Malad. Sang, Hop. Saint-Louis, 2 place du Dr. Fournier, 75475 Paris Cedex 10, France) *Biomed. Express, 23(10), 414-418 (1975)* En;en,fr.

776-U2 **DNA release as a direct measure of microbial killing by phagocytes.** Friedlander,A.M. (Dep. Med., Univ. California Sch. Med., San Diego, CA 92103, USA) *Infect. Immun., 22(1), 148-154 (1978)* En;en.

777-U2 **A simple technique for measuring leucocyte chemotaxis in reversible Boyden chambers.** Arvilommi,H.; Laatikainen,A. (Public Health Lab., SF-40620 Jyvaskyla 62, Finland) *Z. Immunitatsforsch. Immunobiol., 153(2), 179-182 (1977)* En;en.

778-U2 **Simple sensitive microbioassay for adenine arabinoside and hypoxanthine arabinoside in human plasma.** Bryson,Y.J.; Sweetman,L.; Connor,J.D. (Dep. Pediatr., Univ. California at Los Angeles, Los Angeles, CA 90024, USA) *Antimicrob. Agents Chemother., 14(6), 909-915 (1978)* En;en.

779-U2 **Improved detection of immune complexes in human and mouse serum using a microassay adaptation of the C1q binding test.** June,C.H.; Contreras,C.E.; Perrin,L.H.; Lambert,P.H. (WHO Immunol. Res. and Training Cent., Cent. Transfus., Hop. Cantonal, 1211 Geneva 4, Switzerland) *J. Immunol. Methods, 31(1-2), 23-39 (1979)* En;en.

780-U2 **Computerization of a bioassay: quantitation of slow reacting substance of anaphylaxis (SRS-A).** Fleisch,J.H.; Zaborowsky,B.R.; Cerimele,B.J.; Spaethe,S.M. (Lilly Res. Lab., Eli Lilly and Co., MC905, Indianapolis, IN 46206, USA) *J. Pharmacol. Exp. Ther., 209(2), 238-243 (1979)* En;en.

781-U2 **A simple method to demonstrate an autoimmune-response following vaccination and infection. Evaluation of autohaemolysin-forming cells in the peripheral blood by means of a plaque assay.** Huber,H.C.; Hochstein-Mintzel,V. (Bayer, Landesimpfanstalt, Am Neudeck 1, 8 Munchen 95, GFR) *Fortschr. Med., 95(3), 123-126 (1977)* De;de,en.

782-U2 **Surface analysis by bacterial adherence to virus-infected cells.** Huang,A.S.; Okorie,T.G. (Div. Infect. Dis., Child Hosp. Med. Cent., 300 Longwood Ave., Boston, MA 02115, USA) *J. Infect. Dis., 140(2), 147-151 (1979)* En;en.

783-U2 **A direct hemolytic plaque assay on polylysine-coated coverslips for scanning electron microscopy of antibody forming cells.** Pan,S.-H.; Friedman,H. (Dep. Microbiol., Albert Einstein Med. Cent., Philadelphia, PA, USA) *J. Immunol. Methods, 17(3-4), 329-336 (1977)* En;en.

784-U2 **Application of a galactosidase immunosorbent test to carcinoembryonic antigen in plasma.** Weltman,J.K.; Frackelton,A.R.,Jr. (Lab. Exp. Immunol. and Biophys., Miriam Hospital-Brown Univ., Providence, RI 02906, USA) *Cancer Res., 36(8), 2850-2853 (1976)* En;en.

785-U2 **Detection of antibody-forming cells and humoral antibody to horseradish peroxidase.** Burke,M.; Harris,D.; Stollar,B.D.; Borel,H.; Borel,Y. (Div. Immunol., Dep. Pediatr., Child. Hosp. Med. Cent., Boston, MA 02115, USA) *J. Immunol. Methods, 25(4), 365-374 (1979)* En;en.

786-U2 **A fluorescent immunoassay for the quantification of C-reactive protein.** Siboo,R.; Kulisek,E. (Dep. Microbiol. and Immunol., McGill Univ., Montreal, Que., Canada) *J. Immunol. Methods, 23(1-2), 59-67 (1978)* En;en.

787-U2 **RNA synthesis in mouse spleen cells as a probe of antibody mediated cytotoxicity.** Stephens,T.J.; Warner,C.M. (Dep. Biochem. and Biophys., Iowa State Univ., Ames, IA 50011, USA) *J. Immunol. Methods, 26(1), 75-86 (1979)* En;en.

788-U2 **Particle counting immunoassay (PACIA). III. Automated determination of circulating immune complexes by inhibition of an agglutinating factor of mouse serum.** Cambiaso.C.L.; Sindic,C.; *Masson,P.L. (Unit Exp. Med., ICP-UCL 7430, Ave. Hippocrate 75, B 1200 Brussels, Belgium) *J. Immunol. Methods, 28(1-2), 13-23 (1979)* En;en.

789-U2 **A micro macrophage migration inhibition test for the detection of cellular immunity in vitro.** Allardyce,R.A.; Hunt,J.S.; Stewart,R.J. (Dep. Surg., Univ. Otago, Christchurch Clin. Sch. Med., Christchurch, New Zealand) *J. Immunol. Methods, 27(1), 9-18 (1979)* En;en.

790-U2 **A whole-blood microculture assay of human immunocompetence.** Griffin,J.F.T.; Banwell,K. (Dep. Microbiol., Univ. Otago Med. Sch., Otago, New Zealand) *Proc. Univ. Otago Med. Sch., 56(3), 81-82 (1978)* En.

791-U2 **A simple microassay for detection of antibodies to fetal calf serum and related antigens and its application to the serological definition of human tumor antigens.** Liao,S.K.; Rahman,A.F.R.; Kwong,P.C.; Dent,P.B. (Ontario Cancer Found., Hamilton Clin., Henderson Gen. Hosp., 711 Concession St., Hamilton, Ont. L8V 1C3, Canada) *J. Immunol. Methods, 27(2), 111-125 (1979)* En;en.

792-U2 **A direct assay of granulocytic repopulating ability.** Marsh,J.C.; *Blackett,N.M. (Biophys. Div., Inst. Cancer Res., Sutton, Surrey, UK) *Exp. Hematol., 6(2), 135-140 (1978)* En;en.

793-U2 **Micropore filter assays of human granulocyte locomotion: problems and solutions.** Maderazo,E.G.; Woronick,C.L. (Med. Res. Lab., Dep. Med., Hartford Hosp., Hartford, CT 06115, USA) *Clin. Immunol. Immunopathol., 11(2), 196-211 (1978)* En;en.

794-U2 **A rapid, semi-automated counting procedure for enumeration of antibody-forming cells in gel and nucleated cells in suspension.** Katz,D.H.; Faulkner,M.; Katz,L.R.; Lindh,E.; Leonhardt,C.C.; Herr,K.; Tung,A.S. (Dep. Cell. and Dev. Immunol., Scripps Clin. and Res. Found., La Jolla, CA 92037, USA) *J. Immunol. Methods, 17(3-4), 285-291 (1977)* En;en.

795-U2 **Application of CH50 assay to the detection of circulating immune complexes.** Santoro,F.; Wattre,P.; Capron,A. (Cent. Immunol. et Biol. Parasitaire, Inst. Pasteur de Lille, 20, blvd. Louis-XIV, 59012 Lille Cedex, France) *Pathol. Biol., 25(2), 135-138 (1977)* Fr;en,fr.

796-U2 **A simplified one-step procedure for the simultaneous determination of complement receptor lymphocytes and lymphocytes with membrane-bound immunoglobulins.** Winterleitner,H.; Rotter,M.; Knapp,W. (Inst. Immunol., Univ. Wien, Borschkegasse 8a, A-1090 Wien, Austria) *Acta Haematol., 57(2), 74-80 (1977)* En;en.

797-U2 **Inhibition of antibody-mediated rosette formation by alveolar macrophages: a sensitive assay for metal toxicity.** Hadley,J.G.; Gardner,D.E.; Coffin,D.L.; Menzel,D.B. (Biol. Dep., Pacific N.W. Labs., Richland, WA 99352, USA) *J. Reticuloendothel. Soc., 22(5), 417-425 (1977)* En;en.

798-U2 **An improved fluorescence probe cytotoxicity assay.** Brawn,R.J.; Barker,C.R.; Oesterle,A.D.; Kelly,R.J.; Dandliker,W.B. (Dep. Biochem., Scripps Clin. and Res. Found., La Jolla, CA 92037, USA) *J. Immunol. Methods, 9(1), 7-26 (1975)* En;en.

799-U2 **A microplate culture method for assay of guinea pig mitogenic factor.** Ashworth,L.A.E.; Ford,W.H. (Microbiol. Res. Establ., Porton Down, Salisbury, Wilts., UK) *J. Immunol. Methods, 9(1), 47-57 (1975)* En;en.

800-U2 **Automated fluorescent analysis for cytotoxicity assays.** Horan,P.K.; Kappler,J.W. (Dep. Pathol., Univ. Rochester Sch. Med. and Dent., Rochester, NY 14642, USA) *J. Immunol. Methods, 18(3-4), 309-316 (1977)* En;en.

801-U2 **Rapid lymphocyte immunoreactivity test utilizing [^{3}H]uridine in vitro.** Pienkowski,M.M.; Lyerly,M.M.; Miller,H.C. (Dep. Anat. Michigan State Univ., East Lansing, MI 48824, USA) *J. Immunol. Methods, 24(1-2), 163-173 (1978)* En;en.

802-U2 **Improved electroimmunoassay of factor VIII-related antigen.** Chute,A.; Haddow,J.E.; Ritchie,R.F. (Rheum. Dis. Lab., Maine Med. Cent., Portland, ME 04102, USA) *Clin. Chem., 23(3), 602-603 (1977)* En;en.

803-U2 **A pen smear technique for assays of rosette-forming lymphocytes.** Campbell,A.C.; Waller,C.A. (Dep. Immunol., University Hosp., Queen's Med. Cent., Nottingham NG7 2UH, UK) *J. Immunol. Methods, 26(4), 337-344 (1979)* En;en.

804-U2 **Equilibrium binding assay and kinetic characterization of insulin antibodies.** Goldman,J.; Baldwin,D.; Pugh,W.; Rubenstein,A.H. (Dep. Med., Henry Ford Hosp., 2799 West Grand Blvd., Detroit, MI 48202, USA) *Diabetes, 27(6), 653-660 (1978)* En;en.

805-U2 **Semi-automated method for the enumeration of cytotoxicity results.** Goldrosen,M.H.; Formeister,J.; Holyoke,E.D. (Roswell Park Memorial Inst., Dep. Gen. Surg., Tumor Immunol. Group, 666 Elm St., Buffalo, NY 14263, USA) *J. Immunol. Methods, 11(3-4), 367-369 (1976)* En;en.

806-U2 **Automated histamine analysis for in vitro allergy testing. I. A method utilizing allergen-induced histamine release from whole blood.** Siraganian,R.P.; Brodsky,M.J. (Lab. Microbiol. and Immunol., Natl. Inst. Dent. Res., Bethesda, MD 20014, USA) *J. Allergy Clin. Immunol., 57(6), 525-540 (1976)* En;en.

807-U2 **Agarose microwell assay for cell-mediated cytotoxicity.** Maroudas,N.G. (71 Park Ave. North, London, NW10 ILE, UK) *J. Immunol. Methods, 10(4), 389-392 (1976)* En;en.

808-U2 Electro-immunoassay for properdin. A comparison with the radioimmunoassay. Schrager,M.A.; Chapitis,J.; Rothfield,N.F.; Lepow,I.H. (Div. Rheum. Dis., Dep. Med., Univ. Connecticut Sch. Med., Farmington, CT 06032, USA) *Clin. Immunol. Immunopathol., 5(2), 258-263 (1976)* En;en.

809-U2 Automated rheumatoid factor assay in a population of patients with rheumatic disease. Gall,E.P.; Templin,D.W.; Dito,W.R. (Clin. Immunol. Sect., Univ. Arizona Health Sci. Cent., Tucson, AZ 85724, USA) *Arthritis Rheum., 21(2), 224-228 (1978)* En;en.

810-U2 Electroimmunoassay of Cl inactivator and C4 in hereditary angioneurotic edema (HANE). A simplified diagnostic procedure. Laurell,A.-B.; Martensson,U.; Sjoholm,A. (Inst. Med. Microbiol., Dep. Immunol., Univ. Lund, Lund, Sweden) *Clin. Immunol. Immunopathol., 5(3), 308-313 (1976)* En;en.

811-U2 Automated counting of T cell rosettes. Vandenbark,A.A.; Burger,D.R.; Vetto,R.M. (Surg. Res. Veterans Adm. Hosp., Portland, OR 97207, USA) *J. Immunol. Methods, 10(2-3), 261-270 (1976)* En;en.

812-U2 A computerized semi-automatic evaluation system for quanti:ation of electroimmunoassays. Nilsson,B. (Dep. Clin. Chem., Univ. Lund, Malmo Gen. Hosp., S-214 01 Malmo, Sweden) *Scand. J. Clin. Lab. Invest., 39(1), 99-101 (1979)* En;en.

813-U2 Detection of human ADCC activity using automated flow cytometry. Attallah,A.N.; Noguchi,P.D.; Folks,T.; Noguchi,C.T. (Div. Pathol., Bur. Biol., 8800 Rockville Pike, Bethesda, MD 20205, USA) *Experientia, 36(1), 124-125 (1980)* En;en.

814-U2 A simple method to demonstrate an autoimmune-response following vaccination and infection. Evaluation of autohaemolysin-forming cells in the peripheral blood by means of a plaque assay. Huber,H.C.; Hochstein-Mintzel,V. (Bayer. Landesimpfanstalt. Am. Neudeck 1, 8 Munchen 95, GFR) *Fortschr. Med., 95(3), 123-126 (1977)* De;de,en.

815-U2 A newly designed whole microplate automatic harvester for lymphocyte stimulation assays. Dagan,J.; Eshel,I.; Gazit,E.; lawner,M.; Shoman,J. (Dep. Bioeng. Chaim Sheba Med. Cent., Tel Hashomer, Isral) *J. Immunol. Methods, 30(1), 97-103 (1979)* En;en.

816-U2 Evaluation of the C1q solid-phase binding assay for immune complexes. A clinical and laboratory study. Scullion,M.; Balint,G.; Whaley,K. (Pathol. Dep., Western Infirm., Glasgow G11 6NT, UK) *J. Clin. Lab. Immunol., 2(1), 15-22 (1979)* En;en.

817-U2 Functional assay of the alternative complement pathway of rat serum. Coonrod,J.D.; Jenkins,S.D. (Veterans Adm. Med. Cent., Lexington, KY 40506, USA) *J. Immunol. Methods, 31(3-4), 291-301 (1979)* En;en.

818-U2 A simplified [^{3}H]serotonin release assay for the detection of platelet antibodies. Marmer,D.J.; Bowman,R.P.; Kennedy,P.S. (Univ. Arkansas for Med. Sci., Dep. Pediatr., 4301 West Markham, Little Rock, AR 72201, USA) *Am. J. Hematol., 6(1), 45-50 (1979)* En;en.

819-U2 Technical modifications for antibody-dependent cell-mediated cytotoxicity versus separated T- and B lymphocytes: T and B-cell separation by nylon wool columns. Lowry,R.P.; Person,A.E.; Goguen,J.E.; Carpenter,C.B.; Garovoy,M.R. (Tissue Typing Lab., Dep. Med., Peter Bent Brigham Hosp., Boston, MA 02115, USA) *Transplant. Proc., 10(4), 833-837 (1978)* En.

820-U2 Anti-kappa/lambda immunobead for detection of human B lymphocytes. Loren,A.B.; Hendricks,R.L.; Yokoyama,M.M. (Dep. Pathol. and Microbiol., Univ. Illinois Med. Cent., Chicago, IL 60680, USA) *Immunol. Commun., 7(6), 691-699 (1978)* En;en.

821-U2 Convenient non-chromatographic assays for the microbial deconjugation and 7α-OH bioconversion of taurocholate. MacDonald,I.A.; Bishop,J.M.; Mahony,D.E.; Williams,C.N. (Dep. Med., Dalhousie Univ., Halifax, NS, Canada) *Appl. Microbiol., 30(4), 530-535 (1975)* En;en.

822-U2 A microassay for the determination of hemolytic complement activity in mouse serum. Andrews,B.S.; *Theofilopoulos,A.N. (Dep. Immunopathol., Scripps Clin. and Res. Found., La Jolla, CA 92037, USA) *J. Immunol. Methods, 22(3-4), 273-281 (1978)* En;en.

823-U2 A sensitive fluorescent assay for N-acetyltransferase activity in human lymphocytes from newborns and adults. Meisler,M.H.; Reinke,C. (Dep. Hum. Genet., Univ. Michigan, 1137 E. Catherine St., Ann Arbor, MI 48109, USA) *Clin. Chim. Acta, 96(1-2), 91-96 (1979)* En;en.

824-U2 A simple method for the detection and assay of C1. Folkerd,E.J.; Hughes-Jones,N.C. (MRC Exp. Haematol. Res. Unit, St. Mary's Hosp. Med. Sch., London W2, UK) *J. Immunol. Methods, 31(1-2), 187-189 (1979)* En;en.

825-U2 An electronic method for counting human T and B cell rosettes. Denny,T.N.; Kinsley,K.M.; Benson,J.W.; Bull,R.W. (Dep. Med., B220 Life Sci. Build., Michigan State Univ., East Lansing, MI 48824, USA) *Vox Sang., 37(4), 244-250 (1979)* En;en.

826-U2 Non-radioisotopic homogeneous steroid immunoassays. Kohen,F.; Hollander,Z.; Boguslaski,R.C. (Dep. Hormone Res., Weizmann Inst. Sci., Rehovot, Israel) *J. Steroid Biochem., 11(1, part A), 161-167 (1979)* En;en.

827-U2 A simple method for counting antibody bound labelled steroids. Kandeel,F.R.; Rudd,B.T.; Morris,R.; Butt,W.R. (Dep. Clin. Endocrinol., Birmingham and Midland Hosp. for Women, Showell Green Lane, Sparkhill, Birmingham B11 4HL, UK) *Steroids, 28(6), 755-762 (1976)* En;en.

828-U2 Visualization principles in thin-layer immunoassays (TIA) on plastic surfaces. Elwing,H.; Nilsson,L.-Å.; Ouchterlony,O. (Inst. Med. Microbiol., Univ. Goteborg, Guldhedsgatan 10, S-41346 Goteborg, Sweden) *Int. Arch.*

Allergy Appl. Immunol., 51(6), 757-762 (1976) En;en.

829-U2 Comparison of a semiautomated fluorescent immunoassay system and indirect immunofluorescence for detection of antinuclear antibodies in human serum. Casavant,C.H.; Hart,A.C.; Stites,D.P. (Clin. Immunol. Lab., Dep. Lab. Med., Univ. California at San Francisco, San Francisco, CA 94143, USA) *J. Clin. Microbiol., 10(5), 712-718 (1979)* En;en.

830-U2 An improved method for the in vitro evaluation of monocyte leukotaxis. Campbell,P.B. (Div. Infect. Dis., Dep. Med., Cleveland Metropolitan Gen. Hosp., 3395 Scranton Rd., Cleveland, OH 44109, USA) *J. Lab. Clin. Med., 90(2), 381-388 (1977)* En;en.

831-U2 A modified assay for measuring human urinary carcinoembryonic antigenic (CEA) activity. Wu,J.T.; Madsen,A.C.; Bray,P.F. (Div. Pediatr. Neurol., Univ. Utah Med. Cent., Salt Lake City, UT 84132, USA) *Biochem. Med., 14(3), 305-316 (1975)* En;en.

832-U2 A simple, sensitive assay for determining DNA in mononuclear phagocytes and other leukocytes. Cookson,S.L.; Adams,D.O. (Dep. Pathol., Duke Univ. Med. Sch., Durham, NC 27710, USA) *J. Immunol. Methods, 23(1/2), 169-173 (1978)* En;en.

See: 37, 55, 58, 83, 174, 204, 206, 275, 277, 313, 1005, 1062, 1317, 1407, 1408, 1439, 1458, 1461, 1495, 1553, 1588, 1633, 1936, 2198, 22 2325, 14, 2093, 2190, 3262

Typing techniques

833-U2 Typing of *Proteus* strains by proticine production and sensitivity. Senior,B.W. (Dep. Bacteriol., Univ. Dundee, Dundee DD1 4HN, UK) *J. Med. Microbiol., 10(1), 7-17 (1977)* En;en.

834-U2 [Bacteriocin typing of *P. rettgeri*]. Chezzi,C.; Branca,G.; Manzara,S.; Scanu,L.; Tarquini,C. (Ist. Microbiol., Univ. Cattolica S. C., Roma, Italy) *Ig. Mod., 42(11), 1186-1194 (1979)* It;en,it.

835-U2 Modified procedure for pyocin typing of *Pseudomonas aeruginosa*. Bauernfeind,A.; Petermuller,C.; Burrows,J.R. (Max von Pettenkofer Inst., Univ. Munchen, Munchen, GFR) *Zentralbl. Bakteriol. Parasitenkd. Infektionskr. Hyg., I Abt. A, 240(2), 271-278 (1978)* De;de,en.

836-U2 Bacteriocin typing by leakage of ultraviolet light-absorbing material. Farkas-Himsley,H.; Pagel,A. (Dep. Microbiol. and Parasitol., Univ. Toronto, Toronto, Ont. M5S 1A1, Canada) *Infect. Immun., 16(1), 12-19 (1977)* En;de,en.

837-U2 Single phage-typing set for differentiating salmonellae. Gershman,M. (Dep. Microbiol., Univ. Maine, Orono, ME 04473, USA) *J. Clin. Microbiol., 5(3), 302-314 (1977)* En;en.

838-U2 Automation of *Salmonella typhi* phage typing. Farmer,J.J.,III; Hickman,F.W.; Sikes,J.V. (Bacteriophage-Bacteriocin Lab., Cent. Dis. Control, Atlanta, GA 30333, USA) *Lancet, 2(7939), 737-790 (1975)* En;en.

839-U2 Bacteriophage typing of *Shigella sonnei*. Pruneda,R.C.; *Farmer,J.J.,III (Bacteriophage-Bacteriocin Lab., Cent. for Dis. Control, Atlanta, GA 30333, USA) *J. Clin. Microbiol., 5(1), 66-74 (1977)* En;en.

840-U2 Device for mass phage typing of staphylococci. Semenikhina,T.K.; Bugaev,B.M.; Kokojlo,A.N. (Munic. Sanit. Epidemiol. Stn., Artenovsk, USSR) *Mikrobiol. Zh., 40(4), 512-513 (1978)* Uk;en.

841-U2 Device for mass phage typing of staphylococci. Semenikhina,T.K.; Bugaev,B.M.; Kokojlo,A.N. (Munic. Sanit. Epidemiol. Stn., Artemovsk, USSR) *Mikrobiol. Zh., 40(4), 512-513 (1978)* Uk;en.

842-U2 Attempts to simplify the phage-typing methodology of staphylococci and to increase their typability. Popovici,M.; Greceanu,V. (Cantacuzino Inst., Bucharest, Romania) *Arch. Roum. Pathol. Exp. Microbiol., 38(3-4), 273-278 (1979)* En;en,fr.

843-U2 Improved lysotyping scheme for group C streptococci with new phage preparations. Vereanu,A.; Mihalcu,F. (Cantacuzino Inst., Bucharest, Romania) *Arch. Roum. Pathol. Exp. Microbiol., 38(3-4), 265-272 (1979)* En;en,fr.

844-U2 Polyvalent diagnostic cholera bacteriophage. Drozhevkina,M.S.; Alexandrova,I.K.; Kharitonova,T.I. (Anti-Plague Inst., Rostov-on-Don, USSR) *Zh. Mikrobiol. Epidemiol. Immunobiol., 53(6), 82-84 (1976)* Ru;en,ru.

845-U2 Mechanized technique for phage typing and determination of antibiograms. Guinee,P.A.M.; Jansen,W.H.; van Schuylenburg,A.; van Leeuwen,W.J. (Natl. Inst. Public Health, PO Box 1, Bilthoven, Netherlands) *Zentralbl. Bakteriol. Mikrobiol. Hyg., I Abt. A, 246(2), 276-284 (1980)* En;de,en.

846-U2 A comparison of three methods of pyocine typing by pyocine production. Lyne,L.E. (McMaster Univ. Med. Technol., 37(4), 114-122 (1975) *Can. J. Med. Technol., 37(4), 114-122 (1975)* En;en.

847-U2 *Klebsiella* serotyping by counter-current immunoelectrophoresis. Palfreyman,J.M. (Public Health Lab., Coventry, UK) *J. Hyg., 81(2), 219-225 (1978)* En;en.

848-U2 A new serotyping method for *Klebsiella* species: evaluation of the technique. Riser,E.; Noone,P.; Bonnet,M.L. (Bacteriol. Dep., Middlesex Hosp. Med. Sch., Riding House St., London W1, UK) *J. Clin. Pathol., 29(4), 305-308 (1976)* En;en.

849-U2 A new serotyping method for *Klebsiella* species: development of technique. Riser,E.;Noone,P.; Poulton,T.A. (Bacteriol. Dep., Middlesex Hosp. Med. Sch., Riding House St., London W1, UK) *J. Clin. Pathol., 29(4), 296-304 (1976)*

En;en.

850-U2 The use of polyvalent sera for the serotyping of mycobacteria within the *Mycobacterium avium-M. intracellulare-M. scrofulaceum* complex. Matthews,P.R.J.; Brown,A.; Collins,P. (ARC, Inst. Res. on Anim. Dis., Compton, Nr. Newbury, Berks, RG16 0NN, UK) *J. Appl. Bacteriol., 46(3), 425-430 (1979)* En;en.

851-U2 Heat-stable somatic antigens of a group of unclassified fluorescent pseudomonads (UFP). Lanyi,B.; Hoadley,A.W.; Ajello,G.W. (Natl. Inst. Hyg., Budapest, Hungary) *Acta Microbiol. Acad. Sci. Hung., 26(2), 111-120 (1979)* En;en.

852-U2 Direct-plate serological grouping of beta hemolytic streptococci from primary isolation plates with the Phadebact Streptococcus Test. Slifkin,M.; Engwall,C.; Pouchet,G.R. (Microbiol. Sect., Dep. Lab. Med., Allegheny Gen. Hosp., Pittsburgh, PA 15212, USA) *J. Clin. Microbiol., 7(4), 356-360 (1978)* En;en.

853-U2 Evaluation of three commercially available test products for serogrouping beta-hemolytic streptococci. Slifkin,M.; Pouchet-Melvin,G.R. (Dep. Lab. Med. Sect. Microbiol., Allegheny Gen. Hosp., Pittsburgh, PA 15212, USA) *J. Clin. Microbiol., 11(3), 249-255 (1980)* En;en.

854-U2 Serological grouping of streptococci by a rapid slide agglutination test. Deschamps,C.; Ortenberg,M.; Perol,Y. (Lab. Cent. Bacteriol.-Virol., Hop. St-Louis, 2 place du Dr-Fournier, 75010 Paris, France) *Med. Mal. Infect., 6(11), 466-470 (1976)* Fr;en,fr.

855-U2 Biochemical typing of salmonellae by a miniaturized system (API-50 E). Ottaviani,F.; Aliani,A. (Ente Osp. Sandrigo, Lab. Anal. Chimico-Clin. e Microbiol., Vicenza, Italy) *Ann. Sclavo Riv. Microbiol. Immunol., 19(4), 626-632 (1977)* It;en.

856-U2 Mathematical express-interpretation of the results of biochemical typing for identification of representatives of the Enterobacteriaceae family. Finn,V.G.; Chaus,V.I. (Inst. Vaccines and Sera, Moscow, USSR) *Zh. Mikrobiol. Epidemiol. Immunobiol., 53(7), 29-33 (1976)* Ru;en,ru.

857-U2 Biotyping of *Escherichia coli* by a simple multiple-inoculation agar plate technique. Buckwold,F.J.; Ronald,A.R.; Harding,G.K.M.; Marrie,T.J.; Fox,L.; Cates,C. (Dep. Clin. Microbiol., Health Sci Cent., Univ. Manitoba, Winnipeg, Manit. R3E 0W3, Canada) *J. Clin. Microbiol., 10(3), 275-278 (1979)* En;en.

858-U2 Use of antiserum agar plates for serogrouping of meningococci. Sivonen,A.; Renkonen,O-V.; Robbins,J.B. Dep. Bacteriol. and Immunol., Univ. Helsinki, Haartmaninkatu 3, SF-00290 Helsinki, Finland) *J. Clin. Pathol., 30(9), 834-837 (1977)* En;en.

859-U2 Rapid method for auxotyping multiple strains of *Neisseria gonorrhoeae*. Short,H.B.; Ploscowe,V.B.; Weiss,J.A.; *Young,F.E. (Dep. Microbiol., Univ. Rochester Sch. Med. and Dent., Rochester, NY 14642, USA) *J. Clin. Microbiol., 6(3), 244-248 (1977)* En;en.

860-U2 Capillary precipitin typing of *Streptococcus pneumoniae*. Russell,H.; Facklam,R.R.; Padula,J.F.; Cooksey,R. (Cent. Dis. Control., Atlanta, GA 30333, USA) *J. Clin. Microbiol., 8(4), 355-359 (1978)* En;en.

861-U2 Grouping of streptococci by means of antibodies adsorbed to staphylococcal protein A. Wautelet,J.; Wauters,G.; Bruynoghe,G. (Lab. Microbiol., Univ. Louvain, 3000 Louvain, Belgium) *Ann. Microbiol., 126B(2), 251-254 (1975)* Fr;en,fr.

862-U2 Simplified α_1-antitrypsin phenotyping by immunofixation of acid-starch gels. Lieberman,J.; Gaidulis,L. (City of Hope Med,. Cent., 1500 E. Duarte Rd., Duarte, CA 91010, USA) *J. Lab. Clin. Med., 87(4), 710-716 (1976)* En;en.

See: 28, 1681, 1682, 1683, 1717, 1718, 1747, 1765, 1767, 1776, 2000, 2077, 2091, 2968, 1774, 1777

Sensitivity testing

863-U2 Evaluation of the performance parameters of a prediluted, quantitative antibiotic susceptibility test device. Tilton,R.C.; Isenberg,H.D. (Univ. Connecticut Sch. Med., Farmington, CT 06032, USA) *Antimicrob. Agents Chemother., 11(2), 271-276 (1977)* En;en.

864-U2 Laboratory evaluation of an automated antimicrobial susceptibility system. Stubbs,K.G.; *Wicher,K. (Dep. Clin. Microbiol. and Immunol., Erie County Lab., 462 Grider St., Buffalo, NY 14215, USA) *Am. J. Clin. Pathol., 68(6), 769-777 (1977)* En;en.

865-U2 A permanent record of antibiotic susceptibility results (disk method) using a Xerox photocopying machine. Lo,J.W.; Tess,B.R. (Mercy Hosp., and Med. Cent., Chicago, IL 60616, USA) *Am. J. Clin. Pathol., 64(2), 271-272 (1975)* En;en.

866-U2 Improved susceptibility disk assay method employing an agar overlay technique. Anderson,B. (Antibiot. Residue Branch, Natl. Cent. Antibiot. Anal, FDA, Washington, DC 20204, USA) *Antimicrob. Agents Chemother., 14(5), 761-764 (1978)* En;en.

867-U2 A broth-disc technique for the assay of antibiotic synergism. Meyer,M.; Hofherr,L. (Dep. Microbiol., Univ. Minnesota, Minneapolis, MN 55455, USA) *Can. J. Microbiol., 25(11), 1232-1238 (1979)* En;en,fr.

868-U2 [A simple method to test the sensitivity of influenza virus to active substances represented by amantadine]. Adamczyk,B. (Virol. Abt. Bezirks-Hyg.-Inst. DDR-1115 Berin-Buch, Wilbergstr. 50, Haus 106, GDR) *Dtsch. Gesundheitswes., 32(31), 1458-1460 (1977)* De;de,en,ru.

869-U2 An automated susceptibility testing system. Bergogne-Berezin,E.; Thabaut,A.; Durosoir,M.J.;

Berthelot,G.; Courtoux,D. (Serv. Cent. Microbiol., Hop. Bichat, 170 blvd. New, 75877 Paris Cedex 18, France) *Med. Mal. Infect., 8(1), 15-19 (1978)* Fr;en,fr.

870-U2 The microdilution antibiotic susceptibility test. *Bacteroides fragilis.* Steingrimsson,O.; Ryan,R.W.; *Tilton,R.C. (Dep. Lab. Med., Univ. Connecticut Health Cent., Farmington, CT 06032, USA) *Am. J. Clin. Pathol., 65(6), 1010-1016 (1976)* En;en.

871-U2 A disc diffusion method for determining antibiotic susceptibility of *Bacteroides melaninogenicus* subsp. *melaninogenicus.* Gant,C.; Fuerst,R. (Texas Woman's Univ., Denton, TX 76204, USA) *Dev. Ind. Microbiol., 20, 759-768 (1979)* En;en.

872-U2 A simple method for determining bacterial susceptibility to aminoglycosides. Hinchliffe,P.M.; Robertson,L.; Farrell,I.D.; Kirkwood,B.R. (Public Health Lab., R. Infirm., Meadow St., Preston, Lancs, PR1 6PS, UK) *J. Med. Microbiol., 10(3), 381-387 (1977)* En;en.

873-U2 Results of two methods for testing bacterial susceptibility to antibiotics: disc method and a new semi-automatic method in liquid medium (ABACR procedure) and comparison with the minimal inhibitory concentration. Veron,M.; Ghnassia,J.C.; Berche,P.; Daoulas-Le Boudelles,F.; Avril,J.L.; Fauchere,J.L.; de Meirleire,F.; Descamps,P. (Lab. Cent. Bacteriol., Hop. Necker-Enfants Mal., 75730 Paris Cedex 15, France) *Ann. Microbiol., 129A(4), 473-502 (1978)* Fr;en,fr.

874-U2 Comparison of two methods of antibiotic sensitivity tests: gel diffusion and an automatic method using ABAC apparatus. I. Variation and dispersion of the two methods. Drugeon,H.; Courtieu,A.L. (Lab. Bacteriol. A, Cent. Hosp., 44035 Nantes Cedex, France) *Ann. Biol. Clin., 36(4), 363-367 (1978)* Fr;en,fr.

875-U2 Microdilution technique for antimicrobial susceptibility testing of *Haemophilus influenzae.* Jorgensen,J.H.; Lee,J.C. (Dep. Pathol., Univ. Texas Health Sci. Cent. at San Antonio, San Antonio, TX 78284, USA) *Antimicrob. Agents Chemother., 8(5), 610-611 (1975)* En;en.

876-U2 Comparison of rapid methods of antimicrobial susceptibility in *Haemophilus influenzae.* Tilton,R.C.; Kwasnik,I.; Ryan,R. (Dep. Lab. Med., Univ. Connecticut, Sch. Med., Farmington, CT 06032, USA) *Ann. Clin. Lab. Sci., 8(1), 70-73 (1978)* En;en.

877-U2 *Haemophilus influenzae* susceptibility testing within four hours by a modification of the Autobac I system. Hosmer,M.E.; Sprunt,K. (Diagnostic Bacteriol. Lab., Babies Hosp., Columbia Presbyterian Med. Cent., New York, NY 10032, USA) *Antimicrob. Agents Chemother., 13(5), 895-896 (1978)* En;en.

878-U2 A method for testing susceptibility of *Haemophilus influenzae* to co-trimoxazole. Machka,K.; Balg,H.; Smith,M.; *Braveny,I. (Inst. Med. Mikrobiol. und Hyg., Tech. Univ., Ismaninger Str. 22, 8000 Munchen 80, GFR) *Immun. Infekt., 6(6), 249-252 (1978)* De;de,en.

879-U2 Rapid ampicillin susceptibility testing for *Haemophilus influenzae.* Barkin,R.M.; *Todd,J.K.; Roe,M.H. (Dir. Infect. Dis., Child. Hosp. Denver, 1056 E. 19th Ave., Denver, CO 80218, USA) *Am. J. Clin. Pathol., 67(1), 100-103 (1977)* En;en.

880-U2 Rapid detection of chloramphenicol resistance in *Haemophilus influenzae.* Slack,M.P.E.; Wheldon,D.B.; Turk,D.C. (Dep. Bacteriol. and Reg. Public Health Lab., Radcliffe Infirm., Oxford OX2 6HE, UK) *Lancet, 2(8052/3), 1366 (1977)* En.

881-U2 Sensitivity of *Haemophilus influenzae* to 5 antibotics and rapid detection of its resistance to ampicillin. Piot,P.; van Dyck,E.; Pattyn,S.R. (Lab. Bacteriol., Inst. Mal. Trop. Prince Leopold, Nationalestraat 155, B-2000 Antwerpen, Belgium) *Pathol. Biol., 25(2), 83-87 (1977)* Fr,en,fr.

882-U2 Susceptibility of *Haemophilus influenzae* to ampicillin as determined by use of a modified, one-minute beta-lactamase test. Escamilla,J. (Navy Environ. and Prev. Med. Unit No. 2, Norfolk, VA 23511, USA) *Antimicrob. Agents Chemother., 9(1), 196-198 (1976)* En;en.

883-U2 Susceptibility of *Haemophilus* sp. to ampicillin, doxycycline and minocycline determined by microdilution technique. Duret,M.; Philippon,A.; Paul,G.; Nevot,P. (Lab. Bacteriol., Cent. Hosp. Univ. Cochin, 24 rue du Fg. Saint-Jacques, 75014 Paris, France) *Med. Mal. Infect., 8(7), 316-322 (1978)* Fr;en,fr.

884-U2 Rapid method for determining the minimum inhibitory concentration of ampicillin for *Haemophilus influenzae.* Ryan,R.; *Tilton,R.C. (Univ. Connecticut Health Cent., Farmington, CT 06032, USA) *Antimicrob. Agents Chemother., 11(1), 114-117 (1977)* En;en.

885-U2 Acidometric agar plate method for ampicillin susceptibility testing of *Haemophilus influenzae.* Park,C.H.; Lopez,J.S.; Cook,C.B. (Dep. Pathol., Fairfax Hosp., Falls Church, VA 22046, USA) *Antimicrob. Agents Chemother., 13(2), 318-320 (1978)* En;en.

886-U2 Studies on bactericidal activities of β-lactam antibiotics on agar plates: the correlation with the antibacterial activities determined by the conventional methods. Masuda,G. (Dep. Med., Sch. Med., Keio Univ., 35-Shinanomachi, Shinjuku-ku, Tokyo, Japan) *J. Antibiot., 29(11), 1237-1240 (1976)* En;en.

887-U2 New technique in gelose applied to the study of bactericidal activity of ampicillin against a strain of *Listeria monocytogenes* and research of a synergism betwen antibiotic and aminoglycosides. / [presented at a Colloquium of la Societe Francaise de Pathologie Infectieuse, held in Paris, France, on 5 Dec 1975]. Yourassowsky,E.; Vander Linden,M.P.; Schoutens,E. (Hop. Univ. Brugmenn, Serv. Biol. Clin., 1020 Bruxelles, Belgium) *Med. Mal. Infect., 6(9), 41-45 (1976)* Fr;en,fr.

888-U2 Comparison of various measures of sensitivity of *M. tuberculosis* to ethambutol. Subbammal,S.; Nair,N.G.K;

Radhakrishna,S.; Tripathy,S.P. (Tuberc. Chemother. Cent. (Indian Counc. Med. Res.), Spur Tank Road, Madras 600031, India) *Tubercle, 59(3), 185-191 (1978)* En;en,es,fr.

889-U2 Sensitivity tests of mycobacteria to INH, SM, EMB and RMP in a semi-liquid serum-Sauton-agar Neubert,R. (Bezirkskrankenhaus Lungenkr., DDR-7031 Leipzig, Nikolai-Rumjanzaw-Str.100, GDR) *Z. Erkr. Atmungsorgane, 144(1), 63-68 (1976)* De;de,en.

890-U2 Antimicrobial susceptibility testing of *Neisseria gonorrhoeae* to penicillin. Palatnick,L. (Health Sci. Cent., Winnipeg, Manit., Canada) *Can. J. Med. Technol., 41(1), 19-25 (1979)* En;en,fr.

891-U2 A method for determining in vitro drug susceptibilities of some Nocardiae and Actinomadurae. Results with 17 antimicrobial agents. Carroll,G.F.; Brown,J.M.; *Haley,L.D. (Mycol. Training Branch, Cent. Dis. Cont., DHEW, Atlanta, GA 30333, USA) *Am. J. Clin. Pathol., 68(2), 279-283 (1977)* En;en.

892-U2 [A rapid micromethod for the determination of antibiotic sensitivity in *Pseudomonas aeruginosa*]. Rykner,G.; Scavizzi,M.R. (Lab. Cent. Bacteriol., Hop. Bicetre, 78 ave. du General-Leclerc, 94 Bicetre, France) *Med. Mal. Infect.,7(5), 248-253 (1977)* Fr;en,fr.

893-U2 Chemically defined antimicrobial susceptibility test medium for *Pseudomonas aeruginosa*. Jorgensen,J.H.; Lee,J.C.; Jones,P.M. (Dep. Pathol., Univ. Texas Health Sci. Cent., San Antonio, TX 78284, USA) *Antimicrob. Agents Chemother., 11(3), 415-419 (1977)* En;en.

894-U2 Five-hour minimal inhibitory concentration test of four antimicrobial agents for *Pseudomonas aeruginosa*. Brown,W.J.; Sautter,R.L. (Hutzel Hosp., Detroit, MI 48201, USA) *Antimicrob. Agents Chemother., 11(6), 1064-1066 (1977)* En;en.

895-U2 On methods of testing the sensitivity of pathogenic micro-organisms: leaflet test or scatter test. Menzel,G.; Hansel,H. (DDR-1242 Bad Saarow, Pieskower Str., Fach 1381, GDR) *Z. Gesamte Hyg. Grenzgeb., 23(9), 675-677 (1977)* De;de,en,ru.

896-U2 Methicillin-resistant *Staphylococcus aureus* susceptibility testing by an automated system, Autobac I. Cleary,T.J.; Maurer,D. (Dep. Pathol., Univ. Miami, Jackson Mem. Hosp. Miami, FL 33152, USA) *Antimicrob. Agents Chemother., 13(5), 837-841 (1978)* En;en.

897-U2 A modified technique for the detection of antibiotic synergism. Anand,C.M.; Paull,A. (Prov. Lab. Public Health, PO Box 2490, Calgary, Alta. T2P 2M7, Canada) *J. Clin. Pathol., 29(12), 1130-1131 (1976)* En.

898-U2 Simple, direct broth-disk method for antibiotic susceptibility testing of *Ureaplasma urealyticum*. Spaepen,M.S.; *Kundsin,R.B. (Dep. Surg., Peter Bent Brigham Hosp., Boston, MA 02115, USA) *Antimicrob. Agents Chemother., 11(2), 267-270 (1977)* En;en.

899-U2 Rapid sensitivity tests. Newsom,S.W.B. (Papworth Hosp., Cambridge CB3 8RE, UK) *J. Antimicrob. Chemother., 3(3), 201-203 (1977)* En.

900-U2 Urinary infection in urology. A rapid technique for confirmation of the diagnosis and the orientation of medical treatment. Perrin,P.; Gille,Y.; Durand,L.; Vincent,P. (Lab. Microbiol., Hop. Antiquaille, 1 rue de l'Antiquaille, 69325 Lyon Cedex 1, France) *J. Urol. Nephrol., 83(7-8), 499-507 (1977)* Fr;en,fr.

901-U2 Rapid antimicrobial susceptibility test using tetrazolium reduction. Bartlett,R.C.; Mazens,M.F. (Div. Microbiol., Dep. Pathol., Hartford Hosp., Hartford, CT 06115, USA) *Antimicrob. Agents Chemother., 15(6), 769-774 (1979)* En;en.

902-U2 Comparative antibiotic sensitivity testing of bacteria with a paper disc and a tablet method. Rollag,H.,Jr.; Midtvedt,T. (Kapt. W. Wilhelmsen og Frues Bakteriol. Inst., Natl. Hosp. Norway, Oslow, Norway) *Chemotherapy, 24(5), 297-300 (1978)* En;en.

903-U2 Sensitivity of bacteria to antibiotics, evaluation of a rapid automatic method. Acar,J.F.; Swift,R.I.; McKie,J.; Quentin,M.C. (Serv. Microbiol. Med., Hop. St. Joseph, 7 rue Pierre Larousse, 75674 Parix Cedex 14, France) *Pathol. Biol., 26(2), 137-144 (1978)* Fr;en,fr.

904-U2 Routine test for sensitivity. Wiedenmann,B. (Inst. Med. Mikrobiol. und Immunol., Univ. Bonn, An der Immenburg 4, 5300 Bonn Endenich, GFR) *Immun. Infekt., 6(6), 209-213 (1978)* De;de,en.

905-U2 Susceptibility testing with the Autobac system. Baars,B.; Opferkuch,W. (Lehrstuhl Med., Mikrobiol., Ruhr Univ., Universitatsstr. 150, 4630 Bochum, GFR) *Immun. Infekt., 6(6), 270-273 (1978)* De;de,en.

906-U2 Semi-quantitative determination of resistance in agar. Marcenac,F.M.L.; Fernandez,A.J.; Herran,I.L.; Civalero,T.R. (Hosp. Escuela 'Gral. Jose de San Martin', Ave. Cordoba 2351, 1120 Buenos Aires, Argentina) *Arzneim.-Forsch., 28(1, part 4), 582-585 (1978)* De;de,en.

907-U2 Determination by micromethod of the minimal inhibitory concentration of several antibiotics. Madier,S.; Fleurette,J. (Lab. Bacteriol., Hop. Cardiol., BP Lyon Monchat, 69394 Lyon Cedex 3, France) *Med. Mal. Infect., 7(4), 203-207 (1977)* Fr;en,fr.

908-U2 Automated antimicrobial susceptibility system. (1) Tilton,R.C. (Dep. Lab. Med., Univ. Connecticut Health Cent., University Hosp., Farmington, CT 06032, USA); (2) Wicher,K.; Stubbs,K.G. (State Univ. New York, Buffalo, NY 14214, USA) *Am. J. Clin. Pathol., 70(2), 310-311 (1978)* En.

909-U2 Development of the method for quantitative estimation of microbial sensitivity to antibiotics using paper discs. Determination of the minimal inhibitory concentration of doxycycline with the help of antibiotic-containing discs. Chaikovskaya,S.M.; Rezvan,S.P.; Rabinovich,A.S.; Galkina,T.P.; Dmitrova,L.I.; Navashin,S.M. (All-Union Res.

Inst. Antibiot., Moscow, USSR) *Antibiotiki, 23(2), 118-122 (1978)* Ru;en.

910-U2 **Direct method to determine the antibiotic susceptibility of rapidly growing blood pathogens.** Wegner,D.L.; Mathis,C.R.; Neblett,T.R. (Div. Bacteriol., Dep. Pathol., Henry Ford Hosp., Detroit, MI 48202, USA) *Antimicrob. Agents Chemother., 9(5), 861-862 (1976)* En;en.

911-U2 **Rapid determination of microbial sensitivity to antibiotics and chemotherapeutic drugs using the peroxidase test.** Feldman,Yu.M.; Leibman,E.T. (Zhitomir Res. Stn., Zhitomir, USSR) *Antibiotiki, 25(2), 109-112 (1980)* Ru.

912-U2 **Historical survey of tests to determine bacterial susceptibility to antimicrobial agents.** Laverdiere,M.; *Sabath,L.D. (Box 219 Mayo Mem. Build., Univ. Minnesota, Minneapolis, MN 55455, USA) *Mt. Sinai J. Med., 44(1), 73-88 (1977)* En;en.

913-U2 **Rapid determination of minimum inhibitory concentrations of antimicrobial agents by the Autobac method: a collaborative study.** Schoenknecht,F.D.; Washington,J.A.,II; Gavan,T.L.; Thornsberry,C. (Dep. Microbiol., Univ. Washington, Seattle, WA 98195, USA) *Antimicrob. Agents Chemother., 17(5), 824-833 (1980)* En;en.

914-U2 **A computer system for handling data from bacterial sensitivity testing.** Midtvedt,T.; Tvete,E.; Hovig,B.; Rollag,H.,Jr. (Kaptein W. Wilhemsem og Frues Bakteriol. Inst., Oslo, Norway) *Comput. Programs Biomed., 6(3), 178-186 (1976)* En;en.

915-U2 **New in vitro model to study the effect of antibiotic concentration and rate of elimination on antibacterial activity.** Grasso,S.; Meinardi,G.; *de Carneri,I.; Tamassia,V. (Dep. Microbiol., Carlo Erba Res. Inst., 20159 Milan, Italy) *Antimicrob. Agents Chemother., 13(4), 570-576 (1978)* En;en.

916-U2 **Comparison of two methods of antibiotic sensitivity test: gel diffusion and automatic method using ABAC apparatus. II. Study on 399 strains isolated in hospital.** Drugeon,H.; Courtieu,A.L. (Lab. Bacteriol. A, Cent. Hosp., 44035 Nantes Cedex, France) *Ann. Biol. Clin., 36(6), 523-526 (1978)* Fr;en,fr.

917-U2 **New aid to management of septicaemia. Rapid direct antibiotic susceptibility testing.** Funnell,G.R.; Guinness,M.D.G. (Dep. Microbiol., Repatriation Gen. Hosp., Concord, NSW 2139, Australia) *Med. J. Aust., 2(10), 514-515 (1979)* En;en.

918-U2 **Automated direct antimicrobial susceptibility testing of microscopically screened urine cultures.** Tilton,R.E.; *Tilton,R.C. (Dep. Lab. Med., Univ. Connecticut Sch. Med., Farmington, CT 06032, USA) *J. Clin. Microbiol., 11(2), 157-161 (1980)* En;en.

919-U2 **Relationship of early readings of minimal inhibitory concentrations to the results of overnight tests.** Lampe,M.F.; Aitken,C.L.; Dennis,P.G.; Forsthe,P.S.; Patrick,K.E.; Schoenknecht,F.D.; *Sherris,J.C. (Dep. Microbiol., Univ. Washington, Seattle, WA 98195, USA) *Antimicrob. Agents Chemother., 8(4), 429-433 (1975)* En;en.

920-U2 **Microbiological diffusion assay. II. Design and applications.** Kavanagh,F. (Eli Lilly and Co., Indianapolis, IN 46206, USA) *J. Pharm. Sci., 64(7), 1224-1229 (1975)* En;en.

921-U2 **Practical anaerobic broth-disk elution susceptibility test.** Jorgensen,J.H.; Alexander,G.A.; Johnson,J.E. (Dep. Pathol., Univ. Texas Health Sci. Cent., San Antonio, TX 78284, USA) *Antimicrob. Agents Chemother., 17(4), 740-742 (1980)* En;en.

922-U2 **Rapid determination of minimum inhibitory concentrations of antimicrobial agents by regression analysis of light scattering data.** McKie,J.E.; *Seo,J.; Arveson,J.N. (Res. and Dev. Dep., Pfizer Diagnostics, Groton, CT 06340, USA) *Antimicrob. Agents Chemother., 17(5), 813-823 (1980)* En;en.

923-U2 **[Bacterial drug sensitivity tests: classic techniques versus automated techniques].** Thabaut,A.; Durosoir,J.L. (Lab. Bacteriol., Hop. Mil. Begin, 69 ave. de Paris, 94 Saint Mande, France) *Med. Mal. Infect., 9(9), 490-495 (1979)* Fr;en,fr.

924-U2 **Simple inoculum standardizing system for antimicrobial disk susceptibility tests.** Barry,A.L; Amsterdam,D.; Coyle,M.B.; Gerlach,E.H.; Thornsberry,C.; Hawkinson,R.W. (Univ. California, Davis, Med. Cent., Sacramento, CA 95817, USA) *J. Clin. Microbiol., 10(6), 910-918 (1979)* En;en.

925-U2 **Susceptibility of gram-negative bacteria to β-lactam antibiotics and rapid characterization of β-lactamase activity.** Gibbs,D.L.; *Thornsberry,C. (Cent. Dis. Control, Public Health Serv., US Dep. Health, Educ., and Welfare, Atlanta, GA 30333, USA) *Curr. Microbiol., 2(4), 239-244 (1979)* En;en.

926-U2 **The microtiter system for sensitivity testing.** Wilkinson,P.J. (Dep. Microbiol., Bristol Infirm., Bristol BS2 8HW, UK) *Immun. Infekt., 6(6), 267-269 (1978)* De;de,en.

927-U2 **Locally-produced discs for the determination of bacterial sensitivity to antibiotics.** Kerchan,L.L.O.; Bazira,S.S. (Dep. Med. Microbiol., Makerere Univ., Kampala, Uganda) *East Afr. Med. J., 52(6), 301-305 (1975)* En;en.

928-U2 **Automatic systems in determination of antibiotic biological activity by agar-diffusion method.** Bartoshevich,Yu.E.; Egorov,A.A.; Goldshtein,V.O.; Jones,I.J.; Luni,P.A. (All-Union Res. Inst. Antibiot., Moscow, USSR) *Antibiotiki, 21(2), 122-127 (1976)* Ru;en,ru.

929-U2 **Modified agar dilution method for rapid antibiotic susceptibility testing of anaerobic bacteria.** Hanson,C.W.; Martin,W.J. (Clin. Microbiol. Lab., UCLA Cent. Health Sci., Los Angeles, CA 90024, USA) *Antimicrob. Agents Chemother., 13(3), 383-388 (1978)* En;en.

930-U2 Simplified method for antimicrobial susceptibility testing of anaerobic bacteria. Fass,R.J.; Prior,R.B.; Rotilie,C.A. (Div. Infect. Dis., Dep. Med., Ohio State Univ. Coll. Med., Columbus, OH 43210, USA) *Antimicrob. Agents Chemother., 8(4), 444-452 (1975)* En;en.

931-U2 Microdilution transfer plate technique for determining in vitro synergy of antimicrobial agents. Dougherty,P.F.; Yotter,D.W.; *Matthews,T.R. (Nucleic Acid Res. Inst., ICN Pharm., Inc., Irvine, CA 92715, USA) *Antimicrob. Agents Chemother., 11(2), 225-228 (1977)* En;en.

932-U2 Autobac 1. Anon. (c/o Autobac Product Manager, Pfizer Diagn. Div., Pfizer, Inc., 253 E. 42nd St., New York, NY 10017, USA) *J. Am. Med. Technol., 39(5), 245 (1977)* En.

933-U2 Microdilution transfer plate technique for determining in vitro synergy of antimicrobial agents. Dougherty,P.F.; Yotter,D.W.; *Matthews,T.R. (Nucleic Acids Res. Inst., ICN Pharmaceuticals, Inc., Irvine, CA 92715, USA) *Antimicrob. Agents Chemother., 11(2), 225-228 (1977)* En;en.

934-U2 Disk agar diffusion susceptibility testing of yeasts. Saubolle,M.A.; *Hoeprich,P.D. (Dep. Intern. Med., Sch. Med., Univ. California-Davis, Davis, CA 95616, USA) *Antimicrob Agents Chemother., 14(4), 517-530 (1978)* En;en.

935-U2 Turbidimetric studies of growth inhibition of yeasts with three drugs: inquiry into inoculum-dependent susceptibility testing, time of onset of drug effect, and implications for current and newer methods. Galgiani,J.N ; Stevens,D.A. (Stanford Univ., San Jose and Stanford, CA94305, USA) *Antimicrob. Agents Chemother., 13(2), 249-254 (1978)* En;en.

936-U2 Impregnated disk method for antifungal antibiotic testing. Boyer,J.M. (Dep. Microbiol., Philadelphia Coll. Osteopath. Med., Philadelphia, PA 19131, USA) *Antimicrob. Agents Chemother., 9(6), 1070-1071 (1976)* En;en.

937-U2 Antimicrobial susceptibility testing of yeasts: a turbidimetric technique independent of inoculum size. Galgiani,J.N.; Stevens,D.A. (Dep. Med., Santa Clara Valley Med. Cent., San Jose, CA 95128, USA) *Antimicrob. Agents Chemother., 10(4), 721-726 (1976)* En;en.

938-U2 Possible in vitro test for screening drugs for activity against *Babesia* and other blood protozoa. Irvin,A.D.; Young,E.R. (ARC Inst. Res. Anim. Dis., Compton, Newbury, Berks., UK) *Nature, 269(5627), 407-409 (1977)* En.

939-U2 Drug sensitivity of *Plasmodium falciparum*. An in-vitro microtechnique. Rieckmann,K.H.; Sax,L.J.; Campbell,G.H.; Mrema,J.E. (Malana Res. Program, 282 Castetter Hall, Univ. New Mexico, Albuquerque, NM 87131, USA) *Lancet, 1(8054), 22-23 (1978)* En;en.

940-U2 Quantitative assessment of antimalarial activity in vitro by a semiautomated microdilution technique. Desjardins,R.E.; Canfield,C.J.; Haynes,J.D.; Chulay,J.D. (Wellcome Res. Lab., Research Triangle Park, NC 27709, USA) *Antimicrob. Agents Chemother., 16(6), 710-718 (1979)* En;en.

941-U2 Biological assay of potential trichmonoacides in vitro using a counter apparatus. Brotherton,J. (Abt. Gynakol. Endokrinol. Sterilitat und Familien-Planung, Klin. Steglitz, Freien Univ. Berlin, Hindenburgdamm 30, 1000 Berlin 45, GFR) *Arzneim. Forsch., 28(2 part 10), 1665-1669 (1978)* En;de,en.

942-U2 A rapid HLA-D matching method using PHA blasts as responding cells (preliminary data on PHA blasts HLA-D typing). Mawas,C.; Charmot,D.; Comoy,A.; Sasportes,M. (Res. Unit on Immunogenet. of Hum. Transplant., Inserum U93, Hop.-St.-Louis, Paris, France) *Tissue Antigens, 8(2), 121-129 (1976)* En;en.

See: 54, 65, 125, 251, 280, 283, 845, 2289, 2291

Centrifugation and filtration

943-U2 A low-speed centrifugation technique for the preparation of grids for direct virus examination by electron microscopy. Narang,H.K.; Codd,A.A. (Publ. Health Lab., Inst. Pathol., Gen. Hosp., Westgate Road, Newcastle upon Tyne, NE4 6BE, UK) *J. Clin. Pathol., 32(3), 304-305 (1979)* En.

944-U2 Rapid screening for plasmid DNA. Hughes,C.; Meynell,G.G. (Biol. Lab., Univ. Kent, Canterbury, Kent CT2 7NJ, UK) *Mol. Gen. Genet., 151(2), 175-179 (1977)* En;en.

945-U2 A rapid and inexpensive procedure for the purification of adenovirions. Lentfer,D.; Conde,C. (Virol. Abt. des Inst. Hyg. und Mikrobiol., Univ. des Saarlandes, Homburg (Saar), GFR) *Arch. Virol., 56(1-2), 189-193 (1978)* En;en.

946-U2 Purification of some legume carlaviruses. Veerisetty,V.; *Brakke,M.K. (Dep. Plant Pathol., Univ. Missouri, Columbia, MO 65201, USA) *Phytopathology, 68(1), 59-64 (1978)* En;en.

947-U2 A rapid method for the quantitative study of RNA from canine distemper virus infected cells. Martin,S.J.; ter Meulen,V. (Dep. Biochem., Med. Biol. Cent., Queen's Univ., Belfast, UK) *J. Gen. Virol., 32(2), 321-325 (1976)* En;en.

948-U2 High-resolution flow-zonal centrifuge system. McAleer,W.J.; Hurni,W.; Wasmuth,E.; Hilleman,M.R. (Div. Virus and Cell Biol. Res., Merck Inst. Ther. Res., West Point, PA 19486, USA) *Biotechnol. Bioeng., 21(2), 317-321 (1979)* En;en.

949-U2 Purification and properties of launaea mosaic virus. Naqvi,Q.A.; Mahmood,K. (Dep. Bot., Aligarh Muslim Univ., Aligarh 202 001, India) *Indian J. Exp. Biol., 15(3), 243-244 (1977)* En;en.

950-U2 **Rapid purification of lymphocytic choriomeningitis virus by density gradient centrifugation in colloidal silica.** Vahro,M.; Slenczka,W. (Heinrich-Pette-Inst. Exp. Virol. und Immunol., Univ. Hamburg, Martinistr. 52, 2000 Hamburg 20, GFR) *J. Gen. Virol., 42(1), 215-218 (1978)* En;en.

951-U2 **Purification of polyhedra from insect extracts contaminated with a small isometric virus.** Scotti,P.D.; Longworth,J.F. (Entomol. Div., Dep. Sci. Indust. Res., Private Bag, Auckland, New Zealand) *J. Invertebr. Pathol., 32(2), 216-218 (1978)* En.

952-U2 **An improved procedure for purifying RNA tumor viruses from malignant tissue.** Gulati,S.C.; Scholes,J.; Weiss,M.; *Baldi,A.; Zeelon,E.; Spiegelman,S. (Inst. Biol. y Med. Exp., Obligado 2490, Capital Federal, Argentina) *Acta Physiol. Lat. Am., 27(4), 183-187 (1977)* En;en.

953-U2 **Simple, effective method for purifying the AS-1 cyanophage.** Barkley,M.B.; *Desjardins,P.R. (Dep. Plant Pathol., Univ. California, Riverside, CA 92521, USA) *Appl. Environ. Microbiol., 33(4), 971-974 (1977)* En;en.

954-U2 **A rapid procedure for the isolation of intact ^{32}P-labeled RNA from *E. coli* bacteriophage R17.** Muller,K.; Noll,H. (Dep. Biochem. and Mol. Biol., Northwestern Univ., Evanston, IL 60201, USA) *Anal. Biochem., 70(1), 274-278 (1976)* En.

955-U2 **A simple method for the purification of poplar mosaic virus and onion yellow dwarf virus by low centrifugal forces.** Ploaie,P.G. (Inst. Plant Prot., Acad. Agric. and Silvicult. Sci., 8, Bd. Ion Ionescu de la Brad, 71592 Bucharest, Romania) *Rev. Roum. Med., Ser. Virol., 30(2), 117-119 (1979)* En;en,fr.

956-U2 **Standardized method of Sendai virus production for biological assays.** Pastoret,P.-P.; Burtonboy,C.; Lamy,M.E.; Van Dijck,M.C.; Schoenoers,F. (Lab. Virol., Fac. Vet. Med., Univ. Liege, 1070 Bruxelles, Belgium) *Acta Virol., 20(5), 429-431 (1976)* En;en.

957-U2 **Cytoplasmic membrane vesicles of *Escherichia coli*. I. A simple method for preparing the cytoplasmic and outer membranes.** Yamato,I.; Anraku,Y.; Hirosawa,K. (Dep. Microb. Chem., Fac. Pharm. Sci., Univ. Tokyo, Hongo, Bunkyo-ku, Tokyo 113, Japan) *J. Biochem., 77(4), 705-718 (1975)* En;en.

958-U2 **Rapid isolation of *Legionella pneumophila* from seeded donor blood.** Dorn,G.I.; Barnes,W.R. (Wadley Inst. Mol. Med., Dallas, TX 75235, USA) *J. Clin. Microbiol., 10(1), 114-115 (1979)* En;en.

959-U2 **Blood culture technique based on centrifugation: clinical evaluation.** Dorn,G.L.; Burson,G.G.; Haynes,J.R. (Dep. Microbiol., Granville C. Morton Cancer and Res. Hosp., Dallas, TX 75235, USA) *J. Clin. Microbiol., 3(3), 258-263 (1976)* En;en.

960-U2 **Bacterial aggregating activity in human saliva: simultaneous determination of free and bound cells.** Golub,E.E.; Thaler,M.; Davis,C.; Malamud,D. (Dep. Biochem., Sch. Dent. Med., Univ. Pennsylvania, Philadelphia, PA 19104, USA) *Infect. Immun., 26(3), 1028-1034 (1979)* En;en.

961-U2 **Results with a centrifugation technique for the NBT test in 104 cases of infection in children.** Braito,A.; Ciravegna,B.; Ferrea,G.; Navone,C.; Schiavoni,S. (Univ. Studi Genova, I. Clin. Mal. Infettive, Genova, Italy) *Minerva Pediatr., 30(20), 1601-1610 (1978)* It;en,it.

962-U2 **Continuous-flow cell cycle fractionation of eukaryotic micro-organisms.** Lloyd,D.; John,L.; Hamill,M.; Phillips,C.; Kader,J.; Edwards,S.W. (Dep. Microbiol., University Coll., Newport Road, Cardiff CF2 1TA, UK) *J. Gen. Microbiol., 99(1), 223-227 (1977)* En.

963-U2 **Bacterial and fungal activities in soil: separation of bacteria and fungi by a rapid fractionated centrifugation technique.** Faegri,A.; Torsvik,V.L.; Goksoyr,J. (Inst. Gen. Microbiol., Univ. Bergen, 5014 Bergen, Norway) *Soil Biol. Biochem., 9(2), 105-112 (1977)* En;en.

964-U2 **A rapid technique for the cleaning and concentration of *Eimeria* oocysts.** Smith,R.R.; *Ruff,M.D. (Dep. Poult. Sci., Univ. Georgia, Livestock-Poultry Build., Athens, GA 30602, USA) *Poult. Sci., 54(6), 2081-2086 (1975)* En;en.

965-U2 **Large-scale isolation of a native deoxyribonucleohistone complex from baker's yeast.** Franco,L.; Lopex-Brana,I. (Dep. Biochem., Fac. Sci., Univ. Complutense, Madrid-3, Spain) *Nucleic Acids Res., 5(10), 3743-3757 (1978)* En;en.

966-U2 **Technique for concentrating nannoplankton from the tertiary rocks of the California Coast and Peninsular Ranges.** Moore,S.W.; Brabb,E.E.; Warren,A.D. (US Geol. Surv., Menlo Park, CA 94025, USA) *J. Res. US Geol. Surv., 5(2), 207-208 (1977)* En;en.

967-U2 **Collection of dinoflagellates and other marine microalgae by centrifugation in density gradients of a modified silica sol.** Price,C.A.; Reardon,E.M.; Guillard,R.R.L. (Waksman Inst. Microbiol., Rutgers Univ., PO Box 759, Piscataway, NJ 08854, USA) *Limnol. Oceanogr., 23(3), 548-553 (1978)* En;en.

968-U2 **A rapid method for reducing transmission of *Haemobartonella* upon syringe-passage of trypanosomes to rodents.** Sogandares-Bernal,F.; Chandler,J. (Dep. Biol., South Methodist Univ., Dallas, TX 75275, USA) *J. Parasitol., 64(3), 547-548 (1978)* En.

969-U2 **A rapid method for concentrating spleen-derived schizonts in *Plasmodium berghei* malaria.** Sterling,C.R. (Dep. Comp. Med., Wayne State Univ., Sch. Med., 540 E. Canfield, Detroit, MI 48201, USA) *J. Parasitol., 64(4), 747-749 (1978)* En.

970-U2 **A simplified technique to concentrate trypanosomes and microfilariae by sedimentation-centrifugation of blood.** Carrie,J.; Timsit,D. (Address not stated) *Med. Trop., 37(5), 589-591 (1977)* Fr;en,fr.

971-U2 *Trypanosoma brucei*: detection of low parasitaemias in mice by a miniature anion-exchanger/centrifugation technique. Lumsden,W.H.R.; Kimber,C.D.; Strange,M. (Dep. Med. Protozool., London Sch. Hyg. and Trop. Med., Keppel St., London WC1E 7HT, UK) *Trans. R. Soc. Trop. Med. Hyg., 17(5), 421-424 (1977)* En;en.

972-U2 A simple method for the isolation of murine peripheral blood lymphocytes. Chi,D.S.; *Harris,N.S. (Shriners Burns Inst., 610 Texas Avenue, Galveston, TX 77550, USA) *J. Immunol. Methods., 19(2-3), 169-172 (1978)* En;en.

973-U2 A simple and rapid method for the isolation of human monocytes. Cochrum,K.; Hanes,D.; Fagan,G.; van Speybroeck,J.; Sturtevant,F.K. (Histocompat. Lab., Dep. Surg. Univ. California, San Francisco, CA 94143, USA) *Transplant. Proc., 10(4), 867-870 (1978)* En.

974-U2 A rapid method for the separation of large and small thymocytes from rats and mice. Salisbury,J.G.; Graham,J.M.; Pasternak,C.A. (Dep. Biochem., St. George's Hosp. Med. Sch., Cranmer Terrace, Tooting, London SW17 0RE, UK) *J. Biochem. Biophys. Methods, 1(6), 341-347 (1979)* En;en.

975-U2 Two small lymphocyte subpopulations in human peripheral blood. I. Purification and surface marker profiles. Hokland,M.; Hokland,P.; Heron,I. (Inst. Med. Microbiol., Bartholin Build., Univ. Aarhus, DK-8000 Aarhus C, Denmark) *Int. Arch. Allergy Appl. Immununol., 57(5), 435-443 (1978)* En;en.

976-U2 A rapid one-step procedure for purification of mononuclear and polymorphonuclear leukocytes from human blood using a modification of the Hypaque-Ficoll technique. Ferrante,A.; Thong,Y.H. (Dep. Paediatr., Univ. Adelaide, Adelaide Child. Hosp., Adelaide 5006, Australia) *J. Immunol. Methods, 24(3-4), 389-393 (1978)* En;en.

977-U2 A simplified technique for obtaining purified peripheral lymphocytes from chickens. Archambault,L.P.; Dunlop,W.R.; Cucchiara,F.P.; Collins,W.M.; Corbett,A.C. (Dep. Anim. Sci., Univ. New Hampshire, Durham, NH 03824, USA) *Poult. Sci., 55(5), 1972-1973 (1976)* En;en.

978-U2 New centrifugation blood culture device. Dorn,G.L.; Smith,K. (Dep. Microbiol., Leland Fikes Res. Inst., Div. Wadley Inst. Mol. Med., Dallas, TX 75235, USA) *J. Clin. Microbiol., 7(1), 52-54 (1978)* En;en.

979-U2 Evaluation of a new system ('Corvac') for separating serum from blood for routine laboratory procedures. Spencer,W.W.; Nelson,G.H.; Konicki,K.A. (Clin. Chem. Lab., St. Elizabeth Med. Cent., Dayton, OH 45408, USA) *Clin. Chem., 22(7), 1012-1016 (1976)* En;en.

980-U2 A rapid and sensitive technique for the detection of Fc receptors on macrophages. Schroit,A.J.; Kedar,E.; Gallily,R. (Lautenberg Cent. Gen. and Tumor Immunol., Hebrew Univ., Hadassah Med. Sch., Jerusalem, Israel) *J. Immunol. Methods, 12(1-2), 163-170 (1976)* En;en.

981-U2 New centrifugation blood culture device. Dorn,G.L.; Smith,K. (Dep. Microbiol., Leland Fikes Res. Inst., Div. Wadley Inst. Mol. Med., Dallas, TX 75235, USA) *J. Clin. Microbiol., 7(1), 52-54 (1978)* En;en.

982-U2 A rapid single centrifugation step method for the separation of erythrocytes, granulocytes and mononuclear cells on continuous density gradients of Percoll. Segal,A.W.; Fortunato,A.; Herd,T. (Dep. Haematol., University Coll. Hosp. Med. Sch., University St., London WC1, UK) *J. Immunol. Methods, 32(3), 209-214 (1980)* En;en.

983-U2 A simple technique for testing the in vitro response of rabbit lymphocytes to PHA and allogeneic cells. Cornu,P.; Gratwohl,A.; Schmid,E.; Speck,B. (Div. Hematol., Kantonsspital Basel, CH-4031 Basel, Switzerland) *Experientia, 35(2), 281-283 (1979)* En;en.

984-U2 A rapid method for the separation of functional lymphoid cell populations of human and animal origin on PVP-silica (Percoll) density gradients. Kurnick,J.T.; Ostberg,L.; Stegagno,M.; Kimura,A.K.; Orn,A.; Sjoberg,O. (Natl. Jewish Hosp. and Res. Cent., 3800 East Colfax Ave., Denver, CO 80206, USA) *Scand. J. Immunol., 10(6), 563-573 (1979)* En;en.

985-U2 Large scale preparation of leukocyte concentrates and further purification of lymphocytes and granulocytes with the IBM 2991 cell processor. Brutel de la Riviere,A.; Verhoef-Karssen,P.R.; de Wit,J.J.F.M.; Loos,J.A.; Prins,H.K. (Cent. Lab., Netherlands Red Cross Blood Transfus. Serv., Amsterdam, Netherlands) *Transfusion, 17(5), 509-512 (1977)* En;en.

986-U2 Isolation of human peripheral blood lymphocytes: modification of a double discontinuous density gradient of Ficoll-Hypaque. Goldrosen,M.H.; Gannon,P.J.; Lutz,M.; Holyoke,E.D. (Tumor Immunol. Group, Dep. Gen. Surg., Roswell Park Mem. Inst., 666 Elm St., Buffalo, NY 14263, USA) *J. Immunol. Methods, 14(1), 15-17 (1977)* En;en.

987-U2 Immunochemical determination of human immunoglobulins with a centrifugal analyzer. Finley,P.R.; Williams,R.J.; Byers,J.M.,III (Dep. Pathol., Arizona Med. Cent., Tucson, AZ 85724, USA) *Clin. Chem., 22(7), 1037-1041 (1976)* En;en.

988-U2 Separation of T and B cells from human peripheral blood by centrifugal elutriation. Griffith,O.M. (Beckman Instruments, Inc., Spinco Div., 1117 California Ave., Palo Alto, CA 94304, USA) *Anal. Biochem., 87(1), 97-107 (1978)* En;en.

989-U2 Ultracentrifugation in virus diagnosis. Liebermann,H. (DDR-2201 Insel Riems, GDR) *Arch. Exp. Veterinarmed., 32(3), 427-434 (1978)* De;de,en,ru.

990-U2 Simple determination of virus molecular weights by sedimentation equilibrium. Rubinstein,R.; Durham,A.C.H. (Virus Res. Unit, Univ. Cape Town Med. Sch. Observatory, Cape Town 7925, S. Africa) *Anal. Biochem., 81(2), 447-449 (1977)* En.

991-U2 Purification of virions and nucleocapsids of herpes simplex virus by means of metrizamide and sodium metrizoate gradients. Blomberg,J.; Bjorck,E.; Olofsson,S.; Berg,G.; *Lycke,E. (Univ. Goteborg, Inst. Med. Microbiol., Dep. Virol., Guldhedsgaten 10B, S-413 146 Goteborg, Sweden) *Arch. Virol., 50(4), 271-278 (1976)* En;en.

992-U2 A rapid method for the isolation of intracytoplasmic membranes from *Rhodopseudomonas sphaeroides* using an air-driven ultracentrifuge. Shepherd,W.D.; *Kaplan,S. (Dep. Microbiol., Univ. Illinois, Urbana, IL 61801, USA) *Anal. Biochem., 91(1), 194-198 (1978)* En;en.

993-U2 Isolation of stable aggregates of IgG by zonal ultracentrifugation in sucrose gradients containing albumin. Knutson,D.W.; Kijlstra,A.; Lentz,H.; van Es,L.A. (Res. and Med. Serv., VA Wadsworth Hosp. Cent., Los Angeles, CA 90073, USA) *Immunol. Commun., 8(3), 337-345 (1979)* En;en.

994-U2 Automatic membrane-filtration system for the 'on-demand' supply of large volumes of sterile medium in continuous culture. Larsen,V.F.; Spivey,M.; Holdom,R.S. (Dep. Appl. Microbiol., Univ. Strathclyde, R. Coll., Glasgow G1 1XW, UK) *Biotechnol. Bioeng., 18(10), 1433-1443 (1976)* En;en.

995-U2 Reevaluation of the filter method for measuring anti-double-stranded DNA antibody. Kimura,M.; Andoh,T. (Dep. Intern. Med., Inst. Med. Sci., Univ. Tokyo, Shirokanedai, Minato-ku, Tokyo 108, Japan) *Jap. J. Exp. Med., 48(1), 77-79 (1978)* En.

996-U2 A general method for the rapid separation and specific detection of antigenic material by immunoelectrofiltration using multispecific antisera. Self,C.H.; van der Riet,F.de St.J. (Dep. Biol., Open Univ., Walton Hall, Milton Keynes, Bucks MK7 6AA, UK) *J. Immunol. Methods., 13(3-4), 235-239 (1976)* En;en.

997-U2 Concentration of viruses from large volumes of tap water using pleated membrane filters. Farrah,S.R.; Gerba,C.P.; Wallis,C.; *Melnick,J.L. (Dep. Virol. and Epidemiol., Baylor Coll. Med., Houston, TX 77025, USA) *Appl. Environ. Microbiol., 31(2), 221-226 (1976)* En;en.

998-U2 Regeneration of pleated filters used to concentrate enteroviruses from large volumes of tap water. Farrah,S.R.; Gerba,C.P.; Goyal,S.M.; Wallis,C.; *Melnick,J.I. (Dep. Virol. and Epidemiol., Baylor Coll. Med., Houston, TX 77030, USA) *Appl. Environ. Microbiol., 33(2), 308-311 (1977)* En;en.

999-U2 Concentration of enteroviruses from estuarine water. Farrah,S.R.; Goyal,S.M.; Gerba,C.P.; Wallis,C.; *Melnick,J.L. (Dep. Virol. and Epidemiol., Baylor Coll. Med., Houston, TX 77030, USA) *Appl. Environ. Microbiol., 33(5), 1192-1196 (1977)* En;en.

1000-U2 A simple method for concentration of enteroviruses and rotaviruses from cell culture harvests using membrane filters. Farrah,S.R.; Goyal,S.M.; Gerba,C.P.; Conklin,R.H.; Wallis,C.; Melnick,J.L.; DuPont,H.L. (Dep. Virol. and Epidemiol., Baylor Coll. Med., Houston, TX 77030, USA) *Intervirology, 9(1), 56-59 (1978)* En;en.

1001-U2 Arenavirus concentration by molecular filtration. Gangemi,J.D.; Connell,E.V.; Mahlandt,B.G.; Eddy,G.A. (United States Army Med. Res. Inst. Infect. Dis., Fort Detrick, Frederick, MD 21701, USA) *Appl. Environ. Microbiol., 34(3), 330-332 (1977)* En;en.

1002-U2 Characterization of DNA damages by filtration through nitrocellulose filters: a simple probe for DNA-modifying agents. Kuhnlein,U.; Tsang,S.S.; Edwards,J. (Environ. Carcinog. Unit, British Columbia Cancer Res. Cent., 601 West 10th Ave., Vancouver, BC, Canada) *Mutat. Res., 64(3), 167-182 (1979)* En;en.

1003-U2 Purification and characterization of mouse serum interferon. Liu,J.-L.; Peng,S.-Y. (Sch. Med. Technol. and Dep. Clin. Pathol., National Taiwan Univ. Hosp., Taiwan, China) *Chin. J. Microbiol., 8(2), 111-119 (1975)* En;ch,en.

1004-U2 Rapid method for the purification of bacteriophage lambda by gel filtration. Ahmad,M.; Srivastava,B.S.; Agarwala,S.C. (Cent. Drug Res. Inst., Lucknow, India) *Indian J. Exp. Biol., 15(5), 387-388 (1977)* En;en.

1005-U2 The use of membrane filtration and marine agar 2216E to enumerate marine heterotrophic bacteria. Patrick,F.M. (Fish. Res. Div., PO Box 19062, Wellington, New Zealand) *Aquaculture, 13(4), 369-372 (1978)* En;en.

1006-U2 Efficiency of the Seinhorst filter for the recovery of *Eimeria tenella* oocysts, from feces. McCallister,G.; Cowgill,L.M. (Div. Biol. Sci., Mesa Coll., Grand Junction, CO 81501, USA) *Proc. Helminthol. Soc. Wash., 44(2), 218-219 (1977)* En.

1007-U2 Preliminary visual screening of soil samples for the presumptive presence of *Histoplasma capsulatum*. Guar,P.K.; Lichtwardt,R.W. (Food Res. Inst., Univ. Wisconsin, Madison, WI 53706, USA) *Mycologia, 72(3), 259-269 (1980)* En;en.

1008-U2 A simple purification method for the preparation of solubilized chlorophyllase from *Chlorella protothecoides*. Shioi,Y.; Tamai,H.; Sasa,T. (Div. Biol., Miyazaki Med. Coll., Kiyotake, Miyazaki 889-16, Japan) *Anal. Biochem., 105(1), 74-79 (1980)* En;en.

1009-U2 Comparison of phytoplankton numbers in the samples collected by the double filtration method and the standard precipitation method. Sukhanova,I.N.; Rat'kova,T.N. (P.P. Shirshova Inst. Oceanogr., Acad. Sci., USSR) *Okeanologiya, 17(4), 691-693 (1977)* Ru;en,ru.

1010-U2 A simple method for separating *Toxoplasma gondii* from infected mouse peritoneal exudate cells. Masihi,K.N. (Robert Koch-Inst., Abt. Parasitol., D-1000 Berlin 65, Nordufer 20, GFR) *Zentralbl. Bakteriol. Parasitenkd. Infektionskr. Hyg., I Abt. A, 233(4), 556-560 (1975)* En;en,de.

1011-U2 **Lymphocyte purification from normal human peripheral blood by filtration through columns of cotton wool.** Barr,R.D.; Perry,S. (Haematol. Unit, Dep. Pathol., Univ. Aberdeen, Foresterhill, Aberdeen, AB9 2ZD, UK) *Scand. J. Haematol., 17(4), 300-304 (1976)* En;en.

1012-U2 **A rapid technique for the isolation of human IgD myeloma proteins employing Ultragel AcA34.** Jefferis,R. (Dep. Exp. Pathol., Univ. Birmingham Med. Sch., Birmingham B15 2TJ, UK) *J. Immunol. Methods, 9(3/4), 231-234 (1976)* En;en.

1013-U2 **A micro-procedure for quantitative precipitin tests.** Malinowski,K.; Manski,W. (Dep. Microbiol. and Ophthalmol., Coll. Phys. and Surg., Columbia Univ., New York, NY 10032, USA) *J. Immunol. Methods, 21(1-2), 167-177 (1978)* En;en.

1014-U2 **Concentration of bacteria in water using the ultrafiltration method.** Trinel,P.A.; Leclerc,H. (INSERM, U146, Ecotoxicol. Microb., Inst. Pasteur, CERTIA, 59650 Villeneuve d'Ascc, France) *Ann. Microbiol., 127B(2), 201-212 (1976)* Fr;en,fr.

1015-U2 **[An ultrafiltration method of purifying mould lipase].** Grigor'eva,T.A.; Rozhanskaya,T.I.; Selezneva,A.A.; Dubyaga,V.P. (All Union Res. Inst. Antibiot. and Medically Useful Enzymes, Leningrad, USSR) *Khim. Farm. Zh., 11(5), 115-117 (1977)* Ru.

1016-U2 **Human transfer factor prepared by dialysis, utlrafiltration and gel chromatography: biological activity in local transfer of skin sensitivity.** Lawton,J.W.M.; Darg,C.; Pepper,D.; Kay,A.B. (Dep. Pathol., Univ. Hong Kong, Hong Kong) *J. Immunol. Methods, 16(2), 119-129 (1977)* En;en.

1017-U2 **Tryptose-sulfite-cycloserine-egg yolk agar membrane filter method for the detection of *Clostridium perfringens* in water.** Hirn,J.; Raevuori,M. (Coll. Vet. Med., Hameentie 57, 00550 Helsinki 55, Finland) *Water. Res., 12(7), 451-453 (1978)* En;en.

1018-U2 **Two-temperature membrane filter method for enumerating fecal coliform bacteria from chlorinated effluents.** Green,B.L.; Clausen,E.M.; *Litsky,W. (Dep.Environ. Sci., Univ. Massachusetts, Amherst, MA 01003, USA) *Appl. Environ. Microbiol., 33(6), 1259-1264 (1977)* En;en.

1019-U2 **A comparison of membrane filters for counting *Thermoactinomyces* endospores in spore suspensions and river water.** Al-Diwany,L.J.; Unsworth,B.A ; Cross,T. (Postgrad. Sch. Stud. in Biol. Sci., Univ. Bradford, Bradford BD7 1DP, UK) *J. Appl. Bacteriol., 45(2), 249-258 (1978)* En;en.

1020-U2 **Rapid membrane filtration-epifluorescent microscopy technique for direct enumeration of bacteria in raw milk.** Pettipher,G.L.; Mansell,R.; McKinnon,C.H.; Cousins,C.M. (Natl. Inst. Res. Dairying, Shinfield, Reading RG2 9AT, UK) *Appl. Environ. Microbiol., 29(2), 423-429 (1980)* En;en.

1021-U2 **The use of gradocol membrane filter in rapid diagnostic bacteriology. A preliminary communication.** Biswas,S.K. (Dep. Pathol. and Bacteriol., BS Med. Coll., Bankura, West Bengal, India) *Indian J. Med. Sci., 31(4), 68-71 (1977)* En;en.

1022-U2 **Method for determining bioburden of surgical gloves.** Lammerding,A.M.; *Day,D.F. (Dep. Microbiol., Univ. Guelph, Guelph, Ont. N1G 2W1, Canada) *J. Clin. Microbiol., 7(5), 497-498 (1978)* En;en.

1023-U2 **A comment on the membrane filter technique for estimation of length of fungal hyphae in soil.** Sundman,V.; Sivela,S. (Univ. Helsinki, Dep. Gen. Microbiol., Malminkatu 20, SF-00100 Helsinki 10, Finland) *Soil Biol. Biochem., 10(5), 399-401 (1978)* En;en.

1024-U2 **Preparation and purification of merozoites of *Eimeria tenella*.** Stotish,R.L.; Wang,C.C. (Merck Inst. Ther. Res., Rahway, NJ 07065, USA) *J. Parasitol., 61(4), 700-703 (1975)* En;en.

1025-U2 **Single step separation of human T and B cells using AET treated SRBC rosettes.** Saxon,A.; Feldhaus,J.; Robins,R.A. (Immunobiol. Group, Dep. Microbiol. and Immunol., UCLA Sch. Med., Los Angeles, CA 90024, USA) *J. Immunol. Methods, 12(3,4), 285-288 (1976)* En;en.

See: 1060, 1130, 1187, 1343, 1432, 2262, 3268, 3273

Staining techniques

1026-U2 **Preparation of cytochrome c peroxidase from baker's yeast.** Nelson,C.E.; Sitzman,E.V.; Kang,C.H.; *Margoliash,E. (Dep. Biochem. and Mol. Biol., Northwestern Univ., Evanston, IL 60201, USA) *Anal. Biochem., 83(2), 622-631 (1977)* En;en.

1027-U2 **Simple methods for early detection of grapevine leafroll disease.** von der Brelie,D.; Nienhaus,F. (Inst. Pflanzenkr., Univ. Bonn, Nussallee 9, D-5300 Bonn 1, GFR) *Z. Pflanzenkr. Pflanzenschutz., 85(12), 745-747 (1978)* En;de,en.

1028-U2 **Histochemistry of hepatitis B antigen (HB_s-Ag) in routine diagnostic procedures; practical use and interpretation of findings of orcein staining.** von Seebach,H.B. (Pathol. Inst., Stadtischen Krankenhauses, D-6680 Neunkirchen, GFR) *Leber Magendarm, 8(2), 78-82 (1978)* De;de,en.

1029-U2 **Histochemical demonstration of HBsAg in paraffin sections of biopsy material using different staining methods: comparative evaluation.** Furlan,C.; De Bac,F. (Ist. Malattie Infett., Univ. Roma, Roma, Italy) *Boll. Soc. Ital. Biol. Sper., 54(3), 258-263 (1978)* It.

1030-U2 **A trichome stain for the intrahepatic localization of the hepatitis B surface antigen (HBsAg).** Gubetta,L.; Rizzetto,M.; Crivelli,O.; Verme,G.; Arico,S. (Dep. Pathol., Osp. Mauriziano Umberto I, Cs. Turati 46, 10128 Turin, Italy) *Histopathology, 1(4), 277-288 (1977)* En;en.

1031-U2 A spectrophotometric method of selecting orcein dyes for histological demonstration of hepatitis B surface antigen. Iwamoto,T.; Yamashita,K.; Iijima,S. (Dep. Clin. Pathol., Hiroshima Natl. Railway Hosp., Futabanosato, Hiroshima 730, Japan) *Acta Histochem. Cytochem., 12(1), 63-68 (1979)* En;en.

1032-U2 Methods for detection of hepatitis B surface antigen in paraffin sections of liver: a guideline for their use. Sumithran,E. (Dep. Pathol., Med. Fac., Univ. Malaya, Kuala Lumpur, Malaysia) *J. Clin. Pathol., 30(5), 460-463 (1977)* En;en.

1033-U2 [Identification of HBsAg in hepatic tissue through orcein and Gormori's aldehyde fuchsin techniques.] Sarno,E.N.; de Carvalho,F.G.; Alvariz,F.G.; Soares,M.C.G.; Fonseca,F.de B. (Fac. Cienc. Med., Univ. Estado do Rio de Janeiro, Rio de Janeiro, Brazil) *Rev. Inst. Med. Trop. Sao Paulo, 20(1), 1-5 (1978)* Pt;en,pt.

1034-U2 Selective staining of hepatitis B surface antigen in thick epoxy sections of liver. Shamoto,M.; Nishio,H.; Katoh,Y.; Senba,H.; Ito,M. (Dep. Pathol., Fujita-Gakuen Univ. Sch. Med., Toyoake, Japan) *Stain Technol., 52(5), 285-289 (1977)* En;en.

1035-U2 Morphological components of herpesvirus. II. Preservation of virus during negative staining procedures. Vernon,S.K.; Lawrence,W.C.; Cohen,G.H.; Durso,M.; Rubin,B.A. (Dep. Biol. Product Dev., Wyeth Lab., Inc., Philadelphia, PA 19101, USA) *J. Gen. Virol., 31(2), 183-191 (1976)* En;en.

1036-U2 Rapid identification of virus infections. Coleman,D.V. (Dep. Exp. Pathol., St. Mary's Hosp. Med. Sch., London W2, UK) *Br. Med. J., 2(6040), 875-876 (1976)* En.

1037-U2 Sensitivity and specificity of diagnostic tests for genital infection with *Herpesvirus hominis*. Brown,S.T.; Jaffe,H.W.; Zaidi,A.; Filker,R.; Herrmann,K.L.; Lylerla,H.C.; Jove,D.F.; Budell,J.W. (Technical Information Serv., Cent. Dis. Control. Atlanta, GA 30333, USA) *Sex. Transmitted Dis., 6(1), 10-13 (1979)* En;en.

1038-U2 Laboratory diagnosis of cutaneous herpes simplex and varicella-zoster infections. Kurtis,B. (Dep. Dermatol., Stanford Univ. Med. Cent., Stanford, CA 94305, USA) *Trans. St. John's Hosp. Dermatol. Soc., 61(1-2), 35-43 (1975)* En.

1039-U2 A staining technique for smears containing inclusion body viruses. Palmieri,J.R.; Kelderman,W.; Sullivan,J.T. (Inst. Med. Res., Jalan Pahang, Kuala Lumpur 02-14, Malaysia) *J. Invertebr. Pathol., 31(2), 264 (1978)* En.

1040-U2 Application of the Papanicolaou stain to routine histological examination of placenta. Difruscio,D.; deSa,D.J. (Dep. Pathol., MUMC, 1200 Main St. West, Hamilton, Ont. L8S 4J9, Canada) *J. Clin. Pathol., 30(7), 682-683 (1977)* En.

1041-U2 A rapid, simple method for staining bacterial flagella. Mayfield,C.I.; Inniss,W.E. (Dep. Biol., Univ. Waterloo, Waterloo, Ont. N2L 3G1, Canada) *Can. J. Microbiol., 23(9), 1311-1313 (1977)* En;en.

1042-U2 Simplified silver-plating stain for flagella. West,M. *Burdash,N.M.; Freimuth,F. (Div. Clin. Microbiol., Dep. Lab. Med., Univ. South Carolina, Charleston, SC 29403, USA) *J. Clin. Microbiol., 6(4), 414-419 (1977)* En;en.

1043-U2 Description of a method of fixation for bacteria of particular use in the preparation of ultrathin sections of *Bdellovibrio*. Faoro,F. (Gruppo Virus e Virosi delle Piante, Unita di Milano, c/o Ist. Patol. Veg., Univ. Milano, Milano, Italy) *Ann. Microbiol. Enzimol., 25(I-III), 101-110 (1975)* It;en,it.

1044-U2 Sensitivity of immunoperoxidase and immunofluorescence staining for detecting chlamydia in conjunctival scrapings and in cell culture. Woodland,R.M.; El-Sheikh,H.; Darougar,S.; Squires,S. (Virus Lab., Dep. Clin. Ophthalmol., Inst. Ophthalmol., Judd St., London, UK) *J. Clin. Pathol., 31(11), 1073-1077 (1978)* En;en.

1045-U2 Optical demonstration of chlamydiae in cell cultures by means of fluorochrome 33258 H. Rolly,H.; Schachner,B. (Virol. Lab., Hoechst AG, Postfach 80 03 20, D-6230 Frankfurt am Main 80, GFR) *Infection, 7(3), 150-153 (1979)* En;de,en.

1046-U2 Early detection of *Chlamydia trachomatis* using fluorescent, DNA binding dyes. Salari,S.H.; *Ward,M.E. (Univ. Microbiol., South Lab. and Pathol. Block, Southampton Gen. Hosp., Tremona Road, Southampton SO9 4XY, UK) *J. Clin. Pathol., 32(11), 1155-1162 (1979)* En;en.

1047-U2 Easy visualization of *Legionella pneumophila* by 'half-a-Gram' stain procedure. de Freitas,J.L.; Borst,J.; Meenhorst,P.L. (Rijksinst. Volksgezondheid, Bilthoven, Netherlands) *Lancet, 1(8110), 270-271 (1979)* En.

1048-U2 Demonstration of *Legionella pneumophila* by 'half-a-Gram' stain. Boyd,J.F. (Brownice Lab., Ruchill Hosp. and Pathol. Dep., Univ. Glasgow, Western Infirm., Glasgow G20 9NB, UK) *Lancet, 1(8114), 496 (1979)* En.

1049-U2 A simple method of activity staining for phosphate-releasing enzymes in electrophoresis gels. Schafer,H.-J.; Schuerich,P.; Rathgeber,G. (Inst. Biochem., Univ. Mainz, Joh.-Joachim-Bercher-Weg 30, D-6500 Mainz, GFR) *Hoppe-Seyler's Z. Physiol. Chem., 359(10), 1441-1442 (1978)* En;en.

1050-U2 Problems in detection of mycoplasma contamination in cell cultures. Bonissol,C.; Gilbert,M.; Ivanova,L.M. (Unite Bacteriol. Syst., Inst. Pasteur, 75724 Paris Cedex 15, France) *Ann. Microbiol., 129B(2), 245-265 (1978)* Fr;en,fr.

1051-U2 Evaluation of a new technique for the demonstration of gonococci and other micro-organisms in host cells. Sowter,C.; McGee,Z.A. (Anim. Histopathol., Dep. Pathol., Northwick Park Hosp., Watford Road, Harrow HA1 3UJ, UK) *J. Clin. Pathol., 29(5), 433-437 (1976)* En;en.

1052-U2 The evaluation of xylose lysine brilliant green agar medium in the isolation of *Salmonella*. Falcao,D.P.; Suassuna,I.; Suassuna,I.R. (Fac. Cienc. Farm., UNESP, Rua Expedicionarios do Brasil 1621, Campus de Araraquara, 14800 Araraquara, SP, Brazil) *Rev. Saude Publica, Sao Paulo, 13(1), 43-46 (1979)* Pt;en,pt.

1053-U2 A simple improved method to stain *Thiobacillus*. Rodriguez-Leiva,M.; Pichuantes,S. (Catholic Univ. Chile, Inst. Biol. Sci., Lab. Microbiol. and Immunol., Santiago, Chile) *Can. J. Microbiol., 24(6), 756-757 (1978)* En;en,fr.

1054-U2 [Review of laboratory methods used in diagnosis of swine dysentery]. Mazurczak,J.; Teriecki,M.; Klawe,W. (Ul. Nowoursynowska 166, 02-766 Warszawa, Poland) *Med. Weter., 31(5), 283-285 (1975)* Pl.
[Staining; electron microscopy; immunofluorescence; culture]

1055-U2 Novel quality assurance procedure for the Gram strain. Kohn,F.S.; Henneman,S.A. (Schering Corp., PO Box 1288, Madison, WI 53701, USA) *J. Am. Med. Technol., 39(1), 20-21 (1977)* En;en.

1056-U2 A simplified Leifson flagella stain. Clark,W.A. (Cent. Dis. Control, Atlanta, GA 30333, USA) *J. Clin. Microbiol., 3(6), 632-634 (1976)* En;en.

1057-U2 Automatic gram-staining with the Microstainer II. Burdash,N.M.; West,M.E.; Bannister,E.R. (Dep. Lab. Med., Div. Clin. Microbiol., Med. Univ. South Carolina, Charleston, SC 29401, USA) *Health Lab. Sci., 13(1), 20-22 (1976)* En;en.

1058-U2 Automatic Gram staining by a linear conveyor system. Heimer,G.V.; McSwiggan,D.A. (Public Health Lab., Dep. Microbiol., Cent. Middlesex Hosp., London NW10 7NS, UK) *J. Clin. Pathol., 32(12), 1299-1303 (1979)* En.

1059-U2 Staining clinical specimens for acid-fast bacilli by means of a mechanical conveyor system. Heimer,G.V.; Joseph,N.; Taylor,C.E.D. (Public Health Lab. and Dep. Microbiol., Central Middlesex Hosp., Park Royal, London NW10 7NS, UK) *J. Clin. Pathol., 31(2), 185-188 (1978)* En;en.

1060-U2 Direct counts of bacteria by a modified acridine orange method in relation to their heterotrophic activity. Ramsay,A.J. (11A Ashfield Place, Ilam, Christchurch 4, New Zealand) *N. Z. J. Mar. Freshwater Res., 12(3), 265-269 (1978)* En;en.

1061-U2 Identification of gram-negative bacteria in histological sections using Sandiford's counterstain. Leaver,R.E.; Evans,B.J.; Corrin,B. (Dep. Morbid Anat., St. Thomas's Hosp. Med. Sch., London SE1 7EH, UK) *J. Clin. Pathol., 30(3), 290-291 (1977)* En.

1062-U2 [Simple method for evaluation of the bacteriological quality of milk]. Desfleurs,M.; Desfleurs,M. (Inst. Lait, Viandes et Nutr., Univ. Caen, Caen, France) *Tech. Lait., No. 893, 11-13 (1977)* Fr.

1063-U2 Identification of the presence and type of biliary microflora by immediate Gram stains [presented at the Tripartite Meeting, held in Philadelphia, Pennsylvania, USA, 1976]. Keighley,M.R.B.; McLeish,A.R.; Bishop,H.M.; Burdon,D.W.; Quoraishi,A.H.; Oates,G.D.; Dorricott,N.J.; Alexander-Williams,J. (Gen. Hosp., Birmingham B4 6NH, UK) *Surgery, 81(4), 469-472 (1977)* En;en.

1064-U2 Rapid differentiation of meningitis by NBT test. Praharaj,K.C.; Nathsharma,K.C.; Raghavenda Rao,E. (Dep. Pediatr., MKCG Med. Coll., Berhampur (Orissa), India) *Indian Pediatr., 13(9), 697-700 (1976)* En;en.

1065-U2 Application of the Papanicolaou stain to routine histological examination of placenta. Difruscio,D.; deSa,D.J. (Dep. Pathol., MUMC, 1200 Main St. West, Hamilton, Ont. L8S 4J9, Canada) *J. Clin. Pathol., 30(7), 682-683 (1977)* En.

1066-U2 A fluorescence staining method for the examination of microorganisms on natural substrates. Mayfied,C.I. (Dep. Biol., Univ. Waterloo, Waterloo, Ont N2L, 3G1, Canada) *Microsc. Acta, 78(4), 295-299 (1976)* En;de,en.

1067-U2 Slide staining device for use during space flight. Brockett,R.M.; Brady,J.; Day,J.L.; Ferguson,J.K.; Walsh,J.M. (L.B. Johnson Space Cent., NASA, Houston, TX 77058, USA) *Appl. Environ. Microbiol., 33(1), 203-205 (1977)* En;en.

1068-U2 Rapid methenamine silver strain for *Pneumocystis* and fungi. Mahan,C.T.; Sale,G.E. (Div. Oncol., Fred. Hutchinson Cancer Res. Cent., 1124 Columbia St., Seattle, WA 98104, USA) *Arch. Pathol. Lab. Med., 102(7), 351-352 (1978)* En;en.

1069-U2 A staining technique for nuclei of *Rhizoctonia solani* and related fungi. Burpee,L.L.; Sanders,P.L.; Cole,H.,Jr.; Kim,S.H. (Pest Res. Lab. and Dep. Plant Pathol., Pennsylvania State Univ., University Park, PA 16802, USA) *Mycologia, 70(6), 1281-1283 (1978)* En.

1070-U2 Applications of flow cytometry in brewing-biological experiments. Part III: Viability test. Hutter,K.-J. (Fraunhofer-Gesellsch., Inst. Aerobiol., D-5948 Schmallenberg-Grafschaft, GFR) *Brauwissenschaft, 32(1), 13-16 (1979)* De;de,en,fr.

1071-U2 A new fluorescent viability test for fungi cells. Calich,V.L.G.; Purchio,A.; Paula,C.R. (Dep. Microbiol. e Imunol., Inst. Cienc.-Biomed., Univ. Sao Paulo, Sao Paulo, Brazil) *Mycopathologia, 66(3), 175-177 (1979)* En;en.

1072-U2 A simple method for staining semi-thick tissue sections to demonstrate fungal elements. Kuttin,E.S.; Muller,J. (Israel Ref. Lab. Med. Mycol., Israel Inst. Biol. Res., Ness-Ziona, Israel) *Mycopathologia, 68(3), 191-192 (1979)* En;en.

1073-U2 A technique for the fluorescence staining of fungal nuclei. Mogford,D.J. (Dep. Plant Sci., Rhodes Univ., Grahamstown, South Africa) *J.S. Afr. Bot., 45(3), 263-265 (1979)* En;af,en.

1074-U2 Safranin O as a rapid nuclear stain for fungi. Bandoni,R.J. (Dep. Bot., Univ. British Columbia, Vancouver, BC V6T 1W5, Canada) *Mycologia, 71(4), 873-874 (1979)* En.

1075-U2 Rapid haematoxylin staining of lichen nuclei. Renner,B. (Fachber. Biol. Bot., Univ. Marburg, Lahnberge, D-3550 Marburg/Lahn, GFR) *Lichenologist, 9(2), 143-145 (1977)* En;en.

1076-U2 A method for fixing and staining peritrich ciliates. Hazen,T.C.; Smith,G.; Dimock,R.V.,Jr. (Dep. Biol., Wake Forest Univ., Winston-Salem, NC 27109, USA) *Microsc. Acta, 81(1), 15-16 (1978)* En;de,en.

1077-U2 The wet Giemsa method for quick testing of variants in blood and malaria strains. Lillie,R.D.; Donaldson,P.T. (Dep. Pathol., Louisiana State Univ. Med. Cent., New Orleans, LA 70112, USA) *Stain Technol., 54(1), 47-48 (1979)* En.

1078-U2 Modified Grocott's methenamine silver nitrate method for quick staining of *Pneumocystis carinii*. Pintozzi,R.L. (Dep. Pathol., Univ. Illinois Med. Cent., 1853 West Polk St., Chicago, IL 60612, USA) *J. Clin. Pathol., 31(8), 803-805 (1978)* En.

1079-U2 The protargol staining technique: an improved version for *Tetrahymena pyriformis*. Ng,S.F.; *Nelsen,E.M. (Dep. Zool., Univ. Iowa, Iowa City, IA 52242, USA) *Trans. Am. Microsc. Soc., 96(3), 369-376 (1977)* En;en.

1080-U2 Rapid quantitative assessment of *Theileria* infection in ticks. Walker,A.R.; McKellar,S.B.; Bell,L.J.; Brown,C.G.D. (Cent. Trop. Vet. Med., Roslin, UK) *Trop. Anim. Health Prod., 11(1), 21-26 (1979)* En;en,es,fr.

1081-U2 The nitroblue tetrazolium (NBT) reduction test in Wegener's granulomatosis. Niinaka,T.; Watanabe,Y.; Okochi,T.; Hada,T.; Takahashi,Y.; Yamammura,Y. (III Dep. Intern. Med., Osaka University Hosp., Fukushima, Osaka, Japan) *J. Med., 9(2), 99-107 (1978)* En;en.

1082-U2 Rapid identification of monocytes in a mixed mononuclear cell preparation. Tucker,S.B.; Pierre,R.V.; Jordon,R.E. (Dep. Dermatol., Mayo (rad. Sch., Med. Sch. and Found., Rochester, MN 55901, USA) *J. Immunol. Methods, 14(3-4), 267-269 (1977)* En;en.

1083-U2 A rapid, inexpensive and easily quantified assay for phagocytosis and microbicidal activity of macrophages and neutrophils. Simpson,D.W.; Roth,R.; *Loose,L.D. (Inst. Comp. and Human Toxicol., Albany Med. Coll., Union Univ., 47 New Scotland Ave., Albany, NY 12208, USA) *J. Immunol. Methods, 29(3), 221-226 (1979)* En;en.

See: 77, 451, 961, 1204, 1992, 1999, 2078, 55, 3295

Microscopy

1084-U2 Devices which facilitate preparation of biological samples for electron microscopy. Monosov,E.Z. (Address not stated) *Mikrobiologiya, 47(2), 362-366 (1978)* Ru;en,ru.

1085-U2 Electron microscopy in diagnosis of eye, skin and genital lesions. Blaskovic,P.J. (Virus Lab., Ontario Minist. Health, Toronto, Ont., Canada) *J., Can. Med. Assoc., 119(1), 12 (1978)* En.

1086-U2 Electron microscopic analysis of transcription: mapping of initiation sites and direction of transcription. Brack,C (Basel Inst. Immunol., Grenzacherstr. 487, CH-4005 Basel, Switzerland) *Proc. Natl. Acad. Sci. USA, 76(7), 3164-3168 (1979)* En;en.

1087-U2 Determination of the mass of viruses by quantitative electron microscopy. Bahr,G.F.; Engler,W.F.; Mazzone,H.M. (Dep. Cell. Pathol., Armed Forces Inst. Pathol., Washington, DC, USA) *Q. Rev. Biophys., 9(4), 459-489 (1976)* En;en.

1088-U2 Morphological virus diagnosis — use of negative contrast technique for electron-microscopic testing of animal viruses. Solisch,P. (DDR-2201 Insel Riems, GDR) *Arch. Exp. Veterinarmed., 32(4), 465-488 (1978)* De;de,en,ru.

1089-U2 The possibilities of the use of the electron microscope in helping the diagnosis of plant virus diseases. /[presented at the Symposium on Virus Diseases of Tropical Crops, held at the Tropical Agriculture Research Center, Ibaraki-ken 300-21, Japan, Sept 1976]. Kitajima,E.W. (Dep. Biol. Cel., Inst. Cienc. Biol., Univ. Brasilia, 70000 Brasilia, DF, Brazil) *Trop. Agric. Res. Ser., 10, 197-199 (1977)* En.

1090-U2 Electron microscopy for rapid identification of animal viruses in hematoxylin-eosin sections. Bhatnagar,R.; Johnson,G.R.; Christian,R.G. (Alberta Dep. Agric., Vet. Serv. Div., PO Box 8070, Edmonton, Alberta T6H 4P2, Canada) *Can. J. Comp. Med., 41(4), 416-419 (1977)* En;en,fr.

1091-U2 Quantitative electron microscopy of early adenovirus RNA. Westphal,H.; Lai,S.-P. (Lab. Mol. Genet., Natl. Inst. Child Health and Human Dev., NIH, Bethesda, MD 20014, USA) *J. Mol. Biol., 116(3), 525-548 (1977)* En;en.

1092-U2 Preprocessing of electron micrographs of nucleic acid molecules for automatic analysis by computer. Lipkin,L.; Lemkin,P.; Shapiro,B.; Sklansky,J. (Image Processing Unit, Div. Cancer Biol. and Diagn., Natl. Cancer Inst., NIH, Bethesda, MD 20014, USA) *Comput. Biomed. Res., 12(3), 279-289 (1979)* En;en.

1093-U2 Rapid diagnosis of some avian virus diseases. Van Kammen,A.; Spradbrow,P.B. (Vet. Lab., Dep. Primary Industry, PO Box 6372, Boroko, Papua New Guinea) *Avian Dis., 20(4), 748-751 (1976)* En;en.

1094-U2 Rapid diagnosis of barley yellow dwarf virus in plants using serologically specific electron microscopy. Paliwal,Y.C. (Chem. and Biol. Res. Inst., Res. Branch, Agric. Canada, Ottawa, Ont. K1A 0C6, Canada) *Phytopathol. Z., 89(1), 25-36 (1977)* En;de,en.

1095-U2 [Cytomegalic inclusion disease of the newborn. Diagnostic value of electron microscopic study of liver biopsies]. Mallet,E.; Henocq,A.; de Menibus,C.-H.; Buffet-Janvresse,C.; Magard,H. (Serv. Pediatr., CHU, Hop. Charles Nicolle, F 76038 Rouen Cedex, France) *Nouv. Presse Med., 7(12), 1029-1030 (1978)* Fr.

1096-U2 Rapid diagnosis of cytomegalovirus infection in infants by electron microscopy. Lee,F.K.; *Nahmias,A.J.; Stagno,S. (Dep. Pediatr., Emory Univ. Sch. Med., 69 Butler St., SE, Atlanta, GA 30303, USA) *N. Engl. J. Med., 299(23), 1266-1270 (1978)* En;en.

1097-U2 Detection of viruria in cytomegalovirus-infected infants by electron microscopy. Henry,C.; Hartsock,R.J.; Kirk,Z.; Behre,R. (Dep. Lab. Med., Allegheny Gen. Hosp., Singer Mem. Res. Inst., 320 E. North Ave., Pittsburgh, PA 15212, USA) *Am. J. Clin. Pathol., 69(4), 435-439 (1978)* En;en.

1098-U2 Direct diagnosis of herpesvirus in equine foetal organs by means of electron microscopy. Dahle,J.; Petzoldt,K.; von Benten,C. (Inst. Mikrobiol. und Tierseuchen der Tierarztl. Hochs. Hannover, Bischofsholer Damm 15, D-3000 Hannover 1, GFR) *Vet. Microbiol., 4(4), 317-320 (1979)* En;en.

1099-U2 A rapid method for visualization of single-stranded genomic RNAs of animal viruses. Murti,K.G.; Bean,W.J.,Jr.; Hsu,C.H. (Div. Virol., St. Jude Child. Res. Hosp., 332 N. Lauderdale, PO Box 318, Memphis, TN 38101, USA) *J. Ultrastruc. Res., 70(1), 52-57 (1980)* En;en.

1100-U2 Electron microscopy in the rapid diagnosis of orf. Harkness,J.W.; Scott,A.C.; Herbert,C.N. (Minist. Agric., Fish. and Food, Cent. Vet. Lab., New Haw, Weybridge, Surrey, UK) *Br. Vet. J., 133(1), 81-87 (1977)* En;en.

1101-U2 Reliable identification of reovirus-like agent in diarrheal stools. Portnoy,B.L.; Conklin,R.H.; Menn,M.; Olarte,J.; DuPont,H.L. (Univ. Texas Health Sci. Cent., John H. Freeman Bldg., 6400 Cullen St., Houston, TX 77030, USA) *J. Lab. Clin. Med., 89(3), 560-563 (1977)* En;en.

1102-U2 Viral gastroenteritis in children. Holdaway,M.D.; Parkinson,A. (Address not stated) *N.Z. Med. J., 82(547), 176 (1975)* En.

1103-U2 Diagnosis of rotavirus infection by cell culture. Bryden,A.S.; Davies,H.A.; Thouless,M.E.; Flewett,T.H. (Reg. Virus Lab., East Birmingham Hosp., Birmingham B9 5ST, UK) *J. Med. Microbiol., 10(1), 121-125 (1977)* En;en.

1104-U2 Use of electron microscopy and an enzyme-linked immunosorbent assay for the detection of rotaviruses in neonatal calf diarrhea. Payment,P.; Marsolais,G.; Trudel,M.; Fauvel,M.; Lamontagne,L.; Assaf,R.; Marois,P. (Cent. Rech. Virol., Inst. Armand-Frappier, CP 100, Ville de Laval, Quebec H7N 4Z3, Canada) *Can. J. Comp. Med., 43(3), 328-329 (1979)* En;en,fr.

1105-U2 Infantile gastroenteritis caused by rotavirus: comparison of enzyme-linked immunosorbent assay and electron microscopy for rapid diagnosis. Seigneurin,J.M.; Scherrer,R.; Baccard-Longere,M.; Genoulaz,O.; Feynerol,C.; Gout,J.P. (Lab. Virol., CHU Grenoble, BP 217 X, 38043 Grenoble Cedex, France) *Biomedicine, 31(4), 99-104 (1979)* En;en,fr.

1106-U2 Tacaribe-Vero-electron microscopy: model for laboratory diagnosis of arenavirus infection. Blaskovic,P.J.; Mahdy,M.S. (Virus Lab., Ontario Minist. Health, Box 9000, Terminal A, Toronto, Ont., Canada) *J., Can. Med. Assoc., 118(10), 1193-1194 (1978)* En.

1107-U2 Electron microscopical rapid diagnosis of cutaneous viral diseases. Wolff,H.H.; Graser,H. (Dermatol. Univ.-Klin., Fraenlobstr. 9, D-8000 Munchen 2, GFR) *Hautarzt, 28(7), 371-374 (1977)* De;de,en.

1108-U2 Rapid detection of viruses by electron microscopy. Interest of negative staining for the diagnosis of some skin lesions of viral origin. Burtonboy,G.; Lachapelle,J.-M.; Tennstedt,D.; Lamy,M.-E. (Clos-Chapelle-aux-Champs, 30, B 1200 Bruxelles, Belgium) *Ann. Dermatol. Venerol., 105(8-9), 707-712 (1978)* Fr;en,fr.

1109-U2 Rapid diagnosis by electron microscopy of nonbacterial gastroenteritis in children. Blaskovic,P.J.; Kuderewko,O.; McLaughlin,B.; Yong,D.C.; Ball,F.R. (Virus Lab., Ontario Minist. Health, Toronto, Ont., Canada) *J., Can. Med. Assoc., 122(2), 150-152 (1980)* En.

1110-U2 Experimental techniques in the determination of aetiology of acute infantile gastroenteritis. Schoub,B.D.; Jacobs,Y.R.; Robins-Browne,R.M.; Koornhof,H.J.; Lecatsas,G.; Prozesky,O.W. (Dep. Microbiol., Inst. Pathol., PO Box 2034, Pretoria, South Africa) *S. Afr. J. Med. Sci., 41(3), 213-219 (1976)* En;en.

1111-U2 Experimental techniques in the determination of aetiology of acute infantile gastroenteritis. Schoub,B.D.; Jacobs,Y.R.; Robins-Browne,R.M.; Koornhof,H.J.; Lecatsas,G.; Prozesky,O.W. (Dep. Microbiol., Inst. Pathol., PO Box 2034, Pretoria, South Africa) *S. Afr. J. Med. Sci., 41(3), 213-219 (1976)* En;en.

1112-U2 Electron microscopic detection of mycoplasma contaminating cell cultures. Lascano,E.F.; Berria,M.I.; Barrera,O.J.G. (Div. Microsc. Electron., Inst. Nac. Microbiol. Carlos G. Malbran, Avenida Velez Sarsfield 563, 1281 Buenos Aires, Argentina) *Medicina (B. Aires), 38(2), 155-158 (1978)* Es;en,es.

1113-U2 Electron microscopic localization of acid phosphatase in *Tetrahymena pyriformis*: the influence of activities of lysosomal enzymes on fixation and structural preservation. Kolb-Bachofen,V. (Fachber. Biol., Univ. Konstanz, Postfach 7733, D-7750 Konstanz, GFR) *Cytobiologie, 15(1), 135-144 (1977)* En;en.

1114-U2 The identification of human T and B lymphocytes by electron microscopy. Toma,V.A.; Heyns,A.du P.; Retief,F.P. (Dep. Haematol., Fac. Med., Univ. Orange Free State, Bloemfontein, South Africa) *J. Immunol. Methods, 17(1-2), 91-100 (1977)* En;en.

1115-U2 Fluorescence microscopy for detection of *M. leprae* in tissue sections. Jariwala,H.J.; *Kelkar,S.S. (Dep. Microbiol., Grand Med. Coll., Byculla, Bombay 400 008, India) *Int. J. Lepr., 47(1), 33-36 (1979)* En;en,es,fr.

1116-U2 Fluorescent microscopy for tubercle bacilli. Hardas,U.D.; Jahagirdar,V.L.; Pathak,A.A. (Dep. Microbiol., Med. Coll., Nagpur, India) *Clinician, 41(8), 302-304 (1977)* En.

1117-U2 Fluorescent staining for *M. tuberculosis* in tissue sections - comparison with Fite-Faraco procedure. /|presented at the 254th Meeting of the Association of Teching Pathologists, held at the Haffkine Institute, Bombay, India, 25 Jan 1975|. Dave,J.K.; Fernandes,Y.; Gawand,S.C. (Dep. Clin. Pathol., Haffkine Inst., Parel, Bombay 400012, India) *Bull. Haffkine Inst., 4(1), 37-41 (1976)* En;en.

1118-U2 New fluorescent microscopical technique in diagnostic microbiology. Grossgebauer,K.; Kegel,M.; Dann,O. (Inst. Hyg. und Med. Mikrobiol., Freien Univ., 1000 Berlin 45, Hindenburgdamm 27, GFR) *Dtsch. Med. Wochenschr., 101(29), 1098-1099 (1976)* De;de,en.

1119-U2 Use of nuclepore filters for counting bacteria by fluorescence microscopy. Hobbie,J.E.; Daley,R.J.; Jasper,S. (Mar. Biol. Lab., Woods Hole, MA 02543, USA) *Appl. Environ. Microbiol., 33(5), 1225-1228 (1977)* En;en.

1120-U2 Tinopal AN in fluorescent microscopic detection of bacteria within plant tissues. Eng,L.K.; *Cole,A.L.J. (Dep. Bot., Univ. Canterbury, Christchurch, New Zealand) *Stain Technol., 51(5), 277-278 (1976)* En.

1121-U2 Fluorescent feulgen staining of fungal nuclei. Lemke,P.A.; Ellison,J.R.; Marino,R.; Morimoto,B.; Arons,E.; Kohman,P. (Carnegie-Mellon Inst. Res., Carnegie-Mellon Univ., Pittsburgh, PA 15213, USA) *Exp. Cell Res., 96(2), 367-373 (1975)* En;en.

1122-U2 Use of fluorescence microscopy for quantifying phytoplankton, especially filamentous blue-green algae. Brock,T.D. (Dep. Bacteriol., Univ. Wisconsin, Madison, WI 53706, USA) *Limnol. Oceanogr., 23(1), 158-160 (1978)* En;en.

1123-U2 Use of fluorescence microscopy for quantifying phytoplankton, especially filamentous blue-green algae. Brock,T.D. (Dep. Bacteriol., Univ. Wisconsin, Madison, WI 53706, USA) *Limnol. Oceanogr., 23(1), 158-160 (1978)* En;en.

1124-U2 Fluorescence identification of zoochlorellae: a rapid method for investigating alga-invertebrate symbioses. Williamson,C.E. (Biol. Dep., Mount Holyoke Coll., South Hadley, MA 01075, USA) *J. Exp. Zool., 202(2), 187-194 (1977)* En;en.

1125-U2 Rapid detection of malaria and other bloodstream parasites by fluorescence microscopy with 4′6 diamidino-2-phenylindole (DAPI). Hyman,B.C.; Macinnis,A.J. (Dep. Biol., Univ. California, Los Angeles, CA 90024, USA) *J. Parasitol., 65(3), 421-425 (1979)* En;en.

1126-U2 Detection of H-Y antigen by fluorescence microscopy. Galbraith,G.M.P.; Galbraith,R.M.; Faulk,W.P.; Wachtel,S.S. (Dep. Basic and Clin. Immunol. and Microbiol., Med. Univ. South Carolina, Charleston, SC 29401, USA) *Transplantation, 26(1), 25-27 (1978)* En;en.

1127-U2 Interference reflection microscopy of adhesion of *Amoeba proteus*. Opas,M. (Dep. Cell. Biol., M. Nencki Inst. Exp. Biol., Warsaw 02-093, 3 Pasteur Str., Poland) *J. Microsc. (Oxford), 112(part 2), 215-221 (1978)* En;en.

1128-U2 Removal of serum from virus suspensions in preparation for immune electron microscopy. Beesley,J.E. (ARC Inst. Anim. Physiol., Babraham, Cambridge CB2 4AT, UK) *J. Microsc., 111(Part 2), 233-235 (1977)* En;en.

1129-U2 Immuno-electron microscopic detection of adenoviruses concentrated by adsorption with insoluble polyelectrolytes. Lee,V.P. (Prov. Lab. Public Health, Edmonton, Alta., Canada) *Can. J. Med. Technol., 38(5), 129-138 (1976)* En;en,fr.

1130-U2 Enterovirus typing by immune electron microscopy using low-speed centrifugation. Narang,H.K.; Codd,A.A. (Public Health Lab., Inst. Pathol., Newcastle Gen. Hosp., Westgate Road, Newcastle-upon-Tyne NE4 6BE, UK) *J. Clin. Pathol., 33(2), 191-194 (1980)* En.

1131-U2 Morphologically recognizable markers for scanning immunoelectron microscopy. II. An indirect method using T4 and TMV. Kumon,H. (Dep. Virol., Okayama Univ. Med. Sch., Shikatacho, Okayama 700, Japan) *Virology, 74(1), 93-103 (1976)* En;en.

1132-U2 Picornaviruses: rapid differentiation and identification by immune electronmicroscopy and immunodiffusion. Hughes,J.H.; Gnau,J.M.; Hilty,M.D.; Chema,S.; Ottolenghi,A.C.; Hamparian,V.V. (Dep, Med. Microbiol. and Pediatr., Ohio State Univ., Columbus, OH 43210, USA) *J. Med. Microbiol., 10(2), 203-212 (1977)* En;en.

1133-U2 Measurement of antibodies to rabies virus by immunoelectron microscopy. Chaudhary,R.K.; Cho,H.C.; Monette,M.T. (Bur. Virol., Lab. Cent. Dis. Control Health and Welfare Canada, Ottawa, Ont., K1A 0L2, Canada) *Can. J. Microbiol., 25(10), 1209-1211 (1979)* En;en,fr.

1134-U2 A simple method for the inactivation of St. Louis encephalitis virus preparations for immunofluorescent microscopy. Yabrov,A.; Artsob,H.; Spence,L. (Dep. Med. Microbiol., Univ. Toronto, 100 College St., Toronto, Ont. M5G 1L5, Canada) *Can. J. Microbiol., 24(1), 72-74 (1978)* En;en,fr.

1135-U2 Detection of vesicular stomatitis virus (murine leukemia virus) pseudotypes by immunoelectron microscopy.

Chan,J.C.; Hixson,D.C.; Bowen,J.M. (Dep. Mol. Carcinog. and Virol., Univ. Texas System Cancer Cent., M.D. Anderson Hosp., Houston, TX 77030, USA) *Virology, 88(1), 171-176 (1978)* En;en.

1136-U2 A simple technique for immunoelectronmicroscopic localization of Ig on tissue sections with special reference to cutaneous pathology. Schmitt,D.; Germain,D.; Thivolet,J. (Lab. Immunopathol. Cutanee, Hop. Edouard Herriot, F-69374 Lyon Cedex 2, France) *Arch. Dermatol. Res., 259(3), 235-245 (1977)* En;de,en.

1137-U2 [A rapid method for the diagnosis of densonucleosis of *Aedes aegypti* and its use in the detection of susceptible insects.] Lebedeva,O.P.; Topchiy,M.K.; Buchatsky,L.P.; Lebedinets,N.N.; Gonchar,N.M.; Korsh,L.M. (Kiev Univ., Kiev, USSR) *Med. Parazitol. Parasit. Bolezn., 44(5), 612-615 (1975)* Ru;en,ru.

1138-U2 Demonstration of the initial cell in *Streptomyces griseus* by a new microscopic technique. Bisset,K.A. (Dep. Bacteriol., Univ. Birmingham, Birmingham B15 2TJ, UK) *J. Gen. Microbiol., 104(1), 157-159 (1978)* En.

1139-U2 Use of thin, flexible plastic coverslips for microscopy, microcompression, and counting of aerobic microorganisms. Spoon,D.M. (Dep. Biol., Georgetown Univ., Washington, DC 20057, USA) *Trans. Am. Microsc. Soc., 95(3), 520-523 (1976)* En;en.

1140-U2 A direct microscopic ratio-method using polystyrene beads to determine microbial numbers in soil. Peterson,H.L.; Frederick,L.R. (Dep. Agron., PO Box 5248, Mississippi State Univ., Starkville, MI 39762, USA) *Soil Biol. Biochem., 11(1), 77-83 (1979)* En;en.

1141-U2 A direct microscopic ratio-method using polystyrene beads to determine microbial numbers in soil. Peterson,H.L.; Frederick,L.R. (Dep. Agron., PO Box 5248, Mississippi State Univ., Starkville, MI 39762, USA) *Soil Biol. Biochem., 11(1), 77-83 (1979)* En;en.

1142-U2 A simple and rapid method for qualitative and quantitative study of the fungal flora of leaves. Langvad,F. (Dep. Microbiol. and Plant Physiol., Allegaten 70, N-5014 Bergen-Universitetet, Norway) *Can. J. Microbiol., 26(6), 666-670 (1980)* En;en.

1143-U2 A rapid method for reading migration inhibition tests. Doble,A.T.A. (Immuno. Unit, Glaxo Res. Ltd., Greenford, UK) *J. Immunol. Methods, 16(3), 299-300 (1977)* En;en.

1144-U2 Detection of fungi in clinical specimens by phase-contrast microscopy. Roberts,G.D. (Sect. Clin. Microbiol., Mayo Clin. and Mayo Found., Rochester, MN 55901, USA) *J. Clin. Microbiol., 2(3), 261-265 (1975)* En;en.

1145-U2 [A routine method for studying phytopathogenic microorganisms in serial cuttings by the scanning electron microscope]. Ozel,M.; Petzold,H.; Krober,H. (Biol. Bundesanst. Land- und Forstwirtschaft, Inst. Bakteriol., Konigin-Luise-Str. 19, D-1000 Berlin 33, GFR) *Phytopathol. Z., 89(2), 181-186 (1977)* De;de,en.

1146-U2 Principles and techniques of scanning electron microscopy. Volume 5. Biological applications. Hayat,M.A.(ed.) *Publ. by*: Van Nostrand Reinhold Co. Ltd., 450 West 33rd St., New York, NY 10001, US ISBN: 0-442-25692-2 at £18.70. En.

1147-U2 A micromanipulator method to observe the inner structure of diseased leaves by scanning electron microscopy. Kunoh,H.; Ishizaki,H.; Watanabe,T.; Yamada,M.; Nagatani,T. (Lab. Plant Pathol., Fac. Agric., Mie Univ., Tsu-City, 514, Japan) *Plant Dis. Rep., 60(2), 95-97 (1976)* En.

1148-U2 A morphologically recognizable marker for scanning immunoelectron microscopy. I. T_4-bacteriophage. Kumon,H.; Uno,F.; Tawara,J. (Dep. Virol., Okayama Univ. Med. Sch., Shikatacho, Okayama 700, Japan) *Virology, 70(2), 554-557 (1976)* En;en.

1149-U2 Simplified preparation of mycoplasmas, an acholeplasma, and a spiroplasma for scanning electron microscopy. Gallagher,J.E.; Rhoades,K.R. (Natl. Anim. Dis. Cent., Fed. Res., Sci. and Educ. Adm., USDA, Ames, IA 50010, USA) *J. Bacteriol., 137(2), 972-976 (1979)* En;en.

1150-U2 The surface morphology of the phagocytosis of micro-organisms by peritoneal macrophages. Walters,M.N.-I.; Papadimitriou,J.M.; Robertson,T.A. (Dep. Pathol., Univ. Western Australia, rth Med. Cent., Perth, WA, *J. Pathol., 118(4), 221-226 (1976)* En;en.

1151-U2 A method for rapid evaluation of materials for susceptibility to marine biofouling. Zachary,A.; Taylor,M.E.; Scott,F.E.; Colwell,R.R. (Dep. Microbiol., Univ. Maryland, College Park, MD 02742, USA) *Int. Biodeterior. Bull., 14(4), 111-118 (1978)* En;de,en,es,fr.

1152-U2 Chemical analysis of five species of *Aspergillus* by combined scanning electron microscopy and X-ray spectrometry. Thibaut,M.; Ansel,M. (Lab. Parasitol. et Mycol., Paris, France 75006) *Trans. Am. Microsc. Soc., 95(2), 210-214 (1976)* En;en.

1153-U2 A simple method for preparing specimens of *Claviceps purpurea* for scanning electron microscopy. Pokorny,V.; Pazoutova,S.; Rehacek,Z. (Inst. Microbiol., Czechoslovak Acad. Sci., 14200 Prague 4, Czechoslovakia) *Zentralbl. Bakteriol. Parasitenkd. Infektionskr. Hyg., II, 133(2), 188-191 (1978)* En;de,en.

1154-U2 Lichen structure viewed with the scanning electron microscope. Hale,M.E. (Dep. Bot., Natl. Mus. Nat. Hist., Smithsonian Inst., Washington, DC 20560, USA) *In:* Lichenology: Progress and Problems. *Publ.by:* Academic Press Inc. (London) 1976 p.1-15 ISBN 0-12-136750-9 En;en.

1155-U2 A simplified technique to prepare fungal specimens for scanning electron microscopy. Samson,R.A.; Stalpers,J.A.; Verkerke,W. (Centraalbur. Schimmelcult., PO Box 273, Oosterstr. 1, Baarn, Netherlands) *Cytobios, 24(93), 7-11 (1979)* En;en.

1156-U2 A rapid preparation method for scanning electron microscopy of Lugol preserved algae. Bistricki,T.; Munawar,M. (Dep. Fish. Environ., Canada Cent. Inland Waters, PO Box 5050, Burlington, Ont., Canada L7R 4A6) *J. Microsc. (Oxford), 114(2), 215-218 (1978)* En;en.

1157-U2 A technique for preparing ciliated rumen protozoa for scanning electron microscopy. Rittenburg,J.H.; Bayer,R.C.; Stern,M.D. (Univ. Marine, Dep. Anim. and Vet. Sci., Orono, ME 04473, USA) *J. Anim. Sci., 44(4), 710-712 (1977)* En;en.

1158-U2 Practical electron microscopy for biologists. 2nd Edition. Meek,G.A. *Publ. by:* John Wiley and Sons Ltd., Baffins Lane, Chichester, Sussex, UK. 18 Feb 1976. ISBN: 0-471-59031-2 at £13.50 or US $29.70.

1159-U2 A convenient live-box for the microscope stage. Harding,J.P. (Br. Mus. (Nat. Hist.), South Kensington, London, UK) *Microsc., J. Quekett Microsc. Club, 33(2), 70-72 (1976)* En.

1160-U2 Identifying colony types in the case of gonococci. Muller,G. (Hautklin. Poliklin. Bereich Med., Humboldt-Univ., Schumannstr. 20-21, DDR-104 Berlin, GDR) *Z. Gesamte Hyg. Grenzgeb., 24(12), 943-944 (1978)* De;de,en,ru.

1161-U2 The rapid diagnosis of urinary tract infection. A side-room method. Dornfest,F.D. (202 Centrepoint, Loxton Road, Milnerton 7405, South Africa) *S. Afr. Med. J., 56(21), 841-843 (1979)* En;en.

1162-U2 Simplified microscopy for rapid detection of significant bacteriuria in random urine specimens. Barbin,G.K.; *Thorley,J.D.; Reinarz,J.A. (Dep. Med., Univ. Texas Med. Branch, Galveston, TX 77550, USA) *J. Clin. Microbiol., 7(3), 286-291 (1978)* En;en.

1163-U2 A simple technique for observing germ tube formation in *Candida albicans*. Cartwright,R.Y. (Public Health Lab., Guildford, Surrey, UK) *J. Clin. Pathol., 29(3), 267-268 (1976)* En.

1164-U2 Culture and microscopy of microorganisms in frozen sections. Curry,J.; Molokhia,M.M. (Skin Hosp., Univ. Manchester Sch. Med., Manchester M3 3HL, UK) *Sabouraudia, 15(3), 217-219 (1977)* En;en,fr.

1165-U2 Coherent microscopy and matched spatial filtering for real-time recognition of diatom species. Case,S.K.; Almeida,S.P.; Dallas,W.J.; Fournier,J.M.; Pritz,K. (Phys. Dep., Virginia Polytech. Inst. and State Univ., Blacksburg, VA 24061, USA) *Environ. Sci. Technol., 12(8), 940-946 (1978)* En;en.

1166-U2 A simple microscopic method for identifying and quantitating phagocytic cells in vitro. Patterson-Delafield,J.; Lehrer,R.I. (Dep. Bacteriol., Dep. Med., Univ. California, Los Angeles, CA 90024, USA) *J. Immunol. Methods, 18(3-4), 377-379 (1977)* En;en.

1167-U2 A rapid method for determining the percentage of antibacterial phagocytes in a sample population of leukocytes. Janssen,W.A.; Dangerfield,H.G. (U.S. Army Med. Res. Inst. Infect. Dis., Frederick, MD 21701, USA) *J. Reticuloendothel. Soc., 21(5), 299-306 (1977)* En;en.

See: 143, 351, 783, 943, 1020, 2040, 2604, 3010, 3140, 3264, 3265, 3243, 3266, 3279

Chromatography

1168-U2 Construction of a paralytic shellfish toxin analyzer and its application. Buckley,L.J.; *Oshima,Y.; Shimizu.Y. (Dep. Pharmacognosy, Coll. Pharm., Univ. Rhode Island, Kingston, RI 02881, USA) *Anal. Biochem., 85(1), 157-164 (1978)* En;en.

1169-U2 Techniques and instrumentation for preparative immunosorbent separations. Eveleigh,J.W. (Instruments Products Div., E. I. du Pont de Nemours, Experimental Station, Wilmington, DE 19898, USA) *J. Chromatogr., 159(1), 129-145 (1978)* En;en.

1170-U2 A rapid method for the purification of antibody-enzyme conjugates. Page,M.; Audette,M.; Caron,M. (Cent. Rech. Hotel-Dieu de Quebec, 11 Cote du Palais, Quebec, Que, G1R 2J6, Canada) *Can. J. Biochem., 57(3), 286-288 (1979)* En;en,fr.

1171-U2 Rapid purification of plasmid DNAs by hydroxyapatite chromatography. Colman,A.; Byers,M.J.; Primrose,S.B.; Lyons,A. (Dep. Biol. Sci., Univ. Warwick, Coventry, Warwickshire CV4 7AL, UK) *Eur. J. Biochem., 91(1), 303-310 (1978)* En;en.

1172-U2 Purification and characterization of agrocin 84. Thompson,R.J.; Hamilton,R.H.; *Pootjes,C.F. (Dep. Microbiol., Pennsylvania State Univ., University Park, PA 16802, USA) *Antimicrob. Agents Chemother., 16(3), 293-296 (1979)* En;en.

1173-U2 Purification of adenovirus 4 type-specific hemagglutinins for use in diagnostic counterelectrophoresis tests. Hierholzer,J.C.; Tannock,G.A. (Cent. Dis. Control, Respir. Virol. Branch, 7-112, Adanta, GA 30333, USA) *J. Med. Virol., 4(4), 279-290 (1979)* En;en.

1174-U2 The rapid concentration and purification of influenza virus from allantoic fluid. Heyward,J.T.; Klimas,R.A.; Stapp,M.D.; Obijeski,J.F. (Virol. Div., Bur. Lab., Cent. Dis. Control, Public Health Serv., US Dep. Health, Educ. and Welfare, 1600 Clifton Rd., N. E., Atlanta, GA 30333, USA) *Arch. Virol., 55(1-2), 107-119 (1977)* En;en.

1175-U2 Chromatographic isolation of the hemagglutinin polypeptides from influenza virus vaccine and determination of their amino-terminal sequences. Bucher,D.J.; Li,S.S.-L.; Kehoe,J.M.; Kilbourne,E.D. (Dep. Microbiol., Mount Sinai Sch. Med., City Univ. New York, Fifth Ave, and 100th St., New York, NY 10029, USA) *Proc. Natl. Acad. Sci. USA, 73(1), 238-242 (1976)* En;en.

1176-U2 Adsorption chromatography of viruses. Bresler,S.E.; Katushkina,N.V.; Kolikov,V.M.; Potokin,J.L.; Vinogradskaya,G.N. (Leningrad Inst. Nuclear Phys., Leningrad Polytech. Inst., Leningrad, USSR) *J. Chromatogr., 130, 275-280 (1977)* En;en.

1177-U2 A rapid purification of T4 polynucleotide kinase using Blue Dextran-Sepharose chromatography. Nichols,B.P.; Lindell,T.D.; Stellwagen,E.; Donelson,J.E. (Dep. Biochem., Univ. Iowa, Iowa City, IA 52242, USA) *Biochim. Biophys. Acta, 526(2), 410-417 (1978)* En;en.

1178-U2 The rapid purification of T4 DNA ligase from a λT4 *lig* lysogen. Tait,R.C.; *Rodriguez,R.L.; West,R.W.,Jr. (Dep. Genet., Univ. California, Davis, CA 95616, USA) *J. Biol. Chem., 255(3), 813-815 (1980)* En;en.

1179-U2 Separation of biopolymers by liquid sieve chromatography under nonequilibrium conditions. Kolikov,V.M.; Mchedlishvili,B.V.; Lebedev,Yu.Ya.; Krasil'nikov,I.V. (Polytech Inst., Leningrad, USSR) *Kolloidn. Zh., 39(3), 562-567 (1977)* Ru;en,ru.

1180-U2 Rapid separation of immunoglobulin M from immunoglobulin G antibodies for reliable diagnosis of recent rubella infections. Frisch-Niggemeyer,W. (Inst. Virol., Univ. Wien, A-1095 Wien Kinderspitalgasse 15, Austria) *J. Clin. Microbiol., 2(5), 377-381 (1975)* En;en.

1181-U2 [Purification and concentration of tickborne encephalitis virus by adsorption chromatography]. Bresler,S.E.; Kolikov,V.M.; Krasil'Nikov,I.V.; Mamonenko,L.L.; Molodkin,V.M.; Mchedlishvili,B.V.; Pogodina,V.V.; El'bert,L.B. (M.I. Kalinin Leningrad Politech. Inst., Leningrad, USSR) *Dokl. Akad. Nauk SSR, 234(4), 940-942 (1977)* Ru.

1182-U2 Isolation of bacterial and phage proteins by homopolymer RNA-cellulose chromatography. Carmichael,G.G. (Biol. Lab., Harvard Univ., Cambridge, MA 02138, USA) *J. Biol. Chem., 250(15), 6160-6167 (1975)* En;en.

1183-U2 Purification of interferon by adsorption chromatography on controlled pore glass. Edy,V.G.; Braude,I.A.; De Clercq,E.; Billiau,A.; De Somer,P. (Rega Inst. Med. Res., Katholleke Univ. Leuven, Minderbroedersstr. 10, B-3000 Leuven, Belgium) *J. Gen. Virol., 33(3), 517-521 (1976)* En;en.

1184-U2 Argon detector: alternative detection system for gas-liquid chromatographic analysis of short-chain organic acids. Sullivan,N.M.; Mayhew,J.; DiTullio,D.; *Tally,F.P. (Infect. Dis. Sect., Dep. Med., Tufts-New England Med. Cent., Boston, MA 02111, USA) *J. Clin. Microbiol., 8(4), 369-373 (1978)* En;en.

1185-U2 The purification of glutamine synthetase from *Azotobacter* and other procaryotes by blue Sepharose chromatography. Lepo,J.E.; Stacey,G.; Wyss,O.; *Tabita,F.R. (Dep. Microbiol., Univ. Texas at Austin, Austin, TX 78712, USA) *Biochim. Biophys. Acta, 568(2), 4236 (1979)* En;en.

1186-U2 Hydroxyapatite for chromatography. III. Cation and pH effects on fractionation of tRNA for crystallization. Spencer,M.; Neave,E.J.; Webb,N.L. (King's Coll. Dep. Biophys., 26-29 Drury Lane, London WC2B 5RL, UK) *J. Chromatogr., 166(1), 447-454 (1978)* En;en.

1187-U2 A procedure for the rapid, large-scale purification of *Escherichia coli* DNA-dependent RNA polymerase involving Polymin P precipitation and DNA-cellulose chromatography. Burgess,R.R.; Jendrisak,J.J. (McArdle Lab. Cancer Res., Univ. Wisconsin, Madison, WI 53706, USA) *Biochemistry (Wash.), 14(21), 4634-4638 (1975)* En;en.

1188-U2 Purification and properties of the σ subunit of *Escherichia coli* DNA-dependent RNA polymerase. Lowe,P.A.; Hager,D.A.; *Burgess,R.R. (McArdle Lab. Cancer Res., Univ. Wisconsin, Madison, WI 53706, USA) *Biochemistry (Wash.), 18(7), 1344-1352 (1979)* En;en.

1189-U2 A simple procedure for resolution of *Escherichia coli* RNA polymerase holoenzyme from core polymerase. Gonzalez,N.; Wiggs,J.; Chamberlin,M.J. (Dep. Biochem., Univ. California, Berkeley, CA 94720, USA) *Arch. Biochem. Biophys., 182(2), 404-408 (1977)* En;en.

1190-U2 *Escherichia coli* phosphoenolpyruvate dependent phosphotransferase system. Complete purification of enzyme I by hydrophobic interaction chromatography. Robillard,G.T.; Dooijewaard,G.; Lolkema,J. (Dep. Phys. Chem., Univ. Groningen, Paddepoel, Groningen, The Netherlands) *Biochemistry (Wash.), 18(14), 2984-2989 (1979)* En;en.

1191-U2 Simple method for isolation of three DNA-polymerases from *E. coli* using chromatography on DNA-cellulose. Khlebalina,O.I.; Debabov,V.G. (All-Union Insst. Genet. and Selection Ind. Microorganisms, Moscow, USSR) *Biokhimiya, 41(4), 650-654 (1976)* Ru;en,ru.

1192-U2 Retrovirus purification: method that conserves envelope glycoprotein and maximizes infectivity. McGrath,M.; Witte,O.; Pincus,T.; *Weissman,I.L. (Dep. Pathol., Stanford Univ. Med. Cent., Stanford, CA 94305, USA) *J. Virol., 25(3), 923-927 (1978)* En;en.

1193-U2 Purification and some properties of enterotoxin A produced by *Staphylococcus aureus* strain 100 mutant. Yamada,S.; Igarashi,H.; Terayama,T. (Dep. Microbiol., Tokyo Metrop. Res. Lab. Public Health, Tokyo, Japan) *Jap. J. Bacteriol., 31(3), 409-419 (1976)* Ja;en,ja.

1194-U2 Hydrophobic chromatography and fractionation of enzymes from extremely halophilic bacteria using decreasing concentration gradients of ammonium sulfate. Mevarech,M.; Leicht,W.; *Werber,M.M. (Polymer Dep., Weizmann Inst. Sci., Rehovot, Israel) *Biochemistry (Wash.), 15(11), 2382-2387 (1976)* En;en.

1195-U2 An HPLC assay for cephalosphorinase activity. Mehta,R.J.; Fox,M.K.; Newman,D.J.; Nash,C.H. (Microbiol. Dep., Smith Kline and French Lab., Philadelphia, PA 19101, USA) *J. Antibiot., 30(12), 1132-1133 (1977)* En.

1196-U2 Quantitative methods for the gas chromatographic characterization of acidic fermentation by-products of anaerobic bacteria. Bohannon,T.E.; Manius,G.; Mamaril,F.; Li Wen,L.-F. (Quality Control Dep., Hoffman-La Roche Inc., Nutley, NJ 07110, USA) *J. Chromatogr. Sci., 16(1), 28-35 (1978)* En;en.

1197-U2 Analysis of amines and other bacterial products by head-space gas chromatography. Larsson,L.; Marcd,P.A.; Odham,G. (Inst. Med. Microbiol., Univ. Lund, Solvegatan 23, S-223 62 Lund, Sweden) *Acta Pathol. Microbiol. Scand. Ser. B, 86(4), 207-213 (1978)* En;en.

1198-U2 A rapid method for purifying bacterial deoxyribonucleic acid. Gibson,D.M.; Ogden,I.D. (Minist. Agric., Fish. and Food, Torry Research Stn., 135 Abbey Road, Aberdeen AB9 8DG, UK) *J. Appl. Bacteriol., 46(3), 421-423 (1979)* En;en.

1199-U2 Detection of 5-amino-4-imidazole-N-succinocarboxamide ribotide and hypoxanthine accumulation. A simple method for identification of some purine auscotrophs. Bal,J.; Pieniazek,N.J. (Dep. Genet., Warsaw Univ., Al. Ujazdowskie 4, 00-478 Warsaw, Poland) *J. Chromatogr., 169, 474-476 (1979)* En.

1200-U2 An efficient method for the synthesis and purification of trans-[^{14}C]geranylgeranyl pyrophosphate. Gafni,Y.; Schechter,I. (Dep. Biochem., George S. Wise Cent. Life Sci., Tel-Aviv Univ., Tel-Aviv, Israel) *Anal. Biochem., 92(1), 248-252 (1979)* En;en.

1201-U2 Rapid determination of polycyclic aromatic hydrocarbons (PAH) in yeasts grown on n-paraffins and molasses. Santoro,A.; Modica,R.; Paglialunga,S.; *Bartosek,I. (Ist. Ricerche Farm. 'Mario Negri', Via Eritrea 62-20157 Milan, Italy) *Toxicol. Lett., 3(2), 85-93 (1979)* En;en.

1202-U2 Isolation of pure DNA from filamentous fungi by chromatography on hydroxyapatite. Szecsi,A. (Res. Inst. Plant Prot., H-1525 Budapest, POB 102, Hungary) *Acta Microbiol. Acad. Sci. Hung., 25(1), 61-65 (1978)* En;en.

1203-U2 A chromatographic method for estimating fungal growth by glucosamine analysis of diseased tissues. Stahmann,M.A.; Abramson,P.; Wu,L.-C. (Dep. Biochem., Coll. Agric. and Life Sci., Univ. Wisconsin-Madison, Madison, WI 53706, USA) *Biochem. Physiol. Pflanz. BPP, 168(1-4), 267-276 (1975)* En;en.

1204-U2 Aflatoxin determination. Anon. (Anal. Control. Italia, Via Rovani 10, Monza, Italy) *Ind. Aliment., 16(4), 85-86 (1977)* It.

1205-U2 Chromatographic method of a simultaneous semiquantitative measurement of eight mycotoxins in grain. Lvova,L.S.; Kravchenko,L.V.; Shulgina,A.P. (All-Union Res. Inst. Grain, USSR Acad. Med. Sci., Moscow, USSR) *Prikl. Biokhim. Mikrobiol., 15(1), 143-149 (1979)* Ru;en.

1206-U2 Simultaneous extraction and fractionation and thin layer chromatographic determination of 14 mycotoxins in grains. Takeda,Y.; Isohata,E.; Amano,R.; Uchiyama,M.; (Natl. Inst. Hyg. Sci. Div. Food, 1-18-1 Kamiyoga Setagaya, Tokyo 158, Japan) *J. Assoc. Off. Anal. Chem. 62(3), 573-578 (1979)* En;en.

1207-U2 A procedure for the simultaneous measurement of net CO_2-exchange and nitrogenase activity in lichens. Crittenden,P.D.; Kershaw,K.A. (Dep. Bot., Univ. Sheffield, Sheffield S10 2TN, UK) *New Phytol., 80(2), 393-401 (1978)* En;en.

1208-U2 Identification and rapid determination of histamine in fish samples. Cattaneo,P.; Cantoni,C. (Ist Ispezione Alimenti Origine Anim., Univ. Stude di Milano, Fac. Med. Vet., Milan, Italy) *Ind. Aliment., 17(4), 303-307 (1978)* It;en,it.

1209-U2 Purification of proenzymic and activated human C1s free of C1r. Effects of calcium and ionic strength on activated Cls. Arluud,G.J.; Reboul,A.; Meyer,C.M.; Colomb,M.G. (DRF/Biochim., Cent. Etud. Nucl., 85X, 38041 Grenoble-Cedex, France) *Biochim. Biophys. Acta, 485(1), 215-226 (1977)* En;en.

1210-U2 The use of a double gel (G-25/DEAE Sephadex) for one step separation of fluorescein tagged gamma-globulins. Torres-Anjel,M.J.; Auger,J.J.; Riemann,H.P. (Fac. Med. Vet. y Zootec., Univ. Nac. Colombia, Apartado Aereo 11951, Bogota, D.E.1, Colombia) *Rev. Latinoam. Microbiol., 18(1), 37-41 (1976)* En;en,es.

1211-U2 Isolation of serum albumin and immunoglobulins by chromatography on Con-A Sepharose and Sephadex G-150. Hrkal,Z.; Vodrazka,Z. (Inst. Haematol. and Blood Transfusion, 128 20 Prague-2, Czechoslovakia) *J. Chromatogr., 135(1), 193-195 (1977)* En.

1212-U2 Rapid isolation and purification of antibody to Factor VIII by protein A. Lee,H.; Tucker,D.; Allain,J.P. (Lab. d'hemostase et de thrombose exp., Hop. Saint-Louis, 75010 Paris, France) *Thromb. Res., 14(6), 925-930 (1979)* En;en.

1213-U2 An automatic multi-programmed affinity chromatographic system. Folkersen,J.; Teisner,B.; Westergaard,J.; Svehag,S.-E. (Inst. Med. Microbiol., Odense Univ., Odense, Denmark) *J. Immunol. Methods, 23(1/2), 137-147 (1978)* En;en.

1214-U2 A rapid, novel method for the solid-phase derivatization of IgG antibodies for immune-affinity chromatography. Gersten,D.M.; Marchalonis,J.J. (Dep. Pathol., Georgetown Univ., Sch. Med. and Dent., Washington, DC 20007, USA) *J. Immunol. Methods, 24(3-4), 305-309 (1978)* En;en.

1215-U2 Purification of equine infectious anemia virus antigen by affinity chromatography. Sugiura,T.; Nakajima,H. (Equine Infect. Anemia Div., Natl. Inst. Anim. Health, Kodaira, Tokyo 187, Japan) *J. Clin. Microbiol., 5(6), 635-639 (1977)* En;en.

1216-U2 **Purification of hepatitis B surface antigen by affinity chromatography.** Einarsson,M.; Kaplan,L.; Utter,C. (AB Kabi, Res. Dep., Biochem. and Natl. Bacteriol. Lab., Dep. Immunol., Stockholm, Sweden) *Vox Sang., 35(4), 224-233 (1978)* En;en.

1217-U2 **Purification of neuraminidase from influenza viruses by affinity chromatography.** Bucher,D.J. (Dep. Microbiol., Mount Sinai Sch. Med., City Univ. New York, Fifth Ave. at 100th St., New York, NY 10029, USA) *Biochim. Biophys. Acta, 482(2), 393-399 (1977)* En;en.

1218-U2 **Simple affinity procedure for the purification of mammalian viral reverse transcriptases.** Sarngadharan,M.G.; Kalyanaraman,V.S.; Rahman,R.; *Gallo,R.C. (Lab. Tumor Cell Biol., Natl. Cancer Inst., NIH, Bethesda, MD 20205, USA) *J. Virol., 35(2), 555-559 (1980)* En;en.

1219-U2 **Purification of mouse interferon on specifically purified immunoadsorbent.** Hajnicka,V.; Fuchsberger,N.; Borecky,L. (Inst. Virol., Slovak Acad. Sci., 80939 Bratislava 9, Czechoslovakia) *Arch. Immunol. Ther. Exp., 25(5), 619-622 (1977)* En;en.

1220-U2 **Affinity chromatography of mouse interferon: a modified purification procedure utilizing specifically purified antibodies.** Hajnicka,V.; Fuchsberger,N.; Borecky,L. (Inst. Virol., Slovak Acad. Sci., 809 39 Bratislava, Czechoslovakia) *Acta Virol., 20(4), 326-333 (1976)* En;en.

1221-U2 **Purification of human fibroblast interferon by zinc chelate affinity chromatography.** Edy,V.G.; Billiau,A.; de Somer,P. (Rega Inst., Dep. Human Biol., Univ. Leuven, B-3000 Leuven, Belgium) *J. Biol. Chem., 252(17), 5934-5935 (1977)* En;en.

1222-U2 **Simple efficient methods for the isolation of malate dehydrogenase from thermophilic and mesophilic bacteria.** Wright,I.P.; Sundaram,T.K. (Dep. Biochem., Univ. Manchester Inst. Sci. and Technol., Manchester M60 1QD, UK) *Biochem. J., 177(2), 441-448 (1979)* En;en.

1223-U2 **Affinity chromatography of *Anacystis nidulans* ferredoxin-nitrate reductase and NADP reductase on reduced ferredoxin-Sepharose.** Manzano,C.; Candau,P.; Guerrero,M.G. (Dep. Bioquim., Fac. Ciencias y CSIC, Univ. Sevilla, Sevilla, Spain) *Anal. Biochem., 90(1), 408-412 (1978)* En;en.

1224-U2 **Use of immunoadsorbent affinity chromatography to purify Component I of nitrogenase from extracts of *Azotobacter vinelandii.*** Ferrante,J.V.; Nicholas,D.J.D. (Dep. Agric. Biochem., Waite Agric. Res. Inst., Univ. Adelaide, Glen Osmond, SA 5064, Australia) *FEBS Lett., 66(2), 187-190 (1976)* En.

1225-U2 **Affinity chromatography of aminoacyl-transfer ribonucleic acid synthetases. Cognate transfer ribonucleic acid as a ligand.** Clarke,C.M.; Knowles,J.R. (Dep. Biochem., Imperial Coll. Sci. and Technol., London SW7, UK) *Biochem. J., 67(2), 419-428 (1977)* En;en.

1226-U2 **Purification of alcohol dehydrogenase from *Bacillus stearothermophilus* by affinity chromatography.** Runswick,M.J.; Harris,J.I. (deceased) (MRC Lab.Mol. Biol., Hills Road, Cambridge, CB2 2QH, UK) *FEBS Lett., 92(2), 365-367 (1978)* En.

1227-U2 **A rapid method for the purification of β-lactamase from *Bacillus cereus* by affinity chromatography.** Coombe,R.G.; George,A.M. (Dep. Pharm., Univ. Sydney, Sydney, NSW, Australia) *Anal. Biochem., 75(2), 652-655 (1976)* En;en.

1228-U2 **Purification by affinity chromatography of phospholipase C from *Bacillus cereus.*** Little,C.; Aurebekk,B.; Otnaess,A.-B. (Inst. Med. Biol., Univ. Tromso, Tromso, Norway) *FEBS Lett., 52(2), 175-179 (1975)* En.

1229-U2 **Affinity chromatography purification of *Clostridium perfringens* enterotoxin.** Scott,V.N.; *Duncan,C.L (Food Res. Inst., Univ. Wisconsin, Madison, WI 53706, USA) *Infect. Immun., 12(3), 536-543 (1975)* En;en.

1230-U2 **A new affinity column for the purification of thymidine kinase from *Eschericia coli.*** Rohde,W. (Inst. Virol., Justus-Liebig-Univ., Frankfurter Str. 107, D-6300 Giessen/Lahn, GFR) *Anal. Biochem., 80(2), 643-645 (1977)* En.

1231-U2 **Chorismate mutase/prephenate dehydratase from *Escherichia coli* K12. 1. The effect of NaCl and its use in a new purification involving affinity chromatography on Sepharosyl-phenylalanine.** Gething,M.-J.H.; Davidson,B.E.; Dopheide,T.A.A. (Russell Grimwade Sch. Biochem., Univ. Melbourne, Melbourne, Vic., Australia) *Eur.J. Biochem., 71(2), 317-325 (1976)* En;en.

1232-U2 **Purification and characterization of polynucleotide phosphorylase from *Escherichia coli.* Probe for the analysis of 3′ sequences of RNA.** Soreq,H.; Littauer,U.Z. (Dep. Neurobiol., Weizmann Inst. Sci., Rehovot, Israel) *J. Biol. Chem., 252(19), 6885-6888 (1977)* En;en.

1233-U2 **Purification of heat-labile enterotoxin from four *Escherichia coli* strains by affinity immunoadsorbent: evidence for similar subunit structure.** Dagni,Z.; Sack,R.B.; Craig,J.P. (Makor Chem. Ltd., Jerusalem, Israel) *Infect. Immun., 22(3), 852-860 (1978)* En;en.

1234-U2 **Affinity chromatography of N^5-methyltetrahydrofolate-homocysteine methyltransferase on a cobalamin-Sepharose.** Sato,K.; Hiei,E.; Shimizu,S.; Abeles,R.H. (Dep. Food Sci. and Technol., Nagoya Univ., Nagoya 464, Japan) *FEBS Lett., 85(1), 73-76 (1978)* En.

1235-U2 **Nucleotide phosphotransferase of *Escherichia coli*: purification by affinity chromatography.** Brunngraber,E.F.; *Chargaff,E. (Cell Chem. Lab., Roosevelt Hosp., New York, NY 10019, USA) *Proc. Natl. Acad. Sci. USA, 74(8), 3226-3229 (1977)* En;en.

1236-U2 **Combined use of strain construction and affinity chromatography in the rapid, high-yield purification of 6-**

phosphogluconate dehydrogenase from *Escherichia coli*. Wolf,R.E.,Jr.; Shea,F.M. (Dep. Biol. Sci., Univ. Maryland Baltimore County, Catonsville, MD 21228, USA) *J. Bacteriol., 138(1), 171-175 (1979)* En;en.

1237-U2 **Affinity chromatography and inhibition of chorismate mutase-prephenate dehydrogenase by derivatives of phenylalanine and tyrosine.** Smith,G.D.; Roberts,D.V.; Daday,A. (Dep. Biochem., Fac. Sci., Australian Natl. Univ., Canberra, ACT 2600, Australia) *Biochem. J., 165(1), 121-126 (1977)* En;en.

1238-U2 **Preparation of *Escherichia coli* pyruvate oxidase utilizing a thiamine pyrophosphate affinity column.** O'Brien,T.A.; Schrock,H.L.; Russell,P.; Blake,R.,II; Gennis,R.B. (Dep. Chem., Univ. Illinois, Urbana, IL 61801, USA) *Biochim. Biophys. Acta, 452(1), 13-29 (1976)* En;en.

1239-U2 **Affinity chromatography of *Klebsiella* arylsulfatase on tyrosyl-hexamethylenediamino-β-1,3-glucan and an immunoadsorbent.** Murooka,Y.; Yim,M.-H.; Yamada,T.; Harada,T. (Inst. Sci. and Ind. Res., Osaka Univ., Suita, Osaka 565, Japan) *Biochim. Biophys. Acta, 485(1), 134-140 (1977)* En;en.

1240-U2 **An affinity adsorbent containing deoxyguanosine 5'-triphosphate linked to Sepharose and its use for large scale preparation of ribonucleotide reductase of *Lactobacillus leichmannii*.** Hoffmann,P.J.; Blakley,R.L. (Dep. Biochem., Coll. Med., Univ. Iowa, Iowa City, IA 52242, USA) *Biochemistry (Wash.), 14(22), 4804-4812 (1975)* En;en.

1241-U2 **Purification of methylmalonyl-CoA mutase from *Propionibacterium shermanii* using affinity chromatography.** Murthy,V.V.; Jones,E.; Cole,T.W.,Jr.; Johnson,J.,Jr. (Chem. Dep., Talladega Coll., Talledega, AL 35160, USA) *Biochim. Biophys. Acta, 483(2), 487-491 (1977)* En;en.

1242-U2 **Purification of protein A from *Staphylococcus aureus* by affinity chromatography on polyacrylamide activated with thiophosgene.** Rao,A.K.; Swaminathan,B.; Ayres,J.C.; Mendicino,J. (Dep. Biochem., Univ. Georgia, Athens, GA 30602, USA) *FEBS Lett., 83(2), 329-331 (1977)* En.

1243-U2 **A rapid purification of streptokinase by affinity chromatography.** Comp,P.C. (Sect. Exp. Pathol. and Med., Univ. Oklahoma Health Sci. Cent., Oklahoma City, OK 73190, USA) *Thromb. Res., 14(2-3), 405-411 (1979)* En;en.

1244-U2 **Specific method for the purification of *Streptococcus mutans* dextransucrase.** McCabe,M.M.; Smith,E.E. (Div. Oral Biol. and Dep. Biochem., Univ. Miami Sch. Med., Miami, FL 33152, USA) *Infect. Immun., 16(3), 760-765 (1977)* En;en.

1245-U2 **Affinity chromatography of bacterial malate dehydrogenases.** Kay,W.W.; Down,J.A. (Dep. Biochem. and Microbiol., Univ. Victoria, Victoria, BC V8W 2Y2, Canada) *Curr. Microbiol., 1(5), 293-296 (1978)* En;en.

1246-U2 **Affinity chromatography of several proteolytic enzymes on carbobenzoxy-D-phenylalanyl-triethylenetetramine-Sepharose** Fujiwara,K.; Tsuru,D. (Fac. Pharm. Sci., Nagasaki Univ., Nagasaki, Japan) *Int. J. Peptide Protein Res., 9(1) 18-26 (1977)* En;en.

1247-U2 **Purification and properties of pyrocatechuate decarboxylase from *Aspergillus niger*.** Ramachandran,A.; Subramanian,V.; Sugumaran,M.; Vaidyanathan,C.S. (Dep. Biochem., Indian Inst. Sci., Bangalore-560012, India) *FEMS Microbiol. Lett., 5(6), 421-425 (1979)* En.

1248-U2 **Affinity chromatography of homogentisate-1,2-dioxygenase from *Aspergillus niger*.** Sugumaran,C.S.; Vaidyanathan,M. (Dep. Biochem., Indian Inst. Sci., Bangalor-560 012, India) *FEMS Microbiol. Lett., 4(6), 343-347 (1978)* En.

1249-U2 **Chromatographic purification of ribose 5-phosphate ketol isomerase from *Candida utilis* on p-mercuribenzoate-6-aminohexyl-Sepharose 4B adsorbent.** Horitsu,H.; Banno,Y.; Kimura,S.; Shimizu,H.; Tomoyeda,M. (Dep. Agric. Chem., Fac. Agric., Gifu Univ., Kakamigahara, Gifu, Japan) *Agric. Biol. Chem., 43(3), 663-664 (1979)* En.

1250-U2 **Purification of the inhibitor of the 3'-5' cyclic AMP phosphodiesterase of *Dictyostelium discoideum* by affinity chromatography.** Dicou,E.; Brachet,P. (Inst. Pasteur, Dep. Biol. Mol., Unite de Differenciation Cellulaire, 25 rue du Dr. Roux, 75724 Paris Cedex 15, France) *Biochem. Biophys. Res. Commun., 90(4), 1321-1327 (1979)* En;en.

1251-U2 **A novel purification procedure for *Penicillium notatum* phospholipase B and evidence for a modification of phospholipase B activity by the action of an endogenous protease.** Okumura,T.; Kimura,S.; Saito,K. (De. Med. Chem., Kansai Med. Sch., Moriguchi, Osaka 570, Japan) *Biochim. Biophys. Acta, 617(2), 264-273 (1980)* En;en.

1252-U2 **A new method for the purification of RNA polymerase II (or B) from the lower eukaryote *Physarum polycephalum*. The presence of subforms.** Smith,S.S.; Braun,R. (Inst. Allg. Mikrobiol., Univ. Bern, Altenbergrain 21, CH-3013 Bern, Switzerland) *Eur. J. Biochem., 82(1), 309-320 (1078)* En;en.

1253-U2 **Rapid purification of threonyl-tRNA synthetase from *Saccharomyces carlsbergensis* by affinity elution from phosphocellulose.** Yamada,H. (Inst. Mol. Biol., Sch. Sci., Nagoya Univ., Chikusa-ku, Nagoya, Aichi 464, Japan) *J. Biochem., 83(6), 1583-1589 (1978)* En;en.

1254-U2 **Rapid purification and properties of potassium-activated aldehyde dehydrogenase from *Saccharomyces cerevisiae*.** Bostian,K.A.; Betts,G.F. (Rosenstiel Bas. Med. Res. Cent., Brandeis Univ., Waltham, MA 02154, USA) *Biochem. J., 173(3), 773-786 (1978)* En;en.

1255-U2 **Tyrosyl-tRNA synthetase from baker's yeast. Rapid isolation by affinity elution, molecular weight of the enzyme, and determination of essential sulfhydryl groups.** Faulhammer,H.G.; Cramer,F. (Max-Planck-Inst. Exp. Med., Abt. Chem., Gottingen, GFR) *Eur. J. Biochem., 75(2), 561-570 (1977)* En;en.

1256-U2 The purification of orotidine-5′-phosphate decarboxylase from yeast by affinity chromatography. Brody,R.S.; Westheimer,F.H. (James Bryant Conant Lab., Chem. Dep., Harvard Univ., Cambridge, MA 02138, USA) *J. Biol. Chem., 254(10), 4238-4244 (1979)* En;en.

1257-U2 Affinity chromatography of *Anacystis nidulans* ferredoxin-nitrate reductase and NADP reductase on reduced ferredoxin-Sepharose. Manzano,C.; Candau,P.; Guerrero,M.G. (Dep. Bioquim., Fac. Ciencias y CSIC, Univ. Sevilla, Sevilla, Spain) *Anal. Biochem., 90(1), 408-412 (1978)* En;en.

1258-U2 Laurell crossed immunoelectrophoresis and affinity chromatography for the purification of a parasite antigen. Hillyer,G.V.; Cervoni,M. (Dep. Biol., Univ. Puerto Rico, Rio Piedras, PR 00931, USA) *J. Immunol. Methods, 20, 385-390 (1978)* En;en.

1259-U2 Purification of anti-thyroglobulin autoantibodies by affinity chromatography. Hearn,M.T.W.; Adams,D.D.; Hancock,W.S. (Immunopathol. Res. Group, Med. Res. Counc., Dunedin, New Zealand) *Proc. Univ. Otago Med. Sch., 53(2), 45-47 (1975)* En.

1260-U2 Purification of human immunoglobulin M by affinity chromatography on protamine-sepharose. Wichman,A.; Borg,H. (Dep. Biochem. and Mol. Biol., Northwestern Univ., Evanston, IL 60201, USA) *Biochim. Biophys. Acta, 490(2), 363-369 (1977)* En;en.

1261-U2 Purification of native properdin by reversed affinity chromatography and its activation by proteolytic enzymes. Minta,J.O. (Dep. Pathol., Div. Exp. Pathol., Univ. Toronto, Toronto, Ont., Canada) *J. Immunol., 117(2), 405-412 (1976)* En;en.

1262-U2 One step preparation of both human C-reactive protein and Clt. Pontet,M.; Engler,R.; Jayle,M.F. (UER Biomed. Sts-Peres, Serv. Biochim., Lab. Associe n° 87 du CNRS, 45, rue des Sts-Peres, 75270 Paris Cedex 06, France) *FEBS Lett., 88(2), 172-175 (1978)* En.

1263-U2 Heme-octapeptide of cytochrome-c with peroxidase activity coupled to antibody: an alternative method. Gerber,H.A.; Schaffner,T.; Hess,M.W. (Dep. Pathol., Univ. Bern, 3010 Bern, Switzerland) *Immunochemistry, 13(10), 847-848 (1975)* En;en.

1264-U2 Preparation of properdin by affinity chromatography. Ogle,C.K.; Ogle,J.D.; Alexander,J.W. (Paul I, Hoxworth Blood Cent., Univ. Cincinnati Med. Cent., Cincinnati, OH 45267, USA) *Immunochemistry, 14(5), 341-344 (1977)* En;en.

1265-U2 Isolation of porcine secretory immunoglobulin A by affinity chromatography and determination of its component chains. de Buysscher,E.V.; Berman,D.T. (Dep. Vet. Sci., Univ. Wisconsin, Madison, WI 53706, USA) *Am. J. Vet. Res., 36(11), 1659-1661 (1975)* En;en.

1266-U2 Rapid isolation of a venom protein by affinity chromatography on immobilized antibodies. Gubensek,F.; Zunic,D.; Babnik,J. (Dep. Biochem., J. Stefan Inst., PO Box 199/IV, 61001 Ljubljana, Yugoslavia) *Period Biol., 80(Suppl.1), 97-100 (1978)* En;en.

1267-U2 Purification of α-fetoprotein from mouse amniotic fluid by gel-entrapped antibody filtration. Mizejewski,G.J.; Simon,R.; Vonnegut,M. (Birth Defects Inst., Div. Lab. and Res., New York State Dep. Health, Empire State Plaza, Albany, NY 12201, USA) *J. Immunol. Methods, 31(3-4), 333-339 (1979)* En;en.

1268-U2 Protein A reactivity of various mammalian immunoglobulins. Goudswaard,J.; van der Donk,J.A.; Noordzij,A.; van Dam,R.H.; Vaerman,J.-P. (Dep. Immunol., Fac. Vet. Med., Yalelaan 1, de Uithof, Utrecht, Netherlands) *Scand. J. Immunol., 8(1), 21-28 (1978)* En;en.

1269-U2 Elimination of inter-species reactive anti-IgG antibodies by affinity chromatography. Johansson,M.E.; Espmark,J.A. (Dep. Virol., Statens Bakteriol. Lab., S-105 21 Stockholm, Sweden) *J. Immunol. Methods, 21(3-4), 285-293 (1978)* En;en.

1270-U2 Rapid purification of detergent-solubilized HLA antigen by affinity chromatography employing anti-β_2-microglobulin serum. Robb,R.J.; Strominger,J.L.; Mann,D.L. (Biol. Lab., Harvard Univ., Cambridge, MA 02138, USA) *J. Biol. Chem., 251(17), 5427-5428 (1976)* En;en.

1271-U2 Attempt to isolate antierythrocyte antibodies by the method of affinity chromatography. I. IgM-complement antibodies. Simkova,J. (Teaching Hosp., Transfus. Dep., 60000 Brno-Bohunice, Czechoslovakia) *J. Hyg. Epidemiol. Microbiol. Immunol., 22(4), 442-448 (1978)* En;de,en,fr.

1272-U2 Rapid purification of detergent-solubilized Ia antigens by immunoabsorbent chromatography. Clement,L.T.; Kask,A.M.; Shevach,E.M. (Lab. Immunol., Natl. Inst. Allergy and Infect. Dis., Natl. Inst. Health, Bethesda, MD 20014, USA) *Immunochemistry, 15(6), 393-399 (1978)* En;en.

1273-U2 Rapid preparation of peroxidase:anti-peroxidase complexes for immunocytochemical use. Mason,D.Y.; Sammons,R. (Univ. Oxford, Dep. Pathol., Radcliffe Infirm., Oxford, UK) *J. Immunol. Methods, 20, 317-324 (1978)* En;en.

1274-U2 Simple method of purification of bacteriophage lambda by hydroxyapatite column chromatography. Das,S.; Ghosh,S. (Dep. Biochemistry, Bose Inst., 93/1, A.P.C. Road, Calcutta 700 009, India) *Indian J. Biochem. Biophys., 14(1), 65-67 (1977)* En;en.

1275-U2 Simple isolation method and assay for T4 DNA ligase and characterization of the purified enzyme. Knopf,K.W. (Inst. Mol. Genet., Univ. Heidelberg, Heidelberg, GFR) *Eur. J. Biochem., 73(1), 33-38 (1977)* En;en.

1276-U2 Fractionation of Sendai virus RNA by polylysine-Kieselguhr column chromatography. Ayad,S.R.; Delinassios,J.G. (Dep. Biochem., Univ. Manchester, Manchester M13 9PL, UK) *J. Chromatogr., 130, 396-398*

(1977) En.

1277-U2 Isolation of the M polypeptide of Sendai virus (HVJ) with column chromatography. Semba,T.; Hosaka,Y.; Sakiyama,F. (Dep. Prev. Med., Res. Inst., Microb. Dis., Osaka Univ., Yamada-Kami, Suita, Osaka, Japan) *Biken J., 20(2), 77-80 (1977)* En.

1278-U2 Purification of *Bacillus subtilis* RNA polymerase with heparin-agarose. In vitro transcription of φ29 DNA. Davison,B.L.; Leighton,T.; Rabinowitz,J.C. (Dep. Biochem., Univ. California, Berkeley, CA 94720, USA) *J. Biol. Chem., 254(18), 9220-9226 (1979)* En;en.

1279-U2 A rapid column chromatographic method for the isolation of catechol-type siderophores. Robinson,A.V. (Biol. Dep., Battelle, Pacific Northwest Lab., Richland, WA 99352, USA) *Anal. Biochem., 95(2), 364-370 (1979)* En;en.

1280-U2 Separation of deoxythymidine and deoxythymidine nucleotides by column and thin-layer chromatography. Fitt,P.S.; Peterkin,P.I.; Grey,V.L. (Dep. Biochem., Univ. Ottawa, Ottawa, Ont. K1N 9A9, Canada) *J. Chromatogr., 124(1), 137-140 (1976)* En.

1281-U2 Purification of RNA-DNA hybrids by exclusion chromatography. Persson,H.; Perricaudet,M.; Tolun,A.; Philipson,L.; Pettersson,U. (Dep. Microbiol., Biomed. Cent., S-751 23 Uppsala, Sweden) *J. Biol. Chem., 254(16), 7999-8003 (1979)* En;en.

1282-U2 Gas liquid chromatography in the rapid diagnosis of meningitis. Ferguson,I.R.; Tearle,P.V. (Public Health Lab., County Hosp., Hereford HR1 2ER, UK) *J. Clin. Pathol., 30(12), 1163-1167 (1977)* En;en.

1283-U2 Rapid diagnosis of meningitis by gas liquid chromatographic analysis of cerebrospinal fluid lactic acid. Controni,G.; Rodriguez,W.J.; Deane,C.; Ross,S.; Khan,W.; Puig,J.R. (Child. Hosp. Natl. Med. Cent., 2125 13th St., NW, Washington, DC 20003, USA) *Clin. Proc., Child. Hosp. Natl. Med. Cent., 31(9), 194-201 (1975)* En.

1284-U2 Rapid diagnosis of anaerobic infections by direct gas-liquid chromatography of clinical specimens. Gorbach,S.L.; Mayhew,J.W.; Bartlett,J.G.; Thadepalli,H.; Onderdonk,A.B. (Infect. Dis. Sect., Tufts-New England Med. Cent., Boston, MA 02111, USA) *J. Clin. Invest., 57(2), 478-484 (1976)* En;en.

1285-U2 Identification of *Neisseria* by electron capture gas-liquid chromatography of metabolites in a chemically defined growth medium. Morse,C.D.; *Brooks,J.B.; Kellogg,D.S.,Jr. (Cent. Dis. Control, Atlanta, GA 30333, USA) *J. Clin. Microbiol., 6(5), 474-481 (1977)* En;en.

1286-U2 Assessment of technique for rapid detection of *Escherichia coli* and *Proteus* species in urine by head-space gas-liquid chromatography. Hayward,N.J.; Jeavons,T.H. (Microbiol. Dep. Monash Univ. Med. Sch., Prahran, Vic. 3181 Australia) *J. Clin. Microbiol., 6(3), 202-208 (1977)* En;en.

1287-U2 Identification of *Achromobacter* species by cellular fatty acids and by production of keto acids. Dees,S.B.; Moss,C.W. (Cent. Dis. Control, Atlanta, GA 30333, USA) *J. Clin. Microbiol., 8(1), 61-66 (1978)* En;en.

1288-U2 Gas-liquid chromatographic analysis of metabolic products in the identification of Bacteroidaceae of clinical interest. Deacon,A.G.; Duerden,B.I.; Holbrook,W.P. (Dep. Bacteriol., Edinburgh Univ. Med. Sch., Teviot Place, Edinburgh EH8 9AG, UK) *J. Med. Microbiol., 11(2), 81-99 (1978)* En;en.

1289-U2 Rapid determination of dipicolinic acid in the spores of *Clostridium* species by gas-liquid chromatography. Tabor,M.W.; MacGee,J.; Holland,J.W. (Basic Sci. Lab., Veterans Adm. Hosp., Cincinnati, OH 45220, USA) *Appl. Environ. Microbiol., 31(1), 25-28 (1976)* En;en.

1290-U2 Phosphorus pentoxide as a drying agent for bacterial culture extracts analyzed by gas-liquid chromatography. Finne,G.; Matches,J.R. (Inst. Food Sci. and Technol., Coll. Fish., Univ. Washington, Seattle, WA 98195, USA) *J. Assoc. Off. Anal. Chem., 59(3), 602-605 (1976)* En;en.

1291-U2 Determination of penicillin acylase activity of *E. coli* with gas-liquid chromatography. Tochenaya,N.P.; Vagina,I.M.; Sapozhnikov,Yu.M.; Chaikovskaya,S.M. (All-Union Res. Inst. Antibiot., Moscow, USSR) *Antibiotiki, 20(9), 810-813 (1975)* Ru;en.

1292-U2 Head-space gas liquid chromatography for rapid detection of *Escherichia coli* and *Proteus mirabilis* in urine. Coloe,P.J. (Dep. Microbiol., Monash Univ. Med. Sch., Prahran, Vic. 3181, Victoria) *J. Clin. Pathol., 31(4), 365-369 (1978)* En;en.

1293-U2 Development of specific tests for rapid detection of *Escherichia coli* and all species of *Proteus* in urine. Hayward,N.J.; Jeavons,T.H.; Nicholson,A.J.C.; Thornton,A.G. (Microbiol. Dep., Monash Univ. Med. Sch., Prahran, Vic. 3181, Australia) *J. Clin. Microbiol., 6(3), 195-201 (1977)* En;en.

1294-U2 Identification of clinical isolates of mycobacteria with gas-liquid chromatography alone. Tisdall,P.A.; Roberts,G.D.; *Anhalt,J.P. (Dep. Lab. Med., Mayo Clin. and Mayo Found., Rochester, MN 55901, USA) *J. Clin. Microbiol., 10(4), 506-514 (1979)* En;en.

1295-U2 Characterization of ten species of mycobacteria by reaction gas-liquid chromatography. Ohashi,D.K.; Wade,T.J.; Mandle,R.J. (Box 193, Triplef Army Med. Cent., Honolulu, HI 96859, USA) *J. Clin. Microbiol., 6(5), 469-473 (1977)* En;en.

1296-U2 Rapid diagnosis of anaerobic infections by gas-liquid chromatography of clinical material. Phillips,K.D.; Tearle,P.V.; Willis,A.T. (Public Health Lab., Luton and Dunstable Hosp., Lewsey Road, Luton LU4 0DZ, UK) *J. Clin. Pathol., 29(5), 428-432 (1976)* En;en.

1297-U2 Analysis of acetoin and diacetyl in bacterial culture supernatants by gas-liquid chromatography. Lee,S.M.; *Drucker,D.B. (Dep. Bacteriol., Med. Sch., Manchester, M13 9PT, UK) *J. Clin. Microbiol., 2(3), 162-164 (1975)* En;en.

1298-U2 Applicability of pyrolysis/gas-liquid chromatography for the identification of bacteria in sewage treatment plant effluent. Symuleski,R.A.; *Wetzel,D.M. (Dep. Chem. Eng. and Materials Sci., Catholic Univ. America, Washington, DC 20064, USA) *Environ. Sci. Technol., 13(9), 1124-1130 (1979)* En;en.

1299-U2 Rapid diagnosis of anaerobic infections by gas-liquid chromatography. Ladas,S.; Arapakis,G.; Malamou-Ladas,H.; Palikaris,G.; Arseni,A. (Res. Unit Prof. Dep. Med., 'Evangelismos' Hosp., Athens, Greece) *J. Clin. Pathol., 32(11), 1163-1167 (1979)* En;en.

1300-U2 Rapid diagnosis of meningitis with use of selected clinical data and gas-liquid chromatographic determination of lactate concentration in cerebrospinal fluid. Gastrin,B.; Briem,H.; Rombo,L. (Cent. Hosp., S-631 88 Eskilstuna, Sweden) *J. Infect. Dis., 139(5), 529-533 (1979)* En;en.

1301-U2 Rapid differentiation of the major causative agents of bacterial meningitis by use of frequency-pulsed electron capture gas-liquid chromatography: analysis of amines. Brooks,J.B.; Kellogg,D.S.,Jr.; Shepherd,M.E.; Alley,C.C. (Cent. Dis. Control, Atlanta, GA 30333, USA) *J. Clin. Microbiol., 11(1), 52-58 (1980)* En;en.

1302-U2 Clinical evaluation of a simple, rapid procedure for the presumptive identification of anaerobic bacteria. Holland,J.W.; Gagnet,S.M.; Lewis,S.A.; Stauffer,L.R. (Res. Surg. Bacteriol. Lab., Dep. Surg., Univ. Cincinnati, Coll. Med., Cincinnati, OH 45267, USA) *J. Clin. Microbiol., 5(4), 416-426 (1977)* En;en.

1303-U2 Rapid diagnosis of anaerobic empyema by direct gas-liquid chromatograpy of pleural fluid. Thadepalli,H.; Gangopadhyay,P.K. (Charles Drew Postgrad. Med. Sch., 1621 East 120th St., Los Angeles, CA 90059, USA) *Chest, 77(4), 507-513 (1980)* En;en.

1304-U2 Rapid diagnosis of anaerobic infections by direct gas-liquid chromatography of clinical material. Wust,J. (Inst. Med. Mikrobiol., Univ. Zurich, Postfach CH-8028 Zurich, Switzerland) *Schweiz. Med. Wochenschr., 110(10), 362-369 (1980)* De;de,en.

1305-U2 Rapid presumptive identification of anaerobes in blood cultures by gas-liquid chromatography. Sondag,J.E.; Ali,M.; *Murray,P.R. (Clin. Microbiol. Lab., Barnes Hosp. and Washington Univ. Sch. Med., St. Louis, MO 63110, USA) *J. Clin. Microbiol., 11(3), 274-277 (1980)* En;en.

1306-U2 Modification of the gas-liquid chromatography procedure and evaluation of a new column packing material for the identification of anaerobic bacteria. Hauser,K.J.; Zabransky,R.J. (Microbiol. Res. Lab., Dep. Pathol., Mt. Sinai Med. Cent., Milwaukee, WI 53201, USA) *J. Clin. Microbiol., 2(1), 1-7 (1975)* En;en.

1307-U2 Quantitation of patulin pathway metabolites using gas-liquid chromatography. Ehman,J.; Gaucher,G.M. (Dep. Chem., Biochem. Group, Univ. Calgary, Alta. T2N 1N4, Canada) *J. Chromatogr., 132(1), 17-26 (1977)* En;en.

1308-U2 Determination of fatty acid composition of *Bacillus cereus* and related bacteria: a rapid gas chromatographic method using a glass capillary column. Niskanen,A.; Kiutamo,T.; Raisanen,S.; Raevuori,M. (Food Res. Lab., Tech. Res. Cent. Finland, SF-02150 Espoo 15, Finland) *Appl. Environ. Microbiol., 35(2), 453-455 (1978)* En;en.

1309-U2 Identification of some Enterobacteriaceae by gas chromatography of volatile metabolites formed in a standardized growth medium. van Vuuren,H.J.J.; Toerien,D.F.; Lategan,P.M. (Dep. Microbiol., Univ. Orange Free State, POB 339, Bloemfontein 9300, South Africa) *S. Afr. J. Sci., 74(10), 387-388 (1978)* En.

1310-U2 Gas chromatographic characterization of mycobacteria: analysis of fatty acids and trifluoroacetylated whole-cell methanolysates. Larsson,L.; Mardh,P.-A. (Inst. Med. Microbiol., Univ. Lund, S-223 62 Lund, Sweden) *J. Clin. Microbiol., 3(2), 81-85 (1976)* En;en.

1311-U2 Gas chromatographic method for measuring free lysine in the products of microbiological synthesis. Bosenko,A.M.; Damberg,B.E. (Inst. Microbiol., Latvian Acad. Sci., Riga, USSR) *Prikl. Biokhim. Mikrobiol., 14(3), 450-454 (1978)* Ru;en.

1312-U2 Rapid gas-chromatographic method for identification of metabolic products of anaerobic bacteria. Rizzo,A.F. (State Vet. Med. Inst., SF-00100 Helsinki 10, Finland) *J. Clin. Microbiol., 11(4), 418-421 (1980)* En;en.

1313-U2 Acetylene reduction assay for nitrogenase activity: gas chromatographic determination of ethylene per sample in less than one minute. David,K.A.V.; Apte,S.K.; Banerji,A.; *Thomas,J. (Biol. Group, Bhabha Atomic Res. Cent., Trombay, Bombay 400 085, India) *Appl. Environ. Microbiol., 39(5), 1078-1080 (1980)* En;en.

1314-U2 Detection of alcohols and volatile fatty acids by head-space gas chromatography in identification of anaerobic bacteria. Larsson,L.; Mardh,P.-A.; Odham.G. (Dep. Tech. Analyt. Chem., Univ. Lund, S-220 07 Lund, Sweden) *J. Clin. Microbiol., 7(1), 23-27 (1978)* En;en.

1315-U2 Rapid methods for the analysis of the fatty acids of fermenting wort and beer (C_6-C_{10}) and yeast (C_6-C_{18}). Taylor,G.T.; Kirsop,B.H. (Brewing Res. Found., Nutfield, Redhill, Surrey, UK) *J. Inst. Brew., 83(2), 97-99 (1977)* En;en.

1316-U2 High performance liquid affinity chromatography (HPLAC) and its application to the separation of enzymes and antigens. Ohlson,S.; Hansson,L.; Larsson,P.-O.; Mosbach,K. (Div. Biochem., Chem. Cent., Univ. Lund, PO Box 740, S-220 07 Lund 7, Sweden) *FEBS Lett., 93(1), 5-9 (1978)* En.

1317-U2 Rapid assay for tryptophanase using reversed-phase high-performance liquid chromatography. Krstulovic,A.M.; Matzura,C. (Chem. Dep., Manhattanville Coll., Purchase, NY 10577, USA) *J. Chromatogr., 176(2), 217-224 (1979)* En;en.

1318-U2 Gentamicin C-component ratio determination by high-pressure liquid chromatography. Anhalt,J.P.; Sancilio,F.D.; McCorkle,T. (Mayo Clin. and Mayo Found., Rochester, MN 55901, USA) *J. Chromatogr., 153(2), 489-493 (1978)* En;en.

1319-U2 High-performance liquid chromatography of sterigmatocystin and other metabolites of *Aspergillus versicolor*. Kingston,D.G.I.; Chen,P.N.; Vercellotti,J.R. (Dep. Chem., Virginia Polytech. Inst. and State Univ., Blackburg, VA 24061, USA) *J. Chromatogr., 118(3), 414-417 (1976)* En.

1320-U2 High pressure liquid chromatographic determination of patulin in apple juice. Stray,H. (Food Inspect. Processed Fruits and Veg., Minist. Agric., Gladengvn 3B, Oslo 6, Norway) *J. Assoc. Off. Anal. Chem., 61(6), 1359-1362 (1978)* En;en.

1321-U2 High performance liquid chromatography of natural products. I. Separation of cephalosporin C derivatives and cephalosporin antibiotics; isolation of cephalosporin C from fermentation broth. Miller,R.D.; Neuss,N. (Lilly Res. Lab., Indianapolis, IN 56206, USA) *J. Antibiot., 29(9), 902-906 (1976)* En;en.

1322-U2 Analysis of extracts from *Psilocybe semilanceata* mushrooms by high-pressure liquid chromatography. White,P.C. (Metrop. Police Forensic Sci. Lab., 109 Lambeth Rd., London SE1 7LP, UK) *J. Chromatogr., 169, 453-456 (1979)* En.

1323-U2 Analysis of cellulase proteins by high-performance liquid chromatography. Bissett,F.H. (Food Sci. Lab., US Army Natick Res. and Dev. Command, Natick, MA 01760, USA) *J. Chromatogr., 178(2), 515-523 (1979)* En;en.

1324-U2 Application of high performance liquid chromatography to the quantity analysis of aflatoxins and mycotoxins in press-cakes and by-products. Blanc,M.; Midler,O.; Karleskind,A. (Lab. Wolff, 198 rue Sigmund Freud (ex-avenue du Belvedere), 75019 Paris, France) *Oleagineux, 31(11), 495-499 (1976)* Fr;en,es,fr..

1325-U2 A high-pressure liquid chromatographic method for analysis of carbohydrates and polyols from lichens. Gordy,V.; Baust,J.G.; Hendrix,D.L. (Dep. Biol., Univ. Houston, Houston, TX 77004, USA) *Bryologist, 81(4), 532-538 (1978)* En;en.

1326-U2 High pressure liquid chromatographic determination of aflatoxins in peanut products. Pons,W.A.,Jr.; Franz.A.O.,Jr. (USDA, Sci. and Educ. Admin., Southern Reg. Res. Cent., New Orleans, LA 70179, USA) *J. Assoc. Off. Anal. Chem., 61(4), 793-800 (1978)* En;en.

1327-U2 Determination of aflatoxins in peanut products using reverse phase HPLC. Hurst,W.J.; Toomey,P.B. (Hershey Foods Res. Lab., PO Box 54, Hershey, PA 17033, USA) *J. Chromatogr. Sci., 16(8), 372-376 (1978)* En;en.

1328-U2 High performance liquid chromatography of aflatoxins in cottonseed products. Pons,W.A.,Jr.; Franz,A.O.,Jr. (Southern Reg. Res. Cent., ARS, USDA, New Orleans, LA 70179, USA) *J. Assoc. Off. Anal. Chem., 60(1), 89-95 (1977)* En;en.

1329-U2 A clean-up procedure fo HPLC analysis of aflatoxins in agricultural commodities. Lansden,J.A. (Natl. Peanut Res. Lab., USDA, ARS, South. Reg., Georgia-South Carolina Area, Dawson, GA 31742, USA) *J. Agric. Food Chem., 25(4), 969-971 (1977)* En;en.

1330-U2 Determination of zearalenone in cereals by high-pressure liquid-chromatography (HPLC). Schweighardt,H.; Bohm,J.; Leibetseder,J. (Inst. Ernahr. Veterinarmed. Univ. Wien, Linke Bahngasse 11, A-1030 Wien, Austria) *Z. Tierphysiol. Tierernahr. Futtermittelkd., 41(1), 39-47 (1978)* De;de,en.

1331-U2 On-column chromatographic extraction of aflatoxin M_1 from milk and determination by reversed phase high performance liquid chromatography. Winterlin,W.; Hall,G.; Hsieh,D.P.H. (Dep. Environ. Toxicol., Univ. California, Davis, CA 95616, USA) *Anal. Chem., 51(11), 1873-1874 (1979)* En.

1332-U2 A quantitative method for the separation of chlorophylls a and b from phytoplankton pigments by high pressure liquid chromatography. Jacobsen,T.R. (Dep. Zool., Univ. Georgia, Athens, GA 30602, USA) *Mar. Sci. Commun., 4(1), 33-47 (1978)* En;en.

1333-U2 Purification of the ichthyotoxic component of *Gymnodinium breve* (red tide dinoflagellate) toxin by high pressure liquid chromatography. Kim,Y.S.; Padilla,G.M. (Dep. Physiol. and Pharmacol., Duke Univ. Med. Cent., Durham, NC 27710, USA) *Toxicon, 14(5), 379-387 (1976)* En;en.

1334-U2 The determination of phytoplankton pigments by high-performance liquid chromatography. Abaychi,J.K.; *Riley,J.P. (Dep. Oceanogr., Univ. Liverpool, POB 147, Liverpool L69 3BX, UK) *Anal. Chim. Acta, 107, 1-11 (1979)* En;en.

1335-U2 The determination of phytoplankton pigments by high-performance liquid chromatography. Abaychi,J.K.; Riley,J.P. (Dep. Oceanogr., Univ. Liverpool, PO Box 147, Liverpool L69 3BX, UK) *Anal. Chim. Acta, 107(1), 1-11 (1979)* En;en.

1336-U2 An assay of sterigmatocystin and related metabolites of *Aspergillus versicolor* by high speed liquid chromatography. Ito,K.; Yamane,A.; Hamasaki,T.; Hatsuda,Y. (Food Technol. Inst., Tottori Prefect., Sakaiminato, Tottori 684, Japan) *Agric. Biol. Chem., 40(10), 2099-2100 (1976)* En.

1337-U2 The use of ion exchange chromatography for demonstration of rubella-specific IgM antibodies. Nagy,G.; Mezey,I. (Natl. Inst. Hyg., H 1966 Budapest, PO Box 69, Hungary) *Acta Microbiol. Acad. Sci. Hung., 24(3), 189-194 (1977)* En;en.

1338-U2 Simple ion exchange batch technique for detection of IgM antibodies against rubella and flavivirus antigens. Nagy,G.; Mezey,I.; Molnar,E. (Natl. Inst. Hyg., H-1966 Budapest, PO Box 69, Hungary) *Acta. Microbial. Acad. Sci. Hung., 24(4), 317-321 (1977)* En;en.

1339-U2 Purification of proteins from the 50S ribosomal subunit of *Escherichia coli* by ion-exchange chromatography. Zimmermann,R.A.; Stoffler,G. (Dep. Biochem., Univ. Massachusetts, Amherst, MA 01002, USA) *Biochemistry (Wash.), 15(9), 2007-2017 (1976)* En;en.

1340-U2 Stabilization and purification of ornithine transcarbamylase from *Neisseria gonorrhoeae*. Powers,C.N.; *Pierson,D.L. (Dep. Microbiol. and Immunol., Baylor Coll. Med., Houston, TX 77030, USA) *J. Bacteriol., 141(2), 544-549 (1980)* En;en.

1341-U2 A rapid purification procedure for L-asparaginase from *Vibrio succinogenes*. Abuchowski,A.; Kafkewitz,D.; Davis,F.F. (Dep. Biochem., Rutgers Univ., New Brunswick, NJ 08903, USA) *Prep. Biochem., 9(3), 205-211 (1979)* En;en.

1342-U2 Simple method for purifying choleragenoid, the natural toxoid of *Vibro cholerae*. Mekalanos,J.J.; *Collier,R.J.; Romig,W.R. (Dep. Bacteriol., Univ. California, Los Angeles, CA 90024, USA) *Infect. Immun., 16(3), 789-795 (1977)* En;en.

1343-U2 A simplified procedure for the isolation of bacterial polypeptide elongation factor EF-Tu. Leberman,R.; Antonsson,B.; Giovanelli,R.; Guariguata,R.; Schumann,R.; Wittinghofer,A. (Eur. Mol. Biol. Lab., Postfach 10.2209, 69 Heidelberg, GFR) *Anal. Biochem., 104(1), 29-36 (1980)* En;en.

1344-U2 Chromatographic estimation of fungal mass in plant materials. Wu,L.-C.; Stahmann,M.A. (Dep. Plant Pathol., Coll. Agric. and Life Sci., Univ. Wisconsin-Madison, Madison, WI 53706, USA) *Phytopathology, 65(9), 1032-1034 (1975)* En;en.

1345-U2 Purification of alpha-2 macroglobulin by immunoadsorbent chromatography. McEntire,J.E. (Dep. Hum. Biol. Chem. and Genet., Div. Biochem., Univ. Texas Med. Branch, Galveston, TX 77550, USA) *J. Immunol. Methods, 24(1-2), 39-45 (1978)* En;en.

1346-U2 Separation of anti-dinitrophenyl antibodies according to affinity by fractionated elution from immunoadsorbent using thiocyanate. Shimizu,F.; Mossmann,H.; Himmelspach,K.; Takamiya,H.; *Vogt,A. (Abt. Immunol., Zent. Hyg. Univ., D-7800 Freiburg i Br., GFR) *Res. Exp. Med., 173(2), 165-171 (1978)* En;en.

1347-U2 Rapid differentiation of *Streptomyces* from *Nocardia* by liquid chromatography. Tisdall,P.A.; *Anhalt,J.P. (Dep. Lab. Med., Mayo Clin., and Mayo Found., Rochester, MN 55901, USA) *J. Clin. Microbiol., 10(4), 503-505 (1979)* En;en.

1348-U2 Concentration and purification of enteroviruses by membrane chromatography. Henderson,M.; Wallis,C.; *Melnick,J.L. (Dep. Virol. and Epidemiol., Baylor Coll. Med., Houston, TX 77030, USA) *Appl. Environ. Microbiol., 32(5), 689-693 (1976)* En;en.

1349-U2 Quantitative thin-layer chromatographic determination of the macrolide antibiotic turimycin H and some of its degradation products by densitometric in situ measurement with the image analyser 'Quantimet 720'. Fricke,H.; Muhlig,P.; Kuhne,C. (Zentralinst. Mikrobiol. und Exp. Ther. Jena, Akad. Wissenschaft. DDR, Beuthenbergstr. 11, 69 Jena, GDR) *J. Chromatogr., 153(2), 495-506 (1978)* De;en.

1350-U2 Selectivity of RNA chain initiation in vitro. 1. Analysis of RNA initiations by two-dimensional thin-layer chromatography of 5'-triphosphate-labeled oligonucleotides. Miller,J.S.; *Burgess,R.R. (McArdle Lab. Cancer Res., Univ. Wisconsin, Madison, WI 53706, USA) *Biochemistry (Wash.), 17(1), 2054-2059 (1978)* En;en.

1351-U2 Thin-layer chromatography of mycolic acids and its application for the characterization of BCG strains. / [presented at the 3rd International Symposium on Allergenic and immunogenic mycobacterial antigens held in Plzen, Czechoslovakia, on 11-13 Sept 1974]. Michalec,C.; Mara,M. (Lab. Protein Metab., Fac. Gen. Med., Charles Univ., Prague 2, U nemocnice 5, Czechoslovakia) *J. Hyg. Epidemiol. Microbiol., Immunol., 19(4), 467-470 (1975)* En.

1352-U2 Assay of activity of ribonucleotide reductase by thin-layer chromatography (on ECTEOLA-cellulose). Antoshkina,N.V.; Iordan,E.P.; Vorobyova,L.I. (Moscow State Univ., Moscow, USSR) *Prikl. Biokhim. Mikrobiol., 14(2), 295-299 (1978)* Ru;en.

1353-U2 Separation of glycerylphosphorylinositol mannosides by thin-layer chromatography. Khuller,G.K.; Banerjee,B. (Dep. Biochem., Postgrad. Inst. Med. Educ. and Res., Chandigarh, India) *J. Chromatogr., 150(2), 518-520 (1978)* En.

1354-U2 Feasibility of thin layer chromatography as an inexpensive alternative to gas-liquid chromatography for the identification of some anaerobic gram positive non-sporing rods. Palasuntheram,C.; Drucker,D.B.; Tuxford,A.F. (Med. Res. Inst., Colombo, Sri Lanka) *J. Appl. Bacteriol., 42(3), 451-453 (1977)* En;en.

1355-U2 [A thin layer chromatographic purification method for the determination of aflatoxins B_1, B_2, G_1 and G_2]. Arnold,H. (Migros-Genossenschafts-Bund, Fleisch-Laboratorium, CH 1784 Courtepin, Switzerland) *Fleischwirtschaft, 55(7), 985 (1975)* De;de,en.

1356-U2 Comparison of two methods for analysis of aflatoxin M_1 in milk. Bodine,A.B.; Wayne Davis,R.; Janzen,J.J.; Crawford,F.M. (Dep. Dairy Sci., Clemson Univ.,

Clemson, SC 29631, USA) *J. Dairy Sci., 60(3), 450-452 (1977)* En;en.

1357-U2 [A thin layer chromatographic purification method for the determination of aflatoxins B_1, B_2, G_1 and G_2]. Arnold,H. (Migros-Genossenschafts-Bund, Fleisch-Laboratorium, CH 1784 Courtepin, Switzerland) *Fleischwirtschaft, 55(7), 985 (1975)* De;de,en.

1358-U2 A convenient thin layer chromatographic method for the detection of cyclochlorotine and simatoxin, toxins from *Penicillium islandicum* Sopp. Ghosh,A.C.; Manmade,A.; Bousquet,A.; Townsend,J.M.; Demain,A.L. (SISA Inc., and Sheehan Inst. Res., Inc., Cambridge, MA 02138, USA) *Experientia, 34(6), 819-820 (1978)* En;en.

1359-U2 [Rapid determination of ergosterol and total minor sterols in the unsaponifiable fraction of mould mycelium]. Kreuzig,F. (Forsch. and Entwickl. Biochem. G.m.b.H., A-6250 Kundl, Austria) *Z. Anal. Chem., 281(2), 135-137 (1976)* De;de,en.

1360-U2 Note on a two-dimensional tlc procedure for determining aflatoxins in corn. Alexander,R.J.; Baur,M.C. (Krause Milling Co., Milwaukee, WI 53201, USA) *J. Am. Assoc. Cereal Chem., 54(3), 699-704 (1977)* En.

1361-U2 Nondestructive distinction between aflatoxin B_1 and ethoxyquin in thin-layer chromatography. Issaq,H.J.; Barr,E.W.; Zielinski,W.L.,Jr. (NCI Frederick Cancer Res. Cent., PO Box B, Frederick, MD 21701, USA) *J. Chromatogr., 132(1), 115-120 (1977)* En;en.

1362-U2 Rapid thin layer chromatographic method for determining aflatoxin B_1, ochratoxin A, and zearalenone in corn. Balzer,I.; Bogdanic,C.; Pepeljnjak,S. (Inst. Food Sci. and Technol., Zagreb, Yugoslavia) *J. Assoc. Off. Anal. Chem., 61(3), 584-585 (1978)* En;en.

1363-U2 A simple thin-layer chromatography method for separating the ergot alkaloids. Hassan,S.M. (Pharm. Chem. Dep., Fac. Pharm., Mansura Univ., Mansura, Egypt) *Pharmazie, 33(4), 237 (1978)* En.

1364-U2 An improved semi-quantitative method for the estimation of aflatoxin M_1 in liquid mil'. Patterson,D.S.P.; Roberts,B.A. (Cent. Vet. Lab., Weybridge, Surrey, KT15 3NB, UK) *Food Cosmet. Toxicol., 13(5), 541-542 (1975)* En.

1365-U2 Rapid and simple determination of aflatoxin M_1 in milk in the low parts per 10^{12} range. Gauch,R.; Leuenberger,U.; Baumgartner,E. (Kantoles Lab., PO Box, CH-3000 Bern 9, Switzerland) *J. Chromatogr., 178(2), 543-549 (1979)* En;en.

1366-U2 [Chlorobenzidine as reagents for thin layer chromatographic determination of patulin in apple juice]. Meyer,R.A. (Bezirkshygieneinspektion Dresden, DDR-802, Dresden, Reichenbachstr. 71/73, Hyg.-Inst., GDR) *Nahr., Chem. Biochem. Mikrobiol. Technol., 21(1), 85-86 (1977)* De.

1367-U2 Low substrate concentration determination for nutrient limited growth kinetic studies. II. ^{14}C-glucose analysis by glucose oxidase thin layer chromatography. Robertson,B.R.; *Button,D.K. (Inst. Mar. Sci., Univ. Alaska, Fairbanks, AK 99701, USA) *Mar. Sci. Commun., 3(1), 35-43 (1977)* En;en.

1368-U2 Rapid purification of an RNA tumor virus and proteins by high-performance steric exclusion chromatography on porous glass bead columns. Darling,T.; Albert,J.; Russell,P.; Albert,D.M.; *Reid,T.W. (Dep. Ophthalmol. and Visual Sci., Yale Univ. Sch. Med., 333 Cedar St., New Haven, CT 06510, USA) *J. Chromatogr., 131, 383-390 (1977)* En;en.

1369-U2 Simplified methods for purification of staphylococcal enterotoxins A and C and preparation of anti-enterotoxin sera. Shinagawa,K.; Kunita,N.; Sakaguchi,G. (Osaka Prefectural Inst. Public Health, Univ. Osaka Prefecture, Osaka, Japan) *Jap. J. Bacteriol., 30(6), 683-692 (1975)* Ja;en,ja.

1370-U2 Rapid chromatographic method for determination of mycotoxins. Burzynska,H.; Maciejska,K. (00-791 Warszawa, ul. Chocimska 24, Poland) *Rocz. Panstw. Zakl. Hig., 26(6), 671-674 (1975)* Pl;en,pl,ru.

1371-U2 Purification of *Aspergillus niger* α-amylase. Allam,A.M.; Khalil,N.A.; Dabbous,M.K.; (Lab. Microbiol., Chem., Natl. Res. Cent., Cairo, Egypt) *Zentralbl. Bakteriol. Parasitenkd. Infektionskr. Hyg., II, 131(6), 510-511 (1976)* En;de,en.

1372-U2 Isolation of DNA from yeast by chromatography on hydroxyapatite. Ivanov,I.G.; Venkov,P.V.; Markov,G.G. (Inst. Biochem., Bulgarian Acad. Sci., Sofia 13, Bulgaria) *Prep. Biochem., 5(3), 219-228 (1975)* En;en.

1373-U2 A simplified method of isolation of pure HLA antigens and beta$_2$-microglobulin. Rieva,C.; Wikler,M. (Hop. Universitaire St. Pierre, Serv. Immunol. et Transfus. Sanguine, 322 rue Haute, 1000 Bruxelles, Belgium) *Acta Clin. Belg., 31(Suppl. 8), 27-32 (1976)* En;en,fr,nl.

1374-U2 Preparation of an R3 reagent (serum depleted of the third component of human complement (C3)) by immunoadsorption. Application to the hemolytic assay of human C3. Fontaine,M.; Joisel,F.; Dumouchel,L. (INSERM, Unite 78, 543 Chemin de la Breteque, 76230 Bois-Guillaume, France) *J. Immunol. Methods, 33(2), 145-158 (1980)* En;en.

1375-U2 Improved cultural methods for *Tritrichomonas foetus*. Wosu,L.O. (Dep. Trop. Vet. Sci., James Cook Univ., Townsville, Queensl., Australia) *Vet. Microbiol., 2(1), 89-93 (1977)* En;en.

See: 971, 1003, 1016, 1511, 1870, 3268, 3300

Fluorometry

1376-U2 Fluorometric determination of the quality of FITC conjugates. Korting,H.J.; Klagge,E.; Warweg,U.; Hindersin,P. (Virol. Lab., Neurol. Psychiatr. Clin., Med.

Acad. Erfurt, 74 Nordhauser Str. Erfurt, GDR) *Rev. Roum. Med., Ser. Virol., 28(1), 41-43 (1977)* En;en,fr.

1377-U2 Computer-operated scanning analysis of FITC-labelled measles virus in cell cultures. Korting,H.J.; Klagge,E.; Warweg,U.; Porstmann,I.; Weber,P. (Virol. Lab., Neurol.-Psychiatric Clin., Med. Acad. Erfurt, 74 Nordhauser Str., 50 Erfurt, GDR) *Rev. Roum. Med., Ser. Virol., 30(3), 185-188 (1979)* En;en,fr.

1378-U2 A specific method for determination of free tryptophan and endogenous tryptophan in *Escherichia coli*. Bliss,R.D. (Dep. Biol., Univ. California, Riverside, CA 92521, USA) *Anal. Biochem., 93(2), 390-398 (1979)* En;en.

1379-U2 Quantification of L-glutamine using *Escherichia coli* glutamate synthase; a sensitive fluorometric assay. Miller,R.E. (Div. Endocrinol. and Metab., Dep. Med., Case Western Reserve Univ., Veterans Adm. Hosp., Cleveland, OH 44106, USA) *Anal. Biochem., 75(1), 91-99 (1976)* En;en.

1380-U2 Rapid filter method for the microfluorometric analysis of DNA. Cattolico,R.A.; Gibbs,S.P. (Dep. Bot., Univ. Washington, Washington, Seattle, WA 98195, USA) *Anal. Biochem., 69(2), 572-582 (1975)* En;en.

1381-U2 A simple device for assuring high reproducibility in fluorometric measurement of deoxyribonucleic acid in yeast cells. Doi,A.; Doi,K. (Inst. Sci. and Ind. Res., Osaka Univ., Suita, Osaka, Japan) *Agric. Biol. Chem., 40(4), 813-814 (1976)* En.

1382-U2 A fluorometric rapid screen method for aflatoxin in peanuts. Davis,N.D.; Guy,M.L.; Diener,U.L. (Dep. Bot., Plant Pathol. and Microbiol., Auburn Univ. Agric. Exp. Stn., Auburn, AL 36830, USA) *J. Am. Oil Chem. Soc., 57(3), 109-110 (1980)* En;en.

1383-U2 A rapid photofluorometric technique for quantitative analysis of aflatoxin B_1 to be used in animal studies. Llewellyn,G.C.; Carlton,W.W.; Robbers,J.E.; Hansen,W.G. (Dep. Biol., Virginia Commonw. Univ., Richmond, VA 23284, USA) *Dev. Ind. Microbiol., 16, 352-355 (1974)* En;en.

1384-U2 A cytofluorometric method of counting trypanosomes. Mills,J.N.; Valli,V.E.O. (Univ. Guelph, Dep. Pathol., Ontario Vet. Coll., Guelph, Ont., Canada N1G 2W1) *Tropenmed. Parasitol., 29(1), 95-100 (1978)* En;de,en.

1385-U2 A fluorometric technique for measuring sinking rates of freshwater phytoplankton. Titman,D. (Dep. Zool., Univ. Michigan, Ann Arbor, MI 48104, USA) *Limnol. Oceanogr., 20(5), 869-875 (1975)* En;en.

1386-U2 Fluorimetric determination of chlorophyll a and phaeophytin a. Neveux,J. (Lab. Arago, F 66650 Banyuls-sur-Mer, France) *Ann. Inst. Oceanogr., Paris, 52(2), 165-174 (1976)* Fr;fr,en.

1387-U2 Automated analysis of antigen-stimulated lymphocytes. Cassidy,M.; Yee,C.; Costa,J. (Lab. Pathol., Natl. Cancer Inst., Natl. Inst. Health, Bethesda, MD 20014, USA) *J. Histochem. Cytochem., 24(1), 373-377 (1976)* En;en.

1388-U2 A microfluorometric assay for immunoglobulin class and subclass levels in murine serum. Haaijman,J.J.; Brinkhof,J. (Inst. Exp. Gerontol. TNO, 151 Lange Kleiweg, Rijswijk (ZH), Netherlands) *J. Immunol. Methods, 14(3-4), 213-224 (1977)* En;en.

1389-U2 Rapid multiparameter analysis of cell stimulation in mixed lymphocyte culture reactions. Traganos,F.; Gorski,A.J.; Darzynkiewicz,Z.; Sharpless,T.; Melamed,M.R. (Invest. Cytol. Lab., Mem. Sloan-Kettering Cancer Cent., New York, NY 10021, USA) *J. Histochem. Cytochem., 25(7), 881-887 (1977)* En;en.

1390-U2 Flow microfluorometric quantitation of the blastogenic response of lymphocytes. Cram,L.S.; Comez,E.R.; Thoen,C.O.; Forslund,J.C.; Jett,J.H. (Biophys. Instrum. Group, Los Alamos Sci. Lab., Univ. California, Los Alamos, NM 87545, USA) *J. Histochem. Cytochem., 24(1), 383-387 (1976)* En;en.

1391-U2 Automated analysis of antigen-stimulated lymphocytes. Cassidy,M.; Yee,C.; Costa,J. (Lab. Pathol., Natl. Cancer Inst., Natl. Inst. Health, Bethesda, MD 20014, USA) *J. Histochem. Cytochem., 24(1), 373-377 (1976)* En;en.

1392-U2 Rapid fluorometric assay for cerebrospinal fluid immunoglobulin G. Menonna,J.; Galantowicz,D.; Dowling,P.; Cook,S. (Neurol. Serv., Veterans Adm. Hosp., East Orange, NJ 07019, USA) *Neurology, 27(5), 481-483 (1977)* En;en.

1393-U2 Fluorimetric assay of acid lipase in human leukocytes. Kelly,S.; Bakhru-Kishore,R. (New York State Dep. Health, Birth Defects Inst., Albany, NY 12201, USA) *Clin. Chim. Acta, 97(2-3), 239-242 (1979)* En;en.

1394-U2 A microfluorometric assay of the lysosomal arylsulfatases in leukocytes. Bakhru-Kishore,R.; Kelly,S. (New York State Dep. Health, Birth Defects Inst., Albany, NY 12201, USA) *Clin. Chim. Acta, 75(3), 483-485 (1977)* En.

See: 1168

Turbidimetry, nephelometry and colorimetry

1395-U2 Practical colorimeter for direct measurement of microplates in enzyme immunoassay systems. Clem,T.R.; Yolken,R.H. (Biomed. Eng. and Instrument. Branch, Div. Res. Serv., Natl. Inst. Health, Bethesda, MD 20014, USA) *J. Clin. Microbiol., 7(1), 55-58 (1978)* En;en.

1396-U2 Practical colorimeter for direct measurement of microplates in enzyme immunoassay systems. Clem,T.R.; Yolken,R.H. (Biomed. Eng. and Instrument. Branch, Div. Res. Serv., Natl. Inst. Health, Bethesda, MD 20014, USA) *J. Clin. Microbiol., 7(1), 55-58 (1978)* En;en.

1397-U2 An automatic colorimeter and its use in evaluating the growth response of an anaerobic general faty acid auxotroph to cis- and trans-octadecenoic acids. Hazlewood,G.P.; Reynolds,M.J.; Dawson,R.M.C.; Gunstone,F.D. (Dep. Biochem., Inst. Anim. Physiol., Babraham, Cambridge CB2 4AT, UK) *J. Appl. Bacteriol., 47(2), 321-325 (1979)* En;en.

1398-U2 Practical colorimeter for direct measurement of microplates in enzyme immunoassay systems. Clem,T.R.; Yolken,R.H. (Biomed. Eng. and Instrument. Branch, Div. Res. Serv., Natl. Inst. Health, Bethesda, MD 20014, USA) *J. Clin. Microbiol., 7(1), 55-58 (1978)* En;en.

1399-U2 Investigation of ribonuclease-catalysed kinetics by a micro-calorimetric method. Tribout,M.; Paredes,S.; Leonis,J. (Univ. Libre de Bruxelles, Chim. Gen. 1, Fac. sci., Ave. F.D. Roosevelt 50, 1050 Bruxelles *Biochem. J., 153(1), 89-91 (1976)* En;en.

1400-U2 Colorimetric analysis of immunogenic impurities in acetylsalicylic acid. Bundgaard,H. (R. Danish Sch. Pharm., Dep. Pharmaceut., 2 Universitetsparken, DK 2100 Copenhagen, Denmark) *J. Pharm. Pharmacol., 28(7), 544-547 (1976)* En;en.

1401-U2 Simple and rapid colorimetric method for the microdetermination of alpha amylase. Attia,R.M.; Ali,S.A. (Microb. Enzyme Unit, Microbiol. Res. Div., Agric. Res. Cent., Gamma St., Giza, Egypt) *Zentralbl. Bakteriol. Parasitenkd. Infektionskr. Hyg., II, 132(3), 193-195 (1977)* En;de,en.

1402-U2 Rapid enumeration of fecal coliforms in water by a colorimetric β-galactosidase assay. Warren,L.S.; *Benoit,R.E.; Jessee,J.A. (Dep. Biol., Virginia Polytech. Inst. and State Univ., Blacksburg, VA 24061, USA) *Appl. Environ. Microbiol., 35(1), 136-141 (1978)* En;en.

1403-U2 Rapid assay for the determination of bacterial resistance to the lethal activity of serum. Moll,A.; *Cabello,F.; Timmis,K.N. (Dep. Microbiol., New York Med. Coll., Valhalla, NY 10595, USA) *FEMS Microbiol. Lett., 6(5), 273-276 (1979)* En.

1404-U2 Calorimetric identification of several strains of lactic acid bacteria. Fujita,T.; Monk,P.R.; Wadso,I. (Inst. Appl. Microbiol., Univ. Tokyo, Bunkyu-ku, Tokyo, Japan) *J. Dairy Res., 45(3), 457-463 (1978)* En;en.

1405-U2 Rapid, colorimetric test for the determination of hippurate hydrolysis by group B *Streptococcus*. Edberg,S.C.; Samuels,S. (Montefiore Hosp. and Med. Cent., Div. Microbiol. and Immunol., Dep. Pathol., New York, NY 10467, USA) *J. Clin. Microbiol., 3(1), 49-50 (1976)* En;en.

1406-U2 Design and testing of a calorimeter for microbiological uses. Fujita,T.; Nunomura,K.; Kagami,I.; Nishikawa,Y. (Inst. Appl. Microbiol., Univ. Tokyo, Tokyo, Japan) *J. Gen. Appl. Microbiol., 22(1), 43-50 (1976)* En;en.

1407-U2 Colorimetric assay of penicillin amidase activity using phenylacetyl-aminobenzoic acid as substrate. Szewczuk,A.; Siewinski,M.; *Slowinska,R. (Biochem. Lab., Inst. Immunol. and Exp. Ther., Pol. Acad. Sci., Wroclaw, Poland) *Anal. Biochem., 103(1), 166-169 (1980)* En;en.

1408-U2 Bioassay of nystatin bulk material by flow microcalorimetry. Beezer,A.E.; Newell,R.D.; Tyrrell,H.J.V. (Chem. Dep., Chelsea Coll., Manresa Road, London SW3 6LX, UK) *Anal. Chem., 49(1), 34-37 (1977)* En;en.

1409-U2 Rapid test for determining the intracellular rhodanese activity of various bacteria. Vandenbergh,P.A.; Bawdon,R.E.; Berk,R.S. (Dep. Immunol. and Microbiol., Wayne State Univ. Sch. Med., Detroit, MI 48201, USA) *Int. J. Syst. Bacteriol., 29(4), 339-344 (1979)* En;en.

1410-U2 A rapid method for the detection of nonionic surfactant-degrading micro-organisms. Cook,K.A. (Shell Res. Ltd., Shell Biosci. Lab., Sittingbourne Res. Cent., Sittingbourne, Kent ME9 8AG, UK) *J. Appl. Bacteriol., 44(2), 299-303 (1978)* En;en.

1411-U2 Simple and rapid colorimetric method for the microdetermination of alpha amylase. Attia,R.M.; Ali,S.A. (Microbiol. Enzyme Unit, Microbiol. Res. Div., Agric. Res. Cent., Gamma St., Giza, Egypt) *Zentralbl. Bakteriol. Parasitenkd. Infektionskr. Hyg., II, 132(3), 193-195 (1977)* En;de,en.

1412-U2 Automated colorimetric determination of cellulase activity in fermenation samples using a ferricyanide-molybdoarsenic acid reagent. Holm,K.A. (Novo Industri A/S, Anal. Lab., Enzymes Div., Novo Alle, DK 880 Bagsvaerd, Denmark) *Anal. Biochem., 84(2), 522-532 (1978)* En;en.

1413-U2 Improved direct colorimetric method for the determination of total serum globulins. Neeley,W.E.; Pollack,M.; Cupas,C.A. (Div. Biochem., Walter Reed Army Inst. Res., Washington, DC 20012, USA) *Clin. Biochem., 8(4), 273-278 (1975)* En;en.

1414-U2 Colorimetric determination of succinic acid using yeast succinate dehydrogenase. Valle,A.B.F.; Panek,A.D.; Mattoon,J.R. (Fed. Univ. Rio de Janeiro, Inst. Chem., Rio de Janeiro, Brazil) *Anal. Biochem., 91(2), 583-599 (1978)* En;en.

1415-U2 Immunoglobulin assay by automated nephelometry. Buschhausen,J.C. (McMaster Univ. Med. Cent., Hamilton, Ont., Canada) *Can. J. Med. Technol., 37(4), 97-113 (1975)* En;en.

1416-U2 Immunonephelometric assays. Ackerman,E.; Rosevear,J.W. (Health Computer Sci. Dep. Lab. Med. and Pathol., Univ. Minnesota, Minneapolis, MN 55455, USA) *Med. Prog. Technol., 6(2), 81-90 (1979)* En;en.

1417-U2 Ammonium sulphate enhancement of the antigen-antibody reaction. Manual and automated nephelometric studies. Stakenburg,J. (Lab. Chem. Pathol., Ziekenhuis St. Antoniushove, Burg, Banninglaan 1, Leidschendam, Netherlands) *Clin. Chim. Acta, 91(3), 251-261 (1979)* En;en.

1418-U2 A rapid and sensitive method using immunological nephelometric analysis to evaluate the gel-filtration method

(application to the diagnosis of rubella). Sire,J.; Denoyel,G.A.; Colle,A.; Manuel,Y. (Lab. Rech. sur les Proteines et Macromolecules dans les Liquides Biol., Inst. Pasteur, 77, rue Pasteur, 69365 Lyon Cedex 2, France) *Pathol. Biol., 25(6), 425-428 (1977)* En;en,fr.

1419-U2 Serologic-nephelometric method for the detection of allergic reactions to small molecular weight drugs using particulate instead of soluble polystyrene sulfonate-Na as a helper substance. Hoigne,R.; Rink,S.; Patrizzi,R.; Spiess,J.; Bavaud,C.; Spring,M.; Waefler-Tschudi,L. (Zieglerspital, Postfach 2600, CH-3001 Bern, Switzerland) *Schweiz. Med. Wochenschr., 107(32), 1125-1133 (1977)* De;de,en.

1420-U2 An evaluation of the Hyland laser nephelometer PDQ system for the measurement of immunoglobulins. Whicher,J.T.; Perry,D.E.; Hobbs,J.R. (Specific Protein Ref. Unit, Dep. Chem. Pathol., Westminster Hosp., London, UK) *Ann. Clin. Biochem., 15(part 2), 77-85 (1978)* En;en.

1421-U2 Use of laser nephelometry in the measurement of specific antibody to lamb's quarters antigen. Caplin,J.A.; Capriles Hulett,A.; Walters,G.; Neiburger,R.; Merriwether,S.; Straub,D.L.; Neiburger,J.B.; DeYarman,K.; *Dockhorn,R.J. (5300 West 94th St. Terrace, Prairie Village, KS 66207, USA) *Ann. Allergy, 41(4), 236-237 (1978)* En;en.

1422-U2 Immunochemical determination of human apolipoprotein B by laser nephelometry. Fievet-Desreumaux,C.; Dedonder-Decoopman,E.; Fruchart,J.C.; Dewailly,P.; *Sezille,G. (Lab. Physiopathol. Lipides (ERA, CNRS No. 070497), Clin. Med. Gen. A, Fac. Med., Place Verdun, 59045 Lille Cedex, France) *Clin. Chim. Acta, 95(2), 405-408 (1979)* En;en.

1423-U2 Hyland laser nephelometer compared with radial immunodiffusion assay for quantitation of individual immunoglobulins, and an assessment of the excess-antigen. Roberts,N.B.; Nance,L.A. (Dep. Clin. Pathol., Liverpool Area Health Auth. (Teaching), Ashton St., Liverpool L3 5RT, UK) *Clin. Chem., 25(8), 1509-1510 (1979)* En.

1424-U2 Use of laser nephelometry in the measurement of serum proteins. Deaton,C.D.; Maxwell,K.W.; Smith,R.S.; Creveling,R.L. (Hyland Div. Travenol Lab., Inc., POB 2214, 3300 Hyland Ave., Costa Mesa, CA 92628, USA) *Clin. Chem., 22(9), 1465-1471 (1976)* En;en.

1425-U2 Rapid automated turbidimetric assay for chlortetracycline hydrochloride using *Leuconostoc mesenteroides* as the test organism. Mueller,D.L.; Reed,S.J.; Barkate,J.A. (Ralston Purina Co., Microbiol Dep., St. Louis, MO 63188, USA) *J. Assoc. Off. Anal. Chem., 62(1), 160-167 (1979)* En;en.

1426-U2 Semiautomated turbidimetric bioassay for the ionophore A23187. Westhead,J.E. (Lilly Res. Lab., Eli Lilly and Co., Indianapolis, IN 46206, USA) *Antimicrob. Agents Chemother., 11(5), 916-918 (1977)* En;en.

1427-U2 A rapid microbiological method of assaying cefoxitin. Rogers,D.T.; Russell,A.D. (Welsh Sch. Pharm., Univ. Wales, Inst. Sci. and Technol., Cathays Park, Cardiff, CF1 3BU, UK) *J. Antimicrob. Chemother., 5(2), 230-231 (1979)* En.

1428-U2 Rapid microbiological assay for chlorhydroxyquinoline that uses a cryogenically stored inoculum. Cosgrove,R.F. (Int. Dev. Lab., E. R. Squibb and Sos, Moreton, Wirral, Merseyside L46 1QW, UK) *Antimicrob. Agents Chemother., 11(5), 848-851 (1977)* En;en.

1429-U2 Semielectronic turbidimeter for automated monitoring of bacterial growth in test tubes. Marcelis,J.H.; Versteeg,H.; Mansvelt Beck,H.J.; Vinke,D. (Lab. Microbiol., State Univ., Catharijnesingel 59, Utr. Netherlands) *Appl. Environ. Microbiol., 39(2), 281-284 (1980)* En;en.

1430-U2 Spectrodensitometric determination of trichothecene mycotoxins with 4-(p-nitrobenzyl)pyridine on silica gel thin-layer chromatograms. Takitani,S.; Asabe,Y.; Kato,T.; Suzuki,M.; Ueno,Y. (Fac. Pharm. Sci., Sci. Univ. Tokyo, 12 Ichigaya-funagawara-machi, Shinjuku-ku, Tokyo 162, Japan) *J. Chromatogr., 172, 335-342 (1979)* En;en.

1431-U2 A turbidimetric assay method for studying antibody-induced lysis of hapten-coated erythrocytes. Cofrancesco,E.; Vigilante,A.; Revoltella,R. (Lab. Biol. Cell. Consiglio Naz. Ric. (CNR) Via G. Romagnosi 18/A, 00196 Roma, Italy) *Ric. Clin. Lab., 6(2), 111-120 (1976)* En;en.

1432-U2 Immunochemical determination of human immunoglobulins: use of kinetic turbidimetry and a 36-place centrifugal analyzer. Finley,P.R.; Williams,R.J.; Lichti,D.A.; Griffith,F.; Thies,A.C. (Dep. Pathol., Univ. Arizona, Health Sci. Cent., Tucson, AZ 85724, USA) *Clin. Chem., 25(4), 526-530 (1979)* En;en.

1433-U2 Kinetic immunoturbidimetry: the estimation of albumin. Spencer,K.; Price,C.P. (Dep. Chem. Pathol., Southampton Gen. Hosp., Southampton, UK) *Clin. Chim. Acta, 95(2), 263-276 (1979)* En;en.

1434-U2 An automated turbidimetric rate method for immunoglobulin assays. Malkus,H.; Buschbaum,P.; Castro,A. (Dep. Pathol., Univ. Miami, Miami, FL 33152, USA) *Clin. Chim. Acta, 88(3), 523-530 (1978)* En;en.

1435-U2 An optical method for the detection and quantitation of antibodies to glycosphingolipids and other antigens. Hamers,M.N.; Donker-Koopman,W.E.; Reijngoud,D.-J.; Schram,A.W.; Tager,J.M. (Lab. Biochem., Univ. Amsterdam, BCP Jansen Inst., Plantage Muidergracht 12, Amsterdam, Netherlands) *Immunochemistry, 15(2), 97-105 (1978)* En;en.

1436-U2 Some conditions of the turbidimetric method application for determining the yeast concentration. Miskiewicz,T. (Inst. Technol. Przem. Chem. i Spozyw. AE, Wróclaw, Poland) *Przem. Ferment. Rolny, 22(3), 17-19 (1978)* Pl;en,pl,ru.

1437-U2 Microbiological plate and turbidimetric assays of chlortetracycline in feeds: collaborative study. Ragheb,H.S. (Dep. Biochem., Purdue Univ., West Lafayette, IN 47907,

USA) *J. Assoc. Off. Anal. Chem., 60(5), 1119-1124 (1977)* En;en.

1438-U2 **Microcalorimetric measurement of normal and adenovirus infected HeLa cells.** Ljungholm,K.; Wadso,I.; Kjellen,L. (Thermochem. Lab., Chem. Cent., Univ. Lund, Box 740, S 220 7 Lund, Sweden) *Acta Pathol. Microbiol Scand. Ser B, 86(3), 121-124 (1978)* En;en.

1439-U2 **Semiautomated turbidimetric microbiological assay for cefazaflur.** Pitkin,D.H.; *Actor,P.; Baldinus,J.G.; Post,A.; Weisbach,J.A. (Res. and Dev. Div., Smith-Kline and French Lab., Philadelphia, PA 19101, USA) *J. Antibiot., 31(4), 359-363 (1978)* En;en.

1440-U2 **Evaluation of two commercially availabel laser nephelometers and their recommended methods for measurement of serum immunoglobulin G.** Walsh,R.L.; Coles,M.E.; Geary,T.D. (Clin. Chem. Div., Inst. Med. and Vet. Sci. Box 14 Rundle St. PO, Adelaide, S.A. 5000, Australia) *Clin. Chem., 25(1), 151-156 (1979)* En;en.

1441-U2 **Nephelometric method for determination of rheumatoid factor.** Virella,G.; Waller,M.; Fudenberg,H.H. (Dep. Basic and Clin. Immunol. and Microbiol., Med. Univ. South Carolina, Charleston, SC 29401, USA) *J. Immunol. Methods, 22(3-4), 247-251 (1978)* En;en.

See: 738, 935, 937, 2530, 3008, 3057, 3068, 3298

Spectrophotometry and spectroscopy

1442-U2 **Microtiter assay for interferon: microspectrophotometric quantitation of cytopathic effect.** McManus,N.H. (Dep. Med., Dartmouth-Hitchcock Med. Cent., Hanover, NH 03755, USA) *Appl. Environ. Microbiol., 31(1), 35-38 (1976)* En;en.

1443-U2 **Rapid spectrophotometric determination of *Acholesplasma laidlawii* membranes.** Dresdner,G. (Dep. Biophys., Arrhenius Lab., Stockholm Univ., S 106 91 Stockholm, Sweden) *Can. J. Microbiol., 24(9), 1093-1097 (1978)* En;en,fr.

1444-U2 **A spectrophotometric procedure for quantitation of antibody directed to bacterial antigens.** Iacono,V.J.; *Taubman,M.A.; Smith,D.J.; Moreno,E.C. (Dep. Immunol., Forsyth Dent. Cent., Boston, MA 02115, USA) *Immunochemistry, 13(3), 235-243 (1976)* En;en.

1445-U2 **A simple method for the determination of poly-β-hydroxybutyric acid in microbial biomass.** Juttner,R.-R.; Lafferty,R.M.; *Knackmuss,H.-J. (Inst. Mikrobiol., Abt. Bakterienphysiol., D-3400 Gottingen-Weende, Grisebachstr. 8, GFR) *Eur. J. Appl. Microbiol., 1(3), 233-237 (1975)* En;en.

1446-U2 **A spectrophotometric rapid kinetic study of reactions catalysed by coenzyme-B_{12}-dependent ethanolamine ammonia-lyase.** Hollaway,M.R.; White,H.A.; Joblin,K.N.; Johnson,A.W.; Lappert,M.; Wallis,O.C. (Dep. Biochem., Univ. Coll. London, Gower St., London WC1E 6BT, UK) *Eur. J. Biochem., 82(1), 143-154 (1978)* En;en.

1447-U2 **Phospholipase C assay using p-nitrophenylphosphorylcholine together with sorbitol and its application to studying the metal and detergent requirement of the enzyme.** Kurioka,S.; Matsuda,M. (Dep. Biochem., Jikei Univ. Sch. Med., Minato-ku, Tokyo 105, Japan) *Anal. Biochem., 75(1), 281-289 (1976)* En;en.

1448-U2 **Use of umbelliferone derivatives in the study of enzyme activities of mycobacteria.** Grange,J.M. Clark,K. (Dep. Microbiol., Cardiothorac. Inst., Fulham Road, London SW3 6HP, UK) *J. Clin. Pathol., 30(2), 151-153 (1977)* En;en.

1449-U2 **Purification, new assay, and properties of coenzyme A transferase from *Peptostreptococcus elsdenii*.** Tung,K.K.; *Wood,W.A. (Dep. Biochem., Michigan State Univ., East Lansing, MI 48824, USA) *J. Bacteriol., 124(3), 1462-1474 (1975)* En;en.

1450-U2 **A spectrophotometric assay for demethiolating activity.** Laakso,S.; Nurmikko,V. (Dep. Biochem., Univ. Turku, 20500 Turku 50, Finland) *Anal. Biochem., 72(1/2), 600-605 (1976)* En;en.

1451-U2 **Comparison of two methods for evaluating the biodegradation of hydrocarbons in vitro.** Fusey,P.; Oudot,J. (Lab. Crypt., Mus. Natl. Hist. Nat., Lab. Associe CNRS, 12 rue Buffon Paris V°, France) *Mater. Org., 11(4), 241-151 (1976)* Fr;de,en,es,fr.

1452-U2 **Rapid spectrophotometric differentiation between glutathione-dependent and glutathione-independent gentisate and homogentisate pathways.** Crawford,R.L.; Frick,T.D. (Univ. Minnesota, Freshwater Biol. Inst., Navarre, MN 55392, USA) *Appl. Environ. Microbiol., 34(2), 170-174 (1977)* En;en.

1453-U2 **A simple method for the determination of poly-β-hydroxybutyric acid in microbial biomass.** Juttner,R.-R.; Lafferty,R.M.; Knackmuss,H.-J. (Inst. Mikrobiol., Gesellschaft Strahlen- und Umweltforsch. mbH, Munchen, in Gottingen, GFR) *Eur. J. Appl. Microbiol., 1(3), 233-237 (1975)* En;en.

1454-U2 **A simple procedure for determination of bacteriolytic activity in biological fluids.** Bratlid,D. (Dep. Paediatr., University Hosp., Univ. Tromso, N-9012 Tromso, Norway) *Acta Pathol. Microbiol. Scand., Ser. C, 85(1), 17-20 (1977)* En;en.

1455-U2 **First practical assay for soluble nitrous oxide reductase of denitrifying bacteria and a partial kinetic characterization.** Kristjansson,J.K.; Hollocher,T.C. (Dep. Biochem., Brandeis Univ., Waltham, MA 02254, USA) *J. Biol. Chem., 255(2), 704-707 (1980)* En;en.

1456-U2 **A spectrophotometric assay for gentamicin.** Williams,J.W.; Langer,J.S.; *Northrop,D.B. (Sch. Pharm., Univ. Wisconsin, Madison, WI 53706, USA) *J. Antibiot., 28(12), 982-987 (1975)* En;en.

1457-U2 Methodologies for the determination of adenosine phosphates. Bostick,W.D.; Ausmus,B.S. (Oak Ridge Natl. Lab., Oak Ridge, TN 37830, USA) *Anal. Biochem., 88(1), 78-92 (1978)* En;en.

1458-U2 Spectrophotometric assay for amikacin using purified kanamycin acetyltransferase. Scarbrough,E.; Williams,J.W.; *Northrop,D.B. (Sch. Pharm., Univ. Wisconsin, Madison, WI 53706, USA) *Antimicrob. Agents. Chemother., 16(2), 221-224 (1979)* En;en.

1459-U2 Automated spectrophotometric determination of o-diphenoloxidase activity in common mushroom *Agaricus bisporus* L. Varoquaux,P.; Sarris,J.; Albagnac,G. (Stn. Technol. rod. Veg., Cent. Rech. Dijon, INRA, 7 rue Sully, 21034 Dijon Cedex, France) *Ann. Technol. Agric., 26(4), 461-472 (1977)* Fr;en,fr.

1460-U2 An analysis of *Porphyridium* absorption bands with a digital spectrophotometer. Leclerc,J.C. Hoarau,J.; Guerin-Dumartrait,E. (deceased) (Lab. Physiol. Cell. Veg., Associe au CNRS, Univ. Paris-Sud, Centre d'Orsay, 91405-Orsay, France) *Photochem. Photobiol., 22(1-2), 41-48 (1975)* En;en.

1461-U2 1st communication on mycotoxin detection - the routine aflatoxin detection. Kneist,S.; Koch,H.A. (Mykol. Abt., Hautklin. Med. Akad. Erfurt, DDR-50 Erfurt, Klement-Gottwald-Str. 34, GDR) *Z. Gesamte Hyg. Grenzgeb., 23(16), 417-419 (1977)* De;de,en,ru.

1462-U2 Measurements of electron transport activities in marine phytoplankton. Kenner,R.A.; Ahmed,S.I. (Dep. Oceanogr., Univ. Washington, Seattle, WA 98105, USA) *Mar. Biol., 33(2), 119-127 (1975)* En;en.

1463-U2 Spectrophotometric determination of lymphocyte mediated sheep red blood cell hemolysis in vitro. Simpson,M.A.; Gozzo,J.J. (Grad. Div. Med. Lab. Sci., Coll. Pharm. and Allied Health Prof., Northeastern Univ., Boston, MA 02115, USA) *J. Immunol. Methods, 21(1-2), 159-165 (1978)* En;en.

1464-U2 Mass spectral measurement of residual oxygen in the atmosphere of an anaerobic culture chamber. Tabaqchali,S.; Peach,S.; Snedden,W. (Dep. Med. Microbiol., St. Bartholomew's Hosp., London EC1, UK) *J. Clin. Microbiol., 8(3), 339-341 (1978)* En;en.

1465-U2 Determination of molecular weights by fluctuation spectroscopy: application to DNA. Weissman,M.; Schindler,H.; *Feher,G. (Dep. Phys., Univ. California, San Diego, La Jolla, CA 92093, USA) *Proc. Natl. Acad. Sci. USA, 73(8), 2776-2780 (1976)* En;en.

1466-U2 Determination of molecular weights by fluctuation spectroscopy: application to DNA. Weissman,M.; Schindler,H.; *Feher,G. (Dep. Phys., Univ. California, San Diego, La Jolla, CA 92093, USA) *Proc. Natl. Acad. Sci. USA, 73(8), 2776-2780 (1976)* En;en.

1467-U2 Use of proton correlation NMR spectroscopy in the study of living cells: anaerobic metabolism of *Escherichia coli*. Ogino,T.; Arata,Y.; Fujiwara,S.; Shoun,H.; Beppu,T. (Dep. Chem., Univ. Tokyo, Hongo, Tokyo 113, Japan) *J. Magn. Resonance, 31(3), 523-526 (1978)* En.

1468-U2 Microbial petroleum degradation: application of computerized mass spectrometry. Walker,J.D.; Colwell,R.R.; Petrakis,L. (Dep. Microbiol., Univ. Maryland, College Park, MD 20742, USA) *Can. J. Microbiol., 21(11), 1760-1767 (1975)* En;en,fr.

1469-U2 Infra-red method for quantitative determination of hydrocarbons in the yeast suspension. Erokhova,V.P.; Arevshatian,A.A. (All-Union Res. Inst. Biosynthesis Protein Compounds, Moscow, USSR) *Prikl. Biokhim. Mikrobiol., 13(3), 459-464 (1977)* Ru;en,ru.

1470-U2 Mass spectral confirmation of aflatoxins. Haddon,W.F.; Masri,M.S.; Randall,V.G.; Elsken,R.H.; Meneghelli,B.J. (Western Reg. Res. Cent., ARS, USDA, Berkeley, CA 94710, USA) *J. Assoc. Off. Anal. Chem., 60(1), 107-113 (1977)* En;en.

1471-U2 High resolution ^{31}P nuclear magnetic resonance studies of intact yeast cells. Salhany,J.M.; Yamane,T.; Shulman,R.G.; Ogawa,S. (Dep. Biochem. and Cardiovascular Cent., Univ. Nebraska Med. Cent., Omaha, NB 68105, USA) *Proc. Natl. Acad. Sci. USA, 72(12), 4966-4970 (1975)* En;en.

1472-U2 Laser light scattering spectroscopic immunoassay for mouse IgA. von Schulthess,G.K.; Cohen,R.J.; Sakato,N.; Benedek,G.B. (Dep. Phys., Cent. Materials Sci. and Eng., Massachusetts Inst. Technol., Cambridge, MA 02139, USA) *Immunochemistry, 13(12), 955-962 (1976)* En;en.

1473-U2 Pyrolysis-spectrometry: automated identification of microbes. Maugh,T.H.,II (Address not stated) *Science (Wash.), 194(4272), 1403-1405, 1447 (1976)* En.

1474-U2 A spectrophotometric technique for recording uptake of an organomercurial by mycelial fungi. Greenaway,W.; Ward,S.; Rajan,A.K.; Whatley,F.R. (Bot. Sch., South Parks Road, Oxford OX1 3RA, UK) *J. Gen. Microbiol., 107(1), 31-35 (1978)* En;en.

1475-U2 Microbial petroleum degradation: application of computerized mass spectrometry. Walker,J.D.; Colwell,R.R.; Petrakis,L. (Dep. Microbiol., Univ. Maryland, College Park, MD 20742, USA) *Can. J. Microbiol., 21(11), 1760-1767 (1975)* En;en,fr.

See: 729, 1013, 1031, 1152

UV absorbance

1476-U2 Homogeneity of *Pasteurella avicida* bacterin. Kirschbaum,J.; Dunham,J.M. (Squibb Inst. Med. Res., New Brunswick, NJ 08903, USA) *J. Pharm. Sci., 65(3), 433-436 (1976)* En;en.

Electrophoresis and immunoelectrophoresis

1477-U2 A convenient apparatus and tracking dye for anaerobic analytical polyacrylamide-gel electrophoresis. McKenna,C.E.; Nguyen.H.T.; Huang,C.W.; Smith,S.A. (Chem. Biol. Dev. Lab., Dep. Chem., Univ. Southern California, University Park, Los Angeles, CA 90007, USA) *Anal. Biochem., 83(2), 337-345 (1977)* En;en.

1478-U2 Electrophoretic desorption of immunoglobulins from immobilized protein A and other ligands. Morgan,M.R.A.; Johnson,P.M.; Dean,P.D.G. (Dep. Biochem., Univ. Liverpool, POB 147, Liverpool L69 3BX, UK) *J. Immunol. Methods, 23(3-4), 381-387 (1978)* En;en.

1479-U2 An immunochemical method for the detection of protein on disc electrophoresis column. Tamai,Y.; Kuwata,S.; Takakuwa,M. (Dep. Agric. Chem., Fac. Agric., Ehime Univ., Matsuyama 790, Japan) *Agric. Biol. Chem., 40(2), 445-446 (1976)* En.

1480-U2 Cell electrophoresis research directed toward clinical cytodiagnosis. Boltz,R.C.,Jr.; Todd,P.; Gaines,R.A.; Milito,R.P.; Docherty,J.J.; Thompson,C.J.; Notter,M.F.D.; Richardson,L.S.; Mortel,R. (618 Life Sci. Build., University Park, PA 16802, USA) *J. Histochem. Cytochem., 24(1), 16-23 (1976)* En;en.

1481-U2 A simple apparatus and technique for crossed antigen-antibody electrophoresis. Kelkar,S.S.; Jad,C.Y. (Dep. Microbiol., Med. Coll., Aurangabad, India) *Indian J. Med. Res., 64(11), 1691-1694 (1976)* En;en.

1482-U2 Use of proteins covalently bound to agar gel for specific sorption in electrophoresis. Kusovleva,O.B.; Nikolaeva,A.I. (N. F. Gamaleya Inst. Epidemiol. and Microbiol., Acad. Med. Sci. USSR, Moscow, USSR) *Vopr. Med. Khim., 22(3), 413-417 (1976)* Ru;en,ru.

1483-U2 An improved technique for immunofixation of electrophoretograms. Karinkanta,H.H.; Nieminen,E.J. (Cent. Hosp. Vaasa, Cent. Lab., Hietalahdenkatu 2-4, 65130 Vaasa 13, Finland) *Clin. Chem., 24(9), 1639-1641 (1978)* En;en.

1484-U2 'Immunosubtraction' electrophoresis: a simple method for identifying specific proteins producing the cellulose acetate electrophoretogram. Aguzzi,F.; Poggi,N. (Ente Osp. Broni e Stradella, Lab. Anal., 27049 Stradella, Pavia, Italy) *Boll. Ist. Sieroter. Milan, 56(3), 212-216 (1977)* En;en,it.

1485-U2 Simple agarose gel electrophoretic method for the identification and characterization of plasmid deoxyribonucleic acid. Meyers,J.A.; Sanchez,D.; Elwell,L.P.; *Falkow,S. (Dep. Microbiol., Univ. Washington Sch. Med., Seattle, WA 98195, USA) *J. Bacteriol., 127(3), 1529-1537 (1976)* En;en.

1486-U2 Simple agarose gel electrophoretic method for the identification and characterization of plasmid deoxyribonucleic acid. Meyers,J.A.; Sanchez,D.; Elwell,L.P.; *Falkow,S. (Dep. Microbiol., Univ. Washington Sch. Med., Seattle, WA 98195, USA) *J. Bacteriol., 127(3), 1529-1537 (1976)* En;en.

1487-U2 Dye titrations of covalently closed supercoiled DNA analyzed by agarose gel electrophoresis. DeLeys,R.J.; Jackson,D.A. (Dep. Microbiol., Univ. Michigan Med. Sch., Ann Arbor, MI 48109, USA) *Biochem. Biophys. Res. Commun., 69(2), 446-454 (1976)* En;en.

1488-U2 Identification and characterization of large plasmids in *Rhizobium meliloti* using agarose gel electrophoresis. Casse,F.; Boucher,C.; Julliot,J.S.; Michel,M.; Denarie,J. (Lab. Genet. Microorg., INRA, Route de Saint-Cyr, 78000 Versailles, France) *J. Gen. Microbiol., 113(2), 229-242 (1979)* En;en.

1489-U2 A simple purification of avian myeloblastosis virus reverse transcriptase for full-length transcription of 35S RNA. Myers,J.C.; Ramirez,F.; Kacian,D.L.; Flood,M.; *Spiegelman,S. (Inst. Cancer Res., Coll. Physicians and Surgeons, Columbia Univ., 701 W. 168th St., New York, NY 10032, USA) *Anal. Biochem., 101(1), 88-96 (1980)* En;en.

1490-U2 A simple method for isolating pure RNA preparations after electrophoresis in polyacrylamide gel. Dolja,V.V.; Negruk,V.I.; Novikov,V.K.; Atabekov,J.G. (Dep. Virol. and Lab. Bioorg. Chem., Moscow State Univ., Moscow 117234, USSR) *Anal. Biochem., 80(3), 502-506 (1977)* En;en.

1491-U2 Correlation of apparent molecular weight and antigenicity of viral proteins: an SDS-PAGE separation followed by acrylamide-agarose electrophoresis and immunoprecipitation. Webb,K.S.; Mickey,D.D.; Stone,K.R.; Paulson,D.F. (Dep. Surg., Div. Urol., Duke Univ. Med. Cent., Box 3062, Durham, NC 27710, USA) *J. Immunol. Methods, 14(3,4), 343-353 (1977)* En;en.

1492-U2 Simultaneous isolation of major viral proteins in one step. Schetters,J.; McLeod,B. (Univ. Munich, Poliklin., Pettenkoferstr. 8, 8000 Munich 2, GFR) *Anal. Biochem., 98(2), 329-334 (1979)* En;en.

1493-U2 Purification and electrophoretic assay of T4-induced polynucleotide ligase for the in vitro construction of recombinant DNA molecules. Moore,S.K.; James,E. (Dep. Chem., Univ. South Carolina, Columbia, SC 29208, USA) *Anal. Biochem., 75(2), 545-554 (1976)* En;en.

1494-U2 Oligonucleotide mapping of picornavirus RNAs by two-dimensional electrophoresis. Frisby,D.P.; Newton,C.; Carey,N.H.; Fellner,P.; Newman,J.F.E.; Harris,T.J.R.; Brown,F. (Searle Res. Lab., High Wycombe, Buckinghamshire, UK) *Virology, 71(2), 379-388 (1976)* En;en.

1495-U2 Comparison of tomato bioassay and slab gel electrophoresis for detection of potato spindle tuber viroid in potato. Schumann,G.L.; Thurston,H.D.; Horst,R.K.; Kawamoto,S.O.; Nemoto,G.I. (Dep. Plant Pathol., Cornell Univ., Ithaca, NY 14853, USA) *Phytopathology, 68(9), 1256-*

1259 (1978) En;en.

1496-U2 Potato spindle tuber disease: procedures for the detection of viroid RNA and certification of disease-free potato tubers. Morris,T.J.; Smith,E.M. (Dep. Plant Pathol., Univ. California, Berkeley, CA 94720, USA) *Phytopathology, 67(2), 145-150 (1977)* En;en.

1497-U2 The influence of intramolecular disulfide bonds on the structure and function of Semliki Forest virus membrane glycoproteins. Kaluza,G.; Pauli,G. (Inst. Virol., Fachber. Humanmed., Justus-Liebig-Univ. Giessen, Frankfurter Str. 107, 6300 Giessen, GFR) *Virology, 102(2), 300-309 (1980)* En;en.

1498-U2 A simple large-scale method for separating closed circular form DNA by gel electrophoresis. Tanabe,N.; Hidaka,H.; Watanabe,S.; Oda,T. (Dep. Biochem., Cancer Inst., Okayama Univ. Med. Sch., Okayama 700, Japan) *Acta Med. Okayama, 32(6), 379-385 (1978)* En;en.

1499-U2 Use of polyacrylamide gel electrophoresis in diagnosis of farm animal mycoplasmas. Paroz,P.; Krawinkler,M.; Nicolet,J. (Univ. Bern, Vet. Bakteriol. Inst., Postfach 2735, 3001 Bern, Switzerland) *Zentralbl. Veterinarmed., B, 24(8), 668-677 (1977)* De;de,en,es,fr.

1500-U2 Application of polyacrylamide gel electrophoresis to the characterization and identification of *Arthrobacter* species. Seiter,J.A.; *Jay,J.M. (Dep. Biol., Wayne State Univ., Detroit, MI 48202, USA) *Int. J. Syst. Bacteriol., 30(2), 460-465 (1980)* En;en.

1501-U2 Preparative polyacrylamide gel electrophoresis purification of *Clostridium perfringens* enterotoxin. Enders,G.L.,Jr.; *Duncan,C.L. (Campbell Inst. Food Res., Camden, NJ 08101, USA) *Infect. Immun., 17(2), 425-429 (1977)* En;en.

1502-U2 A method to identify individual proteins in four different two-dimensional gel electrophoresis systems: application to *Escherichia coli* ribosomal proteins. Madjar,J.; Michel,S.; Cozzone,A.J.; Reboud,J.P. (Lab. Biochim. Med., Univ. Lyon 1, 43, Bd. du 11 Novembre, F 696211 Villeurbanne, France) *Anal. Biochem., 92(1), 174-182 (1979)* En;en.

1503-U2 Electrophoretic mobility of membrane fragments on a sucrose gradient. Application to isolated purple membrane fragments from *Halobacterium halobium*. Eisenbach,L.; *Eisenbach,M. (Dep. Membr. Res., Weizmann Inst. Sci., Rehovot, Israel) *Anal. Biochem., 92(1), 228-232 (1979)* En;en.

1504-U2 [Study of biospecific complexes by electrophoresis: application to antibiotics specific for bacterial ribosome]. le Goffic,F.; Moreau,N.; Edery,M.; Capmau,M.-L. (CERCOA-CRNS, 2-8 rue Henry-Dunant, 94320 Thiais, France) *C.R. Hebd. Seances Acad. Sci., Paris, Ser. D., 283(5), 551-553 (1976)* Fr;fr,en.

1505-U2 A rapid method for determining the base composition of DNA using unpurified bacterial extracts. Gilpin,M.L.; Dale,J.W. (Dep. Microbiol., Univ. Surrey, Guildford, Surey GU2 5XH, UK) *Microbios Lett., 12(45), 31-35 (1980)* En;en.

1506-U2 Quantitative estimation of newly synthesized radioactive proteins: use of minicolumns of antibody embedded in agarose gel or immobilized on Sepharose. Proffitt,R.T.; Sankaran,L.; Pogell,B.M. (Hematol. Div., Jewish Hosp. St. Louis, 216 South Kingshighway, St. Louis, MO 63110, USA) *Anal. Biochem., 88(1), 236-244 (1978)* En;en.

1507-U2 A simple method for the classification and typing of monoclonal immunoglobulins. Cejka,J.; Kithier,K. (Dep. Immunochem., Child. Hosp. Michigan, Detroit, MI 48201, USA) *Immunochemistry, 13(7), 629-631 (1976)* En;en.

1508-U2 [Application of preparative methods in the liquid phase to the separation of blood cells]. Herve,P.; Masse,M.; Lenys,R.; Peters,A. (Cent. Transfus. Sang., Place St.-Jacques, 25000 Besancon, France) *Rev. Fr. Transfus. Sang., 18(4), 439-456 (1975)* Fr;en,fr.

1509-U2 Simplified cell microelectrophoretic method applied to the macrophage electrophoretic mobility test for cancer diagnosis. Fike,R.M.; van Oss,C.J. (Immunochem. Lab., Dep. Microbiol., State Univ. New York at Buffalo, Buffalo, NY 14214, USA) *Immunol. Commun., 8(2), 173-184 (1979)* En;en.

1510-U2 Determination of human apolipoprotein E by electroimmunoassay. Curry,M.D.; McConathy,W.J.; *Alaupovic,P.; Ledford,J.H.; Popovic,M. (Lipoprotein Lab., Oklahoma Med. Res. Found., Oklahoma City, OK 73104, USA) *Biochim. Biophys. Acta, 439(2), 413-425 (1976)* En;en.

1511-U2 Purification of Fab and Fc fragments from human serum immunoglobulin G by electrophoresis and affinity chromatography. De La Farge,F.; Valdiguie,P. (Lab. Biochim., Fac. Med. Rangueil, Chemin du Vallon, 31400 Toulouse, France) *J. Chromatogr., 123(1), 247-250 (1976)* En.

1512-U2 Counter immunoelectrophoresis in the detection of anti-DNA antibodies. Gava,O.; Riboldi,P.S.; Sala,G. (Osp. Gen. Reg. Milano-Niguardo Ca' Granda, piazza Ospedale Maggiore 3, Div. Med., 'Brera', Milano, Italy) *Osp. Magg., 73(5), 303-307 (1978)* It;en,it.

1513-U2 Radioimmunoelectrophoretic determination of α-fetoprotein. Determination of standard inhibition curves. Nesterova,E.N. (Oncol. Sci. Cent., USSR Acad. Med. Sci., Moscow, USSR) *Byull. Eksp. Biol. Med., 84(7), 124-126 (1977)* Ru;en,ru.

1514-U2 Performance comparison of the Corning thin-film agarose and the Hyland thick-film agar methods for immunoelectrophoresis. Gerson,B.; LaBrie,J.; Copeland,B.E. (Dep. Pathol., New England Deaconess Hosp., 185 Pilgrim Rd., Boston, MA 02215, USA) *Clin. Chem., 24(9), 1634-1635 (1978)* En;en.

1515-U2 Counterimmunoelectrophoresis in diagnosis of infectious disease. Rytel,M.W. (Sect. Infect. Dis., Milwaukee County Gen. Hosp., Milwaukee, WI, USA) *Hosp. Pract., 10(10), 75-82 (1975)* En;en.

1516-U2 A micromodification of the electroimmunoassay. Prause,J.U.; Sjostrom,H.; Noren,O.; Josefsson,L. (Dep. Ophthalmol., Rigshosp., Blegdamsvej 9, DK-2100 Copenhagen 0, Denmark) *Anal. Biochem., 85(2), 564-571 (1978)* En;en.

1517-U2 A technical improvement for crossed immunoelectrophoresis. Kuusi,N. (Cent. Public Health Lab., Mannerheimintie 166, 00280 Helsinki 28, Finland) *J. Immunol. Methods, 31(3-4), 361-364 (1979)* En;en.

1518-U2 Disc-crossed immunoelectrophoresis. A simple 'laying-on' technique permitting the use of commercially available agarose. Ekwall,K.; Soderholm,J.; Wadstrom,T. (Statens Bakteriol. Lab., S-105 21 Stockholm, Sweden) *J. Immunol. Methods, 12(1,2), 103-115 (1976)* En;en.

1519-U2 A review of countercurrent immunoelectrophoresis. Murillo,J.L.;Titelbaum,J.A. (Newark Beth Israel Med. Cent., 201 Lyons Ave., Newark, NJ 07112, USA) *J. Med. Soc. New Jersey, 76(5), 371-375 (1979)* En;en.

1520-U2 Counterimmunoelectrophoresis assay for detection of adenovirus antigen. Mankikar,S.D.; *Middleton,P.J.; Petric,M. (Hosp. Sick Child., Toronto, Ont., Canada) *J. Clin. Microbiol., 10(2), 253-255 (1979)* En;en.

1521-U2 Detection of adenovirus antibody by counterimmunoelectrophoresis. Petric,M.; *Middleton,P.J.; Mankikar,S.D. (Hosp. Sick. Child., Toronto, Ont., Canada) *J. Clin. Microbiol., 10(2), 256-258 (1979)* En;en.

1522-U2 Application of counterimmunoelectrophoresis in rapid detection of cytomegalovirus antibodies. Niebojewski,R.A.; Torres,F.G.; *Rytel,M.W. (Sect. Infect. Dis., Dep. Med., Med. Coll. Wisconsin, Milwaukee County Gen. Hosp., 8700 W. Wisconsin Ave., Milwaukee, WI 53226, USA) *Am. J. Clin. Pathol., 68(3), 343-346 (1977)* En;en.

1523-U2 Rapid detection of dengue viral antigens by counterimmunoelectrophoresis. Churdboonchart,V.; *Bhamarapravati,N.; Harisdangkul,V.; McNair Scott,R.; Futrakul,P.; Chiengson,R.; Nimmanitaya,S. (Dep. Pathol., Fac. Med., Ramathibodi Hosp., Rama VI Road, Bangkok, Thailand) *Am. J. Clin. Pathol., 71(1), 102-108 (1979)* En;en.

1524-U2 Quantitation of influenza vaccine hemagglutinin by immunoelectrophoresis. Mayner,R.E.; Blackburn,R.J.; Barry,D.W. (Div. Virol., Bur. Biol., Food and Drug Adm., 8800 Rockville Pike, Bethesda, MD 20014, USA) *Dev. Biol. Stand., 39, 169-178 (1977)* En;en.

1525-U2 [Evaluation of ovalbumin by counterimmunoelectrophoresis in influenza vaccine]. Barme,M. (Inst. Pasteur, 25 rue du Dr. Roux, 75015 Paris, France) *Dev. Biol. Stand., 39, 213-217 (1977)* Fr;en,fr.

1526-U2 Quantification of influenza virus structural proteins using rocket immunoelectrophoresis. Oxford,J.S.; Schild,G.C. (Div. Virol., Natl. Inst. for Biol. Stand. and Control, Holly Hill, Hampstead, London NW3, UK) *J. Gen. Virol., 38(1), 187-193 (1977)* En;en.

1527-U2 Contraimmunoelectrophoresis with the infleunza antigens. II. Comparative studies on the temporary variation of antibodies tested by contraimmunoelectrophoresis and complement fixation. Cahill,J.C.R.; Mogdasy,M.C.; de Peluffo,M.H.; Somma-Moreira,R. (Dep. Bacteriol. y Virol., Fac. Med. Univ. Republic, Montevideo, Uruguay) *Rev. Latinoam. Microbiol., 18(2), 107-110 (1976)* Es;en,es.

1528-U2 Diagnosis of California La Crosse virus infection by counterimmunoelectrophoresis. Linsey,H.S.; Thompson,W.H.; Obijeski,J.F. (Bur. Lab., Cent. for Dis. Control, Atlanta, GA 30333, USA) *J. Clin. Microbiol., 7(6), 603-608 (1978)* En;en.

1529-U2 Determination of measles virus-specific nucleocapsid antibodies by means of counterimmunoelectrophoresis. Nordal,H.J.; Vandvik,B.; Norrby,E. (Inst. Immunol. and Rheumatol., Rikshosp. Univ. Hosp., F. Qvamsgt. 1, N-OSLO 1, Norway) *Acta Pathol. Microbiol. Scand., Sect. B, 83(5), 425-432 (1975)* En;en.

1530-U2 Counter-immunoelectro-osmophoresis for the detection of infantile gastroenteritis virus (orbigroup) antigen and antibody. Middleton,P.J.; Petric,M.; Hewitt,C.M.; Szymanski,M.T.; Tam,J.S. (Dep. Virol., Hosp. for Sick Child., Toronto, Canada) *J. Clin. Pathol., 29(3), 191-197 (1976)* En;en.

1531-U2 Counter immunoelectrophoresis in rapid diagnosis of pox virus infections. Meher-Homji,K.M.; Bennett,G.; Pearson,L. (Sch. Public Health and Trop. Med., Univ. Sydney, NSW 2006, Australia) *Med. J. Aust., 1(25), 790 (1975)* En;en.

1532-U2 Pseudorabies: adaptation of the countercurrent immunoelectrophoresis for the detection of antibodies in porcine serum. Papp-Vid,G.; Dulac,G.C. (Anim. Pathol. Div., Health Anim. Branch, Agric. Canada, Anim. Dis. Res. Inst. (E), Box 11300, Stn. H, Ottawa, Ont. K2H 8P9, Canada) *Can. J. Comp. Med., 43(2), 119-124 (1979)* En;en,fr.

1533-U2 Analysis of murine C-type virus structural proteins by rocket and crossed immunoelectrophoresis. Geogiades,J.A.; Billiau,A.; Verbruggen,R.; De Somer,P. (Dep. Hum. Biol., Rega Inst., Univ. Leuven, B-3000, Leuven, Belgium) *J. Immunol. Methods, 17(1, 2), 177-187 (1977)* En;en.

1534-U2 Identification of antigens within the so-called vaccinia L-S complex by means of quantitative immunoelectrophoresis. Prante,P.H.; *Haukenes,G.; Gurvin,I. (Mikrobiol. Avd., MFH-bygget, N-5016 Haukeland sykehus, Norway) *Acta Pathol. Microbiol. Scand., Sect. B, 84(5), 293-299 (1976)* En;en.

1535-U2 Rapid virus diagnosis by electrosyneresis. Gaspar,A.; Edlinger,E. (Unite de Diagn. Virol. et

Rickettsiales, Inst. Pasteur, 75724 Paris Cedex 15, France) *Ann. Microbiol., 129A(4), 545-552 (1978)* Fr;en,fr.

1536-U2 Discontinuous counter-immunoelectrophoresis in the study of viruses. Tomori,O.; Oseni,G. (Virus Res. Lab., Univ. Ibadan, Ibadan, Nigeria) *Intervirology, 10(2), 102-114 (1978)* En;en.

1537-U2 Counterimmunoelectrophoresis for detection of microbial antigens; increased sensitivity with dextran-containing gels. Siber,G.R.; Skrapriwsky,P. (Dep. Clin. Microbial., Sidney Farber Cancer Inst., Boston, MA 02115, USA) *J. Clin. Microbiol., 7(4), 392-393 (1978)* En;en.

1538-U2 Methodology and applications of countercurrent immunoelectrophoresis in microbiology. / [from a lecture presented at the Association of Clinical Pathologists Meeting, held at York University, York, UK, on 8 April 1976]. Moody,G.J. (Chem. Dep., Univ. Wales Inst. Sci. and Technol., Cardiff, UK) *Lab. Pract., 25(9), 575-580 (1976)* En;en.

1539-U2 Crossed immunoelectrophoresis of membrane proteins from *Acholeplasma laidlawii* and *Spiroplasma citri*. Wroblewski,H.; Johansson,K.-E.; Burlot,R. (Lab. Biol. Cell., Fac. Sci. Biol., Complexe Beaulieu, BP 25A, 35031 Rennes Cedex, France) *Int. J. Syst. Bacteriol., 27(2), 97-103 (1977)* En;en.

1540-U2 [Rapid and simultaneous screening and quantitation of antitetanus and anti-HBs antibodies using Laurell's method]. Mesnier,F.; Piquet,Y.; Moulinier,J. (Ser. Immunochim., CTS de Bordeaux, BP 24, 33035 Bordeaux Cedex, France) *Rev. Fr. Transfus. Immuno-Hematol., 20(4), 593-601 (1977)* Fr;en,fr.

1541-U2 Rapid detection and quantitation of *Clostridium perfringens* enterotoxin by counterimmunoelectrophoresis. Naik,H.S.; *Duncan,C.L. (Dep. Food Microbiol. and Toxicol., Food Res. Inst., Univ. Wisconsin, Madison, WI 53706, USA) *Appl. Environ. Microbiol., 34(2), 125-128 (1977)* En;en.

1542-U2 Quantification of tetanus antitoxin in human sera. I. Counter-immunoelectrophoresis. Winsnes,R. (Vaccine Dep., Natl. Inst. Public Health, Postuttak Oslo 1, Norway) *Acta Pathol. Microbiol. Scand., Ser. B, 87(3), 191-195 (1979)* En;en.

1543-U2 Rapid determination of *Corynebacterium diphtheriae* toxigenicity by counterimmunoelectrophoresis. Thompson,N.L.; *Ellner,P.D. (Dep. Microbiol., Columbia Univ. Coll. Physicians and Surgeons, New York, NY 10032, USA) *J. Clin. Microbiol., 7(5), 493-494 (1978)* En;en.

1544-U2 Countercurrent immunoelectrophoresis (CIEP) in the diagnosis of childhood meningitis. Krishnan,C.; Wylie,J.S. (Dep. Bacteriol., Hosp. Sick Child., Toronto, Ont., Canada) *Indian Pediatr., 15(9), 703-706 (1978)* En;en.

1545-U2 An improved method for detection of *Haemophilus influenzae* b antigen in cerebrospinal fluid. Pifer,L.; Elliott,S.; Woodard,T.; Woods,D.; Hughes,W.T. (St. Jude Child. Res. Hosp., 332 North Lauderdale, Memphis, TN 38101, USA) *J. Pediatr., 92(2), 227-229 (1978)* En.

1546-U2 Discontinuous counter current immunoelectrophoresis in *Haemophilus* meningitis (a case report). Ayyagari,A.; Iyer,R.V.; Agarwal,K.C.; Kumar,L. (Dep. Microbiol. and Pediatr., Postgrad. Inst. Med. Educ. and Res., Chandigarh-160011, India) *Indian Pediatr., 14(11), 937-941 (1977)* En;en.

1547-U2 Counter current immunoelectrophoresis in the diagnosis of *Haemophilus influenzae* meningitis in children. Ayyagari,A.; Kumar,L.; Agarwal,K.C.; Sharma,M.; Pande,D. (Dep. Microbiol., Postgrad. Inst. Med. Educ. and Res., Chandigarh-160012, India) *Indian J. Med. Res., 70(Aug), 168-172 (1979)* En;en.

1548-U2 Detection of bacteraemia by countercurrent immunoelectrophoresis. Simpson,R.A.; Speller,D.C.E. (Dep. Bacteriol., Bristol R. Infirm., Bristol BS2 8HW, UK) *Lancet, 1(8023), 1206 (1977)* En.

1549-U2 Counterimmunoelectrophoresis in the diagnosis of human leptospirosis. Terpstra,W.J.; Schoone,G.J.; Ligthart,G.S. (R. Trop. Inst., Amsterdam, Netherlands) *Zentralbl. Bakteriol. Parasitenkd. Infektionskr. Hyg., I Abt. A, 244(2-3), 285-290 (1979)* En;de,en.

1550-U2 Antigenic relationships among species of *Mycobacterium* studied by fused rocket immunoelectrophoresis. Chaparas,S.D.; Brown,T.; Hyman,I. (Bur. Biologics, FDA, Bethesda, MD 20014, USA) *Int. J. Syst. Bacteriol., 28(4), 547-560 (1978)* En;en.

1551-U2 Antimyobacterial antibodies in diffuse lepromatous leprosy detected by counterimmunoelectrophoresis. Rojas-Espinosa,O.; Estrada-Parra,S.; Serrano-Miranda,E.; Saul,A.; Latapi,F. (Dep. Immunol., Esc. Nac. Cienc. Biol., IPN, AP 4-870, Mexico 4, DF, Mexico) *Int. J. Lepr., 44(4), 448-452 (1976)* En;en,es,fr.

1552-U2 Rapid detection of mycoplasma antibody. Goldschmidt,B.L.; Menonna,J.P.; *Dowling,P.C.; Cook,S.D. (V.A. Hosp., Neurol. Serv. (180), East Orange, NJ 07019, USA) *J. Immunol., 117(3), 1054-1055 (1976)* En.

1553-U2 Rapid identification of specific etiology in meningitis. McCracken,G.H.,Jr. (Dep. Pediatr., Univ. Texas Health Sci. Cent. at Dallas, 5323 Harry Hines Blvd., Dallas, TX 75235, USA) *J. Pediatr., 88(4), 706-708 (1976)* En.

1554-U2 Difficulties in diagnosing meningococcal meningitis. Lewis,L.S. (Health Cent., Narberth, Dyfed, UK) *Br. Med. J., 1(6172), 1217 (1979)* En.

1555-U2 Detection of *Pseudomonas* antigen by counterimmunoelectrophoresis. Adams,M.B.; Aguilar-Torres,F.G.; Rytel,M.W. (Dep. Surg., Med. Coll. Wisconsin, Milwaukee, WI 53201, USA) *Surg. Forum, 27, 3-5 (1976)* En;en.

1556-U2 Simultaneous detection of *Salmonella typhi* antigen and antibody in serum by counter-

immunoelectrophoresis for an early and rapid diagnosis of typhoid fever. Gupta,A.K.; Rao,K.M. (Dep. Microbiol., Def. Res. and Dev. Establ., Gwalior-474002, India) *J. Immunol. Methods, 30(4), 349-353 (1979)* En;en.

1557-U2 **Counterimmunoelectrophoresis in staphylococcal enterotoxin identification.** Azzi,A.; Gorzoni,S. (Univ. Padova, Ist. Ig., Verona, Italy) *Ann. Slavo Riv. Microbiol. Immunol., 19(2), 293-299 (1977)* It;en.

1558-U2 **Counterimmunoelectrophoresis for the rapid diagnosis of group B streptococcal infections.** Typlin,B.L.; Koranyi,K.; Azimi,P.; Hilty,M.D. (Dep. Pediatr., Michael Reese Hosp. and Med. Cent., 29th Street and Ellis Ave., Chicago, IL 60616, USA) *Clin. Pediatr., 18(6), 366-369 (1979)* En.

1559-U2 **[The identification of streptococcal groups. II. Comparative study of the precipitation reaction in liquid and gel media].** Bergogne-Berezin,E.; Slim,A. (Lab. Cent. Microbiol., Hopital Bichat, 170 blvd. Ney, 75877 Paris Cedex 18, France) *Ann. Biol. Clin., 34(6), 415-421 (1976)* Fr;en,fr.

1560-U2 **A comparison of four methods for the serotyping of group B streptococci.** Triscott,M.X.; Davis,G.H.G. (Dep. Micriobiol., Univ. Queensland, ST. Lucia, Queensl. 4067, Australia) *Aust. J. Exp. Biol. Med. Sci., 57(5), 521-527 (1979)* En;en.

1561-U2 **Rapid specific identification of group D streptococci by counterimmunoelectrophoresis.** Portas,M.R.; Hogan,N.A.; *Hill,H.R. (Div. Clin. Immunol., Dep. Pediatr., Univ. Utah, Salt Lake City, Ut 84132, USA) *J. Lab. Clin. Med., 88(2), 339-344 (1976)* En;en.

1562-U2 **Rapid identification of beta haemolytic streptococci groups A, B, C and G by counter immunoelectrophoresis.** Gopaul,D.L.; McIntosh,M.A. (Microbiol. Dep. St. Joseph's Hosp., London, Ont., Canada) *Can. J. Med. Technol., 40(3), 93-95 (1978)* En;en.

1563-U2 **Rapid recognition of group B streptococci by pigment production and counterimmunoelectrophoresis.** Merritt,K.; Treadwell,T.L.; Jacobs,N.J. (Dep. Surg., Dartmouth Med. Sch., Hanover, NH 03755, USA) *J. Clin. Microbiol., 3(3), 287-290 (1976)* En;en.

1564-U2 **Counterimmunoelectrophoresis of pneumococcal antigens: improved sensitivity for the detection of types VII and XIV.** Anhalt,J.P.; Yu,P.K.W. (Sect. Clin. Microbiol., Dep. Lab. Med., Mayo Clin. and Mayo Found., Rochester, MN 55901, USA) *J. Clin. Microbiol., 2(6), 510-515 (1975)* En;en.

1565-U2 **Counterimmunoelectrophoresis of Reiter treponeme axial filaments as a diagnostic test for syphilis.** Neil,E.E.; *Hardy,P.H.,Jr. (WHO Collaborating Cent. Ref. and Res., Treponematoses, Dep. Microbiol., Johns Hopkins Univ. Sch. Med., Baltimore, MD 21205, USA) *J. Clin. Microbiol., 8(2), 148-152 (1978)* En;en.

1566-U2 **Comparison of the counter-immunoelectrophoresis technique with the Reiter protein and three other serological tests as a first line test for syphilis.** Banffer,J.R.J.; Raghoenath,S.; Hulst,A.M. (Municipal Lab. Epidemiol. Bacteriol., c/o Erasmus Univ., Rotterdam, Netherlands) *J. Clin. Microbiol., 2(5), 361-367 (1975)* En;en.

1567-U2 **Counter-immunoelectrophoresis in the serological diagnosis of syphilis.** Tringali,G. (Ist. Ig., Univ. Palermo, Via del Vespro 133, 90127 Palermo, Italy) *Arch. Roum. Pathol. Exp. Microbiol., 36(1), 37-41 (1977)* En;en,fr,ru.

1568-U2 **Rapid detection of specific treponemal antibodies by counterimmunoelectrophoresis using an extract of *Treponema pallidum*.** Ghinsberg,R.; Grunbaum,B.W.; Blumstein,G. (Dr. Ephraim Rappaport Natl. Ref. Cent. for Treponematoses, Public Health Lab., Minist. Health, Tel Aviv, Istael) *Isr. J. Med. Sci., 13(6), 557-560 (1977)* En;en.

1569-U2 **Antigenic relationship of *Yersinia enterocolitica* type 9 with other gram-negative species.** /[presented at the International Symposium on Yersinia, Pasteurella and Francisella, held in Malmo, Sweden, 10-12 April 1972]. Diaz,R. (INRA, F-37 Nouzilly, France) *Contrib. Microbiol. Immunol., 2, 157-158 (1973)* En.

1570-U2 **Counterimmunoelectrophoresis of blood cultures. Temporal relationship of positive counterimmunoelectrophoresis.** Fossieck,B.,Jr.; Fedorko,J. (Natl. Cancer Inst./Vet. Adm. Med. Oncol. Sect., Veterans Adm. Hosp., 50 Irving Street, NW, Washington DC 20422, USA) *Am. J. Clin. Pathol., 71(3), 326-329 (1979)* En;en.

1571-U2 **Immunoelectroosmophoresis (IEOP) for detection of bacterial antigens in cerebrospinal fluid.** Sillanpaa,M.; Vaha-Eskeli,E.; Willman,K. (Dep. Paediatr., Univ. Turku, SF-20520 Turku 52, Finland) *Scand. J. Infect. Dis., 7(2), 113-115 (1975)* En;en.

1572-U2 **Counterimmunoelectrophoresis in the diagnosis of bacterial meningitis.** Colding,H.; Lind,I. (Dep. Clin.Bacteriol., Inst. Med. Microbiol., University of Copenhagen, 2100 Copenhagen, Denmark) *J. Clin. Microbiol., 5(4), 405-409 (1977)* En;en.

1573-U2 **Rapid bacteriological diagnosis by counter-immunoelectrophoresis.** Montplaisir,S.; Bahous,J. (Dep. Microbiol. et Immunol., Lab. Sero-Immunol., Hop. Ste.-Justine, 3175 Chemin Cote Ste-Catherine, Montreal, Que. H3T 1C5, Canada) *Union Med. Can., 106(10), 1402-1408 (1977)* Fr;en,fr.

1574-U2 **Methodology and applications of counter-currentimmunoelectrophoresis in microbiology.** Moody,G.J. (Chem. Dep., Univ. Wales Inst. Sci. and Technol., Cardiff, UK) *Lab. Pract., 25(9), 575-580 (1976)* En;en.

1575-U2 **[Rapid diagnosis of infections caused by groups B and D streptococci by counterimmunoelectrophoresis].** Geslin,P.; LeGrand,P.; Lemoine,J.-L.; Squinazi,F.; Brioude,R.; Leraillez,J.; Gibert,C.; Tremolieres,F. (Serv. Microbiol., Cent. Hosp. Intercommunal Creteil, 40 ave. de Verdun, F 94010 Creteil, France) *Nouv. Presse Med., 6(45), 4207-4208 (1977)* Fr.

1576-U2 [A technique for the detection of anti-*Aspergillus fumigatus* antibodies in intrinsic allergic alveolitis]. Rottoli,L.; Rottoli,P.; Lalumera,M. (Ist. Tisiol. e Malattie Apparato Respiratorio, Univ. Studi, Siena, Italy) *Boll. Soc. Ital. Biol. Sper., 54(8), 729-732 (1978)* It.

1577-U2 Efficacy of counterimmunoelectrophoresis in the rapid diagnosis of allergic bronchopulmonary aspergillosis. Khan,Z.U.; Sandhu,R.S. (Natl. Inst. Commun. Dis., 22-Sham Nath Marg. Delhi-110054, India) *Indian J. Med. Res., 68, 599-602 (1978)* En;en.

1578-U2 Counterimmunoelectrophoresis in the detection of antibodies against *Coccidioides immitis*. Aguilar-Torres,F.G.; Jackson,L.J.; Ferstenfeld,J.E.; Pappagianis,D.; Rytel,M.W. (Sect. Infect. Dis., Dep. Med., Med. Coll. Wisconsin, Milwaukee, WI 53201, USA) *Ann. Intern. Med., 85(6), 740-744 (1976)* En;en.

1579-U2 Counterimmunoelectrophoresis as a rapid screening test in coccidioidomycosis. Aguilar-Torres,F.G.; Rytels,M.W.; Pappagianis,D.; ackson,L. (Sect. Infect. Dis., Dep. Med., Med. Coll. Wisconsin, Milwaukee Cty. Gen. Hosp., Milwaukee, WI 53226, USA) *Rev. Invest. Clin., 30(3), 247-251 (1978)* En;en,es.

1580-U2 Detection of cryptococcal polysaccharide using counterimmunoelectrophoresis. /[presented in part at the Annual meeing of the American Society for Microbiology, held in Atlantic City, NJ, USA, on 7th May 1976]. Maccani,J.E. (Bacteriol. Lab., VAH, 54th St. and 48th Ave., South, Minneapolis, MN 55417, USA) *Am. J. Clin. Pathol., 68(1), 39-44 (1977)* En;en.

1581-U2 Counter immunoelectrophoresis as a rapid screening test for amoebic liver abscess. Tosswill,J.H.C.; Ridley,D.S.; Warhurst,D.C. (Dep. Serol., Hosp. Trop. Dis., 4 St. Pancras Way, London NW1 0PE, UK) *J. Clin. Pathol., 33(1), 33-35 (1980)* En;en.

1582-U2 [Counter immunoelectrophoresis C.I.E.P. in the diagnosis of visceral leishmaniasis]. Mansueto,S.; Picone,D.; Di Rosa,S.; La Cascia,C (Ist. Clin. Med. 1°, Piazza delle Cliniche 2, 90127 Palermo, Italy) *Boll. Ist. Sieroter. Milan., 57(5), 623-630 (1978)* It;en,it.

1583-U2 Detection of plasmacytoma of the nasal cavity by immunoelectrophoresis of nasal washing fluid. de Gast,G.C.; Weits,J.; Ockhuizen,T.; Annyas,A.A.; Hoedemaeker,P.J. (Dep. Med., Univ. Groningen, Groningen, Netherlands) *J. Laryngol. Otol., 90(8), 785-787 (1976)* En;en.

1584-U2 Evaluation of counter immuno-electrophoresis crossed electro-immunodiffusion and agar gel diffusion for immunodiagnosis of human hydatid disease. Ardehali,S.; Kohanteb,D.J.; Gerami,S.; Behfourouz,N.; Rezai,H.R.; Vaez-Zadeh,K. (Dep. Microbiol., Med. Sch., Pahlav Univ., Shiraz, Iran) *Trans. R. Soc. Trop. Med. Hyg., 71(6), 481-485 (1977)* En;en.

1585-U2 Quantitative counterimmunoelectrophresis for urinary fibrinogen-related antigens. Ruiz-Reyes,G.; Tabe-Tabe,D.; Exaire-Murad,E. (Lab. Clin., Puebla, Blvd. Presidente Diaz Ordaz 808, Colonia Anzures, Puebla, Mexico) *Am. J. Clin. Pathol., 67(2), 174-176 (1977)* En;en.

1586-U2 Immunoelectrophoretic quantitation of alpha-fetoprotein. Pietrogrande,M.; Roffi,L.; Bassi,A.S.; Vergani,C. (Ist. Clin. Med. III, Univ. Milano, Via Pace 15, 20122 Milano, Italy) *Clin. Chem., 23(2, Part 1), 178-179 (1977)* En;en.

1587-U2 Rapid, sensitive detection of myoglobinemia by improved counterimmunoelectrophoresis in cases of acute myocardial infarction. Hiramori,K.; Sumiyoshi,T.; Motegi,S.; Honda,T.; Kimata,S.; Hirosawa,K.; Kawai,H.; Kondo,A.; Iwaasa,M.; Miyoshi,K. (Natl. Cardiovasc. Cent., 5-125 Fujishiro-dai, Suita City, Osaka 565, Japan) *Am. Heart J., 96(2), 187-190 (1978)* En;en.

1588-U2 Crossed immunoelectrophoresis and electroimmunoassay of human IgG subclasses. Oxelius,V.A. (Dep. Pediatr., Univ. Lund, S-221 85 Lund, Sweden) *Acta Pathol. Microbiol. Scand. Ser. C, 86(3), 109-116 (1978)* En;en.

1589-U2 Identification of antibodies to nuclear acidic antigens by counterimmunoelectrophoresis. Kurata,N.; *Tan,E.M. (Scripps Clin. and Res. Found., 476 Prospect St., La Jolla, CA 92037, USA) *Arthritis Rheum., 19(3), 574-580 (1976)* En;en.

1590-U2 Detection of circulating carcinoembryonic antigen (CEA) by counter-immunoelectrophoresis. Hamada,S.; Ishikawa,N.; Imura,T.; Torizuka,K.; Matsumoto,Y.; Honjo,K.; Mayumi,T. (Radioisot. Res. Cent., Kyoto Univ., Kyoto, Japan) *Clin. Chim. Acta, 66(3), 365-370 (1976)* En;en.

1591-U2 Detection of alpha-fetoprotein by supplementation counterimmunoelectrophoresis (CIE). Chan,S.H.; Heng,S.H.; Simons,M.J. (Dep. Microbiol., Fac. Med., Univ. Singapore, Singapore) *J. Immunol. Methods, 29(2), 191-196 (1979)* En;en.

1592-U2 Comparison between immunofixation and crossed immunoelectrophoresis for the detection of C3 activation products. de Vecchi,A.; Montagnino,G.; Massaro,L.; Constantino,A.; Bencini,P.L.; Tarantino,A. (Div. Nefrol., Osp. Maggiore Policlin., Via Commenda 15, 20122 Milano, Italy) *Clin. Chim. Acta, 97(1), 27-32 (1979)* En;en.

1593-U2 Plasma protein profiling: the diagnostic evaluation of disorders in plasma protein composition by a new immunoelectrophoretic method. Scherer,R.; Ruhenstroth-Bauer,G. (Univ. Munchen, Dep. Dermatol., D-8000 Munchen 2, Frauenlobstr. 9, GFR) *Clin. Chim. Acta, 66(3), 417-433 (1976)* En;en.

1594-U2 Rapid method for screening of immunoglobulins in porcine fetuses, using rocket immunoelectrophoresis. Application of an interspecies reaction between human and porcine μ-chain. Dalsgaard,K.; Overby,E.; Metzger,J.J.; Basse,A. (State Vet. Inst. Virus Res., Lindholm, DK-4471 Kalvehave, Denmark) *Acta Vet. Scand., 20(3), 313-320 (1979)* En;da,en.

1595-U2 An assessment under routine conditions of an immunoelectrophoretic assay for thyroxine-binding globulin. Attwood,E.C.; Probert,D.E. (Dep. Clin. Chem., County Hosp., Hereford, UK) *Clin. Biochem., 11(6), 226-229 (1978)* En;en.

1596-U2 Counterimmunoelectrophoresis for rapid serodiagnosis of human hydatid disease. Mahajan,R.C.; Ganguly,N.K.; Sharma,S.; Chandnani,R.E.; Chitkara,N.L. (Dep. Microbiol., Post-grad. Inst. Med. Educ. and Res., Chandigarh, India) *Indian J. Med. Res., 64(8), 1173-1176 (1976)* En;en.

1597-U2 Detection of complement activation by counterimmunoelectrophoresis (CIE). Arroyave,C.M.; Tan,E.M. (Div. Allergy and Immunol., Scripps Clin. and Res. Found., La Jolla, CA 92037, USA) *J. Immunol. Methods, 13(2), 101-112 (1976)* En;en.

1598-U2 Screening test for complement activation by counterimmunoelectrophoresis. Arroyave,C.M.; Taylor,D.G.; Gallup,P.; Nakamura,R.M. (Dep. Rheumatol., Univ. Colorado Med. Cent., 4200 East Ninth Ave., Denver, CO 80220, USA) *Am. J. Clin. Pathol., 69(4), 440-445 (1978)* En;en.

1599-U2 Crossed immunoelectrophoresis from sodium dodecyl sulfate-polyacrylamide gels into antibody-containing agarose: an improved method for evaluation of the number of immunochemical determinants in polypeptides, with reference to spectrin. Kirkpatrick,F.H.; Rose,D.J. (Dep. Radiat. Biol. and Biophys., Univ. Rochester Sch. Med. and Dent., Rochester, NY 14642, USA) *Anal. Biochem., 89(1), 130-135 (1978)* En;en.

1600-U2 Counterimmunoelectrophoresis in diagnosis of hydatid disease. Pinon,J.M. (Lab. Parasitol., Hop. Reims, 51100, France) *Lancet, 1(7954), 310 (1976)* En.

1601-U2 Some methods for microbiological assay. Board,R.G.; Lovelock,D.W. (eds.) *Publ. by:* Academic Press Inc., (London) Ltd., 24-28 Oval Road, London NW1 7DX, UK May 1975 ISBN 0-12-108240-7 at £7.80 En.

1602-U2 Improved method of assay for choline. Tanguay,A.E.; Blanchard,M.H. (American Cyanamid Co., Pearl River, NY 10965, USA) *Appl. Environ. Microbiol., 31(3), 446-447 (1976)* En;en.

1603-U2 Use of Semliki Forest virus to identify lipid mediated antiviral activity and anti-alphavirus immunoglobulin A in human milk. Welsh,J.K.; Skurrie,I.J.; *May,J.T. (Microbiol. Dep., La Trobe Univ., Bundoora, Vic. 3083, Australia) *Infect. Immun., 19(2), 395-401 (1978)* En;en.

1604-U2 Gentamicin assay by enzymatic adenylation and the application of a double osmotic shock procedure to prepare gentamicin adenine mono-nucleotide transferase. Dankert,J.; Woltjes,J. (Lab. Med. Microbiol., Div. Hosp. Infect., Univ. Hosp., Univ. Groningen, Oostersingel 59, Netherlands) *Zentralbl. Bakteriol. Parasitenkd. Infektionskr. Hyg., I Abt. A, 239(1), 113-123 (1977)* En;de,en.

1605-U2 A rapid microbiological procedure using *Bacillus stearothermophilus* for the assay of antibacterial drugs. Wahlig,H.; Holt,R.J. (Dep. Chemother., E. Merck, Darmstadt, GFR) *J. Clin. Pathol., 29(9), 858-861 (1976)* En.

1606-U2 Microbiological assay utilizing an automatic zone scanner. Geigert,J.; Hansen,D.; McDowell,C.; Merrill,R.; Ward,C. (Cetus Corp., 600 Bancroft Way, Berkeley, CA 94710, USA) *Dev. Ind. Microbiol., 17, 153-156 (1976)* En;en.

1607-U2 Simple assay for 5-fluorocytosine in the presence of amphotericin B. Kauffman,C.A.; Carleton,J.A.; Frame,P.T. (Div. Infect. Dis., Dep. Intern. Med., Cincinnati VA Hosp., Cincinnati, OH 45220, USA) *Antimicrob Agents Chemother., 9(3), 381-383 (1976)* En;en.

1608-U2 A simple titration assay for anti-concanavalin-A sera. Schnebli,H.P.; Bachi,T. (Friedrich Miescher-Inst., PO Box 273 CH-4002 Basel, Switzerland) *Experientia, 31(10), 1246-1247 (1975)* En;en.

See: 1172, 1258, 1437, 1647, 1746, 1749, 1763, 1780, 1783, 1815, 2109, 2434, 2648, 3241, 3253, 3266, 3281, 3283, 3287, 3286, 3297, 1774, 32 3256, 3294, 3243, 3254, 3263

Electro-extraction

1609-U2 Isolation of viruses by electro-extraction of infected plants Polson,A. (Dep. Microbiol., Univ, Cape Town, Rondebosch, South Africa) *Prep. Biochem., 7(3-4), 207-215 (1977)* En;en.

Haemolysis, lysis and immunolysis

1610-U2 New tests for characterization of mumps virus antibodies: hemolysis inhibition, single radial immunodiffusion with immobilized virions, and mixed hemadsorption. Norrby,E.; Grandien,M.; Orvell,C. (Dep. Virol., Karolinska Inst. Sch. Med., S-105-21 Stockholm, Sweden) *J. Clin. Microbiol., 5(3), 346-352 (1977)* En;en.

1611-U2 Study on single radial haemolysis for the detection of antibody against Ranikhet disease. Chandra Chowdhury,S.P.; Bhattacharyya,A.K.; Das Gupta,P.; Sen Gupta,D.N. (Amalgamated Res. Scheme, Directorate Vet. Serv., West Bengal, Belgachia Campus, Calcutta-700037, India) *Indian J. Anim. Health, 18(1), 75-77 (1979)* En;en.

1612-U2 Screening for rubella antibodies by a single radial haemolysis test. Degre,M.; Rollag,H.,Jr. (Kapt W Wilhelmsen og Frues Bakteriol. Inst., Univ. Oslo, Rikshospitalet, Oslo, Norway) *NIPH Ann., 2(2), 25-30 (1979)* En;en.

1613-U2 Single radial haemolysis for the assay of antibodies to some haemagglutinating arboviruses. Duca,M.; Duca,E.; Ionescu,L.; Abdalla,H. (Fac. Med., Inst. Med. and Pharm., 6600 Iasi, Romania) *Bull. WHO, 57(6), 937-942 (1979)* En;en,fr.

1614-U2 Immune lysis of Sindbis virus Stollar,V. (Dep. Microbiol., Rutgers Med. Sch., Coll. Med. and Dent. New Jersey. *Virology, 66(2), 620-624 (1975)* En;en.

1615-U2 Rapid lysis of vaccinia virus on neutral sucrose gradients with release of intact DNA. Parkhurst,J.R.; Heidelberger,C. (Med. Coll. Wisconsin, Dep. Microbiol., Milwaukee, WI 53233, USA) *Anal. Biochem., 71(1), 53-59 (1976)* En;en.

1616-U2 Rapid Tween 80 hydrolysis test for mycobacteria. Cox,F.R.; Slack,C.E.; Cox,M.E.; Pruden,E.L.; Martin,J.R. (Dep. Pathol., Schumpert Med. Cent., Shreveport, LA 71101, USA) *J. Clin. Microbiol., 7(1), 104-105 (1978)* En;en.

1617-U2 Gentle lysis of *Staphylococcus aureus* at low temperature. Flowers,R.S.; *Martin,S.E.; Ordal,Z.J. (Dep. Food Sci., Univ. Illinois, Urbana, IL 61801, USA) *Appl. Environ. Microbiol., 33(5), 1215-1217 (1977)* En;en.

1618-U2 Improved lysis of group N streptococci for isolation and rapid characterization of plasmid deoxyribonucleic acid. Klaenhammer,T.R.; *McKay,L.L.; Baldwin,K.A. (Dep. Food Sci. and Nutr., Univ. Minnesota, St. Paul, MN 55108, USA) *Appl. Environ. Microbiol., 35(3), 592-600 (1978)* En;en.

1619-U2 A method for improved lysis of some gram-negative bacteria. Schwinghamer,E.A. (CSIRO Div. Plant Ind., Canberra, ACT 2601, Australia) *FEMS Microbiol. Lett., 7(2), 157-162 (1980)* En.

1620-U2 [Modification of the local hemolysis method for detection of cells producing antibody to virus antigens]. Vasiliev,A.V.; Strelnikova,L.P.; Syurin,V.N. (K.I. Skryabina Vet. Acad., Moscow, USSR) *Vopr. Virusol., No. 4, 491-493 (1976)* Ru;en.

1621-U2 Quantitation of antibody to non-hemagglutinating viruses by single radial hemolysis: serological test for human coronaviruses. Hierholzer,J.C.; Tannock,G.A. (Respir. Virol. Branch, Bur. Lab., Cent. Dis. Control., Atlanta, GA 30333, USA) *J. Clin. Microbiol., 5(6), 613-620 (1977)* En;en.

1622-U2 Single-radial-haemolysis: a new method for the assay of antibody to influenza haemagglutinin. Applications for diagnosis and seroepidemiologic surveillance of influenza. Schild,G.C.; Pereira,M.S.; Chakraverty,P. (WHO Collaborating Cent. for Influenza, Natl. Inst. for Med. Res., Mill Hill, London NW7 1AA, UK) *Bull. WHO, 52(1), 43-50 (1975)* En;en,fr.

1623-U2 Comparison of immunological methods in the diagnosis of influenza. Aymard,M. (Lab. Virol., Cent. Natl. de la Grippe, Fac. Med., 8, ave. Rockefeller, 69373 Lyon Cedex 2, France) *Bull. Inst. Pasteur, 75(4), 309-321 (1977)* En;en.

1624-U2 Determination of mumps and influenza antibodies by haemolysis-in-gel. Vaananen,P.; Hovi,T.; Helle,E.-P.; Penttinen,K. (Viral Prod. Dep., Orion Diagn., PO Box 19, 00101 Helsinki 10, Finland) *Arch. Virol., 52(1-2), 91-99 (1976)* En;en.

1625-U2 Measurement of antibody to influenza virus neuraminidase by single radial hemolysis in agarose gels. Callow,K.A.; Beare,A.S. (MRC Common Cold Unit, Salisbury, Wiltshire SP2 8BW, UK) *Infect. Immun., 13(1), 1-8 (1976)* En;en.

1626-U2 Measurement of parainfluenza-3 virus antibody by the single radial hemolysis technique. Probert,M.; Russell,S.M. (Virol. Dep., Wellcome Res. Lab., Langley Ct., Beckenham, Kent BR3 3BS, UK) *J. Clin. Microbiol., 2(3), 157-161 (1975)* En;en.

1627-U2 Application of the passive haemolysis test for the determination of rubella virus antibodies. Skaug,K.; Orstavik,I.; Ulstrup,J.C. (Microbiol. Lab., Ulleval Hosp., Oslo 1, Norway) *Acta Pathol. Microbiol. Scand., Sect. B, 83(4), 367-372 (1975)* En;en.

1628-U2 Hemolysis-in-gel test for the demonstration of antibodies to rubella virus. Strannegard,O.; Grillner,L.; Lindberg,I.-M. (Dep. Virol., Inst. Med. Microbiol., Univ. Goteborg, Goteborg, Sweden) *J. Clin. Microbiol., 1(6), 491-494 (1975)* En;en.

1629-U2 Evaluation of the single radial haemolysis (SRH) technique for rubella antibody measurement. Russell,S.M.; Benjamin,S.R.; Briggs,M.; Jenkins,M.; Mortimer,P.P.; Payne,S.B. (Wellcome Res. Lab., Langley Court, Beckenham, Kent, UK) *J. Clin. Pathol., 31(6), 521-526 (1978)* En;en.

1630-U2 Detection of antibodies to rubella virus by hemolysis in gel test compared with usual methods. Champsaur,H.; Dussaix,E.; Tournier,P. (Lab. Microbiol., Hop. Bicetre, 78, Rue du General Leclerc, 94270 Le Kremlin-Bicetre, France) *Pathol. Biol., 26(9-10), 553-557 (1978)* Fr;en,fr.

1631-U2 Experience with radial haemolysis for rubella antibody screening. Shafi,M.S.; Jordan,S.M.; Mortimer,P.P. (Public Health Lab. and Dep. Microbiol., Cent. Middlesex Hosp., Park Royal, London NW10 7NS, UK) *J. Med. Microbiol., 12(1), 131-136 (1978)* En;en.

1632-U2 Hemolysis-in-gel test in immunity surveys and diagnosis of rubella. Vaananen,P.; Vaheri,A. (Dep. Virol., Univ. Helsinki, Haartmaninkatu, 3, SF-00290, Helsinki 29, Finland) *J. Med. Virol., 3(4), 245-252 (1979)* En;en.

1633-U2 Characterization of a screening test for diphtherial toxin antigen produced by individual plaques of corynebacteriophages. Welkos,S.L.; *Holmes,R.K. (Uniformed Serv. Univ. Health Sci., Bethesda, MD 20205, USA) *J. Clin. Microbiol., 9(6), 693-698 (1979)* En;en. [passive immune haemolysis].

1634-U2 Radial passive immune hemolysis assay for detection of heat-labile enterotoxin produced by individual colonies of *Escherichia coli* or *Vibrio cholerae*. Bramucci,M.G.; *Holmes,R.K. (Dep. Microbiol., Uniformed Serv. Univ. Health Sci., Bethesda, MD 20014, USA) *J. Clin. Microbiol., 8(2), 252-255 (1978)* En;en.

1635-U2 A single radial haemolysis technique for the measurement of antibody to *Mycoplasma bovis* in bovine sera. Howard,C.J.; Collins,J.; Gourlay,R.N. (ARC, Inst. Res. Anim. Dis., Compton, Nr. Newbury, Berks. RG16 0NN, UK) *Res. Vet. Sci., 23(1), 128-130 (1977)* En;en.

1636-U2 Plate haemolysis test for detection of gonococcal antibodies - a preliminary communication. Ganguly,N.K.; Sharma,S.; Arora,L.D.; Kumar,B.; *Kaur,S.; Chitkara,N.L.; Mahajan,R.C. (Dep. Microbiol., Postgrad. Inst. Med. Educ. and Res., Chandigarh, India) *Indian J. Med. Res., 66(1), 23-26 (1977)* En;en.

1637-U2 Hemolysis inhibition by dextrose: a simple screening method for identifying group A streptococcus. Tuason-Mendoza,M.; Toledo,S.P.; Antonio-Velmonte,M.; Tupasi,T.E. (Dep. Med., Univ. Philippines Coll. Med.-Philippine Gen. Hosp., Manila, Philippines) *Philipp. J. Intern. Med., 17(1), 51-57 (1979)* En;en.

1638-U2 A simple one-step hemolytic assay for C2 with C2-deficient human serum. Ngan,B.-Y.; Gelfand,E.W.; Minta,J.O. (Dep. Pathol., Univ. Toronto, Toronto, Ont., Canada) *J. Immunol., 118(3), 736-741 (1977)* En;en.

1639-U2 A radial hemolysis method in agarose for the functional assay of properdin. Minta,J.O.; Vetvutanapibul,W. (Dep. Pathol. Univ. Toronto, Toronto, Ont. M5S 1A8, Canada) *J. Immunol. Methods, 19(2-3), 153-167 (1978)* En;en.

1640-U2 The quantitation of alternative pathway complement function by timed lysis assay. Jones,B.M. (Dep. Pathol., Queen Mary Hosp., Hong Kong) *J. Immunol. Methods, 29(3), 287-292 (1979)* En;en.

1641-U2 A new haemolytic assay for the second component of complement (C2) in human serum. Naish,P.F.; Barratt,J.; Collins,C. (Dep. Nephrol., North Staffordshire Royal Infirmary, Princes Road, Stoke-on-Trent, UK) *Clin. Exp. Immunol., 25(3), 487-489 (1976)* En;en.

See: 1862, 1882, 3260, 3270, 3259, 3271, 3258

Immunodiffusion

1642-U2 Specific double diffusion microtechnique for the diagnosis of aspergillosis and paracoccidioidomycosis using monospecific antisera. Yarzabal,L.A.; de Albornoz,M.B.; de Cabral,N.A.; Santiago,A.R. (Mycol. Sect. PAHO Panamerican Cent Res. and Training Leprosy and Trop. Dis., Inst. Nac. Dermatol., Apartado 4043, Caracas 101, Venezuela) *Sabouraudia, 16(1), 55-62 (1978)* En;en,es.

1643-U2 Immunodiffusion using tissue sections. Taylor,A.N. (Dep. Microscopic Anat., Baylor Coll. Dent., Dallas, TX 75246, USA) *J. Immunol. Methods, 24(3-4), 377-381 (1978)* En;en.

1644-U2 A simple method for quantifying IgM of low molecular weight. Romero,O.(deceased); Yantorno,C.; Romero-Piffiguer,M. (Catedra Immunol. y Serol., Fac. Ciencias Quimicas, Univ. Nacl. Cordoba, Cordoba, Argentina) *J. Immunol. Methods, 24(1-2), 135-139 (1978)* En;en.

1645-U2 Quantitative immuno-recognition by a radial diffusion technique. Gaffney,P.J.; Mahmoud,M.; Kirkwood,T.B.L. (Natl. Inst. Biol. Stand. and Control, Holly Hill, Hampstead, London NW3 6RB, UK) *J. Immunol. Methods, 14(1), 25-29 (1977)* En.

1646-U2 Light scattering assay of haemolytic antibody (a rapid and sensitive non-flow procedure for quantitation of complement fixing activity). Gancevici,G.; Ghitescu,L.; Calugareanu-Flonta, L.M.; Patru,I. (Cantacuzino Inst., Bucharest, Romania) *Arch. Roum. Pathol. Exp. Microbiol., 35(4), 335-362 (1976)* En;en,fr,ru.

1647-U2 [Critical study of the immunoenzymatic reactions revealed by immunoprecipitation on cellulose acetate. Advantages of ELIEDA (enzyme-linked immunoelectrodiffusion assay) and ELIDEPA (enzyme-linked immuno-double-electrophoresis assay) in parasitology]. Pinon,J.-M.; Charpentier,O.; Dropsy,G. (Lab. Parasitol., Hop. Maison-Blanche, 51100 Reims, France) *C.R. Hebd. Seances Acad. Sci., Paris Ser. D. 286(6), 449-502 (1978)* Fr;en,fr.

1648-U2 Improvement of Mancini's method of simple radial immunodiffusion. Svetlyshev,S.D.; Kodkind,G.Kh. (Inst. Poliomyelitis and Virus Encephalitides, Acad. Med. Sci. USSR, Moscow, USSR) *Vopr. Virusol., No. 6, 747-750 (1975)* Ru;en.

1649-U2 Comparative study of four diagnostic methods of enzootic bovine leukosis. Mammerickx,M.; Burny,A.; Dekegel,D.; Ghysdael,J.; Kettmann,R.; Portetelle,D. (Dep. Virol., Inst. Natl. Rech. Vet., 99, Groeselenberg, 1180 Bruxelles, Belgium) *Zentralbl. Veterinarmed., B, 24(9), 733-740 (1977)* En;de,en,es,fr.

1650-U2 Large scale serological detection in Belgium of enzootic bovine leukosis. Mammerickx,M.; Otte,J.; Rase,F.; Braibant,E.; Portetelle,D.; Burny,A.; Dekegel,D.; Ghysdael,J. (Dep. Virol., Inst. Natl. Rech. Vet., 99 Groeselenberg, 1180 Bruxelles, Belgium) *Zentralbl. Veterinarmed., B, 25(5), 416-424 (1978)* En;de,en,es,fr.

1651-U2 [Application of the gel immunodiffusion method to the experimental diagnosis of *Herpesvirus bovis* infection (infectious rhinotracheitis in the calf and balanoposthitis in the bull)]. Charton,A.; Faye,P.; Le Layec,C. (Lab. Rech. (INRA), Chaire Pathol. Betail, Ec. Vet., Alfort, France) *Bull. Acad. Vet., 48(2-3), 71-76 (1975)* Fr;fr.

1652-U2 An immunodiffusion test for detection of bovine viral diarrhea virus antibodies in bovine serum. Hopkinson,M.F.; Hart,L.T.; Seger,C.L.; Larson,A.D.; Fulton,R.W. (Dep. Vet. Sci., Agric. Exp. Stn., Louisiana State Univ., Baton Rouge, LA 70803, USA) *Am. J. Vet. Res., 40(8), 1189-1191 (1979)* En;en.

1653-U2 Microimmunodiffusion test for detection of antibodies to infectious bovine rhinotracheitis virus in bovine serum. Lejeune,J.M.; Hart,L.T.; Larson,A.D.; Seger,C.L. (Dep. Microbiol., Louisiana State Univ., Agric. Exp. Stn., Baton Rouge, LA 70803, USA) *Am. J. Vet. Res., 38(4), 459-463 (1977)* En;en.

1654-U2 Use of glycoprotein antigen in the immunodiffusion test for bovine leukemia virus antibodies. Miller,J.M.; van der Maaten,M.J. (Natl. Anim. Dis. Cent., North Cent. Reg., US Dep. Agric., Agric. Res. Serv., PO Box 70, Ames, IA 50010, USA) *Eur. J. Cancer, 13(12), 1369-1375 (1977)* En;en.

1655-U2 Rapid diagnosis of citrus tristeza virus infections by sodium dodecyl sulfate-immunodiffusion procedures. Garnsey,S.M.; Gonsalves,D.; Purcifull,D.E. (US Hortic. Res. Lab., Fed. Res. Sci. and Educ. Adm., USDA, Orlando, FL 32803, USA) En;en.

1656-U2 Testing for e antigen and e antibody by immunodiffusion. Tedder,R.S. (Dep. Virol., Middlesex Hosp. Med. Sch., London W1, UK) *J. Med. Virol., 3(1), 51-57 (1978)* En;en.

1657-U2 A simple double immunodiffusion test for typing influenza viruses. Dowdle,W.R.; Galphin,J.C.; Coleman,M.T.; Schild,G.C. (WHO Collaborating Cent. for Influenza, Virol. Div., Cent. for Dis. Control, Atlanta, GA 30333, USA) *Bull. WHO, 51(3), 213-218 (1974)* En;en,fr.

1658-U2 Application of an improved single-radial-immunodiffusion technique for the assay of haemagglutinin antigen content of whole virus and subunit influenza vaccines. Wood,J.M.; Schild,G.C.; Newman,R.W.; Seagroatt,V. (Natl. Inst. Biol. Stand. and Control, Hampstead Lab., Holly Hill, London NW3 6RB, UK) *Dev. Biol. Stand., 39, 193-200 (1977)* En;en.

1659-U2 A single-radial-immunodiffusion technique for the assay of influenza haemagglutinin antigen: proposals for an assay method for the haemagglutinin content of influenza vaccines. Schild,G.C.; Wood,J.M.; Newman,R.W. (Virol. Div., Natl. Inst. for Biol. Stand. and Control, Holly Hill, Hampstead, London NW3 6RB, UK) *Bull. WHO, 52(2), 223-232 (1976)* En;en,fr.

1660-U2 An improved single-radial-immunodiffusion technique for the assay of influenza haemagglutinin antigen: application for potency determinations of inactivated whole virus and subunit vaccines. Wood,J.M.; Schild,G.C.; Newman,R.W.; Seagroatt,V. (Natl. Inst. for Biol. Standards and Control, Hamstead Lab., Holly Hill, London NW3 6RB, UK) *J. Biol.Stand., 5(3), 237-247 (1977)* En;en.

1661-U2 Application of the single radial diffusion test for assay of antibody to influenza type A viruses. Mostow,S.R.; Child,G.C.; Dowdle,W.R.; Wood,R.J. (Cleveland Metropolitan Gen. Hosp., Cleveland, OH 44109, USA) *J. Clin. Microbiol., 2(6), 531-540 (1975)* En;en.

1662-U2 Detection of lily symptomless virus by immunodiffusion. Simmonds,D.H.; Cumming,B.G. (Dep. Biol., ELBA, Carleton Univ., Ottawa, Ont. K1S 5B6, Canada) *Phytopathology, 69(11), 1212-1215 (1979)* En;en.

1663-U2 Immunodiffusion test for ovine progressive pneumonia. Cutlip,R.C.; Jackson,T.A.; Laird,G.A. (Natl. Anim. Dis. Cent., USDA, ARS, POB 70, Ames, IA 50010, USA) *Am. J. Vet. Res., 38(7), 1081-1084 (1977)* En;en.

1664-U2 Evaluation of the agar-gel immunodiffusion test for the detection of precipitating antibodies against progressive pneumonia virus of sheep. Molitor,T.W.; Schipper,I.A.; Berryhill,D.L.; Light,M.R. (Dep. Vet. Sci., North Dakota State Univ., Fargo, ND 58102, USA) *Can. J. Comp. Med., 43(3), 280-287 (1979)* En;en,fr.

1665-U2 Evaluation of the single radial diffusion technique for detection of nuclear polyhedrosis virus (NPV) infection in *Heliothis zea*. Scott,H.A; Yearian,W.C.; Young,S.Y. (Virol. and Biocontrol Lab., Dep. Plant Pathol., Univ. Arkansas, Fayetteville, AR 72701, USA) *J. Invertebr. Pathol., 28(2), 229-232 (1976)* En;en.

1666-U2 Results of serological identification of potato virus X by means of the radial immuno-diffusion test. Fuchs,R.; Richter,J. (Inst. Phytopathol. Aschersleben, Akad. Landwirtschaftswiss., Aschersleben, GDR) *Potato Res., 18(3), 378-384 (1975)* En;en.

1667-U2 Agar-gel immunodiffusion assay for pseudorabies virus antibody. Smith,P.C.; Stewart,W.C. (Dep. Rural Practice, Coll. Vet. Med., Univ. Tennessee, Knoxville, Knoxville, TN 37901, USA) *J. Clin. Microbiol., 7(5), 423-425 (1978)* En;en.

1668-U2 Development and evaluation of a microimmunodiffusion test for detection of antibodies to pseudorabies virus in swine serum. Gutekunst,D.E.; Pirtle,E.C.; Mengeling,W.L. (Natl. Anim. Dis. Cent., Agric. Res. Serv., US Dep. Agric., PO Box 70, Ames, IA 50010, USA) *Am. J. Vet. Res., 39(2), 207-210 (1978)* En;en.

1669-U2 [Titration of serum and rabies immunoglobulins by the method of radial immunodiffusion]. Mouillot,L.; Chaniot,S.; Piat,A.; Netter,R. (Min. Sante, Lab. Natl. de la Sante, 25, Boulevard Saint-Jacques, Paris 75014, France) *Dev. Biol. Stand., 40, 243-246 (1978)* Fr;en.

1670-U2 The immunodiffusion test for the detection of rota virus in fecal material and intestine. Van Opdenbosch,E.; Dekegel,D.; Wellemans,G. (Natl. Inst. Diergeneeskundig Onderzoek, Groeselenberg 99, 1080 Brussels, Belgium) *Vlaams Diergeneeskd. Tijdschr., 47(4), 286-291 (1978)* Nl;en,fr,nl.

1671-U2 Single radial immunodiffusion test for detecting antibodies against surface antigens of intracellular and extracellular vaccinia virus. Prakash,V.J.; *Norrby,E.; Payne,L. (Dep. Virol., Karolinska Inst., Sch. Med., Stockholm 1, Sweden) *J. Gen. Virol., 35(3), 463-472 (1977)* En;en.

1672-U2 Staphylococcal protein A; its preparation and an application to rubella serology. Mallinson,H.; Roberts,C.;

White,G.B.B. (Reg. Public Health Lab., Fazakerley Hosp., Liverpool, UK) *J. Clin. Pathol., 29(11), 999-1002 (1976)* En;en.

1673-U2 Identification of *Brucella abortus* antibodies in cattle serum by single radial diffusion. Iannelli,D.; Diaz,R.; Bettini,T.M. (Inst. Anim. Prod., Univ. Naples, Naples, Italy) *J. Clin. Microbiol., 3(2), 203-205 (1976)* En;en.

1674-U2 Laboratory diagnosis of canine pyometra. Kivisto,A.-K.; Vasenius,H.; Sandholm,M. (Dep. Med., Coll. Vet. Med., F-00 550 Helsinki 55, Finland) *Acta Vet. Scand., 18(3), 308-315 (1977)* En;en,sv.

1675-U2 A simple method for typing of acidic polysaccharide K antigens of *Escherichia coli.* Kauser,B. (Inst. Med. Microbiol., Univ. Goteborg, Guldhedsgatan 10, S-413 46 Goteborg, Sweden) *FEMS Microbiol Lett., 1(5), 285-288 (1977)* En.

1676-U2 The quantitation of ribosome-bound *Escherichia coli* ribosomal proteins L7L12 by radial immunodiffusion. Morrissey,J.J.; Caldwell,P.; Weissbach,H.; Brot,N. (Natl. Heart and Lung Inst., NIH, Bethesda, MD 20014, USA) *Anal. Biochem., 75(1), 53-57 (1976)* En;en.

1677-U2 An application of electroimmunodiffusion technique for the detection of staphylococcal enterotoxin B in contaminated foods. Pedretti,C.; Cicognani,G.; Vertemati,R.; *Caserio,G. (Ist. Ispezione Aliment. Orig. Anim., Univ. Milano, Milano, Italy) *Ind. Aliment., 16(9), 125-131 (1977)* It;en,it.

1678-U2 Collaborative study of the serological identification of staphylococcal enterotoxins by the microslide gel double diffusion test. /[presented at the 89th Annual Meeting of the AOAC, held in Washington, DC, USA, 13-16 Oct 1975]. Bennett,R.W.; McClure,F. (Div. Microbiol., FDA, Washington, DC 20204, USA) *J. Assoc. Off. Anal. Chem., 59(3), 594-601 (1976)* En;en.

1679-U2 [Rapid detection and specific identification of streptococci by immunoelectrodiffusion]. Guinet,R.; Andre,J.; Gille,Y.; Guillermet,F. (Dep. Microbiol., Inst. Pasteur, 77 rue Pasteur, F 69365 Lyon Cedex 2, France) *Nouv. Presse Med., 6(22), 1980 (1977)* Fr.

1680-U2 Immunodiffusion method for identification of group A streptococci. Lyampert,I.M.; Borodiyuk,N.A.; Asoskova,T.K.; Rassokhina,I.I. (Lab. Streptococcal Infect., Inst. Epidemiol. and Microbiol., Acad. Med. Sci., Moscow 123098, USSR) *Infect. Immun., 19(3), 961-965 (1978)* En;en.

1681-U2 Modified method for serological identification of group B streptococci. Dillon,H.C.; Pass,M.A.; Buchanan,B. (Dep. Pediatr., Univ. Alabama in Birmingham, Sch. Med., University Station, Birmingham, AL 35294, USA) *J. Clin. Microbiol., 7(6), 599-600 (1978)* En;en.

1682-U2 Grouping streptococci by means of electrosyneresis (electroimmunodiffusion). Guinet,R. (Inst. Pasteur Lyon, Dep. Microbiol., 77 rue Pasteur, 69365 Lyon Cedex 2, France) *Rev. Inst. Pasteur Lyon, 10(2), 207-208 (1977)* Fr.

1683-U2 [A simple and rapid technique for grouping streptococci]. Laban,P.; Chastel,C. (Dep. Bacteriol.-Virol., Fac. Med., BP 815, 29279 Brest Cedex, France) *Ann. Biol. Clin., 34(3), 229-234 (1976)* Fr;en,fr.

1684-U2 The rapid detection and specific identification of bacterial antigens by electro-immunodiffusion in 80 cases of purulent meningitis. Denis,F.; Samb,A.; Chiron,J.-P.; Sow,A.; Diop Mar,I. (Lab. Bacteriol.-Virol., Fac. Med., Dakar, Senegal) *Nouv. Presse Med., 6(37), 3391-3396 (1977)* Fr;en,fr.

1685-U2 [The application of simple radial immunodiffusion in the quantitative determination of *Candida* antigens]. Schwartze,G.; Friedrich,E. (Klin. und Poliklin fur Hatukr., Ber. Med., Martin-Luther-Univ. Halle, Ernst-Kromayer-Str.5-8, DDR-402 Halle, GDR) *Mykosen, 19(5), 141-147 (1976)* De;de,en.

1686-U2 A new specific quantitation-in-gel method differentiating commercial human serum standards intended for RID analyses. Wadsworth,C. (Dep. Immunol. Inst. Med. Microbiol., Guldhedsgatan 10, S-413 46 Gothenburg, Sweden) *Scand. J. Immunol., 6(1-2), 97-107 (1977)* En;en.

1687-U2 Trial of immunodiffusion on cellulose acetate. Application to the estimate of immunoglobulin A and serum transferrin. Results in alcoholism and cirrhosis. Barres,C.; de Lauture,D.; Zeitoun,C.; Besancon,F. (Inst. Hydrol., Hop. Cochin, 27, rue du fg Saint-Jacques, 75674 Paris Cedex 14, France) *Pathol. Biol., 25(9), 673-676 (1977)* Fr;en,fr.

1688-U2 [Application of radial immunodiffusion to the determination of the titre of a monospecific antiserum]. Durand,G.; Blou,D.; Feger,J. (Dep. Biochim., Fac. Pharm., Univ. Montreal, Montreal, Que., Canada) *Clin. Biochem., 9(4), 195-197 (1976)* Fr;en,fr.

1689-U2 [Quantitative determination of immunoglobulin E with a simple radial immunodiffusion technique. 2. Clinical findings]. Bergmann,K.-C.; Bergmann,I.; Barthelmes,H.; Dehnert,I.; Wallenstein,G. (Forschungsinst. Lungenkrankh. und Tuberk., 1115 Berlin-Buch, Karower Str. 11, GDR) *Dtsch. Gesundheitswes., 31(43), 2048-2054 (1976)* De;de,en,ru.

1690-U2 Assay of horse urinary kallikrein by radial immunodiffusion. Jose,M.; Almeida Prado,B.; Sampaio,C.A.M. (Dep. Biochem. and Pharmacol., Esc. Paulista de Med., Caixa Postal 20320, Sao Paulo, Brazil) *Cienc. Cult., 30(9), 1131-1132 (1978)* En.

1691-U2 A simple method for demonstrating gel-precipitating human anti-immunoglobulin antibodies. Schalen,C.; Christensen,P. (Inst. Med. Microbiol., Univ. Lund, Solvegatan 23, S-223 62 Lund, Sweden) *Acta Pathol. Microbiol. Scand. Ser. C, 85(6), 480-482 (1977)* En;en.

1692-U2 A simple immunoenzymatic method for measuring IgE in human sera. Guesdon,J.L.; Thierry,R.; Avrameas,S. (Unite Immunocytochim., Dep. Biol. Mol., Inst.

Pasteur, 25 Rue du Dr. Roux, 75015 Paris, France) *Clin. Exp. Immunol., 25(1), 180-183 (1976)* En;en.

1693-U2 Identification of specific antibodies to extractable nuclear antigens by passive immunodiffusion. Kozin,F.; Fowler,M. (Div. Rheumatol., Dep. Med., Med. Coll. Wisconsin, Milwaukee County Med. Complex, 8700 West Wisconsin Ave., Milwaukee, WI 53226, USA) *Am. J. Clin. Pathol., 71(4), 437-440 (1979)* En;en.

1694-U2 Confirmation by radial immunodiffusion of derepression of the fourth component of complement in guinea pigs fed aflatoxin or rubratoxin. Thurston,J.R.; Richard,J.L. (Natl. Anim. Dis. Cent., Agric. Res. Sci. and Educ. Adm., US Dep. Agric., POB 70, Ames, IA 50010, USA) *Am. J. Vet. Res., 40(8), 1206-1208 (1979)* En;en.

1695-U2 The two-cross immunodiffusion technique: diffusion coefficients and precipitating titers of IgG in human serum and rabbit serum antibodies. Pokric,B.; Pucar,Z. ('Ruder Boskovic' Inst., POB 1016, 41001 Zagreb, Yugoslavia) *Anal. Biochem., 93(1), 103-114 (1979)* En;en.

See: 1132, 1423, 1559, 1584, 1610, 2427, 2550, 2565, 3263, 3294, 3298, 41, 1560, 1722, 3282, 3245

Immunoprecipitation

1696-U2 Curve-fit calculations with a desk-top calculator interfaced with an automated immunoprecipitin reaction method, for determination of plasma proteins in urine. Hemmingsen,L.; Neirup,H.; Skaarup,P. (Cent. Lab., Dep. Clin. Chem. Cent. Hosp., Nykobing Falster DK-4800 Denmark) *Clin. Chem., 23(6), 1189-1190 (1977)* En.

1697-U2 A modified quantitative precipitin test which is rapid, sensitive and reproducible. Raney,J.P.; McLennan,B.D. (Dep. Biochem., Univ. Saskatchewan, Saskatoon, Sask. S7N 0W0, Canada) *J. Immunol. Methods, 29(1), 65-70 (1979)* En;en.

1698-U2 A new method of immunoprecipitation in gel, rapid, easy and highly sensitive-radial-immuno-electro-osmophoresis. (RIEOP). Llopis,F.; Sanchez-Cuenca,J.M.; Gonzalez-Molina,A. (Ciud. Sanit. Seguridad Social 'La Fe' Cent. Invest., Unid. Inmunol. Exp., Avenida Alferez Provisional, 21, Valencia 9, Spain) *Allergol. Immunopathol., 7(1), 19-28 (1979)* Es;en,es.

1699-U2 An immunoprecipitation-dissociated technique for large scale antibody purification and an antigen consumption electroimmunoassay for antibody quantitation. A model study with antibodies to pregnancy zone protein. Folkersen,J.; Teisner,B.; Svendsen,P.; Svehag,S.-E. (Inst. Med. Microbiol., Odense Univ., Odense, Denmark) *J. Immunol. Methods, 23(1-2), 127-135 (1978)* En;en.

1700-U2 Use of gel precipitin test in the diagnosis of cytomegalovirus infections. Fioretti,A.; Pollini,C. (Farmitalia, 20146 Milano, Italy) *Acta Virol., 22(1), 66-69 (1978)* En;en.

1701-U2 Diagnosis of infectious bursitis (Gumboro disease). Wilke,I.; Schmidt,U.; Schobries,H.D. (1503 Potsdam Bornstedt, Pappelallee 2, GDR) *Monatsh. Veterinarmed., 33(17), 672-675 (1978)* De;de,en,ru.

1702-U2 Application of agar-gel precipitin test for the detection of Sendai virus antibody in rat sera. Yachida,S.; Iritani,Y.; Makino,S.; Seko,S. (Aburahi Lab., Shionogi and Co., Ltd., 520-34, Koka-Cho, Shiga, Japan) *Lab. Anim. Sci., 25(4), 434-436 (1975)* En;en.

1703-U2 Precipitation of visna viral proteins by immune sera of rabbits and sheep. Lin,F.H.; Thormar,H. (New York State Inst. Basic Res. Mental Retardation, Staten Island, NY 10314, USA) *J. Virol., 29(2), 536-539 (1979)* En;en.

1704-U2 Rapid purification of mouse L cell interferon labeled with radioactive amino acid by immune precipitation. Yonehara,S.; Iwakura,Y.; *Kawade,Y. (Inst. Virus Res., Kyoto Univ., Kyoto 606, Japan) *Virology, 100(1), 125-129 (1980)* En;en.

1705-U2 Evaluation of the antiserum agar method for the serogroup identification of *Neisseria meningitidis*. Ashton,F.E.; Ryan,A.; Diena,B.B.; Frasch,C.E. (Natl. Neisseria Ref. Cent., Lab. Cent. Dis. Control, Health and Welfare Canada, Ottawa, Ont. K1A OL2, Canada) *Can. J. Microbiol., 25(6), 784-787 (1979)* En;en,fr.

1706-U2 A rapid spot immunoprecipitate assay method applied to quantitating C-reactive protein in pediatric sera. Wadsworth.C. (Dep. Immunol., Inst. Med. Microbiol., Guldhedsgatan 10, S-413 46 Gothenburg, Sweden) *Scand. J. Immunol., 6(12), 1263-1273 (1977)* En;en.

1707-U2 Some problems in the correction for intrinsic light-scattering of samples in specific protein analysis by automated immunoprecipitation. Shenkin,A.; Morrison,B.; Robertson,D.A. (Dep. Biochem. R. Infirm., Glasgow, G4 OSF, UK) *Ann. Clin. Biochem., 14(3), 163-170 (1977)* En;en.

1708-U2 An improved method for isolation of H-2 and Ia alloantigens with immunoprecipitation induced by protein A-bearing staphylococci. Cullen,S.E.; Schwartz,B.D. (Immunol. BrCanch, Natl.ancer Inst. and Lab. Immunol., Natl. Inst. Allergy and Infect. Dis., NIH, Bethesda, MD 20014, USA) *J. Immunol., 117(1), 136-142 (1976)* En;en.

1709-U2 A modification of the automated immune precipitin method for quantitation of human serum immunoglobulins. Keren,D.F.; Frye,R.M.; Datiles,T.B.; Grindon,A.J. (Dep. Pathol., Univ. Michigan, Ann Arbor, MI 48109, USA) *Am. J. Clin. Pathol., 70(1), 41-44 (1978)* En;en.

1710-U2 Immunoglobulins G, A and M in normal and pathologic human sera determined with the spot immunoprecipitate assay (SIA) in multi-assay plates. Wadsworth,C.; Wadsworth,E. (Dep. Clin. Immunol., Inst. Med. Microbiol., Guldhedsgatan 10, 413 46 Gothenburg, Sweden) *Scand. J. Immunol., 9(3), 255-262 (1979)* En;en.

See: 858, 1491, 1938, 3278, 2434, 2630, 3281, 1560, 3286

Agglutination

1711-U2 Application of the latex flocculation serological assay to curly top virus. Mumford,D.L. (Agric. Res. Serv., US Dep.Agric., Crops Res. Lab., Utah State Univ., Logan, UT 84322, USA) *Phytopathology, 67(7), 949-952 (1977)* En;en.

1712-U2 The latex agglutination test in virus diagnostics: identification of poliomyelitis and coxsackie viruses. Siegel,H.; Klein,M.; *Schmidt,W.A.K. (Inst. Med. Microbiol. and Virol., Univ. Dusseldorf, Moorenstr. 5, D-4000 Dusseldorf 1, GFR) *Zentralbl. Bakteriol. Parasitenkd., Infektionskr. Hyg., I Abt. A, 245(3), 271-275 (1979)* En;de,en.

1713-U2 Sensitive detection of Andean potato latent and Andean potato mottle viruses in potato tubers with the serological latex test. Koenig,R.; Bode,O. (Biol. Bundesanstalt, Inst. Viruskr. Pflanzen, Messeweg 11/12, D 3300 Braunschweig, GFR) *Phytopathol. Z., 92(3), 275-280 (1978)* En;de,en.

1714-U2 A leaf dip method for routine identification of plant viruses using the latex agglutination test. Polak,J. (Div. Plant Prot., Res. Inst. Plant Prod., Ruzyne 507, 161 06 Praha 6, Czechoslovakia) *Biol. Plant., 22(3), 237-238 (1980)* En;en.

1715-U2 A rapid automated latex screen for tetanus toxoid antibodies. Booth,J.R.; Nuttall,P.A. (Reg. Transfus. Cent., Longley Lane, Sheffield S5 7JN, UK) *Vox Sang., 34(4), 239-240 (1978)* En;en.

1716-U2 The latex agglutination test as a rapid serological assay for *Corynebacterium sepedonicum*. Slack,S.A.; Sanford,H.A.; Manzer,F.E. (Dep. Plant Pathol., Univ. Wisconsin-Madison, Madison, WI 53706, USA) *Am. Potato J., 56(9), 441-446 (1979)* En;en,fr.

1717-U2 Rapid grouping of beta-hemolytic streptococci by latex agglutination. Lue,Y.A.; Howit,I.P.; *Ellner,P.D. (Dep. Microbiol., Columbia Univ. Coll. Physicians and Surgeons, New York, NY 10032, USA) *J. Clin. Microbiol., 8(3), 326-328 (1978)* En;en.

1718-U2 Rapid grouping of streptococci by latex agglutination after nitrous acid extraction. Guinet,R. (Inst. Pasteur de Lyon, Dep. Microbiol., Lyon, France) *Rev. Inst. Pasteur Lyon, 12(4), 543-544 (1979)* Fr.

1719-U2 Rapid detection of neonatal group B streptococcal infections by latex agglutination. Bromberger,L.I.; Chandler,B.; Gezon,H.; Haddow,J.E. (Found. Blood Res., POB 426, Scarborough, ME 04074, USA) *J. Pediatr., 96(1), 104-106 (1980)* En.

1720-U2 Latex agglutination in diagnosis of bacterial infections, with special reference to patients with meningitis and septicemia. Kaldor,J.; Asznowicz,R.; Buist,D.G.P. (Dep. Pathol., Fairfield Communicable Dis. Hosp., Yarra Bend Road, Fairfield, Vic. 3078, Australia) *Am. J. Clin. Pathol., 68(2), 284-289 (1977)* En;en.

1721-U2 The toxoplasma latex agglutination test in mice and its application to the diagnosis of ovine abortion. Seamon,P.J.; Clegg,F.G. (Minist. Agric., Fish. and Food, ADAS, Sutton Bonington, UK) *Vet. Rec., 101(15), 324-325 (1977)* En;en.

1722-U2 Employment of three immunological tests (latex-agglutination, indirect hemagglutination, double diffusion in gel) for increasing the effectiveness of diagnosis of echinococcosis and alveococcosis. Ballad,N.E.; Zorikhina,V.I. (Inst. Med. Parasitol. and Trop. Med., Minist. Public Health, Moscow, USSR) *Med. Parazitol. Parazit. Bolezn., 47(2), 25-31 (1978)* Ru;en,ru.

1723-U2 Latex particle adherence (LPA) test for identification of adherent and non-adherent leucocytes. Buben,J.; Malkovsky,K.M. (Inst. Mol. Genet., Czechoslovak Acad. Sci., Praha 6, Czechoslovakia) *Folia Biol., 23(5), 359-365 (1977)* En;en,ru.

1724-U2 Quantitation of rheumatoid factor by latex particle aggregation measurement. Dito,W.R. (Lab. Med., Dep. Pathol., Hosp. Scripps Clin., La Jolla, CA 92037, USA) *Am. J. Clin. Pathol., 69(2), 147-155 (1978)* En;en.

1725-U2 The latex particle adherence (LPA) assay for detection of leukocytes with adhesive surface properties. Bubenik,J.; Malkovsky,M.; Suhajova,E. (Inst. Mol. Genet., Czechoslovak Acad. Sci., 166 10 Prague, Czechoslovakia) *Cell. Immunol., 35(1), 217-225 (1978)* En;en.

1726-U2 A new immunochemical tube test for pregnancy, using the principle of the latex agglutination inhibition reaction. I. Basic studies. Schlebusch,H.; Spona,J.; Keller,M.; Hepp,H.; Hoffmann,G.; Keller,P.J. (Abt. Klin. Chem., Univ.-Frauenklin., 5300 Bonn 1, Venusberg, GFR) *Dtsch. Med. Wochenschr., 101(44), 1587-1591 (1976)* De;de,en.

1727-U2 Comparison of a direct latex-agglutination technic with the tanned red cell hemagglutination inhibition immunoassay (TRCHII) for semiquantitation of fibrinogen/fibrin degradation products. Wilson,J.E.,III; Thornton,R.D. (Lab. Cardiopulmonary Res., Dep. Intern. Med., Univ. Texas Southwestern Med. Sch., 5323 Harry Hines Blvd., Dallas, TX 75235, USA) *Am. J. Clin. Pathol., 65(4), 528-532 (1976)* En;en.

1728-U2 Agglutination assay for human opsonic factor using gelatin-coated latex particles. Check,I.J.; Wolfman,H.C.; Coley,T.B.; Hunter,R.L. (Dep. Pathol., Univ. Chicago, Chicago, IL 60637, USA) *J. Reticuloendothel. Soc., 25(4), 351-362 (1979)* En;en.

1729-U2 A simple, rapid micro-latex fixation test. Cicciarelli,J.C.; Terasaki,P.I. Chia,D.; Barrett,E.V.; Mickey,M.R.; Prince,S.; Iwaki,Y.; Shirahama,S. (1000 Veteran Ave., Los Angeles, CA 90024, USA) *Clin. Exp. Immunol., 38(1), 181-185 (1979)* En;en.

1730-U2 Evaluation of a new commercial latex test for detection of rheumatoid factors. Micheli,A. (Div. Rhumatol.,

Hop. Beau-Sejour, CH 1211 Geneve 4, Switzerland) *Schweiz. Med. Wochenschr., 109(7), 220-222 (1979)* Fr;en,fr.

1731-U2 Mechanisation of the slide agglutination test. Scaby,D.A.; Caughey,C. (Plant Pathol. Res. Div., Dep. Agriculture for Northern Ireland, Newforge Lane, Belfast BT9 5PX, UK) *Potato Res., 20(4), 343-344 (1977)* En;en.

1732-U2 The microleucoagglutination test. Majsky,A. (Ustav hematol. a krevni transfuze, Praha, Czechoslovakia) *Cas. Lek. Ces., 116(4), 123-125 (1977)* Cs;cs,en,fr,ru.

1733-U2 Mechanisation of the slide agglutination test. Seaby,D.A.; Caughey,C. (Plant Pathol. Res. Div., Dep. Agriculture for Northern Ireland, Newforge Lane, Belfast BT9 5PX, UK) *Potato Res., 20(4), 343-344 (1977)* En;en.

1734-U2 Staphylococcal agglutination-inhibition reaction: a rapid and simple test for dengue antibodies. Cheong,C.Y.; Hong,T.S.; Eng,A.S. (Dep. Microbiol., Fac. Med., Sepoy Lines, Singapore 3, Republic of Singapore) *Singapore Med. J., 16(3), 194-195 (1975)* En;en.

1735-U2 Identification of *Aerococcus viridans* by means of co-agglutination. Saxegaard,F.; Hastein,T. (Natl. Vet. Inst., PB 8156, Oslo 1, Norway) *Acta. Vet. Scand., 19(4), 604-606 (1978)* En.

1736-U2 Evaluation of the anaplasmosis rapid card agglutination test for detecting experimentally infected elk. Magonigle,R.A.; Eckblad,W.P. (Dep. Vet. Sci. Univ. Idaho Agric. Exp. Stn., Univ. Idaho, Moscow, IL 83843, USA) *Cornell Vet., 69(4), 402-410 (1979)* En;en.

1737-U2 Technique of agglutination reaction for spore-forming bacteria in the microtiter system. Study on *B. anthracis*. Bohm,R.; Strauch,D. (D-7000 Stuttgart 70, Postfach 106/06200, GFR) *Wien. Tierarztl. Monatsschr., 63(6-7), 224-228 (1976)* De;de,en.

1738-U2 A new and easy test for the diagnosis of brucellosis to be used in the field and in small laboratories: the test with a buffered and Bengal Rose-stained antigen. Rey,J.L. (c/o Medicine Tropicale, parc du Pharo, 13007 Marseille, France) *Med. Trop., 37(5), 593-597 (1977)* Fr;en,fr.

1739-U2 A simple technique to differentiate between animals infected with *Yersinia enterocolitica* IX and those infected with *Brucella abortus*. Mittal,K.R.; Tizard,I.R. (Dep. Vet. Microbiol. and Immunol., Ontario Vet. Coll., Univ. Guelph, Guelph, Ont., Canada) *Res. Vet. Sci., 26(2), 248-250 (1979)* En;en.

1740-U2 Development of a rose bengal stained plate-test antigen for the rapid diagnosis of *Brucella canis*. George,L.W.; Carmichael,L.E. (Dep. Large Anim. Med., Coll. Vet. Med., Univ. Georgia, Athens, GA 30602, USA) *Cornell Vet., 68(4), 530-543 (1978)* En;en.

1741-U2 [The Rose Bengal test in the serological diagnosis of human brucellosis. A comparison with three classical techniques]. Geral,M.-F.; Saurat,P.; Lautie,R.; Ganiere,J.-P.; Meignier,B. (Ecole Natl. Vet. Toulouse, Mal. Contagieuses et Zoonoses, 31076 Toulouse Cedex, France) *Rev. Med. Vet., 126(8-9), 1099-1119 (1975)* Fr;de,en,es,fr.

1742-U2 Study of the plate agglutination test with Rose Bengal antigen for the diagnosis of human brucellosis. Cernyseva,M.I.; Knjazeva,E.N.; Egorova,L.S. (Brucellosis Lab., Inst. Microbiol. and Epidemiol., Acad. Med. Sci. USSR, Moscow, USSR) *Bull. WHO, 55(6), 669-674 (1977)* En;en,fr.

1743-U2 Serological methods for detection of the fireblight organisms, *Erwinia amylovora*. Schroder,C.M. (Inst. Phytopathol., Christian-Albrechts-Univ. Kiel, Olshausenstr. 40-60, D-2300 Kiel, GFR) *Z. Pflanzenkr. Planzenschutz., 85(7), 393-403 (1978)* De;de,en.

1744-U2 Serological relationships among strains of *Erwinia chrysanthemi*. Yakrus,M.; Schaad,N.W. (Univ. Georgia, Dep. Plant Pathol., Athens, GA 30602, USA) *Phytopathology, 69(5), 517-522 (1979)* En;en.

1745-U2 Highly-specific agglutinating O- and OK-immunoglobulins for the identification of *Escherichia* in the agglutination test on glass. Ulisko,I.N.; Landsman,N.M.; Goldshmid,V.K.; Romanenko,E.E. (Sci.-Res. Inst. Vaccines and Sera, Moscow, USSR) *Zh. Mikrobiol. Epidemiol. Immunobiol., 54(9), 36-40 (1977)* Ru;en,ru.

1746-U2 Detection of the K99 antigen by means of agglutination and immunoelectrophoresis in *Escherichia coli* isolates from calves and its correlation with enterotoxigenicity. Guinee,P.A.M.; Jansen,W.H.; Agterberg,C.M. (Rijks Inst. Volksgezondheid, PO Box 1, Bilthoven, Netherlands) *Infect. Immun., 13(5), 1369-1377 (1976)* En;en.

1747-U2 Rapid typing of *Escherichia coli* K1 containing bacteria using the co-agglutination technique. Danielsson,D.; Kaijser,B.; Olcen,P. (Dep. Clin. Bact. and Immunol., Cent. Cty Hosp., Overloro, Sweden) *FEMS Microbiol. Lett., 5(2), 123-126 (1979)* En.

1748-U2 Serological identification of the bacterial agent of contagious equine metritis. Rommel,F.A.; Dardiri,A.H.; Sahu,S.P.; Pierson,R.E. (USDA, Sci. and Educ. Adm., Fed. Res., Plum Island Anim. Dis. Cent., Greenport, Long Island, NY 11944, USA) *Vet. Rec., 103(25), 564 (1978)* En.

1749-U2 Serological methods for rapid diagnosis *Haemophilus influenzae, Neisseria meningitis* and *Streptococcus pneumoniae* in crebrospinal fluid: a comparison of co-agglutination, immunofluorescence and immunoelectroosmophoresis. Olcen,P. (Dep. Clin. Bacteriol. and Immunol., Central County Hosp., S-70185 Orebro, Sweden) *Scand. J. Infect. Dis., 10(4), 283-289 (1978)* En;en.

1750-U2 Detection of antibodies to legionnaires disease organism by microagglutination and micro-enzyme-linked immunosorbent assay tests. Farshy,C.E.; Klein,G.C.; Feeley,J.C. (Cent. Dis. Control, Atlanta, GA 30333, USA) *J. Clin. Microbiol., 7(4), 327-331 (1978)* En;en.

1751-U2 [Use of a modified *Leptospira*-microagglutination method in veterinary serological diagnosis]. Bajmocy,E.

(4032 Debrecen, Pf. 174, Hungary) *Magy. Allatorv. Lapja, 30(7-8), 546-548 (1975)* Hu;de,en,hu,ru.

1752-U2 Evaluation of the use of 'thermoresistant' antigen Patoc 1, in the diagnosis of human and animal leptospirosis. Preliminary report. Cinco del Fabro,M.; Dougan,R.; Jelincic,A.; Piacentini,I. (Inst. Microbiol., Univ. Trieste, Via A. Fleming 22, 34127 Trieste, Italy) *Boll. Ist. Sieroter. Milan., 57(6), 707-712 (1978)* En;en,it.

1753-U2 Use of the lectin concanavalin A in the preparation of mesosomal membrane fractions from *Micrococcus lysodeikticus*. Owen,P.; Salton,M.R.J. (Dep. Microbiol., New York Univ. Sch. Med., New York, NY 10016, USA) *Microbios, 13(51-52), 27-39 (1975)* En;en.

1754-U2 A serodiagnostic test for tuberculosis. Nicholls,A.C. (Midhurst Med. Res. Inst., Midhurst, Sussex, UK) *J. Clin. Pathol., 28(11), 850-853 (1975)* En;en.

1755-U2 Rapid diagnosis of *Mycoplasma pneumoniae* infection: a reminder. Macfarlane,J.T.; Neale,I.A. (Radcliffe Infirm., Oxford, UK) *Br. Med. J., 1(6156), 124 (1979)* En.

1756-U2 A rapid slide agglutination test for the confirmation of *Neisseria gonorrhoeae*. Wallace,R.; Charron,F.; Ryan,A.; Ashton,F.E.; Diena,B.B. (Lab. Cent. Dis. Control., Health and Welfare Canada, Ottawa, Ont. K1A 0L2, Canada) *Can. J. Med. Technol., 37(5), 157 (1975)* En;en,fr.

1757-U2 Identification of clinical isolates of *Neisseria gonorrhoeae* by a coagglutination test. Barnham,M.; Glynn,A.A. (Dep. Bacteriol., Wright Fleming Inst., St. Mary's Hosp. Med. Sch., London W2 1PG, UK) *J. Clin. Pathol., 31(2), 189-193 (1978)* En;en.

1758-U2 Serological grouping of meningococci and detection of antigen in cerebrospinal fluid by coagglutination. Eldridge,J.; Sutcliffe,E.M.; Abbott,J.D.; Jones,D.M. (Public Health Lab., Withington Hosp., Manchester M20 8LR, UK) *Med. Lab. Sci., 35(1), 63-66 (1978)* En;en.

1759-U2 Gonorrhea card test: an agglutination test for uncomplicated gonorrhea. Kamei,K.; Chen,N.; Gaafar,H.A. (Div. Lab. and Res., New York State Dep. Health, Albany, NY 12201, USA) *Health Lab. Sci., 15(4), 192-196 (1978)* En;en.

1760-U2 Identification of *Neisseria gonorrhoeae* from primary cultures by a slide agglutination test. Malysheff,C.; Wallace,R.; Ashton,F.E.; Diena,B.B.; Perry,M.B. (Public Health Lab., Ontario Minist. Health, Ottawa, Ont., Canada) *J. Clin. Microbiol., 8(2), 260-261 (1978)* En;en.

1761-U2 Confirmatory identification of *Neisseria gonorrhoeae* by slide coagglutination. Lim,D.V.; Wall,T. (Dep. Biol., Univ. South Florida, Tampa, FL 33620, USA) *Can. J. Microbiol., 26(2), 218-222 (1980)* En;en,fr.

1762-U2 Rapid laboratory identification of *Neisseria gonorrhoeae* by coagglutination. Helstad,A.G.; Bruns,M.K. (State Lab. Hyg., Madison, WI 53706, USA) *J. Clin. Microbiol., 11(6), 753-754 (1980)* En;en.

1763-U2 Rapid serotyping of groups A, B, and C meningococci by rocket-line immunoelectrophoresis and coagglutination. Danielsson,D.; Olcen,P. (Dep. Clin. Bacteriol. and Immunol., Cent. Cty. Hosp., S-70185 Orebro, Sweden) *J. Clin. Pathol., 32(2), 136-142 (1979)* En;en.

1764-U2 Comparison of a slide coagglutination technique with the Minitek system for confirmation of *Neisseria gonorrhoeae*. Hampton,K.D.; Stallings,R.A.; *Wasilauskas,B.L. (Dep. Pathol., Bowan Gray Sch. Med., Wake Forest Univ., Winston Salem, NC 27103, USA) *J. Clin. Microbiol., 10(3), 290-292 (1979)* En;en.

1765-U2 Rapid plate agglutination procedure for serotyping *Pasteurella haemolytica*. Frank,G.H.; Wessman,G.E. (Natl. Anim. Dis. Cent., ARS, USDA, Ames, IA 50010, USA) *J. Clin. Microbiol., 7(2), 142-145 (1978)* En;en.

1766-U2 Comparison of test typing sera for *Pseudomonas aeruginosa*. Kodama,H.; Ishimoto,M. (Dep.Bacteriol., Toyama Inst. Health, 1-15 Otemachi, Toyama 930, Japan) *Jap. J. Exp. Med., 46), 383-391 (1976)* En;en.

1767-U2 Serotyping of *Serratia marcescens*: simplified tube O-agglutination test and comparison with other serological procedures. Traub,W.H.; Fukushima,P.I. (Clin. Microbiol. Lab., Clin. Pathol. Serv. 113*, VA Med. Cent., 4150 Clement St., San Francisco, CA 94121, USA) *Zentralbl. Bakteriol. Parasitenkd. Infektionskr. Hyg., I Abt. A, 244(4), 474-493 (1979)* En;de,en.

1768-U2 Identification of streptococcal groups A, B, C, and G by slide co-agglutination of antibody-sensitized protein A-containing staphylococci. Hahn,G.; *Nyberg,I. (Pharmacia Diagnostics AB, S-751 03 Uppsala, Sweden) *J. Clin. Microbiol., 4(2), 99-101 (1976)* En;en.

1769-U2 Methods for the detection of *S. aureus* in foods. Terplan,G.; Zaadhof,K.-J. (Veterinarstr. 13, 8000 Munchen 22, GFR) *Arch. Lebensmittelhyg., 29(4), 132-135 (1978)* De;de,en.

1770-U2 Single-tube mixed agglutination test for the detection of staphylococcal protein A. Maxim,P.E.; Mathews,H.L.; Mengoli,H.F. (Dep. Microbiol., West Virginia Univ. Med. Cent., Morgantown, WV 26506, USA) *J. Clin. Microbiol., 4(5), 418-422 (1976)* En;en.

1771-U2 [Progress in the identification of pathogenic streptococci. Grouping by coagglutination]. De Maneville,M.; Lebrun,L.; Pillot,J. (Serv. Bacteriol.-Immunol., Hop. Antoine Beclere, 157 rue de la Porte de Trivaux, F 92141 Clamart, France) *Nouv. Presse Med., 6(21), 1882 (1977)* Fr.

1772-U2 Method for rapid detection of group B streptococci by coagglutination. Leland,D.S.; *Lachapelle,R.C.; Wlodarski,F.M. (Dep. Med. Technol., Univ. Vermont, Burlington, VT 05401, USA) *J. Clin. Microbiol., 7(4), 323-326 (1978)* En;en.

1773-U2 Rapid isolation and identification of group B streptococci from selective broth medium by slide coagglutination test. Szilagyi,G.; Mayer,E.; Eidelman,A.I. (Dep. Lab. Med., Hosp. Albert Einstein Coll. Med., Yeshiva Univ., Bronx, NY 10461, USA) *J. Clin. Microbiol., 8(4), 410-412 (1978)* En;en.

1774-U2 Rapid grouping of streptococci by coagglutination. Comparison with counter-immunoelectrophoresis. Guinet,R. (Dep. Microbiol., Inst. Pasteur de Lyon, 77, rue Pasteur, 69365 Lyon Cedex 2, France) *Rev. Inst. Pasteur Lyon, 13(1), 69-71 (1980)* Fr.

1775-U2 Testing the lactic streptococci for agglutinin titre in cows milk. Ellerman,J.A.G. (Dairy Farmers Co-Operative Ltd., Sydney, NSW, Australia) *Aust. J. Dairy Technol., 30(3), 111-113 (1975)* En;en.

1776-U2 Serological grouping of streptococci by a slide coagglutination method. Finch,R.G.; Phillips,I. (Dep. Med. Microbiol., St. Thomas's Hosp. Med. Sch, London, UK) *J. Clin. Pathol., 30(2), 168-170 (1977)* En;en.

1777-U2 Rapid slide coagglutination test for identifying and typing group B streptococci. Kirkegaard,M.K.; Field,C.R. (Dep. Med. Microbiol., Univ. Wisconsin, Madison, WI 53706, USA) *J. Clin. Microbiol., 6(3), 266-267 (1977)* En;en.

1778-U2 The reliability and rapidity of the coagglutination technic and its comparison with precipitin technics in the grouping of streptococci. Koshi,G.; Thangavelu,C.P.; Brahmadathan,K.N. (Dep. Microbiol., Christian Med. Coll. and Hosp., Ida Scudder Road, POB No. 3, Vellore 632004, India) *Am. J. Clin. Pathol., 71(6), 709-712 (1979)* En;en.

1779-U2 Rapid detection of pneumococcal antigens in sputum and blood serum using a coagglutination test. Edwards,E.A.; Kilpatrick,M.E.; Hooper,D. (Biol. Sci. Div., Naval Health Res. Cent., San Diego, CA 92152, USA) *Mil. Med., 145(4), 256-258 (1980)* En;en.

1780-U2 Comparison of slide coagglutination test and countercurrent immunoelectrophoresis for detection of group B streptococcal antigens in cerebrospinal fluid from infants with meningitis. Webb,B.J.; Edwards,M.S.; *Baker,C.J. (Dep. Immunol., Baylor Coll. Med., Houston, TX 77030, USA) *J. Clin. Microbiol., 11(3), 263-265 (1980)* En;en.

1781-U2 Serological grouping of streptococci by slide agglutination. Efstratiou,A.; Maxted,W.R. (Div. Hosp. Infect., Cent. Public Health Lab, Colindale Ave., London NW9 5HT, UK) *J. Clin. Pathol., 32(12), 1228-1233 (1979)* En;en.

1782-U2 A rapid test for the identification of all serotypes of *Vibrio cholerae* (including 'non-agglutinating' vibrios). Sil,J.; Bhattaharyya,F.K. (WHO Collab. Cent. Ref. and Res. on Vibrios, Natl. Inst. Cholera and Enteric Dis., Calcutta 700 016, India) *J. Med. Microbiol., 12(1), 63-70 (1978)* En;en.

1783-U2 Latex agglutination, counterimmunoelectrophoresis, and protein A co-agglutination in diagnosis of bacterial meningitis. Dirks-Go,S.I.S.; *Zanen,H.C. (Bacteriol. Dep., Lab. Hyg., Univ. Amsterdam, Mauritskade 57, Amsterdam-O, Netherlands) *J. Clin. Pathol., 31(12), 1167-1171 (1978)* En;en.

1784-U2 Stained bacterial antigens for use in microagglutination procedures. Eurell,T.E.; Lewis,D.H.; Grumbles,L.C. (Dep. Vet. Microbiol., Texas A and M Univ., College Station, TX 77843, USA) *Prog. Fish Cult., 41(2), 55-57 (1979)* En;en.

1785-U2 Rapid presumptive identification of *Cryptococcus neoformans*. Muchmore,H.G.; Felton,F.G.; Scott,E.N. (Infectious Dis. Sect., Veterans Adm. Hosp., Oklahoma City, OK 73104, USA) *J. Clin. Microbiol., 8(2), 166-170 (1978)* En;en.

1786-U2 Rapid identification of yeasts by serological methods. A combined serological and biological method. Taguchi,M.; Tsukjii,M.; Tsuchiya,T. (Dep. Serol., Kanagawa Prefect. Coll., Med. Technol., Asahi-ku, Yokohama 241, Japan) *Sabouraudia, 17(3), 185-191 (1979)* En;en,fr.

1787-U2 Rapid latex agglutination (RLA) test for the diagnosis of *Babesia argentina*. Lopez V.G.; Todorovic,R.A. (Vet. Parasitol. and Entomol. Prog., Inst. Colombiano Agrepecuario (ICA), Apartado Aereo 67-13, Cali, Colombia) *Vet. Parasitol., 4(1), 1-9 (1978)* En;en.

1788-U2 Determination of agglutinins in the IgM fraction as a simple method for detecting acute toxoplasmosis. Averbach,S.; Averbach,B.; Yanovsky,J.F.; Schmunis,G.A. (Lab. Toxoplasmosis, Hosp. Muniz, Uspallata 2272, Buenos Aires, Argentina) *Medicina, 35(5), 469-476 (1975)* Es;en,es.

1789-U2 Direct agglutination test for diagnosis of *Toxoplasma* infection: method for increasing sensitivity and specificity. Desmonts,G.; *Remington,J.S. (Div. Allergy, Immunol. and Infect. Dis., Palo Alto Med. Res. Found., Palo Alto, CA 94301, USA) *J. Clin. Microbiol., 11(6), 562-568 (1980)* En;en.

1790-U2 Identification of 19-S-antigammaglobulin factors in a modified latex test (rheumatoid factor tube test). Maute,I.; Schmolke,B.; Vorlaender,K.O. (Immunol. Lab., Keithstr. 9/11, 1000 Berlin 30, GFR) *Aktuel. Rheumatol., 3(4), 183-191 (1978)* De;de,en.

1791-U2 A rapid direct latex agglutination test for HCG. Spona,J. (Hormonlabol. Univ.-Frauenklin., Spitalgasse 23, A-1090 Wien, Austria) *Wien. Klin. Wochenschr., 88(9), 289-291 (1976)* De;de,en.

1792-U2 A simple method of screening for antisperm antibodies in the human male. Detection of spermatozoal surface IgG with the direct mixed antiglobulin reaction carried out on untreated fresh human semen. Jager,S.; Kremer,J.; van Slochteren-Draaisma,T. (Fertility Unit, Dep. Obstet. and Gynecol., University Hosp., Groningen, Netherlands) *Int. J. Fertil., 23(1), 12-21 (1978)* En;en.

1793-U2 Simpler technique for serological detection of H-Y antigen on mouse lymphocytes. Tokuda,S.; Arrington,T.; Goldberg,E.H.; Richey,J. (Dep. Microbiol., Sch. Med., Univ.

New Mexico, Albuquerque, NM 87131, USA) *Nature, 267(5610), 433-434 (1977)* En.

1794-U2 Incidence of antispermatozoidal antibodies in the serum of fertile women. Vanderbeeken,Y.; Breedael,E.; Delespesse,G. (17 rue Longue, 1950 Bruxelles, Belgium) *J. Gynecol. Obstet. Biol. Reprod., 8(2), 111-113 (1979)* Fr;en,fr.

1795-U2 Rapid separation of mouse T and B lymphocytes using wheat germ agglutinin. Bourguignon,L.Y.W.; Rader,R.L.; McMahon,J.T. (Dep. Biol., Wayne State Univ., Detroit, MI 48202, USA) *J. Cell Physiol., 99(1), 95-100 (1979)* En;en.

1796-U2 The evaluation of a latex particle agglutination test for the detection of circulating DNA antibody. Stokes,R.P. (Reg. Immunol. Lab., East Birmingham Hosp., Birmingham B9 5ST, UK) *Med. Lab. Sci., 34(3), 241-245 (1977)* En;en.

1797-U2 A simplified mixed agglutination technique for ABO grouping of dried bloodstains using cellulose acetate sheets. Chatterji,P.K. (Cent. Forensic Sci. Lab., New Delhi, India) *J. Forensic Sci. Soc., 17(2-3), 143-144 (1977)* En.

See: 1560, 1876, 2072, 3252, 3280, 3286, 1815, 3241, 3256, 3283, 3285, 3254, 3255, 3281

Haemagglutination and haemagglutination inhibition

1798-U2 Automatized hemmaglutination kinetics. Monnet,A.; Cabadi,Y. (Cent. Hemotypol. CNRS, CHU de Purpan, Ave. Grande Bretagne, F-31300 Toulouse, France) *Vox Sang., 34(4), 227-230 (1978)* En;en.

1799-U2 A rapid method for detection of Bhanja virus in infected ticks. Gaidamovich,S.Ya.; Klisenko,G.A.; Grokhovskaya,I.M. (D.I. Ivanovsky Inst. Virol., Gamaleya St. 16, Moscow 123098, USSR) *Intervirology, 11(5), 188-190 (1979)* En;en.

1800-U2 Detection of Bhanja virus in experimentally infected Ixodidae ticks by the indirect hemagglutination test. Klisenko,G.A.; Gaidamovich,S.Ya.; Grokhovskaya,I.M.; Shcherbakov,S.V. (D.I. Ivanosky Inst. Virol., Acad. Med. Sci., Moscow, USSR) *Med. Parazitol. Parazit. Bolezn., 47(4), 83-86 (1978)* Ru;en.

1801-U2 Indirect hemagglutination test for detection of antibody against bovine leukemia virus. Andersen,T.H. (State Vet. Lab., Dep. Jutland, Hangovej 2, DK-8200 Aarhus N., Denmark) *Ann. Rech. Vet., 9(4), 675-682 (1978)* En;en,fr.

1802-U2 Indirect hemagglutination in the diagnosis and epidemiology of cytomegalovirus. Christodoulopoulou,G.; Havredaki,M. (Biol. Lab., K.P.E. Dimokritos, Athens, Greece) *Acta Microbiol. Hell., 23(5), 321-328 (1978)* Gr;en.

1803-U2 [The indirect haemagglutination reaction as a method of rapid diagnosis of cytomegalovirus infection in man]. Monastireva,L.A.; Demidova,S.A. (Virus Inst., Acad. Med. Sci. USSR, Moscow, USSR) *Vopr. Virusol., No.4, 487-489 (1975)* Ru.

1804-U2 Improved indirect hemagglutination test for cytomegalovirus using human O erythrocytes in lysine. Yeager,A.S. (Dep. Pediatr., Stanford Univ. Sch. Med., Stanford, CA 94305, USA) *J. Clin. Microbiol., 10(1), 64-68 (1979)* En;en.

1805-U2 Detection of Epstein-Barr virus early antigen-D and its antibodies by passive hemagglutination. Ogbern,C.A.; Zajac,B.A. (Dep. Microbiol., Med. Coll. Pennsylvania, Philadelphia, PA 19129, USA) *Int. J. Cancer, 19(2), 150-160 (1977)* En;en,fr.

1806-U2 Reversed hemagglutination test for the detection of Australia antigen. Theodoropoulos,G.; Archimandritis,A.; Stauropoulos,A.; Angelopoulos,B. (Dep. Pathol. Physiol., Med. Sch., Univ. Athens, Athens, Greece) *Transfusion, 15(5), 466 (1975)* En.

1807-U2 Evaluation of passive hemagglutination techniques in detection of Australia antigen (HBsAg). Scalise,G.; Mura,M.S.; Delia,S. (Infect. Dis. Dep., Univ. Sassari, Sassari, Italy) *Transfusion, 17(6), 625-627 (1977)* En.

1808-U2 Demonstration of hepatitis-B (surface) antigen with the Hepatest: a new passive haemagglutination test. Lange,W.; Kohler,H.; Apodaca,J. (Abt. Virol., Robert-Koch-Inst. Bundesgesundheitsamtes, 1000 Berlin 65, Nordufer 20, GFR) *Dtsch. Med. Wochenschr., 101(35), 1273-1276 (1976)* De;de,en.

1809-U2 HB_sAg detection by passive hemagglutination (Hepanosticon-Organon). Advantages and disadvantages in comparison with other methods. Bals,M.G.; Hagiescu,L. (IInd Clin. Infect. Dis., Colentina Hosp., Bucharest, Romania) *Rev. Roum. Med., Ser. Virol., 27(2), 99-101 (1976)* En;en,fr.

1810-U2 [Comparative study of 3 reverse passive haemagglutination techniques to be used in the detection of HBs antigen]. Chevallier,P.; Trepo,D.; Trepo,C.; Sepetjian,M. (Lab. Hygiene, Domaine Rockefeller, 8 ave. Rockefeller, F-69373 Lyon Cedex 2, France) *Nouv. Presse Med., 6(22), 1975 (1977)* Fr.

1811-U2 Screening of HBs antigen on Groupamatic equipments. Chassaigne,M.; Saint-Paul,B. (Cent. Dep. Transfus. Sang., 26 rue du Marechal-Foch, 78011 Versailles, France) *Rev. Fr. Transfus. Immuno-Hematol., 20(3), 531-534 (1977)* Fr.

1812-U2 A comparative study of passive-haemagglutination methods for the detection of hepatitis B surface antigen in routine hospital practice. Withers,M.J.; McCahill,G.V.; Griffiths,P.D.; Heath,R.B.; Pattison,J.R.; Dane,D.S. (Joint Dep. Virol. Med. Coll. London and St. Bartholomew's Hosp., London, UK) *J. Clin. Pathol., 29(8), 732-735 (1976)* En;en.

1813-U2 Detection of hepatitis B surface antigen with the miniature centrifugal fast analyzer. A modified reversed

passive hemagglutination procedure. Wenz,B.; Karmen,A.; Feng,C.S. (Dep. Lab. Med., Albert Einstein Coll. Med., 1300 Morris Park Ave., Bronx, NY 10461, USA) *Vox Sang, 36(4), 228-235 (1979)* En;en.

1814-U2 An improved haemagglutination technique for the detection of hepatitis B antigen. Archer,A.C. (South West Regional Transfus. Cent., Southmead Road, Bristol BS10 5ND, UK) *Med. Lab. Sci., 34(4), 345-350 (1977)* En.

1815-U2 Methods of demonstrating antigen and antibody in hepatitis type B. Novak,J.; Kselikova,M. (Ustav hematol. a krevni transfuze, Praha, Czechoslovakia) *Cas. Lek. Ces., 116(43), 1331-1334 (1977)* Cs;cs,en,fr,ru.

1816-U2 [Elimination of false positives in the detection of Australia antigen by passive haemagglutination]. Tran-My-Hoang; Leplus,R. (Cent. Sect. de Transfus. Sang., Hop. Bicetre, 94-Le Kremlin-Bicetre, France) *Rev. Fr. Transfus., 19(2), 285-295 (1976)* Fr;en,fr.

1817-U2 Determination of e antigen and antibody to e by means of passive hemagglutination method. Takahashi,K.; Fukuda,M.; Baba,K.; Imai,M.; Miyakawa,Y.; *Mayumi,H. (Hepatitis Div., Tokyo Metropolitan Inst. Med. Sci., Bunkyo-ku, Tokyo 113, Japan) *J. Immunol., 119(5), 1556-1559 (1977)* En;en.

1818-U2 Indirect hemagglutination test in typing herpes simplex virus and antibodies. Seth,P.; Prakash,S.S.; Kesavalu,L. (Dep. Microbiol., All India Inst. Med. Sci., Ansari Nagar, New Delhi 110016, India) *Indian J. Med. Res., 68, 887-895 (1978)* En;en.

1819-U2 Rapid diagnosis of herpes simplex by means of the indirect hemagglutination test. Marennikova,S.S.; Nikulina,V.G.; Maltseva,N.N. (Sci.-Res. Inst. Virus Prep., Minist. Public Health, Moscow, USSR) *Vopr. Virusol., No.5, 626-628 (1977)* Ru;en,ru.

1820-U2 [Production of an erythrocyte preparation for the detection of herpes simplex virus antibody in the passive haemagglutination test]. Nikulina,V.G.; Maltseva,N.N. (Address not stated) *Vopr. Virusol., No. 2, 238-240 (1976)* Ru;en.

1821-U2 Stable erythrocyte diagnostic preparation for passive haemagglutination test with herpes simplex virus antigen. Krichevskaya,G.I.; Basova,N.N. (Helmholtz Res. Inst. Eye Dis., 103064, Moscow, USSR) *Acta Virol., 20(5), 418-423 (1976)* En;en.

1822-U2 Classical swine fever: research of specific antibodies by hemagglutination test. Labadie,J.P.; Corthier,G.; Aynaud,J.M.; Renault,L.; Vaissaire,J. (Lab. Vet. Sanders, 17, quai de l'Industrie, 91200 Athis-Mons, France) *Bull. Acad. Vet. Fr., 50(4), 533-542 (1977)* Fr;en,fr.

1823-U2 Employment of the indirect hemagglutination test for rapid diagnosis of influenza outbreaks. Korneeva,E.P.; Schwartsman,Ya.S.; Gorenskaya,R.L.; Schefer,L.F.; Potapenko,L.B.; Zibina,E.A.; Ralitnaya,T.V. (All-Union Sci.-Res. Inst. Influenza, Minist. Public Health, Leningrad, USSR) *Vopr. Virusol., No.5, 628-632 (1977)* Ru;en,ru.

1824-U2 [Modification of the indirect hemagglutination test for the rapid detection and identification of arbovirus antigens in brain suspensions from infected suckling mice]. Nikolaev,V.P. (Med. Acad. Leningrad, USSR) *Vopr. Virusol., No.2, 196-198 (1975)* Ru;en,ru.

1825-U2 Rapid diagnostic techniques for detection of porcine parvovirus infection in mummified foetuses. Joo,H.S.; Donaldson-Wood, C.R.; Johnson,R.H. (Dep. Trop. Vet. Sci., James Cook Univ. North Queensland, Queensl. 4811, Australia) *Aust. Vet. J., 52(1), 51-52 (1976)* En.

1826-U2 Passive hemagglutination miniaturized technique for identification of pox virus antibodies and antigens using formalinized erythrocytes. Mihailescu,R. (Cantacuzino Inst., Bucharest, Romania) *Arch. Roum. Pathol. Exp. Microbiol., 35(3), 231-238 (1976)* En;en,fr,ru.

1827-U2 A passive haemagglutination technique applied to the serological diagnosis of Aujeszky's disease. Labadie,J.P.; Toma,B. (Lab. Vet. Sanders, 17, quai de l'Industrie, 91200 Athis-Mons, France) *Recl. Med. Vet., Ec. Alfort, Paris, 155(1), 47-55 (1979)* Fr;en,es,fr.

1828-U2 Indirect hemagglutination test for pseudorabies antibody detection in swine. Haffer,K.; *Gustafson,D.P.; Kantiz,C.L. (Dep. Vet. Microbiol., Pathol. and Public Health, Sch. Vet. Med., Purdue Univ., West Lafayette, IN 49707, USA) *J. Clin. Microbiol., 11(3), 217-219 (1980)* En;en.

1829-U2 Serological method of determination of rabies antibodies. Madyarova,R.S.; Dulina,A.V.; Krutilina,D.V.; Latypova,R.G.; Morogova,V.M. (Inst. Vaccines and Sera, UFA, USSR) *Zh. Mikrobiol. Epidemiol. Immunobiol., 56(2), 88-90 (1979)* Ru;en.

1830-U2 Detection of rotavirus from faeces by reversed passive haemagglutination method. Sanekata,T.; Yoshida,Y.; Oda,K. (Kamagawa Prefect. Public Health Lab., 52 Nakao-cho, Ashai-ku, Yokohama, Japan) *J. Clin. Pathol., 32(9), 963 (1979)* En.

1831-U2 Simple procedure for the removal of non-specific inhibitors of rubella virus hemagglutination. Allen,R.; *Hedlund,K. (Div. Commun. Dis. and Immunol., Walter Reed Army Inst. Res., Washington, DC 20012, USA) *J. Clin. Microbiol., 2(6), 524-527 (1975)* En;en.

1832-U2 Further evaluation of a rubella passive hemagglutination test. Kilgore,J.M. (Diagn. Lab., Inc., Dep. Microbiol.-Immunol., 2000 Freedom Drive, Charlotte, NC 28208, USA) *J. Med. Virol., 5(2), 131-136 (1980)* En;en.

1833-U2 Experience with an indirect (passive) haemagglutination test for the demonstration of rubella virus antibody. Haukenes,G. (Gade Inst., Dep. Microbiol., Univ. Bergen, Bergen, Norway) *Acta Pathol. Microbiol. Scand. Sect. B, 88(2), 85-87 (1980)* En;en.

1834-U2 Detection of Sindbis and West Nile viruses in the blood of living birds by indirect haemagglutination.

Gaidamovich,S.Ya.; Ismailov,A.Sh.; Klisenko,G.A.; Mirzoeva,N.M. (D.I. Ivanovsky Inst. Virol., USSR Acad. Med. Sci., 123098 Moscow, USSR) *Acta Virol., 22(5), 430 (1978)* En.

1835-U2 Transmissible gastro-enteritis: research of specific antibodies by means of a passive hemagglutination test. Labadie,J.P.; Aynaud,J.M.; Renault,L.; Vaissaire,J. (Lab. Vet. Sanders, 17, Quai de l'Industrie, 91200 Athis-Mons, France) *Bull. Acad. Vet. Fr., 51(1), 53-60 (1978)* Fr;en,fr.

1836-U2 [Application of the passive hemagglutination test (PHAT) for detection of smallpox antibodies after primary smallpox vaccination and revaccination]. Marennikova,S.S.; Shenkman,L.S.; Shelukhina,E.M.; Matsevich,G.R.; Nikulina,V.G.; Noskov,F.S.; Konikova,R.E.; Ertte,A.P. (Moscow Inst. Virus Prep., Minist. Public Health USSR, Moscow, USSR) *Zh. Mikrobiol. Epidemiol. Immunobiol., 53(2), 38-42 (1976)* Ru;en,ru.

1837-U2 A rapid diagnosis of infection caused by varicella-herpes zoster viruses. Maltseva,N.N.; Marennikova,S.S.; Ertte,A.P.; Avdeenko,M.M. (Sci. Res. Inst. Viral Prep.., Minist. Health USSR, Moscow, USSR) *Vopr. Virusol., No.2, 247-251 (1978)* Ru;en,ru.

1838-U2 Direct hemagglutination technique for differentiating *Bacteroides asaccharolyticus* oral strains from nonoral strains. Slots,J.; Genco,R.J. (Dep. Oral Biol. and Periodont, Dis. Clin. Res. Cent., State Univ. New York at Buffalo, Buffalo,Y 14226, USA) *J. Clin. Microbiol., 10(3), 371-373 1979)* En;en.

1839-U2 Indirect hemagglutination test use for the study of ornithosis infection. Report II. Obtaining and approbation of dry ornithosis erythrocytic diagnostic agent. Gorovits,E.S.; Timasheva,O.A. (Inst. Vaccines and Sera, Perm, USSR) *Zh. Mikrobiol. Epidemiol. Immunobiol., 55(2), 68-71 (1978)* Ru;en,ru.

1840-U2 A new criterion for implicating *Clostridium perfringens* as the cause of food poisoning. Dowell,V.R.,Jr.; Torres-Anjel,M.J.; Riemann,H.P.; Merson,M.; Whaley,D.; Darland,G. (Enterobacteriol. Unit, Bacteriol. Sect., Cent. Dis. Control, Atlanta, GA 30333, USA) *Rev. Latinoam. Microbiol., 17(3), 137-142 (1975)* En;en,es.

1841-U2 Testing the system of unidirected reactions for detection of diphtheria antigens and antibodies. Novikova,A.A.; Basova,N.N.; Cherkasova,V.V.; Baraban,P.S.; Minsenzhnikov,A.V.; Raikher,I.I. (Cent. Inst. Epidemiol., Moscow Inst. Epidemiol. and Microbiol., Moscow, USSR) *Zh. Mikrobiol. Epidemiol. Immunobiol., 56(1), 50-54 (1979)* Ru;en.

1842-U2 Evaluation of an indirect haemagglutination kit for the rapid serological diagnosis of *Mycoplasma pneumoniae* infections. Talor,P. (Dep. Pathol., Brompton Hosp., Fulham Road, London SW3 6HP, UK) *J. Clin. Pathol., 32(3), 280-283 (1979)* En;en.

1843-U2 Reverse passive haemagglutination test for the rapid identification of *Neisseria gonorrhoeae* and detection of penicillinase production. Munro,R.; Mallon,R. (Dep. Bacteriol., Inst. Clin. Pathol. and Med. Res., PO Box 60, Wentworthville, Sydney, NSW 2145, Australia) *Br. J. Vener. Dis., 55(6), 404-407 (1979)* En;en.

1844-U2 A new method for strain identification of *Rhizobium trifolii* in nodules. Cloonan,M.J.; Humphrey,B. (Div. Microbiol., Prince Henry Hosp., Little Bay, NSW 2036, Australia) *J. Appl. Bacteriol., 40(1), 101-107 (1976)* En;en.

1845-U2 A passive haemagglutination test for diagnosis of trench fever due to *Rochalimaea quintana*. Cooper,M.D.; Hollingdale,M.R.; *Vinson,J.W.; Costa,J. (Dep. Microbiol., Harvard Sch. Public Health, 665 Huntington Ave., Boston, MA 02115, USA) *J. Infect. Dis., 134(6), 605-609 (1976)* En;en.

1846-U2 [Search for protein A in 1200 strains of *Staphylococcus* (use of a rapid technique of detection by passive haemagglutination with glutaraldehyde-treated erythrocytes)]. Bind,J.L.; Chiron,J.P.; Dennis,F. (Serv. Vet., 46 ave. Gustave Eiffel, 37002 Tours Cedex, France) *C.R. Seances Soc. Biol. Fil., 172(1), 212-215 (1978)* Fr;en,fr.

1847-U2 Staphylococcal protein A: detection with a rapid diagnostic procedure using stable reagent. Poutrel,B.; Lefort,B. (INRA, Stn. Pathol. Reprod., 37380 Nouzilly, France) *Med. Mal. Infect., 10(1), 38-41 (1980)* Fr;en,fr.

1848-U2 [Rapid serology of rheumatic fever and glomerulonephritis]. Pilars de Pilar,C.E.; Friedrich,J.; Spiess,H. (Kinderpoliklin., Univ. Munchen, Pettenhofferstr. 8a, D-8 Munchen 2, GFR) *Monatsschr. Kinderheilkd., 125(2), 65-68 (1977)* De;en,de.

1849-U2 Assay of human anti-exoproteins to group A streptococci by microhaemagglutination: correlation with anti-streptolysin O and various anti-enzymes. Alouf,J.E.; de Saint Martin,J.; Eyquem,A.; Geoffroy,C.; Jacquemot,C.; Duphot,M. (Unite Antigenes Bact., et Serv. Immuno-Hematol. et Immuno-Pathol., Inst. Pasteur, 75724 Paris Cedex 15, France) *Ann. Microbiol., 129A(4), 447-472 (1978)* Fr;en,fr.

1850-U2 Indirect hemagglutination test with *Treponema* antigen for serodiagnosis of syphilis. Petukhova,R.N.; Diyachenko,L.A. (Sanit. and Hyg. Med. Inst., Dep. Microbiol., Leningrad, USSR) *Vestn. Dermatol. Venerol., No.12, 52-54 (1979)* Ru;en,ru.

1851-U2 Screening test for syphilis. A comparison of the *Treponema pallidum* haemagglutination assay with two automated serological tests. Macfarlane,D.E.; Elias-Jones,T.F. (City Lab., 23 Montrose St., Glasgow G1 1RN, UK) *Br. J. Vener. Dis., 53(6), 348-352 (1977)* En;en.

1852-U2 The microhaemagglutination-TP test in syphlitic problem sera. Brathwaite,A.R. (Comprehensive Health Anat., Kingston, Jamaica) *West Indian Med. J., 27(4), 205-210 (1978)* En;en.

1853-U2 [Interest of the passive hemagglutination test (T.P.H.A.) in the diagnosis of syphilis]. Pujol,M.; Morel,C.;

Duhamel,C.; Freymuth,F. (Lab. Microbiol., CHR, 14033 Caen Cedex, France) *Ouest Med., 28(12), 865-868 (1975)* Fr;fr.

1854-U2 [The significance of the treponemal haemagglutination reaction in the modern serology of syphilis]. Alessi,E. (Univ. Milano, Osp. Maggiore 'Policlinico', I Clin. Dermatol., via Pace 9, Milano, Italy) *Osp. Magg., 70(2), 118-123 (1975)* It;en,it.

1855-U2 [Evaluation and clinical significance of *Treponema pallidum* haemagglutination (TPHA) test in syphilis diagnosis]. Kerl,H.; Bayer,U.; Turex,Th.-D. (Univ. Klin. Dermatol. und Venerol., Auenbruggerplatz 8, A-8036 Graz, Austria) *Z. Hautkr., 51(17), 718-726 (1976)* De;de.

1856-U2 Indirect hemagglutination test for detection of *Toxoplasma* antibodies. Fesefeldt,C.; *Braveny,I. (Inst. Med. Mikrobiol. und Hyg., Tech. Univ., Ismaninger Str. 22, 8000 Munchen 80, GFR) *Immun. Infekt., 6(4), 160-165 (1978)* De;de,en.

1857-U2 A new indirect haemagglutination test in the measurement of antitoxoplasma antibodies. Duvina,P.L.; Santalena,G.; Simoni,M.R.; Pieroni,O.; Pieruccini,G. (Ist. Puericult., Univ. Div. Pediatr., Arcispedale S. M. Nouva, I-50121 Firenze, Italy) *Minerva Pediatr., 31(9), 698-698 (1979)* It;en,it.

1858-U2 Indirect hemagglutination using whole mixed antigen for checking toxoplasmosis immunity and for serodiagnosis of human toxoplasmosis, compared with immunofluorescence. Ambroise-Thomas,P.; Simon,J.; Bayard,M. (Parasitol. Dep., CHU Grenoble, 37000 La Tronche, France) *Biomed. Express, 29(7), 245-248 (1978)* En;en,fr.

1859-U2 Capillary indirect hemagglutination: a field method for the diagnosis of African trypanosomiasis. Bone,G.J.; Charlier,J. (Ecole Sante Publique, Univ. Catholique de Louvain, Clos Chapelle aux Champs 4, 1200 Brussels, Belgium) *Ann. Soc. Belge Med. Trop., 55(5), 559-569 (1975)* Fr;en,fr.

1860-U2 A simplified method for reading hemagglutinations on a flat-bottom microtitration plate in the mouse H-2 assay. Moriwaki,K.; Aotsuka,T.; Shiroishi,T. (Natl. Inst. Genet., Shizuoka-ken, Mishima, 411, Japan) *Experientia, 33(5), 673-674 (1977)* En;en.

1861-U2 Assessment of haemagglutination test for α-foetoprotein. Kutsukake,S.; Kondon,T.; Abe,A.; Shimano,K. (Sagamihara Natl. Hospital, Kanagawa, Japan) *Iryo, 31(3), 81-84, 202 (1977)* Ja;en,ja.

1862-U2 Passive hemagglutination and hemolysis tests for the detection of anti-DNA antibody. Sasaki,T.; Ishida,S.; Onodera,S.; Saito,T.; Furuyama,T.; Yoshinaga,K. (Dep. Intern. Med., Tohoku Univ. Sch. Med., 1-1, Seiryo-machi, 980, Sendai, Japan) *J. Immunol. Methods, 22(3-4), 327-337 (1978)* En;en.

1863-U2 Slide haemagglutination test in hydatid disease: a correlative study of diagnostic procedures. Matossian,R.M.; Mamo,A.J.; Dakroub,R. (Dep. Bacteriol. and Virol., Am. Univ. Beirut, Beirut, Lebanon) *J. Clin. Pathol., 29(1), 39-41 (1976)* En;en.

1864-U2 Application of passive hemagglutination for evaluation of antisperm antibodies and a modified Coombs' test for detecting male autoimmunity to sperm antigens. Mathur,S.; Williamson,H.O.; Landgrebe,S.C.; Smith,C.L.; Fudenberg,H.H. (Dep. Basic and Clin. Immunol. and Microbiol., Med. Univ. South Carolina, 171 Ashley Ave., Charleston, SC 29403, USA) *J. Immunol. Methods, 30(4), 381-393 (1979)* En;en.

1865-U2 An evaluation of two new haemagglutination tests for the rapid diagnosis of autoimmune thyroid diseases. Cayzer,I.; Chalmers,S.R.; Doniach,D.; Swana,G. (Wellcome Res. Lab., Beckenham, Kent BR3 3BS, UK) *J. Clin. Pathol., 31(12), 1147-1151 (1978)* En;en.

1866-U2 Determination of antibody to hepatitis B core antigen by means of immune adherence hemagglutination. Tsuda,F.; Takahashi,T.; Takahashi,K.; Miyakawa,Y.; Mayumi,M. (Immunol. Div., Jichi Med. Sch., Tochigi-ken 329-04, Kitazato Inst., Minato-ku, Tokyo 108, Japan) *J. Immunol., 115(3), 834-838 (1975)* En;en.

1867-U2 Immune adherence hemagglutination test applied to the study of herpes simplex and varicella-zoster virus infections. Gillani,A.; *Spence,L. (Dep. Microbiol., Toronto Gen. Hosp., Toronto, Ont., Canada M5G 1L7) *J. Clin. Microbiol., 7(2), 114-117 (1978)* En;en.

1868-U2 Immune adherence hemagglutination: alternative to complement-fixation serology. Lennette,E.T.; Lennette,D.A. (Div. Virol., Joseph Stokes, Jr., Res. Inst., Child. Hosp. Philadelphia, Philadelphia, PA 19104, USA) *J. Clin. Microbiol., 7(3), 282-285 (1978)* En;en.

1869-U2 Elaboration of a HI test for detection of immunoglobulins of various classes. Noskov,F.S.; Konikova,R.E.; Shakhanina,K.L.; Malkina,L.A.; Avdeenko,M.M. (Acad. Mil. Med., Leningrad, USSR) *Zh. Mikrobiol., Epidemiol., Immunobiol., 54(2), 53-57 (1977)* Ru;en,ru.

1870-U2 A procedure for drug screening without the need to transport urines: use of ion exchange papers and hemagglutination inhibition. Alexander,G.J. (Neurotoxicol. Res. Unit, New York State Dep. Ment. Hyg., Bronx, NY, USA) *Clin. Toxicol., 9(3), 435-446 (1976)* En;en.

1871-U2 Laboratory procedures in adenoviruses. IV. The influence of antiglobulin and complement on neutralization and hemagglutination-inhibition. Zimmer,E.; Wigand,R. (Inst. Hyg. und Mikrobiol., Univ. Saarlandes, D-6650 Homburg (Saar), GFR) *Zentralbl. Bakteriol. Parasitenkd. Infektionskr. Hyg., I Abt A, 238(2), 177-183 (1977)* De;de,en.

1872-U2 Some observations on the value of the infectious bronchitis haemagglutination inhibition test in the field. MacPherson,I.; Feest,A. (Midland Poult, Holdings Ltd., The

Grove, Craven Arms, Salop, UK) *Avian Pathol., 7(3), 337-347 (1978)* En;de,en,fr.

1873-U2 **Development of a kit for the assay of haemagglutination inhibition antibodies to flaviviruses using formalinized goose erythrocytes.** Nagarkatti,P.S.; Nagarkatti,M.; Rao,K.M. (Dep. Microbiol., Defence Res. and Dev. Estab., Gwalior-474002, India) *Trans. R. Soc. Trop. Med. Hyg., 74(1), 22-25 (1980)* En;en.

1874-U2 **Some observations on the diagnosis of an outbreak of ectromelia in 1976.** Carthew,P.; Hill,A.C.; Verstraete,A.P. (Med. Res. Council Lab. Anim. Cent., Woodmansterne Rd., Carshalton, Surrey, UK) *Vet. Rec., 100(4), 293 (1977)* En.

1875-U2 **Some observations on the diagnosis of an outbreak of ectromelia in 1976.** Carthew,P.; Verstraete,A.P. (Med. Res. Counc. Lab. Amin. Cent., Woodmansterne Rd., Carshalton, Surrey, UK) *Vet. Rec., 100(14), 293 (1977)* En.

1876-U2 **HBs antigen screening and Groupamatic equipment: inhibition of passive hemagglutination and latex test.** Soulier,J.P.; Garretta,M.; Benamon,D.; Drouet,J.; Courouce-Pauty,A.M. (Cent. Natl. Transfus. Sang., 6, rue Alexandre-Cabanel, 75739 Paris Cedex 15, France) *Rev. Fr. Transfus. Immuno-Hematol., 21(2), 511-521 (1978)* Fr;en.

1877-U2 **[Detection of hepatitis B antigen by passive haemagglutination-inhibition: study with a view to standardisation].** Benamon,D.; Drouet,J.; Garretta,M. (Cent. Natl. Transfus. Sang., 6, rue Alexandre-Cabanel-75739 Paris Cedex 15, France) *Rev. Fr. Transfus., 18(2), 205-210 (1975)* Fr;en,fr.

1878-U2 **Screening of HBs antigen and detection of corresponding antibodies on Groupamatic equipment.** Garretta,M.; Benamon,D.; Drouet,J.; Courouce-Pauty,A.M.; Gener,J.; Muller,A.; Soulier,J.P. (Cent. Natl. Transfusion Sanguine, Ave. des Tropiques, 91400 Orsay, France) *Vox Sang., 33(3), 175-186 (1977)* En;en.

1879-U2 **Development and use of a micro haemagglutination inhibition (HAI) technique, based on Hepatest, for the detection and quantitation of hepatitis B surface antibody (anti-HBs) in blood donors.** Wiseman,I.C. (Dep. Haematol., Jessop Hosp. Women, Sheffield 3, UK) *J. Clin. Pathol., 30(1), 50-53 (1977)* En;en.

1880-U2 **Prevalence of hepatitis-B surface antigen by rapid passive haemagglutination inhibition technique.** Joshi,S.H.; Baxi,A.J.; *Bhatia,H.M. (Blood Group Ref. Cent., Seth G.S. Med. Coll., Parel, Bombay-400012, India) *Indian J. Med. Res., 69, 978-980 (1979)* En;en.

1881-U2 **A simple method for testing antigenic variation of influenza virus antigenicity.** Liang,J.-k. (Kansu Provincial Health and Anti-epidermic Station, Lanchow, China) *Chin. Med. J., 92(1), 37-40 (1979)* En;en.

1882-U2 **Detection of antibodies of influenza virus by single radial haemolysis and haemagglutination inhibition in human sera.** Mancini,G.; D'Ambriosio,E.; Arangio-Ruiz,G. (Ist. Superiore di Sanita, Lab. Mal. Batteriche e Virali, Rome, Italy) *Microbiologica (Bologna), 2(4), 399-403 (1979)* En;en.

1883-U2 **A simple procedure for removing non-specific serum inhibitors of haemagglutination by Japanese encephalitis virus.** Ho,W.K.K.; Shortridge,K.F. (Dep. Biochem., Chinese Univ. Hong Kong, Shatin, New Territories, Hong Kong) *Jap. J. Med. Sci. Biol., 30(1), 25-29 (1977)* En.

1884-U2 **Comparison of macro and micro methods in kaolin and RBC treatment of sera used in the measles hemagglutination-inhibition test.** Ruggeri.F.; Santoro,R.; Rapicetta,M.; Grandolfo,M.F. (Lab. Mal. Batteriche e Virali, Ist. Sup. Sanita, Rome, Italy) *J. Biol. Stand., 6(4), 257-260 (1978)* En;en.

1885-U2 **Detection of early 'significant' antibody responses in paired sera by the use of an automated bovine parainfluenza-3 hemagglutination-inhibition system.** Webert,D.W.; Bech-Nielsen,B.V. (Diagn. Lab., New York State Coll. Vet. Med., Cornell Univ., Ithaca, NY 14850, USA) *Am. J. Clin. Pathol., 70(4), 679-685 (1978)* En;en.

1886-U2 **Simplified method for the rubella haemagglutination inhibition screening test.** Mortimer,P.P.; Jordan,S.M.; Smith,I.V. (Public Health Lab. and Dep. Microbiol., Cent. Middlesex Hosp., Park Royal, Lonodn NW10 7NS, UK) *J. Clin. Pathol., 30(7), 623-625 (1977)* En;en.

1887-U2 **Diagnosis of rubella by demonstrating rubella-specific 19S and 7S antibodies.** Dibbert,H.-J. (Medizinaluntersuchungsanst. Hyg.-Inst., Freien und Hansestadt Hamburg, Gorch-Fock-Wall, D-2000 Hamburg 36, GFR) *Zentralbl. Bakteriol. Hyg., I Abt., A, 234(2), 145-158 (1976)* De;en,de.

1888-U2 **Detection of rubella haemagglutination-inhibition (HAI) and virus-specific IgM antibody using trypsin-treated human group O erythrocytes in the HAI test.** Al-Nakib,W.; Lilley,H. (Dep. Microbiol., Fac. Med., Kuwait Univ., POB 5969, Kuwait) *J. Clin. Pathol., 31(8), 730-734 (1978)* En;en.

1889-U2 **Improved rubella hemagglutination inhibition test: inactivation of non-immunoglobulin hemagglutination inhibitors by phospholipase C.** Iwasa,S.; Hori,M. (Chem. Res. Lab., Cent. Res. Div., Takeda Chem. Ind., Ltd., 17-85, Jusohonmachi 2-chome, Yodogawa-ku, Osaka 532, Japan) *J. Clin. Microbiol., 4(6), 461-466 (1976)* En;en.

1890-U2 **Evaluation of a rubella hemagglutination inhibition test system.** Smith,J.A.; Cummins,A.C. (Div. Microbiol., Mount Sinai Hosp., Toronto, Ontario M5G 1X5, Canada) *J. Clin. Microbiol., 3(1), 5-7 (1976)* En;en.

1891-U2 **A solid-phase immunosorbent technique for the rapid detection of rubella IgM by haemagglutination inhibition.** Krech,U.; Wilhelm,J.A. (Inst. Med. Microbiol., Frohbergstr. 3, CH-9000 St. Gallen, Switzerland) *J. Gen. Virol., 44(2), 281-286 (1979)* En;en.

1892-U2 Rubella virus hemagglutination inhibition IgM antibodies: the method of absorption of IgG by staphylococcal protein A as compared with density gradient ultracentrifugation. Roggendorf,M.; Schneweis,K.E.; Wolff,M.H. (Inst. Med. Mikrobiol. und Immunol., D-5300 Bonn-Venusberg, GFR) *Zentralbl. Bakteriol. Hyg. I. Abt., A, 235(4), 363-372 (1976)* De;de,en.

1893-U2 An alternative method for inactivating heteroagglutinins in human sera applicable to rubella haemagglutination inhibition testing at low dilutions. Mortimer,P.P. (Virus Ref. Lab., Cent. Public Health Lab., Colindale Ave., London NW9 5HT, UK) *J. Clin. Pathol., 29(5), 417-422 (1976)* En;en.

1894-U2 Micro-modification of kaolin treatment of serum for the rubella haemagglutination-inhibition test. Inouye,S. (Cent. Virus Diagn. Lab., Natl. Inst. Health, Musashimurayama-si 19012, Japan) *J. Med. Microbiol., 9(4), 501-502 (1976)* En;en.

1895-U2 Rubella hemagglutination-inhibition test - a simplified technique. Candeias,J.A.N.; Neira,L.; Racz,M.L.; Candeias,N.M.F. (Dep. Microbiol. e Imunol., Inst. Cien. Biomed. USP, 'Setor Saude Publica', Av. Dr. Arnaldo, 715, 01255, Sao Paulo, SP, Brazil) *Rev. Saude Publica, Sao Paulo, 12(4), 516-522 (1978)* Pt;en,pt.

1896-U2 A rubella haemagglutination inhibition test not requiring removal of non-specific inhibitors. 1. Elaboration of the test. Haukenes,G. (Mikrobiol. avd., MFH-bygget, N-5016 Haukeland sykehus, Norway) *Acta Pathol. Microbiol. Scand., Ser. B, 87(6), 385-389 (1979)* En;en.

1897-U2 *S. aureus* protein A for diagnostic test of primoinvasive rubella. Lefebvre,J.Cl.; Passeron,Ch.; Vanderkerkove,M. (Lab. Bacteriol.-Virol., UER Med., Chemin de Vallombrose, 06034 Nice Cedex, France) *Med. Mal. Infect., 7(9), 422-425 (1977)* Fr;en,fr.

1898-U2 Simplified rubella haemagglutination inhibition test not requiring removal of non-specific inhibitors. Haukenes,G. (Dep. Microbiol., Virol. Sect., Gade Inst., Univ. Bergen, N-5016 Haukeland Sykehus, Bergen, Norway) *Lancet, 2(8135), 196-197 (1979)* En.

1899-U2 Evaluation of a simplified sucrose gradient method for the detection of rubella-specific IgM in routine diagnostic practice. Caul,E.O.; Hobbs,S.J.; Roberts,P.C.; Clarke,S.K.R. (Public Health Lab., Myrtle Rd., Kingsdown, Bristol BS2 8EL, UK) *J. Med. Virol., 2(02), 153-163 (1978)* En;en.

1900-U2 Evaluation of a new test system for rubella haemagglutination inhibiting antibodies. Roberts,P.C.; Hobbs,S.J. (Public Health Lab., Myrtle Rd., Kingsdown, Bristol BS2 8EL, UK) *J. Clin. Pathol., 30(11), 1011-1014 (1977)* En;en.

1901-U2 A cell-free haemagglutinating antigen of *Mycoplasma synoviae* and its use in haemagglutination inhibition tests. Cullen,G.A.; Snell,G.C. (Cent. Vet. Lab., Weybridge, Surrey, UK) *J. Biol. Stand., 4(3), 203-207 (1976)* En;en.

1902-U2 Graded haemagglutination inhibition — a new technique applied to the quantitation of human immunoglobulins IgG and IgA. Wolk,M. (Rafa Lab., Ltd., POB 405, Jerusalem, Israel) *Clin. Chem., 25(5), 819 (1979)* En.

1903-U2 Determination of estrogens in pregnancy urine by hemagglutination inhibition reaction. Yanaihara,T.; Iwahara,K.; Okinaga,S.; Arai,K.; Kosuzume,H.; Inaba,H.; Soma,M.; Ogawa,N. (Teikyo Univ. Sch. Med., 11-1 Kaga, 2 Chome, Itabashi-ku, Tokyo 173, Japan) *Am. J. Obstet. Gynecol., 123(7), 700-704 (1975)* En;en.

1904-U2 A simple means of evaluating cellular immunity in the child: the phytohemagglutinin skin test. Bourrier-Reynaud,C.; Toutel,P.; D'Oelsnitz,M.; Mariani,R. (Serv. Pediatr., Hop. Cimiez, 4, Ave. Victoria, 06000 Nice, France) *Rev. Fr. Allergol., 18(4-5), 191-200 (1978)* Fr;en,fr.

1905-U2 Detection of herpes antibody and herpes simplex virus by different methods. Nikulina,V.G.; Khamaganova,A.V. (Moscow Viral Prep. Res. Inst., Moscow, USSR) *Vestn. Dermatol. Venerol., No. 5, 51-55 (1978)* Ru;en.

1906-U2 Observation on rapid diagnosis of porcine parvovirus in mummified foetuses. Joo,H.S.; Johnson,R.H. (Dep. Trop. Vet. Sci., James Cook Univ., North Qeensland, Townsville, Queensland 4811, Australia) *Aust. Vet. J., 53(2), 106-107 (1977)* En.

1907-U2 Hemagglutination by canine parvovirus: serologic studies and diagnostic applications. Carmichael,L.E.; Joubert,J.C.; Pollock,R.V.H. (James A. Baker Inst. anim. Health, New York State Coll. Vet. Med., Cornell Univ., Ithaca, NY 14853, USA) *Am. J. Vet. Res., 41(5), 784-791 (1980)* En;en.

1908-U2 Comparison of passive haemagglutination and haemagglutination-inhibition techniques for detection of antibodies to rubella virus. Birch,C.J.; Glaun,B.P.; Hunt,V.; Irving,L.G.; Gust,I.D. (Virol. Dep., Fairfield Hosp., Queen's Mem.Infect. Dis. Hosp., Yarra Bend Road, Fairfield, Vic. 3078, Australia) *J. Clin. Pathol., 32(2), 128-131 (1979)* En;en.

1909-U2 Micromethod for the HA and HI tests in ornithosis. Sopranenko,L.G.; Terskikh,I.I.; Degtyarev,Yu.L. (Inst. Epidemiol. and Hyg., Acad. Med. Sci. USSR, Moscow, USSR) *Vopr. Virusol., No. 4, 489-491 (1975)* Ru;en.

See: 1612, 1722, 1727, 1911, 1923, 2525, 2642, 3247, 3255, 3256, 3259, 3261, 3271, 3272, 3293, 3297, 2580, 3244, 3245, 3253, 3257, 3258, 32 3267, 3270, 3273, 3294, 3278, 3274

Haemadsorption and haemadsorption inhibition

1910-U2 Titration of African swine fever (ASF) virus. Enjuanes,L.; Carrascosa,A.L.; Moreno,M.A.; Vinuela,E. (Cent. Biol. Mol. (CSIC-UAM), Velazquez 144, Madrid-6, Spain) *J. Gen. Virol., 32(3), 471-477 (1976)* En;en.

1911-U2 Detection of coronavirus in calf faeces with a haemadsorption-elution-haemagglutination assay (HEHA). Van Balken,J.A.M.; De Leeuw,P.W.; Ellens,D.J.; Straver,P.J. (Cent. Vet. Inst., Virol. Dep., Lelystad, Netherlands) *Vet. Microbiol., 3(3), 205-211 (1979)* En;en.

1912-U2 The advantage and adequacy of the mixed hemadsorption test for rabies antibody determination. Grandien,M.; Espmark,A. (Dep. Virol., Natl. Bacteriol. Lab., S-105 21 Stockholm, Sweden) *Dev. Biol. Stand., 40, 237-242 (1978)* En;en.

1913-U2 Use of the mixed haemadsorption technique to demonstrate lectins adsorbed to monolayer cultures. Jonsson,J.; Ostborn,A.; Fagraeus,A.; Skoog,V. (Dep. Immunol., Natl. Bacteriol. Lab., S-105 21 Stockholm, Sweden) *Acta Pathol. Microbiol. Scand. Sect. C, 83(1), 35-42 (1975)* En;en.

1914-U2 [A rapid and simple possibility for the determination of antibodies to influenza virus antigen.] Luther,P.; Bergmann,K.-C. (Abt. Klin. Immunol., Forschungsinst. Lungenkr. und Tuberculose, DDR-115 Berlin-Buch, Karower Str. 11, GDR) *Dtsch. Gesundheitswes., 32(7), 327-328 (1977)* De.

1915-U2 Identification of *Mycoplasma gallisepticum* in the haemadsorption inhibition test. Grebe,H.-H. (Inst. Geflugelkrankheiten, Bunteweg 17, 3000 Hannover, GFR) *Zentralbl. Veterinarmed., B, 24(2), 134-139 (1977)* De;de,en,fr,es.

1916-U2 [The improvement of diagnostic epidemiological work in the field of influenza virus infection]. Adamczyk,B.; Adamczyk,G. (DDR-115 Berlin Wiltbergstr. 50, Haus 106, GDR) *Dtsch. Gesundheitswes., 32(26), 1239-1242 (1977)* De;de,en,ru.

See: 1610, 1950

Complement fixation

1917-U2 [Quantitative structure comparison of protein surfaces by micro-complement fixation.] Haas,H.; Zwilling,R. (Zool. Inst. II, Univ. Heidelberg, Heidelberg, GFR) *Naturwissenschaften, 63(3), 139-143 (1976)* De;en.

1918-U2 Adaptation of the microtechnique complement fixation test for the diagnosis of African horse sickness. Bernard,G. (Lab. Natl. Elev. et Rech. Vet., BP 2057, Dakar-Hann, Senegal) *Rev. Elev. Med. Vet. Pays Trop., 28(4), 451-457 (1975)* Fr;en,es,fr.

1919-U2 Automated complement fixation test for the detection of antibodies against the core of hepatitis B virus (HBc). Coursaget,P.; Maupas,P.; Goudeau,A.; Millman,I. (Lab. Virol., Fac. Med., 2 bis, boulevard Tonelle, 37000 Tours, France) *J. Immunol. Methods, 13(1), 21-27 (1976)* En;en.

1920-U2 Detection of type-specific antibody to herpes simplex virus type 1 and 2 in human sera by complement-fixation tests. Skinner,G.R.B.; Hartley,C.; Whitney,J.E. (Dep. Virol., Med. Sch., Birmingham B15 2TJ, UK) *Arch. Virol., 50(4), 323-333 (1976)* En;en.

1921-U2 Automatic complement fixation reaction applied to the diagnosis of influenza. Coursaget,P.; Maupas,P.; Larbaigt,G.; Raynaud,B.; Grenier,B. (Lab. Microbiol., Univ. Francois-Rabelais, Fac. Med. et Pharm., 2 bis, blvd. Tonnelle, 37000 Tours, France) *Pathol. Biol., 24(3), 227-231 (1976)* Fr;en,fr.

1922-U2 A quantitative micro-complement fixation method in studies of human wart viruses. Hefton,J.M.; Eisinger,M. (Lab. Hum. Tumor Virol., Mem. Sloan-Kettering Cancer Cent., New York, NY 10021, USA) *J. Gen. Virol., 38(1), 179-182 (1977)* En;en.

1923-U2 Laboratory diagnosis of western encephalomyelitis. Sekla,L.H.; Stackiw,W. (Provincial Public Health Lab., 770 Bannatyne Ave., Winnipeg, Manitoba R3E 0W3, Canada) *Can. J. Public Health, 67(Suppl. 1), 33-39 (1976)* En.

1924-U2 Interpretation of complement fixation test results as displayed on Auto-Analyzer charts. Timbs,D.V.; Moxham,J.; McMillan,D.J. (Cent. Brucellosis Lab., Wallaceville Anim. Res. Cent., Private Bag, Upper Hutt, New Zealand) *N. Z. Vet. J., 26(3), 56-59 (1978)* En;en.

1925-U2 Quality control procedures in an automated serological testing laboratory. Liberona,H.E.; Moxham,J.W.; Timbs,D.V. (Cent. Brucellosis Lab., Wallaceville Anim. Res. Cent., Private Bag, Upper Hutt, New Zealand) *N. Z. Vet. J., 26(3), 60, 65-66 (1978)* En;en.

1926-U2 The use of automated complement fixation techniques in the Brucellosis Eradication Scheme. Timbs,D.V.; Moxham,J.W.; Liberona,H.E. (Cent. Brucellosis Lab., Wallaceville Anim. Res. Cent., Private Bag, Upper Hutt, New Zealand) *N.Z. Vet. J., 26(3), 52-56 (1978)* En;en.

1927-U2 The design and operation of the Central Brucellosis Laboratory. Elliott,R.E.W.; Moxham,J.W.; Digby,J.G. (Minist. Agric. and Fish., PO Box 2298, Wellington, New Zealand) *N.Z. Vet. J., 26(3), 46-52 (1978)* En;en.
[complement fixation].

1928-U2 [A complement fixation micromethod applied to the serodiagnosis of contagious epididymitis in rams]. Sanchis,R.; Giauffret,A. (Lab. Rech. Vet., 63 ave. des Alenes, 06000 Nice, France) *Rec. Med. Vet., 152(5), 305-310 (1976)*

Fr;en,es,fr.

1929-U2 Adaptation of a complement fixation test for large scale serological diagnosis of bovine leptospirosis. Hodges,R.T.; Weddell,W. (Wallaceville Anim. Res. Cent., Res. Div., Minist. Agric. and Fish., Private Bag, Upper Hutt, New Zealand) *N. Z. Vet. J., 25(10), 261-262 (1977)* En;en.

1930-U2 Evaluation of automated large-scale screening tests for syphilis. Macfarlane,D.E.; Hare,K.; Elias-Jones,T.F. (The City Lab., Greater Glasgow Health Board, Glasgow G1 1RN, UK) *J. Clin. Pathol., 29(4), 317-321 (1976)* En;en.

1931-U2 A large-scale radiometric micro-quantitative complement fixation test for serum antibody titration. Bengali,Z.H.; Das,S.R.; Levine,P.H. (Clin. Stud. Sect., Lab. Viral Carcinogenesis, Natl. Cancer Inst., NIH, Bethesda, MD 20205, USA) *J. Immunol. Methods, 33(1), 63-77 (1980)* En;en.

1932-U2 A microtechnique of complement fixation applied to the serodiagnosis of contagious epididymitis in rams. du Belier Sanchis,R.; Giauffret,A. (Lab. Rech. Vet., 63 ave des Arenes, 06000 Nice, France) *Recl. Med. Vet., Ec. Alfort, Paris, 152(5), 305-310 (1976)* Fr;en,es,fr.

See: 1741, 1851, 3244, 3258, 3282, 3289, 1868, 3261, 3265, 3266, 3278, 3245, 3274

Neutralization techniques

1933-U2 On a simple virological and serological micro-diagnostic method. Kiessig,R. (Inst. Med. Mikrobiol. und Epidemiol., Med. Akad. Magdeburg, DDR 301 Magdeburg, Leipziger Str. 44, GDR) *Z. Gesamte Hyg. Grenzgeb., 23(6), 389-390 (1978)* De;de,en,ru.

1934-U2 Laboratory procedures in adenoviruses. VI. Neutralization in HeLa cell cultures by macro and micro methods. Tan,L.-K.; *Wigand,R. (Inst. Hyg. and Microbiol., Univ. Saarlandes, D-6650 Hamburg (Saar), GFR) *Zentralbl. Bakteriol. Parasitenkd. Infektionskr. Hyg., I Abt. A, 243(1), 1-15 (1979)* De;de,en.

1935-U2 Serum dilution neutralization test for California group virus identification and serology. Lindsey,H.S.; Calisher,C.H.; Mathews,J.H. (Virol. Div., Bur. Lab., Cent. for Dis. Control., Atlanta, GA 30333, USA) *J. Clin. Microbiol., 4(6), 503-510 (1976)* En;en.

1936-U2 Correlation of cytopathic effect, fluorescent-antibody microneutralization, and plaque reduction test results for determining avian infectious bronchitis virus antibodies. Wooley,R.E.; Brown,J. (Dep. Med. Microbiol., Coll. Vet. Med., Univ. Georgia, Athens, GA 30602, USA) *J. Clin. Microbiol., 5(3), 361-364 (1977)* En;en.

1937-U2 Microculture neutralization test for serodiagnosis of three avian viral infections. Giambrone,J.J. (Poult. Sci. Dep., Alabama Agric. Exp. Stn., Auburn Univ., Auburn, AL 36830, USA) *Avian Dis., 24(1), 284-287 (1980)* En;en.

1938-U2 [Detection of antibody to infectious bronchitis virus of fowl kept on large units]. Birnbaum,H.; Elflein,G.; Frohlich,E.; Heider,G.; Lupcke,W.; Muller-Molenar,K.; Vogel,S. (402 Halle, Freiimfelder Str. 66-68, GDR) *Monatsh. Veterinarmed., 34(22), 851-855 (1979)* De;de,en,ru.

1939-U2 A microtitre serum neutralization test for bovine ephemeral fever virus. Burgess,G.W. (Dep. Vet. Pathol. and Public Health, Massey Univ., Palmerston North, New Zealand) *Aust. J. Exp. Biol. Med. Sci., 52(5), 851-855 (1974)* En;en.

1940-U2 A rapid plaque neutralization test for bluetongue virus. Thomas,F.C.; Samagh,B.S. (Anim. Pathol. Div., Health Anim. Branch, Agric. Canada, Anim. Dis. Res. Inst. (E), POB 11300, Stn. H, Ottawa, Ont. K2H 8P9, Canada) *Can. J. Comp. Med., 43(2), 234-236 (1979)* En;en,fr.

1941-U2 A rapid neutralization test for antibodies to bovine leukemia virus, with the use of rhabdovirus pseudotypes. Zavada,J.; Cerny,L.; Zavadova,Z.; Bozonova,J.; Altstein,A.D. (Inst. Virol., Slovak Acad. Sci., Mlynskadolina, 809 39 Bratislava 9, Czechoslovakia) *J. Natl. Cancer Inst., 62(1), 95-101 (1979)* En;en.

1942-U2 A rapid microneutralization test for antibody determination and serodiagnosis of human coronavirus OC43 infections. Gerna,G.; Cereda,P.M.; Revello,M.G.; Torsellini,G.M.; Costa,J. (Lab. Virus, Ist. Mal. Infet., Univ. Pavia, 27100 Pavia, Italy) *Microbiologica (Bologna), 2(4), 331-344 (1979)* En;en.

1943-U2 Studies on the occurrence of herpes infection in dogs in Western Germany with the help of a rapid neutralization test. Bibrack,B.; Schaudinn,W. (Veterinarstr. 13, 8000 Munchen 22, GFR) *Zentralbl. Veterinarmed., B, 23(5-6), 384-390 (1976)* De;de,en,es,fr.

1944-U2 An immunofluorescence neutralization test in disposable plastic trays for demonstrating classical swine fever virus antibodies: comparison of a micro with a macro method. Witte,K.H. (Staatl. Veterinaruntersuchungsamt Arnsberg, Zur Taubeneiche 10-12, D-5760 Arnsberg 2, GFR) *Zentralbl. Veterinarmed. B., 26(7), 551-560 (1979)* En;de,en,es,fr.

1945-U2 Plaque reduction neutralization test for the serodiagnosis of infectious bursal disease. Yashicha,S.; Iritani,Y. (Aburahi Lab., Shionogi and Co. Ltd., Koka-cho, Shiga 520-34, Japan) *Jap. J. Vet. Sci., 39(1), 1-5 (1977)* En;en.

1946-U2 Sensitive neutralization test for virus antibody. I. Mumps antibody. Sato,H.; Albrecht,P.; Hicks,J.T.; Meyer,B.C.; Ennis,F.A. (Dep. Health, Educ. and Welfare, Public Health Serv., Food and Drug Admin., Bureau Biol., Div. Virol., Bethesda, MD 20014, USA) *Arch. Virol., 58(4), 301-311 (1978)* En;en.

1947-U2 [The use of a microtechnique for the intratypic serodifferentiation of type 1 and 3 poliovirus strains]. Stewien,K.E.; Lacerda,J.P.G.de (Dep. Microbiol. e Immunol., Inst. Cienc. Biomed. USP, Av. Dr. Arnaldo, 715 Sao Paulo,

Brazil) *Rev. Saude Publica, Sao Paulo, 11(1), 135-142 (1977)* Pt;en,pt.

1948-U2 Neutralization and immunofluorescence test (NIF) for the demonstration of antibodies against rabies virus. Frost,J.W. (Staatliches Veterinaruntersuchungsamt, Deutschordenstr. 48, D-6000 Frankfurt 71/Main, GFR) *Zentralbl. Veterinarmed., B, 25(4), 338-340 (1978)* En;de,en.

1949-U2 Use of the hemadsorption phenomenon for determining virus and neutralizing antibody titers of rabies. Minamoto,N.; Kurata,K.; Kaizuka,I.; Sazawa,H. (Natl. Vet. Assay Lab., Kokubunji, Tokyo, 185 Japan) *Infect. Immun., 13(5), 1454-1458 (1976)* En;en.

1950-U2 Evaluation of tests for rabies antibody and analysis of serum responses after administration of three different types of rabies vaccines. Grandien,M. (Dep. Virol., Natl. Bacteriol. Lab., S-105 21 Stockholm, Sweden) *J. Clin. Microbiol., 5(3), 263-267 (1977)* En;en.

1951-U2 Rabies neutralizing antibody determination by the interference inhibition test (IIT) and the mouse neutralization test (MNT). Nicholson,K.G.; Harrison,P.; Turner,G.S. (Med. Res. Counc., Div. Communic. Dis., Clin. Res. Cent., Northwick Park Hosp., Harrow, Middlesex HA1 3UJ, UK) *J. Biol. Stand., 7(3), 253-261 (1979)* En;en.

1952-U2 [A technique on microplates of neutralizing antibodies against transmissible gastroenteritis virus of swine.] Toma,B.; Benet,J.J. (Lab. Mal. Contagieuses, Ec. Natl. Vet., 94701 Maisons-Alfort, France) *Rec. Med. Vet., Ec. Alfort, Paris, 152(9), 565-568 (1976)* Fr;en,es,fr.

1953-U2 Studies on neutralization of varicella-zoster virus and serological follow-up of cases of varicella and zoster. Asano,Y.; Takahashi,M. (Dep. Virol., Res. Inst. for Microb. Dis., Osaka Univ., Suita, Osaka, Japan) *Biken J., 21(1), 15-23 (1978)* En;en.

1954-U2 On a simple virological and serological micro-diagnostic method. Kiessig,R. (Inst. Med. Mikrobiol. und Epidemiol., Med. Akad. Magdeburg, DDR-301 Magdeburg, Leipziger Str. 44, GDR) *Z. Gesamte Hyg. Grenzgeb., 23(6), 389-390 (1977)* De;de,ru,en.

See: 14, 40, 41, 1871, 1916, 1979, 2018, 3257, 3260, 3274

Immunofluorescence

1955-U2 IF-40. A new polystyrene microscope tray for immunofluorescent studies. Krech,U.; Jung,M.; Pyndiah,N.; Price,P.C. (Frohbergstr. 3, CH-9000 St. Gallen, Switzerland) *Zentralbl. Bakteriol. Hyg., I Abt. A, 234(1), 136-140 (1976)* En;de,en.

1956-U2 A useful technique applicable to the immunofluorescence test. Tung,K.S.K. (Univ. New Mexico, Dep. Pathol., Albuquerque, NM 87131, USA) *J. Immunol. Methods, 18(3-4), 391-392 (1977)* En;en.

1957-U2 Application of immunofluorescent staining on paraffin sections improved by trypsin digestion. Huang,S.; Minassian,H.; More,J.D. (Dep. Pathol., McGill Univ. and R. Victoria Hosp., Montreal, Que., Canada) *Lab. Invest., 35(4), 383-390 (1976)* En;en.

1958-U2 Antigen-coupled beads adherent to slides: a simplified method for immunological studies. Streefkerk,J.G.; Deelder,A.M.; Kors,N.; Kornelis,D. (Dep. Histol., Free Univ., van der Boechorstst. 7, Amsterdam, Netherlands) *J. Immunol. Methods, 8(3), 251-256 (1975)* En;en.

1959-U2 Template method for the fluorescent-antibody technique. Karim,K.A.; *Trust,T.J. (Dep. Bacteriol. and Biochem., Univ. Victoria, Victoria, BC V8W 2Y2, Canada) *J. Clin. Microbiol., 5(5), 543-544 (1977)* En;en.

1960-U2 Antibodies to calf serum as a cause of unwanted reaction in immunofluorescence tests. Johansson,M.E.; Bergquist,N.R.; Grandien,M. (Dep. Virol., Statens Bakteriol. Lab., S-105 21 Stockholm, Sweden) *J. Immunol. Methods, 11(3,4), 265-272 (1976)* En;en.

1961-U2 The micro-membrane-fluorescence test: a new semiautomated technique based on the Microtiter system. Schauenstein,K.; Wick,G.; Kink,H. (Inst. Gen. and Exp. Pathol., Univ. Innsbruck, Innsbruck, Austria) *J. Immunol. Methods, 10(2-3), 143-150 (1976)* En;en.

1962-U2 [Laboratory techniques for the rapid diagnosis of viral infections: Memorandum.] Atanasiu,P. (Inst. Pasteur, Paris, France) *Bull. WHO, 56(1), 99-103 (1978)* Fr;fr.

1963-U2 Immunofluorescent diagnosis of acute viral infection. Knight,V.; Brasier,F.; Greenberg,S.B.; Jones,D.B. (Dep. Microbiol. and Immunol., Baylor Coll. Med., Houston, TX 77025, USA) *South. Med. J., 68(6), 764-766 (1975)* En.

1964-U2 Simple detection of fluorescent stained IgM in sucrose gradients: demonstration of virus-specific IgM. Schmitz,H.; Krainick-Riechert,C.M. (Hyg-Inst. Univ. Freiburg, Hermann-Herder-Str. 11, D-78 Freiburg i. Br., GFR) *Intervirology, 3(5-6), 353-358 (1974)* En;en.

1965-U2 Direct fluorescent antibody technique in early and rapid diagnosis of adenovirus pneumonia in children. Fuhsi,C.; Lan,S.; Hao-yan,C.; Hsiu-yun,W.; Yu-chin,S.; Kuei-ying,L. (Virus Res. Lab., Peking Friendship Hosp., Peking, China) *Chin. Med. J., 3(4), 227-232 (1977)* En;en.

1966-U2 Immunofluorescent detection of adenovirus antigen in epidemic keratoconjunctivitis. Schwartz,H.S.; *Vastine,D.W.; Yamashiroya,H.; West,C.E. (Univ. Illinois Eye and Ear Infirm., 1855 W. Taylor St., Chicago, IL 60612, USA) *Invest. Ophthalmol., 15(3), 199-207 (1976)* En;en.

1967-U2 Detection of African malignant catarrhal fever virus antigens in cell cultures by immunofluorescence. Ferris,D.H.; Hamdy,F.M.; Dardiri,A.H. (USDA, Agric. Res. Serv., Plum Island Anim. Dis. Cent., Greenport, Long Island, NY 11944, USA) *Vet. Microbiol., 1(4), 437-448 (1976)* En;en.

1968-U2 Use of the indirect fluorescent antibody method for detecting antibodies to infectious bronchitis virus in chicken serum. Jones,R.C. (Sub-Dep. Avian Med., Univ. Liverpool, 'Leahurst', Neston, Wirral, L64 7TE, UK) *J. Comp. Pathol., 85(3), 473-479 (1975)* En;en.

1969-U2 A comparison of three methods of diagnosis of infectious laryngotracheitis. Meulemans,G.; Halen,P. (Inst. Natl. de Rech. Vet., Groeselenberg 99, 1180, Brussels, Belgium) *Avian Pathol., 7(3), 433-436 (1978)* En;de,en,fr.

1970-U2 The establishment of antibodies against avian encephalomycelitis by means of the application of indirect immunofluorescence. Simic,V.; Jermolenko,G. (Vet. Inst., Beograd, Yugoslavia) *Vet. Glas., 30(7), 619-622 (1976)* Sn;en,ru.

1971-U2 Rapid detection and identification of JC virus and BK virus in human urine by using immunofluorescence microscopy. Hogan,T.F.; Padgett,B.L.; Walker,D.L.; Borden,E.C.; McBain,J.A. (Dep. Hum. Oncol., Univ. Wisconsin Sch. Med., Madison, WI 53705, USA) *J. Clin. Microbiol., 11(2), 178-183 (1980)* En;en.

1972-U2 Fluorescent antibody test for rapid diagnosis of coronaviral enteritis of turkeys (bluecomb). Patel,B.L.; Deshmukh,D.R.; Pomeroy,B.S. (Dep. Vet. Biol., Coll. Vet. Med., Univ. Minnesota, St. Paul, MN 55108, USA) *Am. J. Vet. Res., 63(8), 1265-1267 (1975)* En;en.

1973-U2 Demonstration of antibodies in the central nervous system for the diagnosis of Borna disease. Danner,K. (Inst. Med. Mikrobiol., Infekt.-und Seuchenmed., Veterinarstr. 13, 8000 Munchen 22, GFR) *Zentralbl. Veterinarmed., B, 23(10), 865-867 (1976)* De;de,en.

1974-U2 Procedure for processing of central nervous system tissue for immunofluorescence, light, and electron microscopic evaluation. Williams,R.M.; Krakowka,S.; Koestner,A. (Dep. Vet. Pathobiol., Coll. Vet. Med., Ohio State Univ., 1925 Coffey Road, Columbus, OH 43210, USA) *Am. J. Vet. Res., 39(12), 1946-1949 (1978)* En;en.

1975-U2 Anti-complement immunofluorescence test for antibodies to human cytomegalovirus. Kettering,J.D.; Schmidt,N.J.; Gallo,D.; *Lennette,E.H. (Viral and Rickettsial Dis. Lab., State of California Dep. Health, Berkeley, CA 94704, USA) *J. Clin. Microbiol., 6(6), 627-632 (1977)* En;en.

1976-U2 Serological diagnosis of cytomegalovirus (CMV) infections. Hekkeren,A.C.; Brand Saathof,B. (Virol. Lab., State Inst. for Public Health, Bilthoven, Netherlands) *Ned. Tijdschr. Geneeskd., 121(45), 1817-1818 (1977)* Nl.

1977-U2 Placental fluorescent-antibody studies as test for congenital cytomegalovirus infection. McCaffree,M.A.; Altshuler,G. (Pathol. and Pediatr. Serv., Oklahoma Child. Mem. Hosp., Oklahoma City, OK 73126, USA) *Lancet, 1(8124), 1029-1030 (1979)* En.

1978-U2 Modification of the fluorescing cell assay method for human cytomegalovirus (HCMV). Tyms,A.S. (Dep. Virol., St. Mary's Hosp. Med. Sch., Paddington, London W2 1PG, UK) *Med. Lab. Sci., 35(3), 307-308 (1978)* En.

1979-U2 Use of isolated nuclei in the indirect fluorescent-antibody test for human cytomegalovirus infection: comparison with microneutralization, anticomplement, and conventional indirect fluorescent-antibody assays. Stagno,S.; Reynolds,D.W.; Smith,R.J. (Dep. Pediatr., Med. Cent., Univ. Alabama in Birmingham, AL 35294, USA) *J. Clin. Microbiol., 7(5), 486-489 (1978)* En;en.

1980-U2 Evaluation of anti-complement immunofluorescence test in cytomegalovirus infection. Rao,N.; Waruszewski,D.T.; Armstrong,J.A.; Atchison,R.W.; Ho,M. (Div. Infect. Dis., Dep. Med., Sch. Med., Univ. Pittsburh, Pittsburgh, PA 15261, USA) *J. Clin. Microbiol., 6(6), 633-638 (1977)* En;en.

1981-U2 Indirect fluorescent antibody technic for demonstration of serum antibody in dengue hemorrhagic fever cases. Boonpucknavig,S.; Vuttivirojana,O.; Siripont,J.; Futrakul,P.; Nimmannitya,S. (Dep. Pathobiol., Fac. Sci., Mahidol Univ., Rama VI Rd., Bangkok 4, Thailand) *Am. J. Clin. Pathol., 64(3), 365-371 (1975)* En;en.

1982-U2 A rapid fluorescent focus-inhibition test for determining dengue neutralizing antibody and for identifying prototype dengue viruses. Thacker,W.L.; Lewis,V.J.; Baer,G.M.; Sather,G.E. (Viral Zoonoses Branch, Virol. Div., Bur. Lab., Cent. Dis. Control, Public Health Serv., US Dep. Health, Educ. and Welfare, POB 363, Lawrenceville, GA 30246, USA) *Can. J. Microbiol., 24(12), 1553-1556 (1978)* En;en,fr.

1983-U2 A simple technique for the detection of dengue antigen in mosquitoes by immunofluorescence. Kuberski,T.T.; Rosen,L. (Pacific Res. Sect., Lab. Parasitic Dis., Natl. Inst. Allergy and Infect. Dis., Natl. Inst. Health, Honolulu, HI 96806, USA) *Am. J. Trop. Med. Hyg., 26(3), 533-537 (1977)* En;en.

1984-U2 Epstein-Barr virus early antigen titer by immunofluorescence in microplates. A new semi-automated method based on microtiter system. Favart,A.M.; Lamy,M.E.; Allemeersch,D.; Burtonboy,G.; Vanoverschelde,J. (Lab. Virol., Catholic Univ. Louvain, B-1200 Brussels, Belgium) *Med. Microbiol. Immunol., 166(1-4), 209-217 (1978)* En;en.

1985-U2 The simultaneous observations of immunofluorescence and cellular morphology in viral infections: a new procedure. Takada,K.; Aya,T.; *Osato,T. (Dep. Virol., Cancer Inst., Hokkaido Univ. Sch. Med., 060 Sapporo, N15 W7, Japan) *Med. Microbiol. Immunol., 164(4), 239-246 (1978)* En;en.

1986-U2 Epstein-Barr virus early antigen titer by immunofluorescence in microplates. A new semi-automated method based on microtiter system. Favart,A.M.; Lamy,M.E.; Allemeersch,D.; Burtonboy,G.; Vanoverschelde,J. (Lab. Virol., Catholic Univ. Louvain, B-1200 Brussels, Belgium) *Med. Microbiol. Immunol., 166(1-4), 209-217 (1978)* En;en.

1987-U2 Epstein-Barr virus viral capsid antigen titer by immunofluorescence with microplates: new semiautomated method based on the microtiter system. Lamy,M.E.; Favart,A.M.; Burtonboy,G.; Arana,A. (Lab. Virol., Catholic Univ. Louvain, 1200 Brussels, Belgium) *J. Clin. Microbiol., 6(1), 66-71 (1977)* En;en.

1988-U2 [Study of equine infectious anaemia by indirect immunofluorescence]. Toma,B.; Goret,P. (Chaire des Mal. Contag., Ec. Natl. Vet., 94701 Maisons-Alfort, France) *Rec. Med. Vet., 15(8-9), 499-504 (1975)* Fr;en,es,fr.

1989-U2 Contained indirect viable-cell membrane immunofluorescence microassay for surface antigen analysis of cells infected with hazardous viruses. Cloyd,M.W.; Bigner,D.D. (Dep. Pathol., Duke Univ. Med. Cent., Durham, NC 27710, USA) *J. Clin. Microbiol., 5(1), 86-90 (1977)* En;en.

1990-U2 An immunofluorescence blocking test to detect antibodies against the hepatitis B delta (δ) and core (HB_cAg) antigens. Crivelli,O.; Lavarini,C.; Arico,S.; Rizzetto,M. (Div. Gastroenterol., Osp. Mauriziano Umberto 1-Corso Turati, 46, 10128 Torino, Italy) *Boll. Ist. Sieroter. Milan, 57(2), 173-177 (1978)* En;en,it.

1991-U2 [Detection of HBs antigen by immunofluorescence in skin blood vessels. Preliminary results.] Doutre,M.S.; Beylot,J.; Deminiere,C.; Beylot,C.; Devars,D. (Serv. Dermatol., CHU Bordeaux, Hop. Haut-Leveque, F 33604 Pessac, France) *Nouv. Presse Med., 8(39), 3170 (1979)* Fr.

1992-U2 [Identification of the hepatitis B surface antigen (HBsAg) in the liver in fulminating hepatitis. With reference to twenty cases]. Vitrey,D.; Tempelhoff,G.; Patricot,L.M.; Trepo,C.; Robert,D.; Vauzelle,J.L. (Lab. Anat. Pathol., Hop. Croix-Rousse, F 69317 Lyon Cedex 1, France) *Lyon Med., 238(16), 243-248 (1977)* Fr;en,fr.

1993-U2 Comparison and evaluation of fluorescent antibody techniques in the detection of herpes simplex virus in oral infection. Nitzan,D.W.; Pisanti,S.; Dishon,T. (Dep. Periodontol., Harvard Sch. Dent. Med., 188 Longwood Ave., Boston, MA 02115, USA) *Oral Surg. Oral Med. Oral Pathol., 45(2), 207-213 (1978)* En;en.

1994-U2 Exclusion of false positive reactions due to IgG-binding Fc-receptor(s) in herpesvirus serology. Simon,M. (Natl. Inst. Hyg., H-1966 Budapest, PO Box 69, Hungary) *Acta Microbiol. Acad. Sci. Hung., 24(4), 307-316 (1977)* En;en.

1995-U2 Application of immunofluorescent technique in the cytologic diagnosis of human herpes simplex keratitis. Olding Stenkvist,E.; Goran Brege,K. (Dep. Clin. Cytol., University Hosp., Uppsala, Sweden) *Acta Cytol., 19(5), 411-414 (1975)* En;en.

1996-U2 Diagnosis of herpes simplex virus infection by immunofluorescence. Taber,L.H.; Brasier,F.; Couch,R.B.; Greenberg,S.B.; Jones,D.; Knight,V. (Dep. Pediatr., Baylor Coll. Med., Houston, TX 77030, USA) *J. Clin. Microbiol., 3(3), 309-312 (1976)* En;en.

1997-U2 Sensitivity of the virus isolation and immunofluorescent staining methods in diagnosis of infections with herpes simplex virus. Cho,C.T.; Feng,K.K. (Dep. Pediatr., Univ. Kansas Med. Cent., 39th and Rainbow Blvd., Kansas City, KS 66103, USA) *J. Infect. Dis., 138(4), 536-540 (1978)* En;en.

1998-U2 Immunofluorescent staining for the measurement of antibodies to *Herpesvirus hominis*. Cho,C.T.; Feng,K.K.; Brahmacupta,N.; Liu,C. (Dep. Pediatr., Univ. Kansas Med. Cent., 39th and Rainbow Blvd., Kansas City, KS 66103, USA) *J. Infect. Dis., 132(3), 311-315 (1975)* En;en.

1999-U2 Rapid diagnostic tests for cutaneous eruptions of herpes simplex. Veien,N.K.; Vestergaard,B.F. (Dep. Dermatol., Municipal Hosp., Univ. Copenhagen, Copenhagen, Denmark) *Acta Derm.-Venereol., 58(1), 83-85 (1978)* En;en.

2000-U2 Type specificity and use of human sera for rapid typing of herpes simplex virus isolates by an indirect peroxidase-labeled antibody technique: a comparison with three other methods. Gerna,G.; Ditmore,J.; Chambers,R.W. (Viral Diagn. Serv., Dep. Pathol., Georgetown Univ. Med. Cent., Washington, DC, USA) *Arch. Virol., 54(1-2), 119-130 (1977)* En;en.

2001-U2 Indirect micro-immunofluorescence test for detecting type-specific antibodies to herpes simplex virus. Forsey,T.; Darougar,S. (Virus Lab., Inst. Ophthalmol., Judd St., London WC1H 9QS, UK) *J. Clin. Pathol., 33(2), 171-176 (1980)* En;en.

2002-U2 Measurement of antibodies to *Herpesvirus hominis* by indirect immunofluorescent antibody method. Cho,C.T.; Feng,K.K.; Liu,C. (Dep. Pediatr., Univ. Kansas Med. Cent., 39th and Rainbow Blvd., Kansas City, KS 66103, USA) *Micros. Acta, 79(1), 47-54 (1977)* En;en,de.

2003-U2 Virological studies of smallpox in an endemic area. I. Evaluation of immunofluorescence staining as a rapid diagnostic procedure in the field. Kitamura,T.; Aoyama,Y.; Kurata,T.; Arita,M.; Imagawa,Y. (Div. Poxviruses, Natl. Inst. Health, Shinagawa-ku, Tokyo 141, Japan) *Jap. J. Med. Sci. Biol., 30(5), 215-227 (1977)* En;en.

2004-U2 [Swine fever. I Diagnosis by immunofluorescence using tissue sections and blood smears]. Leaniz,R.; Bermudez,R.; Siri,A.M.; Pereira,J.; Lagos,R.; Carrera,D. (Fac. Vet., Univ. Montevideo, Uruguay) *Gac. Vet., 38(308), 62-69 (1976)* Es;es.

2005-U2 Rapid diagnosis of influenza A infection by immunofluorescence: methodological problems and clinical material. Olding-Stenkvist,E.; Grandien,M. (Dep. Infect. Dis., Univ. Hosp., 5 014 Uppsala 14, Sweden) *Acta Pathol. Microbiol. Scand., Ser. B, 85(5), 296-302 (1977)* En;en.

2006-U2 Rapid diagnosis of influenza A infection by direct immunofluorescence of nasopharyngeal aspirates in adults. Daisy,J.A.; Lief,F.S.; Friedman,H.M. (Infect. Dis. Sect., Dep. Med., Univ. Pennsylvania, PA 19104, USA) *J. Clin. Microbiol., 9(6), 688-692 (1979)* En;en.

2007-U2 Preparation of sera for subtyping of influenza A viruses by immunofluorescence. Johansson,M.E.; Grandien,M.; Arro,L. (Dep. Virol., Natl. Bacteriol. Lab., S-105 21 Stockholm, Sweden) *J. Immunol. Methods, 27(3), 263-272 (1979)* En;en.

2008-U2 Utilization of peroxidase- and FITC-labelled antibody for express diagnosis of influenza. Ketiladze,E.S.; Zhilina,N.N.; Herrmann,H.; Naumova,V.K.; Knyazeva,L.D.(D.I. Ivanovsky Inst. Virol., Acad. Med. Sci. USSR, Moscow, USSR) *Vopr. Virusol., No. 6, 695-699 (1978)* Ru;en,ru.

2009-U2 Comparison of immunofluorescence and immunoperoxidase methods for viral diagnosis at a distance: a WHO collaborative study. Gardner,P.S.; Grandien,M.; McQuillin,J. (Dep. Virol., R. Victoria Infirm., Newcastle-upon-Tyne, UK) *Bull. WHO, 56(1), 105-110 (1978)* En;en,fr.

2010-U2 A rapid method for detecting Junin virus viremia in the guinea pig. Rabinovich,A.; Cossio,P.M.; Carballal,G.; Arana,R.M. (CEMIC, Sanchez de Bustamante 2560, Buenos Aires, Argentina) *Intervirology, 8(6), 360-363 (1977)* En;en.

2011-U2 Serologic diagnosis of human infections with lymphocytic choriomeningitis virus: comparative evaluation of seven methods. Lehmann-Grube,F.; Kallay,M.; Ibscher,B.; Schwartz,R. (Heinrich Pette-Inst., Martinistr. 52, 2000 Hamburg 20, GFR) *J. Med. Virol., 4(2), 125-136 (1979)* En;en.

2012-U2 Rapid detection of measles virus in skin rashes by immunofluorescence. Olding-Stenkvist,E.; Bjorvatn,B. (Dep. Infect. Dis., Univ. Hosp., S-75014 Uppsala, Sweden) *J. Infect. Dis., 134(5), 463-469 (1976)* En;en.

2013-U2 Application of immunofluorescence to a study of measles. McQuillin,J.; Bell,T.M.; *Gardner,P.S.; Downham,M.A.P.S. (Dep. Virol., R. Victoria Infirm., Newcastle-upon-Tyne NE1 4LP, UK) *Arch. Dis. Child., 51(6), 411-419 (1976)* En;en.

2014-U2 The use of trypan blue for counterstaining in the Sepharose bead immunofluorescence test. The application of the test for the demonstration of primate retrovirus-specific antibodies and antigens. Micheel,B. (Cent. Inst. Cancer Res., Acad. Sci. DDR. Dep. Exp. and Clin. Immunol., 1115 Berlin-Buch, GDR) *Acta Biol. Med. Ger., 37(8), K19-K24 (1978)* En;en.

2015-U2 Rapid diagnosis of acute respiratory diseases by means of indirect immunofluorescence test. Baratta,L.; Valeriani,M.; Andreoni,G. (Univ. degli Stud., Clin. Med. Gen. e Ter. Med. 3, Rome, Italy) *Ann. Sclavo, 17(1), 40-49 (1975)* It;it,en,fr.

2016-U2 A rapid method for quantitative assay of poliovirus from water with the aid of the fluorescent antibody technique. Katzenelson,E. (Environ. Health Lab., Hebrew Univ.-Hadassah Med. Sch., PO Box 1172, Jerusalem, Israel) *Arch. Virol., 50(3), 197-206 (1976)* En;en.

2017-U2 Preliminary report on an agar-gel immunodiffusion test for pseudorabies. Pfeiffer,N.E.; Schipper,I.A. (Address not stated) *N. D. Farm Res., 34(6), 21-22 (1977)* En.

2018-U2 Diseases of swine in Minas Gerais. II. Comparison between neutralization and fluorescent antibody test for detecting Aujeszky's disease virus antibody. Apolonio de Oliveira,A.; Reis,R. (Empresa Brasil. Pesqui. Agropecu., Rio de Janeiro, Brazil) *Arq. Esc. Vet., 29(1), 121-127 (1977)* Pt;en,pt.

2019-U2 Antemortem detection of rabies using a skin biopsy technique. Navarro,J.A.; Whittington,E.A. (Coll. Vet. Med., Univ. Philippines, Diliman, Quezon City, Philippines) *Philipp. J. Vet. Med., 14(1), 183-186 (1975)* En;en.

2020-U2 Use of a free viral immunofluorescence assay to detect human reovirus-like agent in human stools. Yolken,R.H.; Wyatt,R.G.; Kalica,A.R.; Kim,H.W.; Brandt,C.D.; Parrott,R.H.; Kapikian,A.Z.; Chanock,R.M. (Lab. Infect. Dis., Natl. Inst. Allergy and Infect. Dis., Natl. Inst. Health, Bethesda, MD 20014, USA) *Infect. Immun., 16(2), 467-470 (1977)* En;en.

2021-U2 Studies on reovirus infection in canine cells by the fluorescent antibody technique. Murakami,T.; Kato,H.; Ohmori,N. (Dep. Vet. Microbiol., Fac. Agric., Iwate Univ., Morioka-shi, Iwate 020, Japan) *Jap. J. Vet. Sci., 37(2), 179-186 (1975)* En;en.

2022-U2 Respiratory syncytial virus infection. Rapid diagnosis in children by use of indirect immunofluorescence. Kaul,A.; Scott,R.; Gallagher,M.; Scott,M.; Clement,J.; Ogra,P.L. (Child. Hosp., 219 Bryant St., Buffalo, NY 14222, USA) *Am. J. Dis. Child., 132(11), 1088-1090 (1978)* En;en.

2023-U2 Rapid immunofluorescence diagnosis of respiratory syncytial virus infections among children in European countries. Orstavik,I.; Grandien,M.; Halonen,P.; Arstila,P.; Mordhorst,C.H.; Hornsleth,A.; Popow-Kraup,T.; McQuillan,J.; *Gardner,P.S. (Div. Microbiol. Reagents and Quality Control, PHLS, Colindale Ave., London BW9 5HT, UK) *Lancet, 2(8184), 32 (1980)* En.

2024-U2 Rapid diagnosis of respiratory virus infections in children. Respiratory syncytial (RS) virus demonstrated by the immunofluorescent technique. Hornsleth,A.; Friis,B.; Mordhorst,C.H.; Ursin Knudsen,F.; Uldall,P.; Andersen,O.; Karup Pedersen,F.; Krasilnikoff,P.A. (Inst. Med. Mikrobiol., Juliane Maries Vej 22, DK-2100 Kobenhavn O, Denmark) *Ugeskr. Laeg., 142(18), 1138-1141 (1980)* Da;da,en.

2025-U2 A note on fluorescent antibody technique for rapid diagnosis of rinderpest. Prabhudas,K.; Sambamurti,B. (Dep. Microbiol., Coll. Vet. Sci., Rajendranagar, India) *Indian J. Anim. Sci., 46(8), 454-457 (1976)* En.

2026-U2 A simple immunofluorescent technique for the detection of human rotavirus. Moosai,R.B.; Gardner,P.S.; Almeida,J.D.; Greenaway,M.A. (Dep. Virol., R. Victoria Infirm. and Univ. Newcastle upon Tyne, Newcastle upon Tyne NE1 4LP, UK) *J. Med. Virol., 3(3), 189-194 (1979)*

En;en.

2027-U2 Identification of rotaviruses. De Silva,L.M.; Marshall,I. (Dep. Microbiol., Bedford Gen. Hosp. (North Wing), Bedford MK40 2NV, UK) *Lancet, 1(8011), 609-610 (1977)* En.

2028-U2 Comparison of methods for immunocytochemical detection of rotavirus infections. Graham,D.Y.; Estes,M.K. (Dep. Med., Baylor Coll. Med., Houston, TX 77030, USA) *Infect. Immun., 26(2), 686-689 (1979)* En;en.

2029-U2 An immunofluorescence assay for human antibodies to Ross River virus. Aaskov,J.G.; Davies,C.E.A. (Queensland Inst. Med. Res., Bramston Terrace, Herston, Queensl. 4006, Australia) *J. Immunol. Methods, 25(1), 37-41 (1979)* En;en.

2030-U2 The application of different cell lines and virus strains for detection of immunofluorescence rubella antibodies. Jankowski,M.; Gut,W.; Imbs,D.; Kantoch,M. (Dep. Virol., Natl. Inst. Hyg., Chocimska 24, 00-791 Warsaw, Poland) *Acta Microbiol. Pol., 9(4), 439-446 (1977)* En;en.

2031-U2 Diagnosis of pre- and postnatal rubella by demonstration of specific IgM-class antibodies by a microplate immunofluorescence test. Dall,V.; Jensen,N.H.; Hansen,U. (Rubella Dep., Statens Seruminst., Amager Blvd. 80, 2300 Copenhagen S, Denmark) *Med. Microbiol. Immunol., 168(1), 73-80 (1980)* En;en.

2032-U2 A rapid assay of Sindbis virus infectivity by counting immunofluorescence foci in *Aedes albopictus* cell culture. Digoutte,J.P.; Tignor,G.H.; Smith,A.L.; Knudson,D.L. (Inst. Pasteur de la Guyane Francaise, INSERM U 79, B.P. 304, 97305 Cayenne, France) *Ann. Microbiol., 127B(4), 573-576 (1976)* Fr;en,fr.

2033-U2 Rapid diagnosis of tick-borne encephalitis by immunofluorescence assay of specific serum IgM antibodies. Frankova,V.; Chyle,M.; Duniewicz,M.; Marhoul,Z.; Kolman,J.M.; Mancal,P. (Dep. Med. Microbiol. and Immunol., Charles Univ., Fac. Med., 100 00 Praha 2, Studnickova 7, Czechoslovakia) *J. Hyg. Epidemiol. Microbiol. Immunol., 22(4), 502-504 (1978)* En;de,en,es,fr.

2034-U2 An indirect fluorescent antibody test for antibodies to transmissible gastroenteritis of swine. Benfield,D.A.; Haelterman,E.O.; *Burnstein,T. (Dep. Vet. Microbiol., Pathol and Public Health, Sch. Vet. Med., Purdue Univ., West Lafayette, IN 47907, USA) *Can. J. Comp. Med., 42(4), 478-482 (1978)* En;en,fr.

2035-U2 The use of immunofluorescence techniques for the laboratory diagnosis of transmissible gastroenteritis of swine. Solorzano,R.F.; Marin,M.; Morehouse,L.G. (Dep. Vet. Microbiol., Vet. Med. Univ. Missouri-Columbia, Columbia, MO 65201, USA) *Can. J. Comp. Med., 42(2), 385-391 (1978)* En;en,fr.

2036-U2 Studies on the improvement of laboratory diagnosis of transmissible gastroenteritis of swine (TGE). Mocsari,E. (1581 Budapest. Pf. 2, Haungary) *Magy. Allatorv. Lapja, 32(8), 508-511 (1977)* Hu;de,en,hu,ru.

2037-U2 Quantitative and rapid assays of togaviruses by immunofluorescence. Enzmann,P.J. (Fed. Res. Inst. Anim. Virus Dis., D-7400 Tubingen, GFR) *Acta Virol., 23(4), 329-334 (1979)* En;en.

2038-U2 Determination of varicella immunity by the indirect immunofluorescence test in urgent clinical situations. Grandien,M.; Appelgren,P.; Espmark,Å.; Hanngren,K. (Dep. Virol., Statens Bakteriol. Lab., Stockholm, Sweden) *Scand. J. Infect. Dis., 8(2), 65-69 (1976)* En;en.

2039-U2 In situ antigen localization with fluorescamine-conjugated antibodies: a new method. Sung,M.T.; Bozzola,J.J.; Richards,J.C. (Dep. Chem. and Biochem., Southern Illinois Univ. at Carbondale, Carbondale, IL 62901, USA) *Anal. Biochem., 84(1), 225-230 (1978)* En;en.

2040-U2 Virus diagnostic procedure in clinical virology. Gardner,P.S. (Virol. Dep., R. Victoria Infirmary, Newcastle upon Tyne, UK) *Experientia, 33(12), 1674-1676 (1977)* En.

2041-U2 Neutralisation of indirect immunofluorescence (NIFI): a specific and quantitative test for detection of viral antigens. Wellemans,G.; van Opdenbosch,F.; de Kegel,D. (Inst. Rech. Vet., Groeselenberg 99, 1180 Bruxelles, Belgium) *Ann. Med. Vet., 123(3), 185-194 (1979)* Fr;en,fr,nl.

2042-U2 Progress in the rapid diagnosis of viral infections: a Memorandum. Anon. (Virus Dis., Div. Commun. Dis., World Health Organization, 1211 Geneva 27, Switzerland) *Bull. WHO, 56(2), 241-244 (1978)* En;en.

2043-U2 Serological identification of *Actinomyces* using fluorescent antibody techniques. Gerencser,M.A.; Slack,J.M. (Dep. Microbiol., Med. Cent., West Virginia Univ., Morgantown, WV 26506, USA) *J. Dent. Res., 55(Special Issue A), A184-A191 (1976)* En;en.

2044-U2 Fluorescent antibody test kit for rapid detection and identification of members of the *Bacteroides fragilis* and *Bacteroides melaninogenicus* groups in clinical specimens. Holland,J.W.; Stauffer,L.R.; Altemeier,W.A. (Res. Surg. Bacteriol. Lab., Dep. Surg., Univ. Cincinnati, Coll. Med., Cincinnati, OH 45267, USA) *J. Clin. Microbiol., 10(2), 121-127 (1979)* En;en.

2045-U2 Identification of *Bacteroides fragilis* by indirect immunofluorescence. Weintraub,A.; Lindberg,A.A.; Nord,C.E. (Dep. Bacteriol., Natl. Bacteriol. Lab., S-105 21 Stockholm, Sweden) *Med. Microbiol. Immunol., 167(4), 223-230 (1979)* En;en.

2046-U2 Evaluation of two fluorescent test kits for detection of selected *Bacteroides* species in clinical specimens. Holst,E.; Oscarson,J.; *Mardh,P.-A. (Inst. Med. Microbiol., Univ. Lund, Solvegatan 23, S-223-62 Lund, Sweden) *Curr. Microbiol., 3(3), 133-136 (1979)* En;en.

2047-U2 Characterization of a polyvalent conjugate of *Bacteroides fragilis* by fluoresccent antibody staining.

Lambe,D.W.Jr. (Dep. Microbiol., Coll. Med., East Tennessee State Univ., Johnson City, TN 37601, USA) *Am. J. Clin. Pathol., 71(1), 97-101 (1979)* En;en.

2048-U2 Single-antigen indirect immunofluorescence test for screening venereal disease clinic populations for chlamydial antibodies. Richmond,S.J.; Caul,E.O. (Public Health Lab., Kingsdown, Bristol BS2 8EL, UK) *In:* Nongonococcal urethritis and related infections. Hobsón,D.; Holmes,K.K. (eds.) *Publ. by:* American Society for Microbiology, 1913 I St., N.W., Washington, DC 20006, USA 1 Aug 1977 p. 259-265 ISBN 0-914826-11-5 En.

2049-U2 Modification of the microimmunofluorescence test to provide a routine serodiagnostic test for chlamydial infection. Treharne,J.D.; Darougar,S.; Jones,B.R. (Virus Lab., WHO Collab. Cent. Ref. and Res. on Trachoma and other Chlamydial Infect., Dep. Clin. Ophthalmol., Inst. Ophthalmol., Univ. London, London, UK) *J. Clin. Pathol., 30(6), 510-517 (1977)* En;en.

2050-U2 Serological responses to chlamydial ocular and genital infections in the United Kingdom and Middle East. Treharne,J.D.; Dines,R.J.; Darougar,S. (Inst. Ophthalmol., Univ. London, London WC1H 9QS, UK) *In:* Nongonococcal urethritis and related infections. Hobson,D.; Holmes,K.K. (eds.). *Publ. by:* American Society for Microbiology, 1913 I St., N.W., Washington, DC 20006, USA 1 Aug 1977 p. 249-258 ISBN 0-914826-11-5 En.

2051-U2 Serodiagnosis of *Chlamydia trachomatis* infection with the micro-immunofuorescence test. Wang,S.P.; Grayston,J.T.; Kuo,C.-C.; Alexander,E.R.; Holmes,K.K. (Sch. Public Health and Community Med., Univ. Washington, Seattle, WA 98195, USA) *In:* Nongonococcal urethritis and related infections. Hobson,D.; Holmes,K.K. (eds.) *Publ. by:* American Society for Microbiology, 1913 I St., N.W., Washington, DC 20006, USA 1 Aug 1977 p.237-248 ISBN 0-914826-11-5 En.

2052-U2 Rapid diagnosis of chlamydial infection of the cervix. Treharne,J.D.; Darougar,S.; Simmons,P.D.; Thin,R.N. (Inst. Ophthalmol., Judd St., London WC1H 9QS, UK) *Br. J. Vener. Dis., 54(6), 403-408 (1978)* En;en.

2053-U2 Simplified serological test for antibodies to *Chlamydia trachomatis*. Thomas,B.J.; Reeve,P.; Oriel,J.D. (Clin. Res. Cent., Watford Road, Harrow HA1 3UJ, UK) *J. Clin. Microbiol., 4(1), 6-10 (1976)* En;en.

2054-U2 Rapid serological test for diagnosis of chlamydial ocular infections. Darougar,S.; Treharne,J.D.; Minassian,D.; El-Sheikh,H.; Dines,R.J.; Jones,B.R. (Inst. Ophthalmol., Judd St., London WC1H 9QS, UK) *Br. J. Ophthalmol., 62(8), 503-508 (1978)* En;en.

2055-U2 Preparation of antigens for microimmunofluorescence testing for antichlamydial antibodies. Nisbet,I.T.; Graham,D.M.; Ng,K.M.; Lunt,P.A. (Dep. Microbiol., Univ. Melbourne, Parkville, Vic. 3052, Australia) *Lancet, 1(8147), 859 (1979)* En.

2056-U2 Immunofluorescence adsorption microtest (mIFAT) in detection of botulin A toxin. Banasiewicz,R.; Reiss,J. (30-901 Krakow, Osrodek Wojskowego Inst. Hig. i Epidemiol. Krakowie, Poland) *Med. Dosw. Mikrobiol., 28(1), 31-35 (1976)* Pl;en,pl,ru.

2057-U2 Fluorescent antibody test in diagnosis of ulcerative enteritis. Berkhoff,G.A.; Kanitz,C.L. (Dep. Vet. Microbiol., Pathol. and Public Health, Sch. Vet. Med., Purdue Univ., West Lafayette, IN 47907, USA) *Avian Dis., 20(3), 525-533 (1976)* En;en.

2058-U2 Identification of *Corynebacterium diphtheriae* by immunofluorescence. Schoenemann,W. (Hyg. Inst. Ruhrgebiets zu Gelsenkirchen, Gelsenkirchen, GFR) *Zentralbl. Bakteriol. Parasitenkd. Infektionskr. Hyg., I Abt. A, 240(2), 258-264 (1978)* De;en,de.

2059-U2 Detection of bovine enteropathogenic *Escherichia coli* by indirect fluorescent antibody technique. Hadad,J.J.; *Gyles,C.L. (Dep. Vet. Microbiol. and Immunol., Ontario Vet. Coll., Univ. Guelph, Guelph, Ont. N1G 2W1, Canada) *Am. J. Vet. Res., 39(10), 1651-1655 (1978)* En;en.

2060-U2 [Detection of *Francisella tularensis* in animal organs at the early stages of development of infection]. Olsufjev,N.G.; Ananova,E.V.; Mescheryakova,I.S.; Savelieva,R.A. (Inst. Epidemiol. and Microbiol., Acad. Med. Sci. USSR, ul. Gamelei 2, Moscow D-98, USSR) *Zh. Mikrobiol. Epidemiol. Immunobiol., 52(10), 13-17 (1975)* Ru;ru,en.

2061-U2 An indirect fluorescent antibody technique for *Haemophilus ducreyi*. Denys,G.A.; Chapel,T.A.; Jeffries,C.D. (Dep. Immunol. and Microbiol., Wayne State Univ. Sch. Med., Detroit, MI 48201, USA) *Health Lab. Sci., 15(3), 128-132 (1978)* En;en.

2062-U2 Immunofluorescence detection of nitrogenase proteins in whole cells. Rennie,R.J. (ARC Unit Nitrogen Fixation, Univ. Sussex, Falmer, Brighton BN1 9QJ, UK) *J. Gen. Microbiol.. 97(2), 289-296 (1976)* En;en.

2063-U2 Rapid definitive diagnosis of Legionnaire's disease. Giglia,A.R.; Morgan,P.N.; Bates,J.H. (4301 West Markham, Little Rock, AR 72201, USA) *Chest, 76(1), 98-99 (1979)* En;en.

2064-U2 Immunofluorescent identification of *Listeria monocytogenes*. Khan,M.A.; Seaman,A.; *Woodbine,M. (Microbiol. Unit, Dep. Appl. Biochem. and Nutr., Fac. Agric. Sci., Univ. Nottingham, Sutton Bonington, Loughborough LE12 5RD, UK) *Zentralbl. Bakteriol. Parasitenkd. Infektionskr. Hyg., I Abt. A, 239(1), 62-69 (1977)* En;de,en.

2065-U2 Detection of *Mycoplasma hominis* and *Mycoplasma orale* in cell cultures by immunofluorescence. Payment,P.; Corbeil,M.; Chagnon,A. (Cent. Rech. Virol., Inst. Armand-Frappier, CP 100, Laval, PQ H7N 4Z3, Canada) *Can. J. Microbiol., 24(6), 689-692 (1978)* En;en,fr.

2066-U2 Simple method for immunofluorescent identification of mycoplasma colonies. Bradbury,J.M.;

Oriel,C.A.; Jordan,F.T.W. (Sub-Dep. Avian Med., Univ. Liverpool, 'Leahurst,' Neston, Wirral L64 7TE, UK) *J. Clin. Microbiol., 3(4), 449-452 (1976)* En;en.

2067-U2 Development of a template for use in immunofluorescent identification of mycoplasmas. Panangala,V.S.; Lein,D.H. (New York State Coll. Vet. Med., Cornell Univ., Ithaca, NY 14853, USA) *J. Clin. Microbiol., 7(4), 399-400 (1978)* En;en.

2068-U2 Mycoplasmas: use of polyvalent antisera for identification by indirect immunofluorescence. Erno,H. (Inst. Med. Microbiol., Univ. Aarhus, DK-8000 Aarhus C, Denmark) *Acta Vet. Scand., 18(2), 176-186 (1977)* En;da,en.

2069-U2 Technical modification of the method for direct and indirect immunofluorescence of unfixed *Mycoplasma* colonies. Moller,B.R. (Inst. Med. Microbiol., Univ. Aarhus, Aarhus, Denmark) *J. Bacteriol., 46(1), 185-188 (1979)* En;en.

2070-U2 *Mycoplasma suipneumoniaae* enzootic pneumoni in the pig. Rapid diagnosis and research of antibodies. Kobisch,M.; Tillon,J.P.; Vannier. (Minist. Agric., Direction de la Qual., Serv. Vet., Stn. Pathol. Porcine, 22440 Ploufragan, France) *Recl. Med. Vet., Ec. Alfort, Paris, 154(10), 847-852 (1978)* Fr;en,es,fr.

2071-U2 Fluorescent antibody techniques in identification of *Neisseria gonorrhoeae* microcolonies grown on membrane filters. Blenk,H.;Junge,W.; Blenk,B. (Med. Untersuchungsstelle 1 Bundeswehr, Lesserstr. 180, D2000 Hamburg 70, GFR) *Zentralbl. Bakteriol. Parasitenkd. Infektionskr. Hyg., I Abt. A, 240(4), 480-488 (1978)* De;de,en.

2072-U2 Laboratory identification of pathogenic *Neisseria* with special regard to atypical strains: an evaluation of sugar degradation, immunofluorescence andn co-agglutination tests. Olcen,P.; Danielsson,D.; Kjellander,J. (Dep. Clin. Bacteriol. and Immunol., Regionsjukhuset, S-701 85 Orebro, Sweden) *Acta Pathol. Microbol. Scand. Ser. B, 86(6), 327-334 (1978)* En;en.

2073-U2 Comparison of methods for rapid immunofluorescent staining of group A streptococci and *Neisseria gonorrhoeae*. Quraishi,M.A.H.; Washington,W.; Rosenthal,S.L. (Room 6C 22, Van Etten Hosp., Bronx Munic. Hosp. Cent., Bronx, NY 10461, USA) *Health Lab. Sci., 14(4), 279-281 (1977)* En;en.

2074-U2 Immunofluorescence studies of *Nitrobacter* populations in soils. Rennie,R.J.; Schmidt,E.L. (Joint FAO/ IAEA Div. At. Energy in Food and Agric. IAEA, PO Box 590, Kartner Ring 11, A-1011, Vienna, Austria) *Can. J. Microbiol., 23(8), 1011-1017 (1977)* En;en,fr.

2075-U2 Adaptation of a microimmunofluorescence test to the study of human *Rickettsia tsutsugamushi* antibody. Robinson,D.M.; Brown,G.; Gan,E.; Huxsoll,D.L. (US Army Med. Res. Unit, Inst. Med. Res., Kuala Lumpur, Malaysia) *Am. J. Trop. Med. Hyg., 25(6), 900-905 (1976)* En;en.

2076-U2 Early diagnosis of Rocky Mountain spotted fever. Oster,C.N. (NIH, NIAID, Lab. Parasit. Dis., Bethesda, MD 20014, USA) *Arch. Intern. Med., 139(4), 400-401 (1979)* En. [immunofluorescence; culture].

2077-U2 Serologic typing of rickettsiae of the spotted fever group by microimmunofluorescence. Philip,R.N.; Casper,E.A.; Burgdorfer,W.; Gerloff,R.K.; Hughes,L.E.; Bell,E.J. (US Dep. Health Educ., and Welfare, Public Health Serv., NIH, Natl. Inst. Allergy and Infect. Dis., Rocky Mountain Lab., Hamilton, MT 59840, USA) *J. Immunol., 121(5), 1961-1968 (1978)* En;en.

2078-U2 Early detection of *Rickettsia tsutsugamushi* in peripheral monocyte cultures derived from experimentally infected monkeys and dogs. Shirai,A.; Sankaran,V.; Gan,E.; Huxsoll,D.L. (US Army Med. Res. Unit, Inst. Med. Res., Kuala Lumpur, Malaysia) *The South East Asian Journal of Tropical Medicine and Public Health, 9(1), 11-14 (1978)* En;en.

2079-U2 A semi-automatic method for the detection of salmonellas in food products. Barrell,R.A.E.; Paton,A.M. (Public Health Lab., Withington Hosp., Manchester M20 8LR, UK) *J. Appl. Bacteriol., 46(1), 153-159 (1979)* En;en.

2080-U2 Synthetic disaccharide-protein antigens for production of specific 04 and 09 antisera for immunofluorescence diagnosis of *Salmonella*. Svenungsson,B.; *Lindberg,A.A. (Dep. Bacteriol., Natl. Bacteriol. Lab., S-105 21 Stockholm, Sweden) *Med. Microbiol. Immunol., 163(1), 1-11 (1977)* En;en.

2081-U2 [Immunofluorescence technique for the detection of *Salmonella* carriers]. Schioppa,F.; Boccia,A.; Montanaro,D. (Address not stated) *Ig. Mod., 72(12), 1324-1331 (1979)* It;en,it.

2082-U2 Evaluation of an automated fluorescent antibody procedure for detection of *Salmonella* in foods and feeds. Munson,T.E.; Schrade,J.P.; Bisciello,N.B.,Jr.; Fantasia,L.D.; Hartung,W.H.; O'Connor,J.J. (FDA, Brooklyn, NY 11232, USA) *Appl. Environ. Microbiol., 31(4), 514-521 (1976)* En;en.

2083-U2 Evaluation of a semiautomated sytem for direct fluorescent antibody detection of salmonellae. Thomason,B.M.; Hebert,G.A.; Cherry,W.B. (Cent. Dis. Control, Atlanta, GA 30333, USA) *Appl. Microbiol., 30(4), 557-564 (1975)* En;en.

2084-U2 Identification and quantitation of *Streptococcus mutans* by the fluorescent antibody technique. Jablon,J.M.; Ferrer,T.; Zinner,D.D. (Div. Oral Biol., Univ. Miami, Miami, FL 33152, USA) *J. Dent. Res., 55(special issue A), A78-A79 (1976)* En.

2085-U2 Rapid fluorescent-antibody method with Bromelase for identification of group A streptococci. Waters,C.A.; Makens,M.A. (Microbiol. Lab., St. Joseph's Hosp., St. Paul, MN 55102, USA) *J. Clin. Microbiol., 5(2), 255-256 (1977)* En;en.

2086-U2 **Identification of group A beta-hemolytic streptococci by the fluorescent antibody method.** Myers,J.B.; *Crum,J.W.; Beaulieu,L.P. (US Med. Component, AFRIMS, APO San Francisco, CA 96346, USA) *Mil. Med., 143(11), 872-874 (1978)* En;en.

2087-U2 **Detection of *Thiobacillus ferrooxidans* in acid mine environments by indirect fluorescent antibody staining.** Apel,W.A.; *Dugan,P.R.; Filppi,J.A.; Rheins,M.S. (Dep. Microbiol., Ohio State Univ., Columbus, OH 43210, USA) *Appl. Environ. Microbiol., 32(1), 159-165 (1976)* En;en.

2088-U2 **Study of fluorescent treponemal antibody test on cerebrospinal fluid using monospecific anti-immunoglobulin conjugates IgG, IgM, and IgA.** Leclerc,G.; Geroux,M.; Birry,A.; *Kasatiya,S. (20045 Chenin Ste-Marie Quest, Ste-Anne-de-Bellevue, Que., H9X 3LZ, Canada) *Br. J. Vener. Dis., 54(5), 303-308 (1978)* En;en.

2089-U2 **Double-staining procedure for the fluorescent treponemal antibody absorption (FTA-ABS) test.** Hunter,E.F.; McKinney,R.M.; Maddison,S.E.; Cruce,D.D. (Cent. Dis. Control, Atlanta, GA 30333, USA) *Br. J. Vener. Dis., 55(2), 105-108 (1979)* En;en.

2090-U2 **Immunofluorescent staining of *Treponema pallidum* and *Treponema pertenue* in tissues fixed by formalin and embedded in paraffin wax.** Al-Samarrai,H.T.; *Henderson,W.G. (Dep. Bacteriol., R. Postgrad. Med. Sch., Hammersmith Hosp., Du Cane Road, London12 0HS, UK) *Br. J. Vener. Dis., 53(1), 1-11 (1977)* En;en.

2091-U2 **Comparison of growth inhibition and immunofluorescence tests in serotyping clinical isolates of *Ureaplasma urealyticum*** Piot,P. (Lab. Bacteriol., Inst. Trop. Geneeskd., Nationalestraat 155, B-2000 Antwerp, Belgium) *Br. J. Vener. Dis., 53(3), 186-189 (1977)* En;en.

2092-U2 **Use of direct and indirect immunofluorescence tests for identification of *Xanthomonas campestris*.** Schaad,N.W. (Dep. Plant Pathol., Univ. Georgia, Georgia Experiment Stn., Experiment, GA 30212, USA) *Phytopathology, 68(2), 249-252 (1978)* En;en.

2093-U2 **[Rapid diagnosis of *Yersinia pestis* infection in the laboratory].** Karimi,Y. (Inst. Pasteur, Teheran, Iran) *Bull. Soc. Pathol. Exot., 71(1), 45-48 (1978)* Fr.

2094-U2 **Counterimmunoelectrophoresis for rapid identification of blood-culture isolates.** Waters,C.; Fossieck,B.,Jr.; *Parker,R.H. (Sect. Infect. Dis., Md. Serv., Veterans Adm. Hosp., 50 Irving St., NW, Washington, DC 20422, USA) *Am. J. Clin. Pathol., 71(3), 330-332 (1979)* En;en.

2095-U2 **Immunofluorescence and site of urinary tract infections.** Courteau,C.; *Montplaisir,S.; Martineau,B.; Pelletier,M. (Hop. Saint-Justine, Lab. Sero-Immunol., 3175 Cote Sainte-Catherine, Montreal, Que. H3T 1C5, Canada) *Union Med. Can., 109(3), 397-407 + 450-452 (1980)* Fr;en,fr.

2096-U2 **New immunofluorescent antibody test for diagnosis of candidiasis using macrophage-engulfed antigens as substrate.** Goudswaard,J.; Virella,G. (Dep. Immunol., Fac. Vet. Med., Utrecht, Netherlands) *J. Immunol. Methods., 20, 357-363 (1978)* En;en.

2097-U2 **An immunofluorescent technique for rapid control of the purity of yeast cultures.** Aries,V.C. (Nestle Prod. Tech. Assistance Co. Ltd., Microbiol. Sect., Ind. Lab., CH-1350, Orbe, Switzerland) *Eur. J. Appl. Microbiol., 2(2), 113-119 (1976)* En;en.

2098-U2 **An appraisal of tests for native DNA antibodies in connective tissue diseases. Clinical usefulness of *Crithidia luciliae* assay.** Chubick,A.; Sontheimer,R.D.; Gilliam,J.N. (Rheumatic Dis. Unit, Dep. Intern. Med., Univ. Texas Health Sci. Cent., Dallas, TX 75235, USA) *Ann. Intern. Med., 89(2), 186-192 (1978)* En;en.

2099-U2 **Immunofluorescence test in the diagnosis of mushroom worker's lung.** Noster,U.; Schulz,K.H.; Hausen,B.M. (Abt. Allergol. und Berufsdermatosen, Univ. Hautklin., 2000 Hamburg 20, Martinistr. 52, GFR) *Dtsch. Med. Wochenschr., 103(15), 655-657 (1978)* De;de,en.

2100-U2 **Rapid determination of the purity of yeast cultures by immunofluorescence and flow cytometry.** Hutter,K.J.; Punessen,U.; Eipel,H.E. (Fraunhofer-Gesellsch. Inst. Aerobiol., D-5948 Schmallenberg, GFR) *J. Inst. Brew., 85(1), 21-22 (1979)* En;en.

2101-U2 **The immunofluorescent staining technique for the detection of wild yeasts — practical problems.** Hammond,J.R.M.; Jones,M. (Res. Lab., Arthur Guinness Son and Co. (Park Royal) Ltd., Park Royal Brewery, London NW10 7RR, UK) *J. Inst. Brew., 85(1), 26-30 (1979)* En;en.

2102-U2 **The application and limitations of immunodiagnostic techniques in parasitic infections.** Thomas,V.; *Ogunba,E.O.; Fabiyi,A. (Dep. Med. Microbiol., Univ. Ibadan, Nigeria) *Afr. J. Med. Med. Sci., 7(2), 107-112 (1978)* En;en,fr.

2103-U2 **Comparison of the dried blood on filter paper and serum techniques for the diagnosis of bovine babesiosis utilizing the indirect fluorescent antibody (IFA) test.** Todorovic,R.; Garcia,R. (Inst. Trop. Vet. Med., College Station, TX 77843, USA) *Tropenmed. Parasitol., 29(1), 88-94 (1978)* En;de,en.

2104-U2 **A combined fluorescent antibody and soil sieving technique to count chlamydospores of *Phytophthora cinnamomi* in soil.** Malajczuk,N.; Bowen,G.D.; Greenhalgh,F.C. (Div. Land Resources Manage., CSIRO, Private Bag, P. O., Wembley, W. A., 6014, Australia) *Soil Biol. Biochem., 10(5), 437-438 (1978)* En.

2105-U2 **Rapid diagnosis of sporotrichosis by immunofluorescent method.** Chuang,T.-Y.; Deng,J.-S.; Young,L.K.; Lu,Y.-C. (Dep. Dermatol., Natl. Taiwan Univ. Hosp., Taipei, Taiwan) *Chin. J. Microbiol., 8(4), 259-261 (1975)* En;ch,en.

2106-U2 Kinetoplast of the flagellate *Crithidia luciliae*: a suitable substrate for the detection of antibodies to native, double-stranded DNA in the indirect immunofluorescence test. Stingl,G.; Knapp,W. (Inst. Immunol., Univ. Wien, Borschkegasse 8A, A-1090 Wien, Austria) *Wien. Klin. Wochenschr., 87(24), 827-829 (1975)* De;de,en.

2107-U2 Comparison of conventional and immunofluorescent techniques for the detection of *Entamoeba histolytica* in rectal biopsies. Gilman,R.; Islam,M.; Paschi,S.; Goleburn,J.; Ahmad,F. (Johns Hopkins Univ. ICMR, Baltimore City Hosp., 4940 Eastern Ave., Baltimore, MD 21224, USA) *Gastroenterology, 78(3), 435-439 (1980)* En;en.

2108-U2 A simply prepared amastigote leishmanial antigen for use in the indirect fluorescent antibody test for leishmaniasis. Shaw,J.J.; Lainson,R. (Inst. Evandro Chagas, 66000 Belem, Para, Brazil) *J. Parasitol., 63(2), 384-385 (1977)* En.

2109-U2 Immunofluorescence and counter immunoelectrophoresis in the diagnosis of kala-azar. Rezai,H.R.; Behforouz,N.; Amirhakimi,G.H.; Kohanteb,J. (Dep. Microbiol., Med. Sch., Pahlavi Univ., Shiraz, Iran) *Trans. R. Soc. Trop. Med. Hyg., 71(2), 149-151 (1977)* En;en.

2110-U2 Detection and measurement of species-specific malarial antibodies by immunofluorescence tests. Manawadu,B.R.; Voller,A. (Dep. Anesthesiol. (B 113), Univ. Colorado Sch. Med., 4200 East Ninth Ave., Denver, CO 80262, USA) *Trans. R. Soc. Trop. Med. Hyg., 72(5), 463-466 (1978)* En;en.

2111-U2 Titration of human serum antibodies to *Toxoplasma gondii* with a simple fluorometric assay. Walls,K.W.; Barnhart,E.R. (Parasitol. Div., Cent. Dis. Control, Public Health Serv., US Dep. Health, Educ. and Welfare, Atlanta, GA 30333, USA) *J. Clin. Microbial., 7(2), 234-235 (1978)* En;en.

2112-U2 A rapid method of IF-ABS with blood for serodiagnosis of urogenital trichomoniasis. Zilman,S.L. (Address not stated) *Vestn. Dermatol. Venerol., No.5, 26-28 (1978)* Ru;en,ru.

2113-U2 Antibodies to double-stranded DNA. A comparison of the indirect immunofluorescent test using *Crithidia luciliae* and the DNA-binding assay. Deegan,M.J.; Walker,S.E.; Lovell,S.E. (Dep. Pathol., Univ. Michigan, Ann Arbor, MI 48109, USA) *Am. J. Clin. Pathol., 69(6), 599-604 (1978)* En;en.

2114-U2 [Chagas' disease: an antigen of easy conservation for the direct immunofluorescence test]. Duarte de Storni,P. (Lab. Cent., Sist. Nac. Integrado Salud, Resistencia, Chaco, Argentina) *Medicina (B. Aires), 37(3), 317 (1977)* Es.

2115-U2 Differentiation by microimmunofluorescence of *T. cruzi* and *T. cruzi*-like strains from *T. conorhini* and *T. rangeli*, using protection from the sponge *Aaptos papillata*. Bretting,H.; Schottelius,J. (Zool. Inst. und Zool. Mus., Univ. Hamburg, Martin-Luther-King-Platz 3, D-2000 Hamburg 13, GFR) *Z. Parasitenkd., 57(3), 213-219 (1978)* De;de,en.

2116-U2 Comparison of three techniques for the detection of antibodies to double-stranded DNA: immunofluorescence on *Trypanosoma gambiense*, immunofluorescence on *Crithidia luciliae* and radioimmunoassay using the Farr technique. Monier,J.C.; Perraud,M.; Picard,F.; Giourd,M. (Lab. Med. Prev., Sante Publ. et Hyg., Univ. Lyon 1, Domaine Rockefeller, 69008 Lyon Cedex 2, France) *Ann. Immunol., 129C(4), 463-474 (1978)* En;en,fr.

2117-U2 [The value of indirect immunofluorescence in diagnosis of human trichinosis (with reference to two recent epidemics)]. Gentilini,M.; Richard-Lenoble,D.; Pinon,J.M.; Niel,G.; Danis,M.; Bouree,P. (Dep. Parasitol. Med. Trop., Groupe Hospitalier Pitie-Salpetriere, 83 blvd de l'Hopital, 75634 Paris, Cedex 13, France) *Nouv. Presse Med., 5(11), 720 (1976)* Fr.

2118-U2 A modification of the immunofluorescence test in vesicular dermatoses. Trofimova,L.Ya.; Antonova,T.N.; Dechko,B.D. (Minist. Public Health USSR, Cent. Dermatol. Inst., Moscow, USSR) *Vestn. Dermatol. Venerol., No. 10, 40-43 (1976)* Ru;en.

2119-U2 *Crithidium luciliae* immunofluorescence test (CL test): a simple and reliable method for detecting anti-native DNA antibodies. Jarzabeck-Chorzelska,M.; Chorzelski,T.; Rzesa,G.; Blaszczyk,M. (Dep. Dermatol., Med. Acad., 02-008 Warsaw, Poland) *Arch. Immunol. Ther. Exp., 26(1-6), 791-794 (1978)* En;en.

2120-U2 A simple fluorescent method to determine complement-mediated liposome immune lysis. Smolarsky,M.; Teltelbaum,D.; Sela,M.; Gitler,C. (Dep. Chem. Immunol., Weizmann Inst. Sci., Rehovot, Israel) *J. Immunol. Methods, 15(3), 255-265 (1977)* En;en.

2121-U2 A simple immunofluorescence test for the detection of platelet antibodies. von dem Borne,A.E.G.Kr.; Verheugt,F.W.A.; Oosterhof,F.; von Riesz,E.; Brutel de la Riviere,A.; Engelfriet,C.P. (Dep. Immunohaematol., Cent. Lab. Netherlands Red Cross Blood Transfus. Ser., POB 9190, Amsterdam, Netherlands) *Br. J. Haematol., 39(2), 195-207 (1978)* En;en.

2122-U2 Wheat grains: a substrate for the determination of gluten antibodies in serum of gluten-sensitive patients. Eterman,K.P.; Hekkens,W.T.J.M.; Pena,A.S.; Lems-Van Kan,P.H.; Feltkamp,T.E.W. (Dep. Autoimmune Dis., Cent. Lab. Netherlands Red Cross Blood Transfus. Serv., Lab. Exp. and Clin. Immunol., Univ. Amsterdam, Amsterdam, Netherlands) *J. Immunol. Mèthods, 14(1), 85-92 (1977)* En;en.

2123-U2 A simplified solid-phase immunofluorescence assay for measurement of serum immunoglobulins. Wang,R.; Merrill,B.; *Maggio,E.T. (Res. and Dev. Dep., Int. Diagnostic Technol., 2551 Walsh Ave., Santa Clara, CA 95050, USA) *Clin. Chim. Acta, 102(2-3), 169-177 (1980)* En;en.

2124-U2 The fluorescein-antifluorescein quenching system as a model for assaying changes in antibody function.

Blakeslee,D. (Div. Cancer Res., Dep. Pathol., Queen's Univ., Kingston, Ont., Canada) *J. Immunol. Methods, 12(1-2), 19-29 (1976)* En;en.

2125-U2 Method and value of immunofluorescence. Optical demonstration of IgM- IgG- and IgA rheumatoid factors. Schulze,A.; Geiler,G. (Pathol. Inst., Karl-Marx-Unv., Leipzig, DDR-701 Leipzig, Liebigstr. 26, GFR) *Z. Rheumatol., 38(1-2), 50-58 (1979)* De;de,en.

2126-U2 A simple immunofluorescence assay for C1q binding. Linder,E. (Clin. Immunol. Sect., Dep. Bacteriol., Univ. Helsinki, SF-00290 Helsinki 29, Finland) *J. Immunol. Methods, 32(3), 239-249 (1980)* En;en.

2127-U2 Abolition of non-specific fluorescent staining of labrocytes (mast cells) with protamine sulfate. Kalland,T.; Doskeland,S.O. (Inst. Anat., Univ. Bergen, Arstadvn, 19, N-5000 Bergen, Norway) *Cell. Mol. Biol., 25(1), 25-29 (1979)* En;en,fr.

2128-U2 The application of the Sepharose bead immunofluorescence test for the detection of allergen-specific IgE and IgG antiodies in pollinosis. Bergman,C.; *Crisci,C.D.; Jinnouchi,H.; Oehling,A. (Dep. Alergol., Clin. Univ. Navarra, Apto. 192, Pamplona, Spain) *Allergol. Immunopathol., 7(4), 283-288 (1979)* En;en,es.

2129-U2 Carcinogen-induced immunoreactivity to antinucleoside antibodies in HeLa cells. Bases,R.; Mendez,F.; Neubort,S.; Liebeskind,D.; Hsu,K.C. (David and Libbie Coleman Radiol. Res. Unit, Dep. Radiol., Albert Einstein Coll. Med., Yeshiva Univ., Bronx, NY 10461, USA) *Exp. Cell Res., 103(1), 175-181 (1976)* En;en.

2130-U2 Antifluorescein affinity columns. Isolation and immunocompetence of lymphocytes that bind fluoresceinated antigens in vivo or in vitro. Scott,D.W. (Div. Immunol., Dep. Microbiol. and Immunol., Duke Univ. Med. Cent., Durham, NC 27710, USA) *J. Exp. Med., 144(1), 69-78 (1976)* En;en.

2131-U2 The detection of granulocyte antibodies in relation to granulocyte transfusion. Verheugt,F.W.A.; Von Dem Borne,A.E.G.Kr.; Prins,H.K.; Engelfriet,C.P. (Dep. Blood Cell Sep., Cent. Lab. Netherlands Red Cross Blood Transfus. Serv., Amsterdam, Netherlands) *Exp. Hematol., 5(Suppl.1), 151-155 (1977)* En.

2132-U2 The detection of platelet isoantibodies by membrane immunofluorescence. van der Schans,G.S.; Veenhoven,W.A.; Snijder,J.A.M.; Nieweg,H.O. (59 Oostersingel, Groningen, Netherlands) *J. Lab. Clin. Med., 90(1), 4-10 (1977)* En;en.

2133-U2 A quantitative assay for low levels of IgM by solid-phase immunofluorescence. Sundeen,J.T.; Krakauer,R.S. (Dep. Immunol., Res. Div., Cleveland Clin. Found., Cleveland, OH 44106, USA) *J. Immunol. Methods, 26(3), 229-244 (1979)* En;en.

See: 829, 848, 849, 1037, 1044, 1858, 1948, 1950, 2258, 2645, 3035, 3243, 3245, 3267, 3273, 3276, 3279, 143, 1741, 1749, 1851, 2642, 264 3280, 3287, 3293, 3264, 3282, 3286, 3294

General diagnostic techniques

2134-U2 Rapid diagnosis in infectious diseases. Rytel,M.J. *Publ.by:* CRC Press, Inc.; 2000 NW 24th St., Boca Raton, FL 33431, USA 1979 250 pp. ISBN: 0-8493-5535-4 At $55.00 in USA, $64.00 outside USA.En.

2135-U2 Leukohematocrit index in the diagnosis of bovine leukosis. Grundboeck,M. (Vet. Inst., Pulawy, Poland) *Vet. Microbiol., 1(2-3), 383-385 (1976)* En;en.

2136-U2 Serologic diagnosis of dengue haemorrhagic fever using filter paper discs and one dengue antigen. Top,F.H.,Jr.; Gunakasem,P.; Chantrasri,C.; Supavadee,J. (Dep. Virus Dis., Walter Reed Army Inst. Res., Washington, DC 20012, USA) *South East Asian J. Trop. Med. Public Health, 6(1), 18-24 (1975)* En;en.

2137-U2 Diagnostic value of buffy coat preparation in dengue hemorrhagic fever. Suvatte,V.; Longsaman,M. (Dep. Pediatr., Fac. Med., Siriraj Hosp., Mahidol Univ., Bangkok, Thailand) *Southeast Asian J. Trop. Med. Public Health, 10(1), 7-12 (1979)* En;en.

2138-U2 [A micromethod adaptation of the Paul-Bunnel-Davidsohn technique]. Beirao Nogueira Catarino,M.M. (Cent. Virol. Appl., Fac. Pharm., Univ. Lisbonne, Lisbonne, Portugal) *Rev. Inst. Pasteur Lyon, 7(2), 189-198 (1974)* Fr;en,fr.

2139-U2 Rapid indexing procedures for detecting yellow speckle disease in grapevines. Mink,G.I.; Parsons,J.L. (Dep. Plant Pathol., Washington State Univ., Irrigated Agric. Res. and Extension Cent., Prosser, WA 99350, USA) *Plant Dis. Rep., 59(11), 869-872 (1975)* En;en.

2140-U2 Antibodies to single-stranded DNA: an aid in diagnosis of viral hepatitis. Villarejos,V.M.; Arquembourg,P.C.; Visona,K.A.; Gutierrez D.,A. (Louisiana State Univ., Int. Cent. Med. Res. and Training, San Jose, Costa Rica) *J. Med. Virol., 2(4), 359-367 (1978)* En;en.

2141-U2 Rapid diagnosis of primary influenza pneumonia Bogart,D.B.; *Liu,C.; Ruth,W.E.; Kerby,G.R.; Williams,C.H. (Univ. Kansas Med. Cent., Kansas City, KS 66103, USA) *Chest, 68(4), 513-517 (1975)* En,en.

2142-U2 Rapid serological diagnosis of outbreaks of influenza. Grist,N.R. (Univ. Dep. Infect. Dis., Ruchill Hosp., Glasgow, UK) *Br. Med. J., 1(6029), 581 (1976)* En.

2143-U2 [Rapid immunological diagnosis of viral infection]. Schmidt,J. (Inst. Med. Mikrobiol. Epidemiol., Ernst-Moritz-Arndt-Univ., DDR-22 Greifswald, Martin-Luther-Str. 6, GDR) *Allerg. Immunol. (Leipzig), 20/21(2), 195-202 (1974/5)* De;de.

2144-U2 Use of inclusions in the rapid diagnosis of virus diseases of red clover. Khan,M.A.; Maxwell,D.P. (Dep. Plant

Pathol., Univ. Wisconsin, Madison, WI 53706, USA) *Plant Dis. Rep., 61(8), 679-683 (1977)* En;en.

2145-U2 Rapid virus diagnosis. Gardner,P.S. (Dep. Virol., R. Victoria Infirm., Newcastle upon Tyne, NE1 4LP, UK) *J. Gen. Virol., 36(1), 1-29 (1977)* En.
[review].

2146-U2 Rapid diagnosis of viral infections. Al-Nakib,W. (Dep. Microbiol., Fac. Med., Univ.Kuwait City, Kuwait) *J. Kuwait Med. Assoc., 13(4), 221-224 (1979)* En.
[review].

2147-U2 A new substrate for the rapid evaluation of enteric microbial overgrowth. Wolgemuth,R.L.; Hanson,K.M.; Zassenhaus,P.H. (Rohm and Haas Res. Lab., Spring House, PA 19477, USA) *Am. J. Dig. Dis., 21(9), 821-826 (1976)* En;en.

2148-U2 [Tuberculin tests and the evaluation of cellular immunity. 2. The intradermal cutaneous tuberculin reaction as a means of evaluating cellular immunity. Optimum delay in reading. Stability or variations in the time]. Papillon,F.; Breau,J.L.; *Chretien,J. (Serv. Voies Respir., Cent. Hosp. Intercommunal, 40 ave. de Verdun, 94010 Creteil, France) *Rev. Fr. Mal. Respir., 4(2), 101-118 (1976)* Fr;en,fr.

2149-U2 Serologic testing for gonorrhea. Koransky,J.R.; Jacobs,N.F.,Jr. (Vener. Dis. Res. Branch, Cent. Dis. Control, 1600 Clifton Road, NE, Atlanta, GA 30333, USA) *Sex. Transmitted Dis., 4(1), 27-31 (1977)* En;en.

2150-U2 Bimodal behaviour studies of gonococci on peripheral cellular elements. Callahan,W.S. (Microbiol. Branch, Dep. Hum. Resourc., Div. Lab., 300 Indiana Ave. NW, Room 6001, Washington, DC 20001, USA) *J. Am. Med. Technol., 39(4), 199 (1977)* En.

2151-U2 Dysentery diagnosis with the aid of the carbon agglomeration test in an epidemic focus in elderly individuals. Evdokimova,T.V. (Leningrad Sanit.-Hyg. Med. Inst., Leningrad, USSR) *Zh. Mikrobiol. Epidemiol. Immunobiol., 53(5), 95-97 (1976)* Ru;en,ru.

2152-U2 The early serological detection of colonisation by *Staphylococcus epidermidis* of ventriculo-atrial shunts. Holt,R. (Queen Mary's Hosp. Child., Carshalton, Surrey, UK) *Infection, 8(1), 8-12 (1980)* En;de,en.

2153-U2 [Report of the application of ipronidazole in swine dysentery in combination with rapid on-the-spot diagnosis]. Stadler,E. (Fachtierarzt Schweine, Gartenstr. 6, 6908 Wiesloch, GFR) *Tierarztl. Umbchau, 31(7), 314-316 (1976)* De;de.

2154-U2 Evaluation of the Syphla Chek test for syphilis in a high-risk population. Nelson,M.; Portoni,E. (Dep. Family Med., Univ. California Coll. Med., Irvine, CA 92717, USA) *Am. J. Public Health, 67(10), 970-972 (1977)* En;en.

2155-U2 An integrated two-test automated syphilis screening system. MacFarlane,D.E. (Greater Glasgow Health Board, City Lab., 23 Montrose St., Glasgow G1 1RN, UK) *J. Clin. Pathol., 32(11), 1189-1191 (1979)* En.

2156-U2 Evaluation of the time necessary for performance of microtests for syphilis. Omarov,M.O.; Nemkaeva,R.M.; Zhodzishsky,I.A.; Vorobeichik,S.M. (Kazakh Dermatol. Inst., Alma Ata, USSR) *Vestn. Dermatol. Venerol., No. 1, 42-44 (1978)* Ru;en.

2157-U2 [Lytic activity of antiplague bacteriophages with respect to some strains of *Escherichia coli*]. Karimi,Y.; Alonso,J.M.; Mollaret,H.H. (Serv. Epidemiol., Inst. Pasteur l'Iran, Teheran, Iran) *Bull. Organ Mond Sante, 53(4), 480-481 (1976)* Fr;fr.

2158-U2 Comparative study of two systems for detecting bacteraemia and septicaemia. Wood,N.G. (Dep. Microbiol., Green Lane Hosp., Auckland, New Zealand) *J. Clin. Pathol., 29(6), 530-533 (1976)* En;en.

2159-U2 A novel, convenient, multiple test system for detection of bacteriuria. Strenkoski,L.F.; Freake,R.; Wiemeri,C.A. (Ames Company, Div. Miles Lab., Inc., 1127 Myrtle St., Elkhart, IN 46514, USA) *Am. J. Med. Technol., 41(10), 353-359 (1975)* En;en.

2160-U2 Evaluation of automatic mastitis detection equipment. Gebre-Egziabher,A.; Wood,H.C.; Robar,J.D.; Blankenagel,G. (Dep. Dairy and Food Sci., Univ. Saskatchewan, Saskatoon, Sask. S7N 2R6, Canada) *J. Dairy Sci., 62(7), 1108-1114 (1979)* En;en.

2161-U2 Early diagnosis of neonatal bacteremia by buffy-coat examination. Faden,H.S. (Div. Infect. Dis., Univ. Utah Med. Cent., 50 N. Medical Dr., Salt Lake City, UT 84132, USA) *J. Pediatr., 88(6), 1032-1034 (1976)* En;en.

2162-U2 An automated test for the detection of significant bacteriuria. Johnston,H.H.; Mitchell,C.J.; Curtis,G.D.W. (Bacteriol. Dep., Gibson Lab., Radcliffe Infirm., Oxford OX2 6HE, UK) *Lancet, 2(7982), 400-402 (1976)* En;en.

2163-U2 Simplified resazurin rennet test for diagnosis of mastitis. Kobayashi,Y. (Dep. Food Technol., Coll. Agric. and Vet. Med., Nihon Univ., Tokyo, Japan) *J. Dairy Sci., 61(5), 592-595 (1978)* En;en.

2164-U2 Rapid diagnosis of septic arthritis by quantitative analysis of joint fluid lactic acid with a Monotest Lactate Kit. Brook,I.; Contronì,G. (Infect. Dis. and Clin. Microbiol. Lab., Child. Hosp. Natl. Med. Cent., George Washington Univ., Washington, DC 20010, USA) *J. Clin. Microbiol., 8(6), 676-679 (1978)* En;en.

2165-U2 [Disc puncture for the bacteriological diagnosis of bacterial spondylodiscitis]. Seignon,B.; Weilbacher,H.; Thorel,J.B.; Bussieres,J.L.; Deshayes,P.; Gougeon,J. (Serv. Rhumatol., CHU, Hop. Sebastopol, 48, rue de Sebastopol, F 51100 Reims, France) *Rev. Rhum. Mal. Osteo-Articulaires, 47(1), 45-47 (1980)* Fr;fr.

2166-U2 The early diagnosis of gram negative septicemia in the pediatric surgical patient. /[presented at the Annual Meeting of the American Surgical Association, held in

Quebec City, Que., Canada, 7-9 May, 1975]. Rowe,M.I.; Buckner,D.M.; Newmark,S. (Div. Pediatr. Surg., Univ. Miami Sch. Med., PO Box 520875, Biscayne Annex, Miami, FL 33152, USA) *Ann. Surg., 182(3), 280-286 (1975)* En;en.

2167-U2 A cow-side test to detect mastitis. Milne,J.R. (Dairy Advisory Off. (Farms), Natl. Dairy Lab., Hamilton, New Zealand) *N.Z. J. Agric., 132(5), 52-55 (1976)* En.

2168-U2 Detection of urinary tract infections in 3- to 5-year-old girls by mothers using a nitrite indicator strip. Kunin,C.M.; DeGroot,J.E.; Uehling,D.; Ramgopal,V. (Veterans Adm. Hosp., 2500 Overlook Terrace, Madison, WI 53705, USA) *Pediatrics, 57(6), 829-835 (1976)* En;en.

2169-U2 ^{86}Rubidium release: a rapid and sensitive assay for amphotericin B. Drazin,R.E.; Lehrer,R.I. (Dep. Med., Cent. Health Sci., Univ. California, Los Angeles, CA 90024, USA) *J. Infect. Dis., 134(3), 238-244 (1976)* En,en.

2170-U2 Methods for detecting *Babesia microti* infection in wild rodents. Etkind,P.; Piesman,J.; Ruebush,T.K.,II; Spielman,A.; Juranek,D.D. (Dep. Trop. Public Health, Harvard Sch. Public Health, 665 Huntington Avenue, Boston, MA 02115, USA) *J. Parasitol., 66(1), 107-110 (1980)* En;en.

2171-U2 [Determination of immunglobulins in saliva; the value of this test in diagnosis of Gougerot-Sjogren syndrome]. Amor,B.; Mach,P.S.; Ghozlan,R.; Delbarre,F. (Inst. Rhumatol., Hop. Cochin, 27 rue du Faubourg Saint-Jacques, F 75674 Paris Cedex 14, France) *Rev. Rheum. Mal. Osteo-Articulaires, 44(7-9), 491-496 (1977)* Fr;fr.

2172-U2 [Prenatal diagnosis of foetal sex]. Faust,J. (Kinderklin. Med. Fak. Tech., Hochschule, 5100 Aachen, Goethestr. 27-29, GFR) *Deutsch. Med. Wochenschr., 101(43), 1576-1577 (1976)* De.

2173-U2 [Studies on the complement binding reaction in equine rhinopneumonitis]. Tatarov,G.; Martinov,S.; Panova,M. (Cent. Vet. Res. Inst., Sofia, Bulgaria) *Vet. Med. Nauk., 16(8), 78-84 (1979)* Bg;en,ru.

2174-U2 C.S.F. lactic acid for differential diagnosis of meningitis. Brook,I.; Rodriguez,W.J.; Controni,G.; Ross,S. (Child. Hosp. Natl. Med. Cent., Washington, DC 20010, USA) *Lancet, 1(8124), 1035 (1979)* En.

2175-U2 The detection of sub-clinical mastitis by electrical conductivity measurements. Davis,J.-G. (Address not stated) *Dairy Ind., 40(8), 286-291 (1975)* En.

2176-U2 A simple, quick technique for determining apple scab infection periods. MacHardy,W.E. (Dep. Bot. and Plant Pathol., Univ. New Hampshire, Durham, NH 03824, USA) *Plant Dis. Rep., 63(3), 199-204 (1979)* En;en.

See: 900, 1032, 1064, 1081, 1085, 1111, 1161, 1299, 1300, 1547, 1549, 1556, 1558, 1673, 1721, 1840, 1850, 1859, 2017, 2063, 2076

Radioisotope techniques

2177-U2 Validation of a simple radiochemical assay measuring hydrolysis of choline-labelled microsomal phosphatidylcholine by phospholipase C - pH dependence. Cater,B.R.; Trivedi,P.; Hallinan,T. (BIBRA, Woodmansterne Road, Carshalton, Surrey, UK) *Biochem. J., 160(3), 475-479 (1976)* En;en.

2178-U2 Automatic treatmental of radioimmunoassay data: an experiment validation of the results. Pilo,A.; Zucchelli,G.C. (CNR Clin. Physiol. Lab., Via P.Savi 8, 56100 Pisa, Italy) *Clin. Chim. Acta, 64(1), 1-9 (1975)* En;en.

2179-U2 Rapid quantitation of membrane antigens. Welsh,K.I.; Dorval,G.; Wigzell,H. (McIndoe Res. Unit, Queen Victoria Hosp., East Grinstead, Sussex, UK) *Nature, 254(5495), 67-69 (1975)* En.

2180-U2 A novel method for delineating antigenic determinants: peptide synthesis and radioimmunoassay using the same solid support. Smith,J.A.; Hurrell,J.G.R.; Leach,S.J. (Russell Grimwade Sch. Biochem., Univ. Melbourne, Parkville, Vic. 3052, Australia) *Immunochemistry, 14(8), 565-568 (1977)* En;en.

2181-U2 Mathematics of the inverse solid-phase radioimmunoassay. Bernath,S.; Zalotai,L. (State Inst. Control Vet. Serobacteriol. Prod., Szallas ut 8, H-1107, Budapest, Hungary) *Acta Microbiol. Acad. Sci. Hung., 25(3), 185-189 (1978)* En;en.

2182-U2 A method for a radioimmunoassay using microtiter plates allowing simultaneous determination of antibodies to two non cross-reactive antigens. Gruss,A.D.; Spier-Michl,I.B.; Gotschlich,E.C. (Dep. Immunol. and Microbiol., Rockefeller Univ., New York, NY 10021, USA) *Immunochemistry, 15(10-11), 777-780 (1978)* En;en.

2183-U2 A screening assay for rapid and quantitative measurement of immunoglobulins and anti-immunoglobulins in unknown sera. Werblin,T.; Mage,R. (Lab. Immunol., Natl. Inst. Allergy and Infect. Dis., Natl. Inst. Health, Bethesda, MD 20014, USA) *J. Immunol. Methods, 16(4), 337-350 (1977)* En;en.

2184-U2 ^{57}Co: a volume marker for the triple-isotope, double-antibody radioimmune assay. Egan,M.L.; Todd,C.W.; Knight,W.S. (Div. Immunol., City Hope Natl. Med. Cent., Duarte, CA 91010, USA) *Immunochemistry, 14(8), 611-613 (1977)* En;en.

2185-U2 A procedure for the preferential radioiodination of small amounts of protein in the presence of excess lipids. Montelaro,R.C.; Bolognesi,D.P. (Dep. Biochem., Louisiana State Univ., Baton Rouge, LA 70803, USA) *Anal. Biochem., 99(1), 92-96 (1979)* En;en.

2186-U2 Computer controlled automated radioimmunoassay. Bagshawe,K.D. (Dep. Med. Oncol., Charing Cross Hosp., Fulham Palace Rd., London W6 8RF,

UK) *Lab. Pract., 24(9), 573-575 (1975)* En;en.

2187-U2 A simple radiolabelled rheumatoid factor binding assay for the measurement of circulating immune complexes. Barratt,J.; Naish,P. (Dep. Nephrol., North Staffordshire R. Infirm., Princes Road, Stock-on-Trent ST4 7LN, UK) *J. Immunol. Methods, 25(2), 137-146 (1979)* En;en.

2188-U2 A competitive labeling method for the determination of the chemical properties of solitary functional groups in proteins. Duggleby,R.G.; *Kaptan,H. (Dep. Biochem., Univ. Ottawa, Ottawa, Ont., Canada) *Biochemistry, 14(23), 5168-5175 (1975)* En;en.

2189-U2 Application of ^{125}I radioimmunoassay to measure inhibition of precipitin reactions using carbohydrate-specific antibodies. Boullanger,P.H.; Nagpurkar,A.; Noujaim,A.A.; Lemieux,R.U. (Dep. Chem., Univ. Alberta, Edmonton, Alta. T6G 2G2, Canada) *Can. J. Biochem., 56(12), 1102-1108 (1978)* En;en,fr.

2190-U2 The 'Southmead system', a simple, fully-automated, continuous-flow system for immunoassays. (Appendix: application to serum thyroxine radioimmunoassays). Ismail,A.A.A.; West,P.M.; Goldie,D.J. (Dep. Clin. Chem., Southmead Gen. Hosp., Westbury-on-Trym, Bristol BS10 5NB, UK) *Clin. Chem., 24(4), 571-579 (1978)* En;en.

2191-U2 Enzyme multiplied immunoassay technique (EMIT). Bagrel,A. (Lab. Pr G, Siest, Cent. Med. Prev., 2, Ave. du Doven Jacques-Parisot, 54500 Vandoeuvre-les-Nancy, France) *Ann. Biol. Clin., 36(4), 383-388 (1978)* Fr;en,fr.

2192-U2 Rationalization of radioimmunologic analyses by employment of a flow-through gamma counter. Dwenger,A.; Friedel,R.; Schweer,H.H.; Trautschold,I. (Inst. Klin. Biochem., Med. Hochsch. Hannover, Hannover, GFR) *Acta Endocrinol., 82(Suppl. 202), 31-33 (1976)* En.

2193-U2 Radioimmunoassay dose interpolation based on the mass action law with antibody heterogeneity. Exact correction for variable mass of labeled ligand. Rodbard,D.; Tacey,R.L. (Endocrinol. and Reprod. Res. Branch. Natl. Inst. Child Health and Human Dev., Natl. Inst. Health, Bethesda, MD 20014, USA) *Anal. Biochem., 90(1), 13-21 (1978)* En;en.

2194-U2 A solid phase radioimmunoassay for pancreatic glucagon. Lundqvist,G.; Edwards,J.; Wide,L. (Dep. Clin. Chem., Univ. Hosp., S-750 14 Uppsala, Sweden) *Upsala J. Med. Sci., 81(2), 65-69 (1976)* En;en.

2195-U2 An introduction to radioimmunoassay. Self,M.; Rees,L.H.; Landon,J. (Dep. Chem. Pathol., St. Bartholomew's Hosp., London EC1, UK) *Med. Lab. Sci., 33(3), 221-228 (1976)* En;en.

2196-U2 [Radioimmunoassays, their value in clinical application]. Adam,W.; Pfannenstiel,P. (Stiftung Deutsch. Klin. Diagn., Fachber. Nuklearmed., Aukammallee 33, 6200 Wiesbaden, GFR) *Med. Welt, 27(37), 1709-1710 (1976)* De.

2197-U2 Radioimmunoassay. Eckert,H.G. (Hoechst AG, 6230 Frankfurt am Main 80, GFR) *Angew. Chem. (Engl. Ed.), 15(9), 525-533 (1976)* En;en.

2198-U2 Radioimmunoassay and bioassay data processing using a logistic curve fitting routine adapted to a desk top computer. Grotjan,H.E.,Jr.; Steinberger,E. (Dep. Reprod. Med. and Biol., Univ. Texas Med. Sch., PO Box 20708, Houston, TX 77025, USA) *Comput. Biol. Med., 7(2), 159-163 (1977)* En;en.

2199-U2 A modified technique for the radioallergosorbent test. Church,J.A.; Bellanti,J.A. (Dep. Pediatr., Georgetown Univ., Sch. Med., Washington, DC 20007, USA) *J. Immunol. Methods, 11(3,4), 385-387 (1976)* En;en.

2200-U2 A simplified approach for processing large numbers of double-antibody radioimmunoassay samples. Shorey,J. (Dep. Int. Med., Univ. Texas Southwestern Med. Sch., Dallas, TX 75235, USA) *J. Lab. Clin. Med., 86(6), 1052-1055 (1975)* En;en.

2201-U2 [Automated data processing of radioimmunoassay or protein competition assay results]. Zunzunegui Pastor,J.R.; Martinez Alonso,V.; Marin Rojas,M.C.; Moreno Gonzalez,J.; Millan Santos,I.; Ruiz Gimeno,M.C.; Alonso Duran,M.; Martinez Sales,V.; Ortiz Berrocal,J. (Serv. Bioestad. y Radioquim. Clin., Puerta de Hierro, Venezuela) *Radiologia, 18(3), 209-220 (1976)* Es;de,en,es,fr.

2202-U2 Quantitative immunoautoradiography at the cellular level. II. Absolute measurements using labeled standard cells as a source of reference. Thiel,E.; Dormer,P.; Ruppelt,P.; Thierfelder,S. (Abt. Immunol., Inst. Hamatol., Ges. Strahlen-und Umweltforsch., D-8 Munchen 2, GFR) *J. Immunol. Methods, 12(3/4), 237-251 (1976)* En;en.

2203-U2 Dried *Staphylococcus aureus* as a rapid immunological separating agent in radioimmunoassays. Ying,S.-Y.; Guillemin,R. (Lab. Neuroendocrinol., Salt Inst. Biol. Stud., La Jolla, CA 92037, USA) *J. Clin. Endocrinol. Metab., 48(2), 360-362 (1979)* En;en.

2204-U2 Methodological principles of the capillary radioimmunoassay. Friedel,R. (Inst. Klin. Biochem., Med. Hochsch., Hannover, GFR) *Acta. Endocrinol., 82(Suppl. 202), 33-35 (1976)* En.

2205-U2 Solid-phase, magnetic particle radioimmunoassay. Nye,L.; Forrest,G.C.; Greenwood,H.; Gardner,J.S.; Jay,R.; Roberts,J.R.; London,J. (Dep. Chem. Pathol., St. Bartholomew's Hosp., London EC1, UK) *Clin. Chim. Acta, 69(3), 387-396 (1976)* En;en.

2206-U2 A modified Clq deviation radioimmunoassay for use in a routine clinical laboratory. Danis,V.; Trent,R.; Clancy,R. (Immunol. Unit, Univ. Sydney, Sydney, N.S.W., Australia) *J. Immunol. Methods, 33(2), 175-182 (1980* En;en.

2207-U2 Observations on the automated calculation of radioimmunoassay results. Challand,G.S.; Spencer,C.A.; Radcliffe,J.G. (Dep. Clin. Biochem, Addenbrooke's Hosp.,

Hills Rd., Cambridge CB2 2QR, UK) *Ann. Clin. Biochem., 13(2), 354-360 (1976)* En;en.

2208-U2 Sensitive radioimmunoassays using partially purified gamma globulins coupled to enzacryl (acrylamide polymer) solid support. Kardana,A.; Squires,M.; Adam,T.; Goka,G.K.; Bagshawe,K.D. (Dep. Med. Oncol., Charing Cross Hosp. (Fulham), Fulham Palace Road, London W6 8RF, UK) *J. Immunol. Methods, 30(1), 47-53 (1979)* En;en.

2209-U2 ^{125}I-labeled anti-2,4-dinitrophenyl (DNP)-antibodies: a universal reagent for solid phase radioimmunoassays. Neurath,A.R.; Strick,N. (Lindsley F. Kimball Res. Inst., New York Blood Cent., New York, NY 10021, USA) *Mol. Immunol., 16(8), 625-628 (1979)* En;en.

2210-U2 [Radioimmunology and clinical biology. Value of radioimmunological assay]. Ferret-Bouin,P.; Croizet,M.; Belabas,M. (Serv. Explor. Fonct. par les Radio-isotopes, CHU Bordeaux, 146, rue Leo-Saignat, 33076 Bordeaux Cedex, France) *Bord. Med., 9(22), 1707-1714 (1976)* Fr.

2211-U2 A simple method for the interpolation of radioimmunoassay data. England,P.; Cain,O. (Res. Dep., R. Maternity Hosp., Rottenrow, Glasgow G4 0NA, UK) *Clin. Chim. Acta, 72(2), 241-244 (1976)* En;en.

2212-U2 Rapid radioimmunoassay for methotrexate in biological fluids. Hendel,J.; Sarek,L.J.; Hvidberg,E.F. (Dep. Clin. Chem., Finsen Inst., 49, Strandboulevarden, DK-2100 Copenhagen O, Denmark) *Clin. Chem., 22(6), 813-816 (1976)* En;en.

2213-U2 Gammaflow: a completely automated radioimmunoassay system. Brooker,G.; Terasaki,W.L.; Price,M.G. (Dep. Pharmacol., Univ. Virginia Sch. Med., Charlottesville, VA 22901, USA) *Science (Wash.), 194(4262), 270-276 (1976)* En.

2214-U2 Optimized radioimmunoassay curve fitting program suitable for use on desktop calculators. van der Sluijs Veer,G.; Kennedy,J.C. (Dep. Clin. Chem., Onze Lieve Vrouwe Gasthuis, le Oosterparkstraat 179, Amsterdam, Netherlands) *Clin. Chim. Acta, 88(1), 27-30 (1978)* En;en.

2215-U2 Radioimmunoassay: a probe for the fine structure of biologic systems. Yalow,R.S. (Veterans Admin. Hosp., Bronx, NY 10468, USA) *Science (Wash.), 200(4347), 1236-1245 (1978)* En.

2216-U2 A semi-automated radioimmunoassay technique employing a nine-channel pipetting and sample handling. Lahteenmaki,P. (Steroid Res. Lab., Dep. Med. Chem., Univ. Helsinki, SF-00170 Helsinki 17, Finland) *Scand. J. Clin. Lab. Invest., 38(4), 393-396 (1978)* En;en.

2217-U2 Automatic pipetting system for radioimmunoassay. Paton,J.S.; Shaw,A.; Giles,C.A. (West of Scotland Health Boards, Dep. CClin. Phys. and Bio-Eng., 11 West Graham St., Glasgow, UK) *Med. Lab. Technol., 32(4), 295-300 (1975)* En;en.

2218-U2 Increasing the precision and speed of the separation step in radioimmunoassays. Merrett,T.G.; Pantin,C.F.A. (RAST Allergy Unit, Benenden Chest Hosp., Beneden, Kent, UK) *Clin. Chim. Acta, 65(1), 131-134 (1975)* En;en.

2219-U2 Enzymatic radioidination of succinyl cyclic AMP tyrosine methyl ester by lactoperoxidase and radioimmunoassay for cyclic AMP. Sato,K.; Miyachi,Y.; Mizuchi,A.; Ohsawa,N.; Kosaka,K. (Third Dep. Intern. Med., Fac. Med., Univ. Tokyo, Bunkyo-ku, Tokyo 113, Japan) *Endocrinol. Jap., 23(3), 251-257 (1976)* En;en.

2220-U2 End-point parameter adjustment on a small desk-top programmable calculator for logit-log analysis of radioimmunoassay data. Hatch,K.F.; Coles,E.; Busey,H.; Goldman,S.C. (Clin. Lab. Dep., Picker Corp., Northford, CT 06472, USA) *Clin. Chem., 22(8), 1383-1389 (1976)* En;en.

2221-U2 Triiodothyronine immunoassay using polyethylene glycol to precipitate antibody-bound hormone. Miyamoto,Y.; Seino,Y.; Matsukura,S.; Imura,H. (III Dep. Intern. Med., Kobe Univ. Sch. Med., 13-14, 7-chome, Kusunoki-cho, Ikuta-ku, Kobe, 650, Japan) *Jap. J. Med., 17(1), 15-21 (1978)* En;en.

2222-U2 Performance of various mathematical methods for computer-aided processing of radioimmunoassay results. Vogt,W.; Sandel,P.; Langfelder,Ch.; Knedel,M. (Inst. Klin. Chem., Klin. Grosshadern, Univ. Munchen, Munchen, GFR) *Clin. Chim. Acta, 87(1), 101-111 (1978)* En;en.

2223-U2 Analysis of radioimmunoassay data: polynomial regression. Shastri,S.; Bright,M.; Wagman,E. (Bergen Pines County Hosp., Paramus, NJ 07652, USA) *Clin. Chem., 24(7), 1290-1291 (1978)* En.

2224-U2 A membrane permeability test for the detection of cell surface antigens. Kurth,R.; Medley,G. (Dep. Tumour Virol., Imperial Cancer Res. Fund Lab., PO Box 123, Lincoln's Inn Fields, London WC2A 3PX, UK) *Immunology, 29(5), 803-812 (1975)* En;en.

2225-U2 Characterization of tumour virus proteins. I. Radioimmunoassay of the P27 protein of avian viruses. Higuchi,T.; August,J.T. (Inst. Quimica, Univ. Sao Paulo, Brasil) *Rev. Bras. Pesquisas Med. Biol., 10(1), 1-14 (1977)* En;en,pt.

2226-U2 Radioimmunoassay for Epstein-Barr virus (EBV)-associated nuclear antigen (EBNA). Binding of iodinated antibodies to antigen immobilized in polyacrylamide gel. Dolken,G.; Klein,G. (Med. Univ.-klin., Hugstetter Str., 55, D-78 Freiburg, GFR) *Eur. J. Cancer, 13(11), 1277-1286 (1977)* En;en.

2227-U2 Detection of hepatitis A viral antigen by radioimmunoassay. Hollinger,F.B.; Bradley,D.W.; Maynard,J.E.; Dreesman,G.R.; Melnick,J.L. (Dep. Virol. Epidemiol., Baylor Coll. Med., Houston, TX 77025, USA) *J. Immunol., 115(5), 1464-1466 (1975)* En;en.

2228-U2 Serodiagnosis of viral hepatitis A: detection of acute-phase immunoglobulin M anti-hepatitis A virus by radioimmunoassay. Bradley,D.W.; Maynard,J.; Hindman,S.H.; Hornbeck,C.L.; Fields,H.A.; McCaustland,K.A.; Cook,E.H.,Jr. (Phoenix Lab. Div., Cent. for Dis. Control, Phoenix, AZ 85014, USA) *J. Clin. Microbiol., 5(5), 521-530 (1977)* En;en.

2229-U2 [Critical study of various radioimmunological techniques for the detection of hepatitis B surface antigen.] Drouet,J. (Cent. Natl. Transfus. Sanguine, Ave. du Canada, 91403 Orsay-Courtabeuf, France) *Rev. Fr. Transfus., 18(2), 217-225 (1975)* Fr;en,fr.

2230-U2 Hepatitis B antibody detection in solid-phase radioimmunoassay. Artaz,C.; Voyat,O.; Rosso-Chioso,M. (Reg. Blood Tranfus. Serv., Maternita, I-11100 Aosta, Italy) *Vox Sang., 31(3), 222-224 (1976)* En;en.

2231-U2 Experiences with solid-phase 'radioimmunoassay' in the rapid diagnosis of hepatitis B. Cameron,C.H. (Dep. Virol., Middlesex Hosp. Med. Sch., London W1, UK) *J. Med. Virol., 3(1), 27-30 (1978)* En.

2232-U2 Rapid detection of hepatitis B surface antigen by double antibody radioimmunoassay. Burrell,C.J.; Leadbetter,G.; Black,S.H.; Hunter,W.M. (Hepatitis Ref. Lab., Dep. Bacteriol., Edinburgh Univ. Med. Sch., Teviot Place, Edinburgh, UK) *J. Med. Virol., 3(1), 19-26 (1978)* En;en.

2233-U2 Detection of antibody to hepatitis Bs-antigen in patients with acute and chronic hepatitis as measured by a modified procedure of the radioimmunoassay Ausria I 125. Muller,R.; Stephan,B.; Kramer,R.; Deicher,H. (Abt. Gastroenterol. und Hepatol., Dep. Inn. Med., Med. Hochsch. Hannover, Karl-Wiechert-Allee 9, D-3000 Hannover, GFR) *Vox Sang., 29(5), 330-337 (1975)* En;en.

2234-U2 Subtyping of hepatitis B surface antigen and antibody by radioimmunoassay. Hoofnagle,J.H.; *Gerety,R.J.; Smallwood,L.A.; Barker,L.F. (Div. Blood and Blood Products, Bur. Biol., 8800 Rockville Pike, Bethesda, MD 20014, USA) *Gastroenterology, 72(2), 290-296 (1977)* En;en.

2235-U2 [Solid-phase radioimmunoassay for the detection of Australia antigen]. Mesnier,F.; Blanquet,P.; *Moulinier,J.; Collignon,G.; Piquet,Y. (Cent. Transfus. Sang. Bordeaux, Place Amelie-Rabat-Leon, 33-Bordeaux, France) *Bord. Med., 8(6), 657-660 (1975)* Fr;de,en,es,fr,it.

2236-U2 Detection of circulating immune complexes in hepatitis by means of a new method employing ^{125}I-antibody: circulating immune complexes in hepatitis. Fresco,G.F. (Ist. Sci. Med. Intern., Viale Benedetto XV, no. 6, I-16132 Genoa, Italy) *Acta Hepato-Gastroenterol., 25(3), 185-188 (1978)* En;de,en.

2237-U2 New technique for solid-phase immunoassay: application to hepatitis B surface antigen. Park,H. (Dep. Pathol., Coll. Med. and Dent. New Jersey, New Jersey Med. Sch., 100 Bergen St. Newark, NJ 07103, USA) *Clin. Chem., 25(1), 178-182 (1979)* En;en.

2238-U2 Application of microtiter solid-phase radioimmunoassay to the determination of hepatitis B e antigen. Miyakawa,Y.; Tsuda,F.; Akahane,Y.; Mayumi,M. (Third Dep. Intern. Med., Fac. Med., Univ. Tokyo, Hongo, Tokyo 113, Japan) *Vox Sang., 36(5), 307-311 (1979)* En;en.

2239-U2 Radioimmunoassay detection of HBsAg in hemoderivatives. Lostia,O.; Costantini,A.; Jori,D.; Lorenzini,P. (Ist. Superiore di Sanita, Lab. Tecnol. Biomed., Viale Regina Elena, 299, 00161 Roma, Italy) *Boll. Ist. Sieroter. Milan, 58(1), 5-24 (1979)* En;en,it.

2240-U2 Radioimmunoassay of hepatitis B core antigen and antibody with autologous reagents. Vyas,G.N.; Roberts,I.M. (Dep. Lab. Med., Univ. California, Sch. Med., San Francisco, CA 94134, USA) *Vox Sang., 33(6), 369-372 (1977)* En;en.

2241-U2 Simple radiometric techniques for rapid detection of herpes simplex virus type 1 in WI-38 cell culture. D'Antonio,N.; *Tsan,M.-F.; Charache,P.; Larson,S.; Wagner,H.N.,Jr. (Johns Hopkins Med. Inst., 615 N. Wolfe St., Baltimore, MD 21205, USA) *J. Nucl. Med., 17(6), 503-507 (1976)* En;en.

2242-U2 Simultaneous radioimmunoassay for specific antibodies to members of the human herpesvirus group. Gehle,W.D.; Smith,K.O.; Fuccillo,D.A.; Perry,A.; Andrese,A.P. (Litton Bionetics, Inc., Kensington, MD 20795, USA) *J. Immunol. Methods, 26(4), 381-389 (1979)* En;en.

2243-U2 Radioimmunoassay for herpes simplex virus. Enlander,D.; Dos Remedios,L.V.; Weber,P.M.; Drew,L. (Dep. Lab. Med., Univ. California Sch. Med., 1001 Portero Ave. and Twenty-Second St., San Francisco, CA, USA) *J. Immunol. Methods, 10(4), 357-362 (1976)* En;en.

2244-U2 Use of *Staphylococcus aureus* for rapid radioimmunoassay of influenza A virus haemagglutinin. Polakova,K.; Simkovicova,M.; Chorvath,B.; Russ,G.; Styk,B.; Sourek,J. (Cancer Res. Inst., Slovak Acad. Sci., 88032 Bratislava, Czechoslovakia) *Acta Virol., 23(2), 107-112 (1979)* En;en.

2245-U2 Solid-phase radioimmunoassay as a method for evaluating antigenic differences in type A influenza viruses. Schieble,J.H.; Cottam,D. (c/o Edwin H. Lennette, Viral and Rickettsial Dis. Lab., State of California Dep. Health, 2151 Berkeley Way, Berkeley, CA 94704, USA) *Infect. Immun., 15(1), 66-71 (1977)* En;en.

2246-U2 Electrophoretic analysis of iodine-labeled influenza virus RNA segments. Bean,W.J.,Jr.; Sriram,G.; Webster,R.G. (Div. Virol. St. Jude Child. Res. Hosp., 332 N. Lauderdale, PO Box 318, Memphis, TN 38101, USA) *Anal. Biochem., 102(1), 228-232 (1980)* En;en.

2247-U2 The use of indirect solid phase radioimmunoassay for investigation of influenza virus antigenic properties. Trushinskaya,G.N.; Yamnikova,S.S.; Slepushkin,A.N.; Fedorova,G.I.; Grigorieva,T.A.; Berezina,O.N. (DI Ivanovsky Inst. Virol., Acad. Med. Sci. USSR, Moscow, USSR) *Vopr. Virusol., No. 4, 403-408 (1980)* Ru;en.

2248-U2 The rapid determination of binding constants for antiviral antibodies by a radioimmunoassay. An analysis of the interaction between hybridoma proteins and influenza virus. Frankel,M.E.; Gerhard,W. (Wistar Inst. Anat. and Biol., 36th Street at Spruce, Philadelphia, PA 19104, USA) *Mol. Immunol., 16(2), 101-106 (1979)* En;en.

2249-U2 Radioimmunoassay for LCM virus antigens and anti-LCM virus antibodies and its application in an epidemiologic survey of people exposed to Syrian hamsters. Blechschmidt,M.; Gerlich,W.; Thomssen,R. (Inst. Hyg., Georg-August-Univ. Gottingen, Dep. Med. Microbiol., Kreuzbergring 57, D-3400 Gottingen, GFR) *Med. Microbiol. Immunol., 163(1), 67-76 (1977)* En;en.

2250-U2 Radioimmunoassay of murine leukemia virus p30 using *Staphylococcus aureus* as immunoadsorbent. Brown,J.P.; Klitzman,J.M.; Hellstrom,I. (Div. Tumor Immunol., Fred Hutchinson Cancer Res. Cent., Seattle, WA 98104, USA) *J. Immunol. Methods, 21(1-2), 23-33 (1978)* En;en.

2251-U2 Simplified radioimmunoassay for viral antigens: use of *Staphylococcus aureus* as an adsorbent for antigen-antibody complexes. Premkumar-Reddy,E.; Devare,S.G.; Vasudev,R.; Sarna,P.S. (Microbiol. Assoc., 5221 River Rd., Bethesda, MD 20016, USA) *J. Natl. Cancer Inst., 58(6), 1859-1861 (1977)* En;en.

2252-U2 In vitro radiolabeling procedure which labels the proteins of Newcastle disease virions with carbon-14. McMillen,J.; Consigli,R.A. (Div. Biol., Kansas State Univ., Manhattan, KS 66506, USA) *Infect. Immun., 12(6), 1411-1414 (1975)* En;en.

2253-U2 Solid-phase microtiter radioimmunoassay for detection of the Norwalk strain of acute nonbacterial, epidemic gastroenteritis virus and its antibodies. Greenberg,H.B.; Wyatt,R.G.; Valdesuso,J.; Kalica,A.R.; London,W.T.; Chanok,R.M.; Kapikian,A.Z. (Natl. Inst. Health, Build. 7, Room 105, Bethesda, MD 20014, USA) *J. Med. Virol., 2(2), 97-108 (1978)* En;en.

2254-U2 A solid-phase radioimmunoassay for polyhedron protein from baculoviruses. Crawford,A.M.; Kalmakoff,J.; Longworth,J.F. (Dep. Mirobiol., Univ. Otago, PO Box 56, Dunedin, New Zealand) *Intervirology, 8(2), 117-121 (1977)* En;en.

2255-U2 Assay of mouse-cell clones for retrovirus p30 protein by use of an automated solid-state radioimmunoassay. Kennel,S.J.; Tennant,R.W. (Biol. Div., Oak Ridge Natl. Lab., Oak Ridge, TN 37830, USA) *Virology, 97(2), 464-467 (1979)* En;en.

2256-U2 Studies on the intracellular reverse transcriptase of RNA tumor viruses. Panet,A. (Dep. Virol., Hebrew Univ.-Hadassah Med. Sch., Jerusalem, Israel) *Isr. J. Med. Sci., 13(7), 731-739 (1977)* En.

2257-U2 Indirect solid-phase microradioimmunoassay for detection of pseudorabies virus antibody in swine sera. Kelling,C.L.; Staudinger,W.L.; Rhodes,M.B. (Dep. Vet. Sci., Inst. Agric. and Nat. Resources, Univ. Nebraska, Lincoln, NE 68583, USA) *Am. J. Vet. Res., 39(12), 1955-1957 (1978)* En;en.

2258-U2 Comparison of rabies humoral antibody titers in rabbits and humans by indirect radioimmunoassay, rapid-fluorescent-focus-inhibition technique, and indirect fluorescent-antibody assay. Lee,T.-K.; *Hutchinson,H.D.; Ziegler,D.W. (Cent. for Dis. Control, Atlanta, GA 30333, USA) *J. Clin. Microbiol., 5(3), 320-325 (1977)* En;en.

2259-U2 A microtiter solid phase radioimmunoassay for detection of the human reovirus-like agent in stools. Kalica,A.R.; Purcell,R.H.; Sereno,M.M.; Wyatt,R.G.; Kim,H.W.; Chanock,R.M.; Kapikian,A.Z. (Lab. Infect. Dis., Natl. Inst. Allergy and Infect. Dis. and Child. Hosp. Natl. Med. Cent. District of Columbia, USA) *J. Immunol., 118(4), 1275-1279 (1977)* En;en.

2260-U2 Simplified radioimmunoassay for detection of human rotavirus in stools. Cukor,G.; Berry,M.K.; Blacklow,N.R. (Div. Infect. Dis., Univ. Massachusetts Med. Sch., 55 Lake Ave. North, Worcester, MA 01605, USA) *J. Infect. Dis., 138(6), 906-910 (1978)* En;en.

2261-U2 Ultrasensitive enzymatic radioimmunoassay: application to detection of cholera toxin and rotavirus. Harris,C.C.; Yolken,R.H.; Krokan,H.; Hsu,I.C. (Human Tissue Stud. Sect., Lab. Exp. Pathol., Natl. Cancer Inst., Natl. Inst. Health, Bethesda, MD 20205, USA) *Proc. Natl. Acad. Sci. USA, 76(10), 5336-5339 (1979)* En;en.

2262-U2 Solid-phase radioimmunoassay of rubella virus immunoglobulin M antibodies: comparison with sucrose density gradient centrifugation test. Meurman,O.H.; Viljanen,M.K.; Granfors,K. (Dep. Virol., Univ. Turku, SF-20520 Turku 52, Finland) *J. Clin. Microbiol., 5(3), 257-262 (1977)* En;en.

2263-U2 A solid phase micro-radioimmunoassay to detect minute amounts of Ig class specific antiviral antibody in a mouse model system. Charlton,D.; *Blandford,G. (Gage Res. Inst., 223 College St., Toronto, Ont., Canada) *J. Immunol. Methods, 8(4), 319-330 (1975)* En;en.

2264-U2 Solid phase indirect radioimmunoassays for rapid diagnosis of Sindbis virus antigen. Enzmann,P.J. (Fed. Res. Inst. Anim. Virus Dis., PO Box 1149, D-7400 Tubingen, GFR) *J. Gen. Virol., 41(3), 641-644 (1978)* En;en.

2265-U2 Solid-phase radioimmunoassay for antibodies to flavivirus structural and nonstructural proteins. Trent,D.W.; Harvey,C.L.; Qureshi,A.; LeStourgeon,D. (Cent. for Dis. Control, Fort Collins, CO 80522, USA) *Infect. Immun., 13(5), 1325-1333 (1976)* En;en.

2266-U2 Detection of SV40 T antigen with labelled antibodies: radioimmunoassay and autoradiography. Flugel,R.M.; Crefeld,T.; Munk,K. (Inst. Virus. Res., German Cancer Res. Cent., Im Neuenheimer Feld 280, 6900 Heidelberg, GFR) *Int. J. Cancer, 19(5), 656-663 (1977)* En;en,fr.

2267-U2 Detection of antibodies to varicella-zoster virus by radioimmunoassay and enzyme immunoassay techniques. Friedman,M.G.; Haikin,H.; Leventon Kriss,S.; Joffe,R.; Goldstein,V.; Sarov,I. (Virol. Unit, Fac. Health Sci. and Soroka Med. Cent., Ben Gurion Univ. Negev, Beer Sheva, Israel) *Med. Microbiol. Immunol., 166(1-4), 177-186 (1978)* En;en.

2268-U2 Solid-phase radioimmunoassay for rapid detection and identification of western equine encephalomyelitis virus. Levitt,N.H.; Miller,H.V.; Eddy,G.A. (US Army Med. Res. Inst. Infect. Dis., Frederick, MD 21701, USA) *J. Clin. Microbiol., 4(4), 382-383 (1976)* En;en.

2269-U2 A sensitive radioimmunoasorbent assay for the detection of plant viruses. Ghabrial,S.A.; Shepherd,R.J. (Plant Pathol. Dep., Univ. Kentucky, Lexington, KY 40546, USA) *J. Gen. Virol., 48(2), 311-317 (1980)* En;en.

2270-U2 Nucleotide sequences in DNA. Sanger,F. (MRC Lab. Mol. Biol., Cambridge, UK) *Proc. R. Soc. Lond., B, 191(1104), 317-333 (1975)* En;en.

2271-U2 Rapid radiorespirometric method for the estimation of bacterial counts of *Cellulomonas* cultures. Jakubick-Neradova,V. (Inst. Strahlentechnol., Bundesforschungsanstalt Ernahrung, Karlsruhe, GFR) *Chem. Mikrobiol. Technol. Lebensm., 4(4), 117-119 (1975)* De;de,en,fr.

2272-U2 A simple radioassay for dihydroorotate dehydrogenase. Smithers,G.W.; Gero,A.M.; O'Sullivan,W.J. (Sch. Biochem., Univ. New South Wales, Kensington, NSW 2033, Australia) *Anal. Biochem., 88(1), 93-103 (1978)* En;en.

2273-U2 High specific activity labelling of typhoid antigen. Patel,M.C.; Ganatra,R.D.; Dhabhuwala,J.B.; Mukherji,S. (Radiat. Med. Cent., Bio-med. group, Tata Mem. Annexe, Jerbai Wadia Road, Parel, Bombay-400012, India) *Int. J. Appl. Radiat. Isotopes, 29(3), 185-186 (1978)* En.

2274-U2 Solid-phase radioimmunoassays for quantitative antibody determination of *Clostridium perfringens* type D ε toxin. Bernath,S. (State Inst. Control. Vet. Serobacteriol. Proc., 1107 Budapest, Szallas u.8., Hungary) *Appl. Microbiol., 30(4), 499-502 (1975)* En;en.

2275-U2 Detection and localization of Epstein-Barr virus-associated early antigens in single cells by autoradiography using ^{125}I-labeled antibodies. Moar,M.H.; Siegert,W.; Klein,G. (Dep. Tumor Biol., Karolinska Inst., S-104 01 Stockholm 60, Sweden) *Intervirology, 8(4), 226-239 (1977)* En;en.

2276-U2 Microtiter solid-phase radioimmunoassay for detection of *Escherichia coli* heat-labile enterotoxin. Greenberg,H.B.; Sack,D.A.; Rodriguez,W.; Sack,R.B.; Wyatt,R.G.; Kalica,A.R.; Horswood,R.L.; Chanock,R.M.; Kapikian,A.Z. (Lab. Infect. Dis., Natl. Inst. Allergy Infect. Dis., NIH, Bethesda, MD 20014, USA) *Infect. Immun., 17(3), 541-545 (1977)* En;en.

2277-U2 Use of glutaraldehyde-fixed *Escherichia coli* cells as a convenient solid support for radioimmune assays. Whei-Yang Kao,W.; Guzman,N.A.; Prockop,D.J. (Dep. Ophthalmol., Sch. Med., Eye and Ear Hosp., Pittsburgh, PA 15213, USA) *Anal. Biochem., 81(1), 209-219 (1977)* En;en.

2278-U2 Inhibition of thymidine phosphorylase in vivo provides a rapid method for switching DNA labeling. Womack,J.E. (Dep. Biochem., Texas A and M Univ., College Station, TX 77843, USA) *Mol. Gen. Genet., 158(1), 11-15 (1977)* En;en.

2279-U2 Radioimmunological screening method for specific membrane proteins. Henning,U.; Schwarz,H.; Chen,R. (Max-Planck-Inst. Biol., D-7400 Tubingen, GFR) *Anal. Biochem., 97(1), 153-157 (1979)* En;en.

2280-U2 Convenient method for detecting $^{14}CO_2$ in multiple samples: application to rapid screening for mutants. Tabor,H.; White Tabor,C.; Hafner,E.W. (Lab. Pharmacol., Natl. Inst. Arthritis, Metab., and Dig. Dis., Bethesda, MD 20014, USA) *J. Bacteriol., 128(1), 485-486 (1976)* En;en.

2281-U2 Measurement of endotoxin. I. Fundamental studies on radioimmunoassay of endotoxin. Kimura,H. (Dep. Surg., Okayama Univ. Med. Sch., Okayama 700, Japan) *Acta Med. Okayama, 30(4), 245-255 (1976)* En;en.

2282-U2 A radioisotope incorporation method for studying the transport and utilization of peptides by *Escherichia coli*. Payne,J.W.; Bell,G. (Dep. Bot., Univ. Durham, Sci. Lab., South Road, Durham, DH1 3LE, UK) *FEMS Microbiol. Lett., 1(2), 91-94 (1977)* En.

2283-U2 The stoichiometry of polypeptide chains in the pyruvate dehydrogenase multienzyme complex of *E. coli* determined by a simple novel method. Bates,D.L.; Harrison,R.A.; Perham,R.N. (Dep. Biochem., Univ. Cambridge, Tennis Court Road, Cambridge CB2 1QW, UK) *FEBS Lett., 60(2), 427-430 (1975)* En.

2284-U2 A simple radioassay for dihydrofolate synthetase activity in *Escherichia coli* and its application to an inhibition study of new pteroate analogs. Ho,R.I.; Corman,L.; Ho,J.; Nair,M.G. (Samuel M. Best Res. Lab., Massachusettes Coll. Pharm., Boston, MA 02115, USA) *Anal. Biochem., 73(2), 493-500 (1976)* En;en.

2285-U2 Radiometric detection of *Haemophilus influenzae* in cerebrospinal fluid specimens. Cox,F.R.; Martin,J.R.; *Cox,M.E.; Stoer,U.H.; Silberman,R. (Dep. Pathol., Schumpert Med. Cent., 915 Margaret Place, Shreveport, LA 71101, USA) *Scand. J. Infect. Dis., 10(1), 93-94 (1978)* En;en.

2286-U2 Radioimmunoassay of capsular polysaccharide antigens of groups A and C meningococci and *Haemophilus influenzae* type b in cerebrospinal fluid. Kayhty,H.; Makela,P.H.; Ruoslahti,E. (Cent. Public Health Lab., Helsinki, Finland) *J. Clin. Pathol., 30(9), 831-833 (1977)* En;en.

2287-U2 Radioimmunoassay for antibodies against *Mycobacterium paratuberculosis* using ^{125}I-labelled PPD. Worsaae,H. (Dep. Forensic and State Vet. Med., R. Vet. and Agric. Univ., Bulowsvej 13, DK-1870 Copenhagen, Denmark) *Acta Vet. Scand., 19(1), 153-155 (1978)* En.

2288-U2 Automatable radiometric detection of growth of *Mycobacterium tuberculosis* in selective media. Reggiardo,Z.; Tigertt,W.D. (G. Middlebrook, Dep. Pathol., 660 W. Redwood St., Baltimore, MD 21201, USA) *Am. Rev. Respir. Dis., 115(6), 1066-1069 (1977)* En;en.

2289-U2 A rapid radiometric method for determining the sensitivity of clinical isolates of *Mycobacterium tuberculosis* to several chemotherapeutic agents. McDaniel,R.E.; Abensohn,M.K.; Spoon,D.R.; Kobayashi,G.S.; MeofMcDaniel,R.E.; Abensohn,M.K.; Spoon,D.R.; Kobayashi,G.S.; Medoff,G.; *Marr,J.J. (Sect. Infect. Dis., Dep. Med., St. Louis Univ. Sch. Med., 1325 S. Grand, St. Louis, MO 63104, USA) *J. Lab. Clin. Med., 89(4), 861-867 (1977)* En;en.

2290-U2 Autoradiographic and metabolic studies of *Mycobacterium leprae*. Khanolkar,S.R.; Ambrose,E.J.; Chulawala,R.G.; Bapat,C.V. (Found. Med. Res., Worli, Bombay, India) *Lepr. Rev., 49(3), 187-198 (1978)* En;en.

2291-U2 Rapid radiometric susceptibility testing of *Mycobacterium tuberculosis*. Kertcher,J.A.; Chen,M.F.; Charache,P.; Hwangbo,C.C.; Camargo,E.E.; McIntyre,P.A.; *Wagner,H.N.,Jr. (Div. Nucl. Med., Johns Hopkins Med. Inst., 615 N. Wolf St., Baltimore, MD 21205, USA) *Am. Rev. Respir. Dis., 117(4, Part 1), 631-637 (1978)* En;en.

2292-U2 Formate oxidation by mycobacteria using a rapid radiometric assay. Deyhle,R.R.; Barton,L.L. (Dep. Biol., Univ. New Mexico, Albuquerque, NM 87131, USA) *Microbios Lett., 2(7-8), 225-229 (1976)* En;en.

2293-U2 A solid phase radioimmunoassay on hydrophobic membrane filters: detection of antibodies to gonococcal antigens. Lambden,P.R.; Watt,P.J. (Dep. Microbiol., Southampton Univ. Med. Sch., South Lab. and Pathol. Block, Southampton Gen. Hosp., Tremona Road, Southampton SO9 4XY, UK) *J. Immunol. Methods, 20, 277-286 (1978)* En;en.

2294-U2 In vitro interaction of *Neisseria gonorrhoeae* type 1 and type 4 with tissue culture cells. Brodeur,B.R.; Johnson,W.M.; Johnson,K.G.; Diena,B.B. (Lab. Cent. Dis. Control, Health and Welfare Canada, Ottawa, Ont. K1A 0L2, Canada) *Infect. Immun., 15(2), 560-567 (1977)* En;en.

2295-U2 Comparison of a radiometric procedure with conventional methods for identification of *Neisseria*. Strauss,R.R.; Holderbach,J.; Friedman,H. (Dep. Microbiol., Albert Einstein Med. Cent., Philadelphia, PA 19141, USA) *J. Clin. Microbiol., 7(5), 419-422 (1978)* En;en.

2296-U2 [Radio-immunological measurement of antimeningococcal antibodies: technique and results of vaccination]. Fayet,M.T.; Triau,R.; Stellmann,C.; Mynard,M.C.; Ayme,G. (Inst. Merieux, 69260 Marcy-l'Etoile, France) *Bull. Soc. Med. Afr. Noire, Langue Fr., 23(3), 299-306 (1978)* En;it.

2297-U2 As assay method for determining extracellular lipases from *Pseudomonas aeruginosa*. Keyser,P.D.; Lacko,A.G.; Christensen,J.N. (Texas Coll. Osteopathic Med., North Texas State Univ., Denton, TX 76203, USA) *J. Am. Osteopath. Assoc., 77(7), 551 (1978)* En.

2298-U2 A radioimmunoprecipitation test with '*Salmonella typhi*'. Mukerji,S.; Dhabuwala,J.B.; Patel,M.G.; Ganatra,R.D.; Sant,M.V. (Haffkine Inst., Bombay, India) *Bull. Haffkine Inst., 6(3), 84-87 (1978)* En;en.

2299-U2 Double antibody solid-phase radioimmunoassay for staphylococcal enterotoxin A. Lindroth,S.; Niskanen,A. (Tech. Res. Cent. Finland, Food Res. Lab., Biologinkuja 1, SF-02150 Espoo 15, Finland) *Eur. J. Appl. Microbiol., 4(2), 137-143 (1977)* En;en.

2300-U2 Detection of staphylococcal enterotoxin B by affinity radioimmunoassay. Niyomvit,N.; Stevenson,K.E.; McFeeters,R.F. (Kasetsart Univ., POB 4-170, Bangkok-4, Thailand) *J. Food Sci., 43(3), 735-739 (1978)* En;en.

2301-U2 Modified radioimmunoassay determination for staphylococcal enterotoxin B in foods. Pober,Z.; *Silverman,G.J. (Food Sci. Lab., US Army Natick Res. and Dev. Command, Natick, MA 01760, USA) *Appl. Environ. Microbiol., 33(3), 620-625 (1977)* En;en.

2302-U2 Double antibody solid-phase radioimmunoassay for staphylococcal enterotoxin A. Lindroth,S.; Niskanen,A. (Tech. Res. Cent. Finland, Food Res. Lab., Biologinkaju 1, SF-02150 Espoo 15, Finland) *Eur. J. Appl. Microbiol., 4(2), 137-143 (1977)* En;en.

2303-U2 Solid-phase radioimmunoassay for detection of staphylococcal antigen in serum of rabbits with endocarditis due to *Staphylococcus aureus*. Wheat,L.J.; Kohler,R.B.; White,A. (Wishard Mem. Hosp., OP 317, 1001 West 10th St., Indianapolis, IN 46202, USA) *J. Infect. Dis., 38(2), 174-180 (1978)* En;en.

2304-U2 Rapid solid-phase radioassay for staphylococcal protein A. Dickie,N.; Konowalchuk,J. (Bur. Microb. Hazards, Food Directorate, Health Prot. Branch, Health and Welfare Canada,Ottawa, Ont. K1A 0L2, Canada) *Can. J. Microbiol., 23(6), 832-834 (1977)* En;en,fr.

2305-U2 An indirect immunoradiometric assay of microbes. Strange,R.E.; Hambleton,P. (Microbiol. Res. Establ., Porton Down, Salisbury SP4 0J4, Wilts., UK) *J. Gen. Microbiol., 93(2), 401-404 (1976)* En.

2306-U2 The rate of microbial methane production determined by the radioisotope technique. Laurinavichus,K.S.; Beljaev,S.S. (Inst. Biochem. and Physiol. Microorg., Acad. Sci. USSR, Moscow, USSR) *Mikrobiologiya, 47(6), 1115-1117 (1978)* Ru;en,ru.

2307-U2 A simple test for flexirubin-type pigments. Fautz,E.; Reichenbach,H. (Gesellschaft Biotechnol. Forschung, Abt. mikrobiol., Mascheroder Weg 1, D-3300

Braunschweig-Stockheim, GFR) *FEMS Microbiol. Lett., 8(2), 87-91 (1980)* En.

2308-U2 Radiometric detection of bacteremia: requirement for terminal subcultures. Strauss,R.R.;Throm,R.; Friedman,H. (Dep. Microbiol., Albert Einstein Med. Cent., Philadelphia, PA 19141, USA) *J. Clin. Microbiol., 5(2), 145-148 (1979)* En;en.

2309-U2 Automatic quantitative radiometric assay of bacterial metabolism. Buddemeyer,E.; Hutchinson,R.; Cooper,M. (Univ. Maryland Hosp., 22 S. Greene St., Baltimore, MD 21201, USA) *Clin. Chem., 22(9), 1459-1464 (1976)* En;en.

2310-U2 A radiochemical assay for glycolytic activity in dental plaque. Williams,R.A.D.; Nuki,K.; Nanda,V.; Schlenker,R. (London Hosp. Med. Coll., Turner St., London E1A 2AD, UK) *J. Periodontal Res., 11(5), 256-261 (1976)* En;en.

2311-U2 Radioimmunoassay procedure for quantitating bacterial antibody in human sera. Sanford,B.A.; Smith,K.O. (Dep. Microbiol., Univ. Texas Health Sci. Cent. at San Antonio, San Antonio, TX 78284, USA) *J. Immunol. Methods, 14(3-4), 313-323 (1977)* En;en.

2312-U2 Determination of antibacterial antibodies in serum by immunoradiometric assay. Hambleton,P.; Strange,R.E. (Microbiol. Res. Estab., Porton Down, Salisbury SP4 0JG, UK) *J. Med. Microbiol., 10(2), 151-159 (1977)* En;en.

2313-U2 A rapid radiometric technique used to trap and quantitate $^{14}CO_2$ evolved by slow-growing microorganisms. Moser,S.A.; Pollack,J.D. (Dep. Microbiol. and Immunol, Tulane Univ. Sch. Med., New Orleans, LA 70112, USA) *Can. J. Microbiol., 24(5), 625-628 (1978)* En;en,fr.

2314-U2 ^{14}C-labeled aflatoxin B_1 prepared with yeastlike cultures of *Aspergillus parasiticus*. Jackson,L.K.; Ciegler,A. (Northern Reg. Res. Lab., ARS, USDA, Peoria, IL 61604, USA) *J. Environ. Sci. Health, 11(4), 317-329 (1976)* En;en.

2315-U2 Determination of antibodies against *Candida albicans* by radio-immuno-assay. Szollosy,E.; Heszler,E.; Palmai,M. (Lungenkrankenhaus Komitats Csongrad, Alkotmany u. 36, 6772 Deszk, Hungary) *Mykosen, 22(1), 11-14 (1979)* De;de,en.

2316-U2 A ^{51}chromium release assay for phagocytic killing of *Candida albicans*. Yamamura,M.; Boler,J.; Valdimarsson,H. (Dep. Immunol., St.ary's Hosp., Med. Sch., London W2 1PG, UK) *J. Immunol. Methods, 13(3, 4), 227-233 (1976)* En;en.

2317-U2 Solid-phase competitive-binding radioimmunoassay for detecting antibody to the M antigen of histoplasmin. Reiss,E.; Hutchinson,H.; Pine,L.; Ziegler,D.W.; Kaufman,L. (Mycol. Div., Cent. Dis. Control, Atlanta, GA 30333, USA) *J. Clin. Microbiol., 6(6), 598-604 (1977)* En;en.

2318-U2 An automated radiometric microassay of fungal growth: quantitation of growth of *T. mentagrophytes*. Qualman,S.J.; *Jones,H.E.; Artis,W.M. (Div. Dermatol., Emory Sch. Med., Room 215, Woodruff Mem. Build., Atlanta, GA 30322, USA) *Sabouraudia, 14(3), 287-297 (1976)* En;de,en.

2319-U2 Radioimmunological differentiation of yeasts. Tripatzis,I. (Inst. Mikrobiol., Med. Hochsch., Karl Wiechart Allee 9, D-3000 Hannover 61, GFR) *Zentralbl. Bakteriol., Parasitenkd., Infektionskr. Hyg., I, 241(4), 525-535 (1978)* De;de,en.

2320-U2 Radiometric method for determining the susceptibility of yeasts to 5-fluorocytosine. Hopfer,R.L.; Mills,K.; Groschel,D. (Dep. Lab. Med., Univ. Texas System Cancer Cent., M.D. Anderson Hosp. and Tumor Inst., Houston, TX 77030, USA) *Antimicrob. Agents Chemother., 15(2), 313-314 (1979)* En;en.

2321-U2 Evaluation of an automatic radiometric system for the detection of fungi in pharmaceutical preparations. Udagawa,S.; Sakabe,F.; Narita,N.; Kurata,H. (Address not stated) *Bull. Natl. Inst. Hyg. Sci., Tokyo, 53(96), 78-83 (1978)* Ja;en.

2322-U2 A stable isotope tracer method to measure silicic acid uptake by marine phytoplankton. Nelson,D.M.; Goering,J.J. (Dep. Biol., Woods Hole Oceanographic Inst., Woods Hole, MA 02543, USA) *Anal. Biochem., 78(1), 139-147 (1977)* En;en.

2323-U2 Recovery of ^{14}C-CO_2 at high flow rates from a CHN analyzer. Taft,J.L. (Chesapeake Bay Inst., Johns Hopkins Univ., Baltimore, MD 21218, USA) *Limnol. Oceanogr., 21(1), 161-164 (1976)* En;en.

2324-U2 Identification of *Leishmania* spp. by radiorespirometry. Decker,J.E.; Schrot,J.R.; Levin,G.V. (Dep. Zool., Morrill Sci. Cent., Univ. Massachusetts, Amherst, MA 01003, USA) *J. Protozool., 24(3), 463-470 (1977)* En;en.

2325-U2 *Trypanosoma brucei*: new radioisotope assay for quantitating cell lysis. Rifkin,M.R. (Rockefeller Univ., New York, NY 10021, USA) *Exp. Parasitol., 46(2), 207-212 (1978)* En;en.

2326-U2 The growth of *Trypanosoma cruzi* in human diploid cells for the production of trypomastigotes. Sanderson,C.J.; Thomas,J.A.; Twomey,C.E.; (Transplantation Biol., Clin. Res. Cent., Watford Road, Harrow, Middx, HA1 3UJ, UK) *Parasitology, 80(1), 153-162 (1980)* En;en.

2327-U2 Rapid heterologous hapten radioimmunoassay for insect moulting hormone. Maroy,P.; Vargha,J.; Horvath,K. (Inst. Genet., Biol. Res. Cent., Hungarian Acad. Sci., H 6701 Szeged, Hungary) *FEBS Lett., 81(2), 319-322 (1977)* En.

2328-U2 A simple and sensitive radioimmunoassay of insect juvenile hormone using an iodinated tracer. Baehr,J.C.; Pradelles,P.; Lebreux,C.; Cassier,P.; *Dray,F. (INSERM, Unit. Radioimmunol. Analytique, Inst. Pasteur, 28 rue du Dr. Roux, F-75724 Paris Cedex 15, France) *FEBS Lett., 69(1),*

123-128 (1976) En.

2329-U2 Rat spleen leucocyte (RSL) radioimmunoassay for the detection and quantification of soluble immune complexes. Al-Khateeb,S.F.; Barkas,T. (Univ. Immunol. Lab., 2 Forrest Road, Edinburgh EH1 2QW, UK) *Clin. Exp. Immunol., 34(3), 429-435 (1978)* En;en.

2330-U2 Radioimmunoassay for urinary albumin. Woo,J.; Floyd,M.; Cannon,D.C.; Kahan,B. (Dep. Pathol. and Lab. Med., Univ. Texas Med. Sch., Houston, TX 77025, USA) *Clin. Chem., 24(9), 1464-1467 (1978)* En;en.

2331-U2 A solid phase radioimmunoassay for detection of serum autoantibodies in systemic lupus erythematosus. Smith,K.O.; Harrington,J.T.; Gehle,W.D. (Dep. Microbiol., Univ. Texas Health Sci. Cent., San Antonio, TX 78284, USA) *J. Immunol. Methods, 15(1), 17-28 (1977)* En;en.

2332-U2 Salivary phenytoin radioimmunoassay. A simple method for the assessment of non-protein bound drug concentrations. Paxton,J.W.; Rowell,F.J.; Ratcliffe,J.G.; Lambie,D.G.; Nanda,R.; Melville,I.D.; Johnson,R.H. (Dep. Materia Med. and Chem., Univ. Glasgow and Radioimmunoassay Unit, Stobhill Hosp., Glasgow, UK) *Eur. J. Clin. Pharmacol., 11(1), 71-74 (1976)* En;en.

2333-U2 A simple radio-immunoassay for the measurement of human and rat IgE levels by ammonium sulfate precipitation. Carson,D.; *Metzger,H.; Bazin,H. (Arthritis and Rheum. Branch, Natl. Inst. Arthritis, Metab. and Dig. Dis., Natl. Inst. Health, Bethesda, MD 20014, USA) *J. Immunol., 115(2), 561-563 (1975)* En;en.

2334-U2 The direct radioimmunoassay technique for the determination of human α_1-fetoprotein. Breborowicz,J.; Majewski,P. (Dep. Pathol. Anat., Inst. Biostructure, Med. Acad., 60-356 Poznan, Poland) *Arch. Immunol. Ther. Exp., 25(3), 303-307 (1977)* En;en.

2335-U2 Simple radioimmunoassay of cortisol in diluted samples of human plasma. Connolly,T.M.; Vecsei,P. (Dep. Pharmacol., Univ. Heidelberg, 6900 Heidelberg, GFR) *Clin. Chem., 24(9), 1468-1472 (1978)* En;en.

2336-U2 A micro immunoassay method for measuring IgG antibodies using staphylococcal protein-A. Moran,D.M.; Dupe,B.E.; Gauntlett,S. (Beecham Pharmaceuticals, Res. Div., Brockham Park, Betchworth, Surrey, UK) *J. Immunol. Methods, 24(1-2), 183-191 (1978)* En;en.

2337-U2 A solid-phase radioassay for the detection of anti-thyroglobulin antibodies in different immunoglobulin classes. Nineham,L.J.; Hay,F.C. (Dep. Immunol., Middlesex Hosp., Med. Sch., Arthur Stanley House, 40-50 Tottenham St., London W1, UK) *Int. Arch. Allergy Appl. Immunol., 51(1), 94-100 (1976)* En;en.

2338-U2 A rapid specific radioimmunoassay for unconjugated estriol in plasma. Brown,T.K.; Brammall,M.A.; Schiller,H.S. (Dep. Obstet. and Gynecol., Univ. Washington, Seattle, WA 98195, USA) *Steroids, 27(4), 459-468 (1976)* En;en.

2339-U2 A simple and rapid radioimmunoassay of triiodothyronine in unextracted serum. Premachandra,B.N. (Veterans Adm. Hosp., Jefferson Barracks and Washington Univ., St. Louis, MO 63125, USA) *J. Nucl. Med., 17(5), 411-416 (1976)* En;en.

2340-U2 Rapid radioimmunoassay of total urinary estriol. Anderson,D.W.; Goebelsmann,U. (Dep. Pathol., Univ. Southern California Sch. Med., 1240 N. Mission Rd., Los Angeles, CA 90033, USA) *Clin. Chem., 22(5), 611-615 (1976)* En;en.

2341-U2 Urine aldosterone radioimmunoassay: validation of a method without chromatography. Brown,R.D.; Swander,A.; McKenzie,J.K. (Mayo Clin., Rochester, MN 55901, USA) *J. Clin. Endocrinol. Metab., 42(5), 894-900 (1976)* En;en.

2342-U2 Introduction of a rapid, simple radioimmunoassay and quality control scheme for thyroxine. Nye,L.; Hassan,M.; Willmott,E.; Landon,J. (Dep. Chem. Pathol., St. Bartholomew's Hosp., London EC1, UK) *J. Clin. Pathol., 29(5), 452-457 (1976)* En;en.

2343-U2 Semi-automatic solid-phase radioimmunoassay for carcinoembryonic antigen. Wang,R.; Sevier,E.D.; Reisfeld,R.A.; *David,G.S. (Scripps Clin. and Res. Found., 10666 N. Torrey Pines Rd., La Jolla, CA 92037, USA) *J. Immunol. Methods, 18(1-2), 157-164 (1977)* En;en.

2344-U2 Polystyrene balls as the solid-phase of a double-antibody radioimmunoassay for human serum albumin. Ziola,B.R.; Matikainen,M.-T.; Salmi,A. (Neurovirol. Study Group, Dep. Virol., Univ. Turku, Kiinamyllnkatu 10, 20520 Turku 52, Finland) *J. Immunol. Methods, 17(3-4), 309-317 (1977)* En;en.

2345-U2 Structure and biological functions of human IgD. XIV. The development of a solid-phase radioimmunoassay for the quantitation of IgD in human sera and secretions. Leslie,G.A.; Teramura,G. (Dep. Microbiol. and Immunol., Univ. Oregon Health Sci. Cent., Portland, OR 97201, USA) *Int. Arch. Allergy Appl. Immunol., 54(5), 451-456 (1977)* En;en.

2346-U2 A simple method for the measurement of low levels of serum IgE. Gamo,T.; Maruyama,G. (Dep. Pediatr., Osaka Police Hosp., 8 Komiya-cho, Tennouji-ku, Osaka 543, Japan) *Clin. Allergy, 7(6), 597-604 (1977)* En;en.

2347-U2 Radioimmunoassay of the anti-hypertensive agent debrisoquin. Dixon,R.; Fahrenholtz,K.E.; Burger,W.; Perry,C. (Res. Div., Hoffman-La Roche Inc., Nutley, NJ 07110, USA) *Res. Commun. Chem. Pathol. Pharmacol., 16(1), 121-129 (1977)* En;en.

2348-U2 Use of highly specific antibodies against 17α-OH-progesterone in a simplified nonchromatographic RIA and in the simultaneous determination of four sex hormones in human plasma. Forest,M.G. (Unite Rech. Endocr. et Metab. chez l'Enfant, Inserm U 34, 29, rue Soeur-Bouvier, F-69322 Lyon Cedex 1, France) *Horm. Res., 7(4-5), 260-273 (1976)* En;en.

2349-U2 Radioimmunoassay of 13,14-dihydro-15-keto-prostaglandin F_{2a}. Youssefnejadian,E.; Brodovcky,H.; Johnson,M.; Craft,I. (Dep. Biochem. Endocrinol., Chelsea Hosp. Women, London SW3, UK) *Prostaglandins, 15(2), 239-253 (1978)* En;en.

2350-U2 A new assay for rapid measurement of MIF levels by ^{3}H-labelled cells in liquid scintillation counting vials: statistical implications for the measurement of migration inhibition. McDaniel,M.C.; Robbins,C.H.; Hokanson,J.A.; Papermaster,B.W. (Basic Sci. Res. Prog., Frederick Cancer Res. Cent., Frederick, MD, USA) *J. Immunol. Methods, 20, 225-239 (1978)* En;en.

2351-U2 Solid-phase radioimmunoassay for morphine, with use of an affinity-purified morphine antibody. Steiner,M.; Spratt,J.L. (Dep. Pharmacol., Univ. Iowa, Iowa City, IA 52242, USA) *Clin. Chem., 24(2), 339-342 (1978)* En;en.

2352-U2 Microfilter paper method for 17α-hydroxyprogesterone radioimmunoassay: its application for rapid screening for congenital adrenal hyperplasia. Pang,S.; Hotchkiss,J.; Drash,A.L.; Levine,L.S.; New,M.I. (Div. Pediatr. Endocrinol., New York Hosp.-Cornell Med. Cent., 525 East 68th St., New York, NY 10021, USA) *J. Clin. Endocrinol. Metab., 45(5), 1003-1008 (1977)* En;en.

2353-U2 Plasma and urinary aldosterone measurement in healthy subjects with a radioimmunoassay kit not requiring chromatography. Demers,L.M.; Sampson,E.; Hayes,A.H.,Jr. (Div. Clin. Pathol., Milton S. Hershey Med. Cent., Pennsylvania State Univ., Hershey, PA 17033, USA) *Clin. Biochem., 9(5), 243-246 (1976)* En;en.

2354-U2 Radioimmunoassay of antihaemophilic factor (factor VIII) antigen. Paulssen,M.M.P.; van de Graaf-Wildschut,M.; Kolhorn,A.; Planje,M.C. (Dep. Clin. Biochem., St. Clara Hosp., Rotterdam, Netherlands) *Clin. Chim. Acta, 63(3), 349-353 (1975)* En;en.

2355-U2 Solid-phase radioimmunoassay for thyroxine in untreated serum. Seth,J.; Rutherford,F.J.; McKenzie,I. (Dep. Clin. Chem., R. Infirmary, Edinburgh, EH3 9YW, UK) *Clin. Chem., 21(10), 1406-1413 (1975)* En;en.

2356-U2 A radioimmunoassay for serum rat thyroglobulin. Physiologic and pharmacological studies. van Herle,A.J.; Klandorf,H.; Uller,R.P. (Dep. Med., Div. Endocrinol., Cent. Health Sci., Univ. California, Los Angeles, CA 90024, USA) *J. Clin. Invest., 56(5), 1073-1081 (1975)* En;en.

2357-U2 The use of amberlite XAD-2 columns for the determination of plasma aldosterone by radioimmunoassay. Roginsky,M.S.; Panetz,A.I.; Gordon,R.D. (Dep. Med., Nassau County Med. Cent., East Meadow, NY 11554, USA) *Clin. Chim. Acta, 63(3), 303-308 (1975)* En;en.

2358-U2 Purification and iodination of antibody for use in an immunoradiometric assay for serum ferritin. Gonyea,L.M. (Dep. Lab. Med. and Pathol., Univ. Minnesota Hosp., Minneapolis, MN 55455, USA) *Clin. Chem., 23(2, Part 1), 234-236 (1977)* En;en.

2359-U2 New, advantageous approach to the direct radioimmunoassay of cortisol. Tilden,R.L. (Clin. Lab. Serv., Veterans Adm. Hosp., Gainesville, FL 32602, USA) *Clin. Chem., 23(2, Part 1), 211-215 (1977)* En;en.

2360-U2 Radioimmunoassay of 17-hydroxyprogesterone in serum with or without thin-layer chromatographic purification. Wickings,E.J. (Med. Klin. und Poliklin., Univ. Dusseldorf, Moorenstr. 5, D-4 Dusseldorf 1, GFR) *Hormone Res., 6(2), 78-84 (1975)* En;en.

2361-U2 Solid-phase immunoradiometric assay of factor-VIII protein. Counts,R.B. (Div. Hematol., 325-9th Ave., Seattle, WA 98104, USA) *Br. J. Haematol., 31(4), 429-436 (1975)* En;en.

2362-U2 Radioimmunoassay of the anticonvulsant agent clonazepam. Dixon,W.R.; Young,R.L.; Ning,R.; Liebman,A. (Dep. Biochem. and Drug Metab., Hoffmann-La Roche Inc., Nutley, NJ 07110, USA) *J. Pharm. Sci., 66(2), 235-237 (1977)* En;en.

2363-U2 Simplified radioimmunoassay for aldosterone using antisera to aldosterone-γ-lactone. Antunes,J.R.; Dale,S.L.; Melby,J.C. (Sect. Endocrinol. and Metab., University Hosp., Boston Univ. Sch. Med., Boston Univ. Med. Cent., Boston, MA 02118, USA) *Steroids, 28(5), 621-630 (1976)* En;en.

2364-U2 Purification and radioimmunoassay of human alpha-1-fetoprotein: the effect of aggregates on the radioimmunoassay. Young,J.; Reid,R.G.; Crawford,J.W. (Dep. Obstet. and Gynaecol., Univ. Dundee, Ninewells Hosp., Dundee, UK) *Clin. Chim. Acta, 69(1), 11-20 (1976)* En;en.

2365-U2 A microassay for the detection of tumour-specific complement-dependent serum cytotoxicity against a chemically induced rat hepatoma. Price,M.R. (Cancer Res. Campaign Lab., Univ. Nottingham, University Park, Nottingham, NG7 2RD, UK) *Transplantation, 25(4), 224-226 (1978)* En.

2366-U2 An isotope-release assay and a terminal-labeling assay for measuring cell-mediated allograft and tumor immunity to small numbers of adherent target cells. Tamerius,J.D.; Garrigues,H.J.; Hellstrom,K.E. (Div. Mol. Immunol., Scripps Clin. and Res. Found., La Jolla, CA 92037, USA) *J. Immunol. Methods, 22(1-2), 1-22 (1978)* En;en.

2367-U2 Radioimmunoassay of 15α-hydroxyestriol (estetrol) in pregnancy serum. Kundu,N.; Grant,M. (Roswell Park Mem. Inst., Sch. Med. and Dent., State Univ. New York at Buffalo, Buffalo, NY 14214, USA) *Steroids, 27(6), 785-796 (1976)* En;en.

2368-U2 Radioimmunoassay of carcinoembryonic antigen (CEA) without extraction and dialysis using solid-phase antibody. Kollmann,G.; Brennan,J. (Dep. Nuclear Med., Div. Radiol., Albert Einstein Med. Cent., Philadelphia, PA 19141, USA) *J. Immunol. Methods, 29(4), 387-394 (1979)* En;en.

2369-U2 **'Release radioimmunoassay': description of a new method and its application to the determination of human chorionic somatomammotropin plasma levels.** Cocola,F.; Orlandini,A.; Barbarulli,G.; Tarli,P.; Neri,P. (ISVT Sclavo, Res. Cent., Via Fiorentina 1, 53100 Siena, Italy) *Anal. Biochem., 99(1), 121-128 (1979)* En;en.

2370-U2 **A simple solid-phase radioimmunoassay for the measurement of IgG secreted in vitro by human lymphocytes.** Romagnani,S.; Del Prete,G.F.; Giudizi,G.M.; Almerigogna,F.; Ricci,M. (Clin. Immunol. Lab., Univ. Florence, Policlin. Careggi, Viale Morgagni, 50134-Firenze, Italy) *J. Immunol. Methods, 29(3), 263-270 (1979)* En;en.

2371-U2 **Characterization and measurement of anti-IgG antibodies in human sera by radioimmunoassay (RIA).** Yamagata,J.; Barnett,E.V.; Knutson,D.W. Nasu,H.; *Chia,D. (1000 Veteran Ave., Los Angeles, CA USA) *J. Immunol. Methods, 29(1), 43-56 (1979)* En;en.

2372-U2 **Measurements of antibodies to tubulin by radioimmunoassay.** Mead,G.M.; Cowin,P.; Whitehouse,J.M.A. (CRC Med. Oncol. Unit, Southampton Gen Hosp., Southampton, UK) *J. Immunol. Methods, 28(3-4), 243-253 (1979)* En;en.

2373-U2 **Radioimmunoassay of T_3, r-T_3 and r-T'_2 in human serum.** Skovsted,L. (Dep. Intern. Med. and Endocrinol., Herlev Hosp., 2730 Copenhagen, Denmark) *Acta Med. Scand., Suppl. 624, 19-24 (1979)* En;en.

2374-U2 **An optimized radioimmunoassay of F VIII related antigen (F VIIIR:AG) in plasma and eluates.** Savidge,G.F.; Carlebjork,G. (Dep. Blood Coagulation Res., Karolinska Inst., Stockholm, Sweden) *Thromb. Res., 14(2-3), 363-376 (1979)* En;en.

2375-U2 **Polymorphonuclear leukocyte phagocytosis: quantitation by a rapid radioactive method.** Miller,D.S.; Beck,S. (Box 2985, Duke Univ. Med. Cent., Durham, NC 27710, USA) *J. Lab. Clin. Med., 86(2), 344-348 (1975)* En;en.

2376-U2 **A simplified isotope release assay for cell-mediated cytotoxicity against anchorage dependent target cells.** Timonen,T.; Saksela,E. (Lab. Pathol., I Dep. Gynecol., Univ. Cent. Hosp., Univ. Helsinki, Helsinki, Finland) *J. Immunol. Methods, 18(1,2), 123-132 (1977)* En;en.

2377-U2 **Fractionation of lymphocyte surface antigens. I. Rapid method for eliminating labeled lipid from cell surface antigens iodinated by the lactoperoxidase catalysed reaction.** Zimmerman,B.; Chapman,M.L. (Dep. Immunol., Res. Inst., Hosp. Sick Child., Toronto, Ont., Canada) *J. Immunol. Methods, 15(2), 183-192 (1977)* En;en.

2378-U2 **Radioassay of granulocyte chemotaxis. Studies on human granulocytes and chemotactic factors.** Gallin,J.I. (Lab. Clin. Invest., Natl. Inst., Allergy and Infect. Dis., Natl. Inst. Health, Bethesda, MD 20014, USA) *Antibiot. Chemother., 19, 146-160 (1975)* En;en.

2379-U2 **Interaction of ^{125}I-labelled complement subcomponents Clr and Cls with protease inhibitors in plasma.** Sim,R.B.; Reboul,A.; Arlaud,G.J.; Villiers,C.L.; Colomb,M.G. (DRF/Biochim., Cent. d'Etudes Nucleaires des Grenoble, 85 X, 38041, Grenoble Cedex, France) *FEBS Lett., 97(1), 111-115 (1979)* En.

2380-U2 **Direct radioimmunoassay of nuclear 3,5,3'-triiodothyronine in rat anterior pituitary.** Larsen,P.R.; Bavli,S.Z.; Castonguay,M.; Jove,R. (Howard Hughes Med. Inst. Lab. and Thyroid Unit, Dep. Med., Peter Bent Brigham Hosp. Boston, MA 02115, USA) *J. Clin. Invest., 65(3), 675-681 (1980)* En;en.

2381-U2 **Rejection and immunosuppression in kidney transplantation.** White,A.G.; de Araujo,A.M.; Anderton,J.L. (South East Scotland Reg. Blood Transfus. Serv., R. Infirm., Edinburgh, UK) *J. Clin. Lab. Immunol., 2(4), 299-302 (1979)* En;en.
[DNA biosynthesis, RNA biosynthesis, radioactive isotopes, blood]

2382-U2 **Simultaneous detection of morphine and barbiturates in urine by radioimmunoassay.** Usate-gui-Gomez,M.; Heveran,J.E.; Cleeland,R.; McGhee,B.; Telischak,Z.; Awdziej,T.; Grunberg,E. (Diagn. Res., Hofmann-La Roche Inc., Nutley, NJ 07110, USA) *Clin. Chem., 21(10), 1378-1382 (1975)* En;en.

2383-U2 **Detection of immune complexes. The use of radioimmunoassays with C1q and monoclonal rheumatoid factor.** Gabriel,A.,Jr.; Agnello,V. (Div. Rheumatol. and Clin. Immunol., Tufts-New-England Med. Cent., Boston, MA 02111, USA) *J. Clin. Invest., 59(5), 990-1001 (1977)* En;en.

2384-U2 **A method for radioiactive antiglobulin testing with ^{125}I labeled anti-IgG.** Jenkins,D.E.; Moore,W.H.; Hawiger,A. (321 22nd Ave., North, Nashville, TN 37203, USA) *Transfusion, 17(1), 16-22 (1977)* En;en.

2385-U2 **Rapid heterologous method of radioimmunochemical determination of luteinizing hormone in the blood serum of rats. II. Specificity of the method.** Antonov,A.S.; Nikiforova,G.P.; Krivosheev,O.G.; Isachenkov,V.A. (Div. Biochem., Minist. Public Health SSSR, Moscow, USSR) *Probl. Endokrinol., 23(3), 80-83 (1977)* Ru;en.

2386-U2 **A rapid radioimmunoassay of human LH and HCG useful in early pregnancy problems.** Koninckx,P.; De Moor,P. (Lab. Exp. Med., Dep. Dev. Biol., Katholicke Univ. Leuven, Rega Inst., Leuven, Belgium) *Europ. J. Obstet. Gynecol. Reprod. Biol., 6(5), 239-241 (1976)* En.

2387-U2 **Evaluation of a new immobilized-antibody radioimmunoassay of plasma cortisol: comparison with two previous solid-phase procedures.** Del Chicca,M.G.; *Clerico,A.; Zucchelli,G.C.; Mariani,G. (Lab. Fisiol. Clin. CNR, Via Savi 8, 5600 Pisa, Italy) *J. Nucl. Med. Allied Sci., 22(2), 97-100 (1978)* En;en.

2388-U2 **A comparison of the 'EMIT' assay with two iodinated radioimmunoassays for diphenylhydantoin.**

Kampa,I.S.; Jarzabek,J.; Hundertmark,J.M. (Valley Hosp., Dep. Pathol. and Sect. Clin. Biochem., Linwood and North Van Dien Ave., Ridgewood, NJ 07451, USA) *Clin. Biochem., 11(4), 167-168 (1978)* En;en.

2389-U2 Specific and sensitive determination of pregnancy-specific β_1-glycoprotein by radioimmunoassay. A new pregnancy test. Grudzinskas,J.G.; Gordon,Y.B.; Jeffrey,D.; *Chard,T. (St. Bartholomew's Hosp. Med. Coll., West Smithfield, London EC1A 7BE, UK) *Lancet, 1(8007), 333 (1977)* En;en.

2390-U2 The measurement of 3-O-methyldopamine in urine and plasma by a rapid and specific radioimmunoassay. Faraj,B.A.; Camp,V.M.; Pruitt,A.W.; Isaacs,J.W.; Ali,F.M. (Dep. Radiol., Div. Nucl. Med., Emory Univ., Atlanta, GA 30322, USA) *J. Nucl. Med., 18(10), 1027-1033 (1977)* En;en.

2391-U2 Simultaneous radioimmunoassay of thyrotropin and thyroxine in human serum. Bluett,M.K.; *Reiter,E.O.; Duckett,G.E.; Root,A.W. (Ed Wright Pediatr. Endocrinol. Res. Lab., All Child. Hosp., 801 Sixth St. South, St. Petersburg, FL 33701, USA) *Clin. Chem., 23(9), 1644-1647 (1977)* En;en.

2392-U2 A new sensitive and precise method for determining testosterone and 11-ketotestosterone in fish plasma by radioimmunoassay. Sangalang,G.B.; Freeman,H.C. (Fisheries and Marine Serv., Environment Canada, Halifax Lab. Halifax, Nova Scotia B3J 2R3, Canada) *Gen. Comp. Endocrinol., 32(4), 432-439 (1977)* En;en.

2393-U2 Radioimmunoassay for nortriptyline and amitryptyline. Aherne,G.W.; Marks,V.; Mould,G.; Stout,G. (Dep. Clin. Biochem., Univ. Surrey, Guildford, Surrey, UK) *Lancet, 1(8023), 1214 (1977)* En.

2394-U2 Solid phase radioimmunoassays using labelled antibodies: a conceptual framework for designing assays. Kalmakoff,J.; Parkinson,A.J.; Crawford,A.M.; Williams,B.R.G. (Dep. Microbiol., Univ. Otago, Dunedin, New Zealand) *J. Immunol. Methods, 14(1), 73-84 (1977)* En;en.

2395-U2 A rapid, sensitive and specific radioimmunoassay for methotrexate. Paxton,J.W.; *Rowell,F.J. (Sch. Pharmacy, Sunderland Polytech., Galen Build., Green Terrace, Sunderland SR1 3SD, UK) *Clin. Chim. Acta, 80(3), 563-572 (1977)* En;en.

2396-U2 Double-antibody radioimmunoassay for factor VIII-related antigen. Green,D.; Reynolds,N. (Hematol. Sect., Dep. Med., Northwestern Univ. Med,. Sch., 303 E. Chicago Ave., Chicago, IL 60611, USA) *Clin. Chem., 23(9), 1648-1653 (1977)* En;en.

2397-U2 Solid-phase radioimmunoassay for detection of antibodies to myelin basic protein. Randolph,D.H.; Kibler,R.F.; Fritz,R.B. (Dep. Microbiol., Emory Univ. Sch. Med., Atlanta, GA 30322, USA) *J. Immunol. Methods, 18(3-4), 215-224 (1977)* En;en.

2398-U2 A simple radioimmunoassay for plasma cortisol and 11-deoxycortisol (17,21-dihydroxy-4-pregnene-3,20-dione). Brown,J.R.; Cavanaugh,A.H.; Farnsworth,W.E. (Veterans Adm. Hosp., 3495 Bailey Ave., Buffalo, NY 14215, USA) *Steroids, 28(4), 487-498 (1976)* En;en.

2399-U2 A solid-state immunoradiometric assay for plasminogen antigen (plasmin antigen). Fukao,K. (Dep. Surg., Inst. Clin. Med., Univ. Tsukuba, Ibaraki, Japan) *Acta Haematol. Jap., 39(6), 890-896 (1976)* Ja;en.

2400-U2 Detection of cytotoxic antibody to erythroblasts. Zaentz,S.D.; Luna,J.A.; Baker,A.S.; *Krantz,S.B. (Dep. Med., Veterans Adm. Hosp., Nashville, TN 37203, USA) *J. Lab. Clin. Med., 89(4), 851-860 (1977)* En;en.

2401-U2 Radioimmunoassay of fibrinopeptide A (double antibody method). Schramm,W.; Schmid,M.; Erhardt,F.W. (Med. Klin. Innenstadt., Univ. Munchen, Hamostaseol, Abt., Ziemssenstr. 1, 8 Munchen 2, GFR) *Fortschr. Med., 95(19), 1291-1298 (1977)* De;de,en.

2402-U2 A critical evaluation of a procedure for measurement of serum bile acids by radioimmunoassay. Mihas,A.A.; Spenney,J.G.; Hirschowitz,B.I.; Gibbon,R.G. (Div. Gastroenterol., Univ. Alabama in Birmingham and Veterans Adm. Hosp., Birmingham, AL 35294, USA) *Clin. Chim. Acta, 76(3), 389-397 (1977)* En;en.

2403-U2 Measurement of absolute amounts of antigen-specific human IgE by a radioallergosorbent test (RAST) elution technique. Schellenberg,R.R.; Adkinson,N.F.,Jr. (Good Samaritan Hosp., 5601 Loch Raven Blvd., Baltimore, MD 21239, USA) *J. Immunol., 115(6), 1577-1583 (1975)* En;en.

2404-U2 Validation of a heterologous radioimmunoassay for the determination of rat thyrotrophic hormone. Garcia,M.D.; Cacicedo,L.; Morreale de Escobar,G. (Dep. Endocrinol. Exp., Inst. G. Maranon, Cent. Invest. Biol. CSIC, Madrid-6, Spain) *Rev. Esp. Fisiol., 32(1), 59-76 (1976)* En;en.

2405-U2 Liquid scintillation vial for radiometric assay of lymphocyte carbohydrate metabolism in response to mitogens. Tran,N.; Wagner,H.N.,Jr. (Dep. Radiol., Univ. California, Irvine Med. Cent., 101 City Dr. S., Orange, CA 92668, USA) *J. Nucl. Med., 19(1), 61-63 (1978)* En;en.

2406-U2 An indirect, quantifiable assay for cytophilic antibody. Kuhn,R.E.; Cassida,G.W. (Dep. Biol., Wake Forest Univ., Winston-Salem, NC 27109, USA) *J. Immunol. Methods, 19(4), 387-393 (1978)* En;en.

2407-U2 A new micromethod for lymphocyte stimulation using whole blood. Eskola,J.; Soppi,E.; Viljanen,M.; Ruuskanen,O. (Dep. Med. Microbiol., Univ. Turku, SF-20520 Turku 52, Finland) *Immunol. Commun., 4(4), 297-307 (1975)* En;en.

2408-U2 A micromethod for stimulation of chicken lymphocytes in vitro using whole blood. Lassila,O.; Eskola,J.; Toivanen,P. (Dep. Med. Microbiol., Turku Univ., SF-20520 Turku 52, Finland) *Clin. Exp. Immunol., 26(3), 641-646*

(1976) En;en.

2409-U2 A microsystem to evaluate the synthesis of [^{3}H]leucine labeled proteins by macrophages. Varesio,L.; Eva,A. (Lab. Immunodiagnosis, Natl. Cancer Inst., NIH, Bethesda, MD 20205, USA) *J. Immunol. Methods, 33(3), 231-238 (1980)* En;en.

2410-U2 A simple and efficient method for measuring the uptake of radiolabelled compounds by leukocytes in whole blood. Pauly,J.L.; Schuller,M.G.; Germain,M.J. (Dep. Med. B, Roswell Park Mem. Inst., New York State Dep. Health, Buffalo, NY 14263, USA) *Int. J. Appl. Radiat. Isotopes, 28(5), 509-512 (1977)* En;de,en,fr,ru.

2411-U2 A radiomicroassay for cytotoxic antibody to human spermatozoa. Quantification by tritiated actinomycin D. Sung,J.S.; Shizuya,H.; Black,D.D.; Mumford,D.M. (Dep. Obstet. and Gynecol., Baylor Coll. Med., Texas Med. Cent., Houston, TX 77030, USA) *Clin. Exp. Immunol., 27(3), 469-477 (1977)* En;en.

2412-U2 Radioimmunological determination of 25-hydroxycholecalciferol. Gemeiner,M. (Inst. Biochem., Veterinarmed. Univ. Wien, Linke, Bahngasse 11, A-1030 Wien, Austria) *Mikrochim. Acta, 2(1-2), 161-173 (1976)*

2413-U2 Radioimmunoassay for aldosterone in plasma without chromatography. Hubl,W.; Buchner,M.; Stahl,F.; Rohde,W. (Bezirkskrankenhaus, Zentrallab., DDR-801 Dresden, Friedrichstr. 41, GDR) *Endokrinologie, 66(3), 292-300 (1975)* En;en.

2414-U2 A rapid method for assessment of a macrophage chemotactant produced by SaD2 fibrosarcoma cells in vitro. Schaub Simon,L.; Patterson,R.; Jones,T.L. (Dep. Surg., Michigan State Univ., East Lansing, MI 48824, USA) *J. Immunol. Methods, 32(2), 195-206 (1980)* En;en.

2415-U2 A microcytotoxicity assay using ^{111}indium oxine release from platelets. Hawker,R.J.; Hawker,L.M. (Dep. Surg., Univ. Birmingham, Queen Elizabeth Hosp. Med. Cent., Edgbaston, Birmingham B15 2TH, UK) *J. Immunol. Methods, 33(1), 45-53 (1980)* En;en.

2416-U2 Preparation of ^{125}I-thyroxine and ^{125}I-triiodothyronine of high specific activity for the radioimmunoassay of serum total T4 and T3. Thurlow,V.R.; Puxley,H.J. (Dep. Chem. Pathol., Lewisham Hosp., High St., Lewisham, SE13 6LH, UK) *Ann. Clin. Biochem., 13(2), 364-368 (1976)* En;en.

2417-U2 Modifications and evaluation of double antibody radioimmunoassay of human carcinoembryonic antigen. Das,S.; Das,B.R.; Terry,W.D. (Adv. Test. and Dev. Lab., Litton Bionet., Inc., 5516 Nicholson Lane, Kensington, MD 20795, USA) *Cancer Res., 36(6), 1954-1961 (1976)* En;en.

2418-U2 Simple solid-phase radioimmunoassays for total triiodothyronine and thyroxine in serum, and their clinical evaluation. Seth,J.; Toft,A.D.; Irvine,W.J. (Dep. Clin. Chem., Univ. Edinburgh, R. Infirm., Edinburgh EH3 9YW, UK) *Clin. Chim. Acta, 68(3), 291-301 (1976)* En;en.

2419-U2 A simple radioimmunological method for the determination of plasma total oestriol during pregnancy. Craig,A. (Searle Sci. Serv., Lane End Rd., High Wycombe, Bucks. HP12 4HL, UK) *Clin. Chim. Acta, 68(3), 277-286 (1976)* En;en.

2420-U2 A column radioimmunoassay method for the determination of digoxin. Boguslaski,R.C.; Denning,C.E. (Ames Res. Lab., Ames Co., Div. Miles Lab., Inc., Elkhart, IN 46514, USA) *Biochem. Med., 14(1), 83-92 (1975)* En;en.

2421-U2 Rapid simultaneous radioimmunoassay for the measurement of triiodothyronine and thyroxine in unextracted human serum. Ljunggren,J.-G.; Persson,B.; Tryselius,M. (Dep. Endocrinol. and Metab. and Surg., Karolinska Sjukhuset, Stockholm, Sweden) *Acta Endocrinol., 81(3), 487-494 (1976)* En;en.

2422-U2 Rapid, column, flow-through radioimmunoassay for insulin. Davis,J.W.; Yoder,J.M.; Adams,E.C. (Ames Res. Lab., Ames Co., Div. Miles Lab., Inc., Elkhart, IN 46514, USA) *Clin. Chim. Acta, 66(3), 379-386 (1976)* En;en.

2423-U2 A rapid radioimmunoassay method of growth hormone in dog plasma. Cocola,F.; Udeschini,G.; Secchi,C.; Panerai,A.E.; Neri.P.; Muller,E.E. (Res. Cent. ISVT Sclavo, 53100 Siena, Italy) *Proc. Soc. Exp. Biol. Med., 151(1), 140-145 (1976)* En;en.

2424-U2 Autoradiographic immunoassay (ARIA): a rapid technique for the semiquantitative mass screening of haptens. Weiler,E.W.; Zenk,M.H. (Lehrstuhl Pflanzenphysiol., Ruhr-Univ. Biochim., D-4630 Bochum, GFR) *Anal. Biochem., 92(1), 147-155 (1979)* En;en.

2425-U2 A rapid, flow-through column radioimmunoassay for human chorionic somatomammotropin. Johnson,P.K.; *Yoder,J.M.; Adams,E.C. (Ames Res. Lab., Ames Co., Div. Miles Labs., Inc., Elkhart, IN 46514, USA) *Am. J. Obstet. Gynecol., 125(1), 45-50 (1976)* En;en.

2426-U2 Direct radioimmunoassay of progesterone in mare plasma. Mathieu,H.P.; Mathieu-Nast,C.; Vrignaud,C. (Lab. Physique Pharmaceutique, UER Pharmacie, Place de la Victoire, 33000 Bordeaux, France) *Steroids, 30(1), 33-39 (1977)* En;en.

2427-U2 Comparison of serum IgE determination by radioimmunoassay and by single radial immunodiffusion. Berrens,L.; Bruynzeel,P. (Div. Exp. Allergy, Dep. Dermatol., Acad. Hosp., Catharijnesingel 101, Utrecht, Netherlands) *Clin. Chim. Acta, 70(3), 337-342 (1976)* En;en.

2428-U2 A method for radioimmunological assay of glisoxepid. Nieuweboer,B.; Gabriel,D.; Lubke,K. (Schering AG, Pharma Forsch, Postfach 65 03 11, 1000 Berlin 65, GFR) *Arzneim.-Forsch., 26(9), 1633-1636 (1976)* De;de,en.

2429-U2 Simplified radioimmunoassay for triiodothyronine: evaluation of a new commercial kit. Balachandran,S.; Moses,D.C.; Sisson,J.C.; Patel,S.R. (Div. Nucl. Med., Buffalo Gen. Hosp., Buffalo, NY 14203, USA) *Clin. Chem., 22(8), 1402-1404 (1976)* En;en.

2430-U2 Radioimmunoassay for human plasma apolipoprotein B. Bedford,D.K.; *Shepherd,J.; Morgan,H.G. (Lipid Res. Clin., Fondren-Brown Build., Methodist Hosp., 6516 Bertner Blvd., Houston, TX 77025, USA) *Clin. Chim. Acta, 70(2), 267-276 (1976)* En;en.

2431-U2 A simultaneous radioimmunoassay for growth hormone and insulin in the plasma of rats and rabbits. Kervran,A.; Rieutort,M.; Guillaume,M. (Univ. Pierre et Marie Curie, Lab. Physiol. Dev., 9, quai Saint-Bernard, 74230 Paris Cedex 05, France) *Diabete Metab., 2(2), 67-72 (1976)* En;en,fr.

2432-U2 A simple radioimmunoassay for the measurement of testosterone glucosiduronate in unextracted urine. Tresguerres,J.A.F.; Lisboa,B.P.; Tamm,J. (2 Med. Klin. und Frauenklin. Univ., D-2 Hamburg 20, Martinstr. 52, GFR) *Steroids, 28(1), 13-23 (1976)* En;en.

2433-U2 Radioimmunoassays for aldosterone and deoxycorticosterone in plasma and urine. Kurtz,A.B.; Bartter,F.C. (Cent. Middlesex Hosp., London NW10, UK) *Steroids, 28(1), 133-142 (1976)* En;en.

2434-U2 A comparison of methods for the immunoassay of serum apolipoprotein B in man. Durrington,P.N.; Whicher,J.T.; Warren,C.; Bolton,C.H.; Hartog,M. (Dep. Med., Bristol R. Infirmary, Bristol BS2 8HW, UK) *Clin. Chim. Acta, 71(1), 95-108 (1976)* En;en.

2435-U2 A radio (^{51}Cr) micro-tube leukocyte adherence inhibition assay: specific tumor-associated immunity in 3 murine tumor systems. Russo,A.J.: Nordin,A.A.; Goldrosen,M.H. (Gerontol. Res. Cent., Baltimore City Hosp., Baltimore, MD 21224, USA) *J. Immunol. Methods, 31(3-4), 259-269 (1979)* En;en.

2436-U2 A solid-phase radioimmunoassay for anti-SNP antibodies. Adam,C.; Verroust,P.; Robitaille,P.; Pontillon,F. (INSERMU64, Hop. Tenon, 4 rue de la Chine, 75790 Paris Cedex 20, France) *J. Immunol. Methods, 27(2), 133-143 (1979)* En;en.

2437-U2 A microprocessor-controlled assay for the estimation of human placental lactogen. Adam,T.; Roulston,J.E.; Bagshawe,K.D. (Dep. Med. Oncol., Charing Cross Hosp. (Fulham), Fulham Palace Road, London W6 8RF, UK) *J. Immunol. Methods, 31(3-4), 323-332 (1979)* En;en.

2438-U2 Extraction of human plasma or sera by heat treatment for a solid-phase radioimmunoassay of carcinoembryonic antigen. Kim,Y.D.; Tomita,J.T.; Schenck,J.R.; Moeller,C.; Weber,G.F.; Hirata,A.A. (Immunol. Lab., Diagn. Div., D-90D, Abbott Lab., North Chicago, IL 60064, USA) *Clin. Chem., 25(5), 773-776 (1979)* En;en.

2439-U2 Studies of the molecular mechanisms of C3b inactivation and a simplified assay of β1H and the C3b inactivator (C3bINA). Gaither,T.A.; Hammer,C.H.; Frank,M.M.(Lab Clin. Invest., Natl. Inst. Allergy and Infect. Dis., Natl. Inst. Health, Bethesda, MD 20014, USA) *J. Immunol., 123(3), 1195-1204 (1979)* En;en.

2440-U2 Detection of IgE insulin antibody with radioallergosorbent test. Nakagawa,S.; Saito,N.; Nakayama,H.; Sasaki,T.; Watanabe,T.; Aoki,S. (Second Dep. Med., Hokkaido Univ. Sch. Med., Kita-15, Nishi-7, Kita-Ku, Sapporo 060, Japan) *Diabetologia, 14(1), 33-38 (1978)* En;en.

2441-U2 Radioactive antiglobulin testing with ^{125}I anti-C4 and anti-C3. Jenkins,D.E.,Jr.; Johnson,R.M.; Moore,W.H. (Nashville Reg. Red Cross Blood Cent., Vanderbilt Univ. Sch. Med., Nashville, TN 37203, USA) *Transfusion, 18(4), 407-416 (1978)* En;en.

2442-U2 Radioimmunoassays for prostaglandins. II. Measurement of prostaglandin E_2 and the 13,14-dihydro-15-keto metabolites of the E and F series. Description of a reliable technique with a universal applicability. Thomas,C.M.G.; van den Berg,R.J.; de Koning Gans,H.J.; Lequin,R.M. (Dep. Obstet. and Gynaecol., St. Radboud Hosp., Catholic Univ., Nijmegen, Netherlands) *Prostaglandins, 15(5), 849-855 (1978)* En;en.

2443-U2 A rapid, automated radioimmunoassay for estimation of plasma renin activity. Roulston,J.E.; Adam,T.; MacGregor,G.A. (Dep. Med., Charing Cross Hosp. (Fulham), Fulham Palace Road, London W6 8RF, UK) *Lab. Pract., 28(8), 835-839 (1979)* En;en.

2444-U2 Second antibody chemically linked to cellulose for the separation of bound and free hormone: an improvement over soluble second antibody in gonadotrophin radioimmunoassay. Koninckx,P.; Bouillon,R.; de Moor,P. (Dep. Ontwikkelingsbiol., Lab. Exp. Geneeskd., K.U. Lueven, Belgium) *Acta Endocrinol., 81(1), 43-53 (1976)* En;en.

2445-U2 Conglutinin binding polyethylene glycol precipitation assay for immune complexes. Macanovic,M.; *Lachman,P.J. (MRC Group Mechanisms Tumour Immunity, Med. Sch., Hills Road, Cambridge, UK) *Clin. Exp. Immunol., 38(2), 274-283 (1979)* En;en.

2446-U2 Comparison of an automated radioimmunoassay method and a competitive protein-binding method for evaluation of serum cortisol. Jekowsky,E.; Trainer,T.D.; *Copeland,B.E. (Dep. Pathol., New England Deaconess Hosp., Boston, MA 02215, USA) *Am. J. Clin. Pathol., 72(2), 146-150 (1979)* En;en.

2447-U2 A rapid radioimmunological evaluation of the androstenone content in boar fat. Andresen,O. (Dep. Reproductive Physiol. and Pathol., Vet. Coll. Norway, PO Box 8146 Dep., Oslo 1, Norway) *Acta Vet. Scand., 20(3), 343-350 (1979)* En;en,no.

2448-U2 Automated radioimmunoassay of nicotin. Castro,A.; Monji,N.; Malkus,H.; Eisenhart,W.; McKennis,H.,Jr.; Bowman,E.R. (Univ. Miami, Sch. Med., Dep. Pathol., R 40, POB 016960, Miami, FL 33101, USA) *Clin. Chim. Acta, 95(3), 473-481 (1979)* En;en.

2449-U2 Rapid in vivo assay of mouse natural killer cell activity. Riccardi,C.; Puccetti,P.; Santoni,A.; Herberman,R.B. (Inst. Pharmacol., Univ. Perugia, 06100 Perugia, Italy) *J. Natl. Cancer Inst., 63(4), 1041-1045 (1979)* En;en.

2450-U2 Comparison of IgE values as determined by different solid phase radioimmunoassay methods. Johansson,S.G.O.; Berglund,A.; Kjellman,N.-I.M. (Dep. Paediatr., Regionsjukhuset, S-581 85 Linkoping, Sweden) *Clin. Allergy, 6(1), 91-98 (1976)* En;en.

2451-U2 [Radioimmunoassay of thyroid-stimulating hormone. Immunological study of the system and its clinical application]. Vandalem,J.L.; Ketelslegers,J.M.; Pirens,G.; Closset,J.; Hennen,G. (Sect. Endocrinol., Dep. Clin. et Semeiol. Med., Inst. Med., Univ. Liege, Liege, Belgium) *Acta Clin. Belg., 30(6), 521-537 (1975)* Fr;en,fr,fl.

2452-U2 Rapid radioimmunoassay of triiodothyronine on Sephadex G-25 by the Ames kit. Howarth,P.J.N.; Marsden,P. (Dep. Chem., Pathol. and Med., King's Coll. Hosp. Med. Sch., London SE5 8RX, UK) *J. Nucl. Med., 17(4), 321-322 (1976)* En;en.

2453-U2 Rapid assay of gonadotrophins: a radio-immunological method in 6 hours. Dotti,C.; Filippi,G.; Franchini,R. (Sez. Radioisot. (1° Serv. Radiol.), Arcispedale Santa Maria Nuova, Via Risorgimento, 80, 42100 Reggio Emilia, Italy) *J. Nucl. Biol. Med., 19(4), 177-185 (1975)* En;en.

2454-U2 New simplified procedures for the determination of progesterone by competitive protein binding and radioimmunoassay. Batra,S. (Dep. Obstet. and Gynecol., Univ. Lund., Lund, Sweden) *J. Steroid Biochem., 7(2), 131-134 (1976)* En;en.

2455-U2 A simple ultrasensitive method for the assay of cyclic AMP and cyclic GMP in tissues. Frandsen,E.K.; Krishna,C. (Sect. Drug-Tiss. Interact., Lab. Chem. Pharmacol., Natl. Heart and Lung Inst., Natl. Inst. Health, Bethesda, MD 20014, USA) *Life Sci., 18(5), 529-542 (1976)* En;en.

2456-U2 Methodological simplifications in radioimmunoassay of urinary aldosterone. Malvano,R.; Orlandini,S.; Cozzani,P.; Duranti,P.; Simonini,N.; Salvetti,A. (Lab. Clin. Physiol. CNR, Via Savi 8, 56100 Pisa, Italy) *Clin. Chim. Acta, 66(3), 331-343 (1976)* En;en.

2457-U2 Determination of plasma aldosterone in children by thin layer chromatography and radioimmunoassay. Parth,K.; Zimprich,H.; Brunel,R. (Ludwig Boltzmann Inst. Paediatr., Endocrinol., Vienna, Austria) *Acta Endocrinol., 81(2), 330-339 (1976)* En;en.

2458-U2 Solid phase radioimmunoassay of apolipoprotein B (apo B) in normal human plasma. Thompson,G.R.; Birnbaumer,M.E.; Levy,R.I.; Gotto,A.M.,Jr. (Med. Res. Counc. Lipid Metab. Unit, Hammersmith Hosp., Ducane Rd., London W12 OHS, UK) *Atherosclerosis, 24(1-2), 107-118 (1976)* En;en.

2459-U2 Radioimmunoassay as an improved method for measurement of serum levels of gentamicin. Longmore,P.; *Atkins,R.C.; Casley,D.; Johnston,C.L. (Renal Unit, Prince Henry's Hosp., St. Kilda Road, Melbourne, Vic. 3004, Australia) *Med. J. Aust., 1(20), 738-740 (1976)* En;en.

2460-U2 On the assessment of validity of steroid radioimmunoassays. Cekan,S.Z. (Swedish Med. Res. Counc., Reprod. Endocrinol. Res. Unit, Karolinska Hosp., 104 01 Stockholm, Sweden) *J. Steroid Biochem., 11(1, part A), 135-141 (1979)* En;en.

2461-U2 Urinary and plasma testosterone glucosiduronate measurement by a simple RIA method. Tresguerres,J.A.F.; Tamm,J. (Catedra Endocrinol. Exp., Fac. Med., Uiv. Complutense, Madrid, Spain) *J. Steroid Biochem., 11(1, part A), 143-146 (1979)* En;en.

2462-U2 Automation of steroid radioimmunoassays for clinical and research purposes. Vihko,R.; Hammond,G.L. (Dep. Clin. Chem., Univ. Oulu, SF-90220 Oulu 22, Finland) *J. Steroid Biochem., 11(1, part A), 125-128 (1979)* En;en.

2463-U2 Histocompatibility typing by cellular radioimmunoassay. Longenecker,B.M.; Singh,B.; Gallatin,M.; Havele,C. (Dep. Immunol., Univ. Alberta, 845E Med. Sci. Build., Edmonton, Alta., T6G 2H7, Canada) *Immunogenetics, 7(3), 201-211 (1978)* En;en.

2464-U2 Rapid and sensitive radioimmunoassays for human myoglobin. Norregaard Hansen,K.; Norgaard-Pedersen,B. (Dep. Clin. Chem., Sonderborg Hosp., DK-6400 Sonderborg, Denmark) *Scand. J. Clin. Lab. Invest., 39(6), 525-531 (1979)* En;en.

2465-U2 Microradioisotopic techniques for the detection of antibody to adherent target cells in tissue culture. Cohen,A.M.; Wood,W.C. (Dep. Surg., Harvard Med. Sch., Boston, MA 02114, USA) *J. Surg. Res., 26(4), 404-410 (1979)* En;en.

2466-U2 Transfer of proteins from gels to diazobenzyloxymethyl-paper and detection with antisera: a method for studying antibody specificity and antigen structure. Renart,J.; Reiser,J.; Stark,G.R. (Inst. Enzimol. CSIC, Fac. Med. Univ. Autonoma, Arzobispo Morcillo s/n, Madrid 34, Spain) *Proc. Natl. Acad. Sci. USA, 76(7), 3116-3120 (1979)* En;en.

2467-U2 Liquid scintillation counting for measurement of tritiated uptake in lymphocyte transformation in vitro: a new direct-suspension procedure. Bader,C.A.; Monet,J.-D.; Assailly,J.; Funck-Brentano,J.L. (INSERM Unite 90, Hop. Necker, 161, rue de Sevres, 75730 Paris Cedex 15, France) *J. Immunol. Methods, 9(3-4), 307-314 (1976)* En;en.

2468-U2 Evaluation of a commercial reagent system ofr the radioimmunoassay fo human thyroid stimulating hormone. Robinson,T.R. (Sci. Dep., St. Lawrence Coll. Appl. Arts and Technol., Kingston, Ont., Canada) *Can. J. Med. Technol., 38(4), 97-113 (1976)* En.

2469-U2 A simplified radioimmunoassay procedure for serum progesterone. Khadempour,M.H.; Laing,I.; *Gowenlock,A.H. (Biochem. Dep., R. Infirm., Manchester M13 9WL, UK) *Clin. Chim. Acta., 82(1-2), 161-171 (1978)* En;en.

2470-U2 Radioimmunoassay for 11-deoxycortisol using iodine-labeled tracer. Sakamoto,N.; Matsukura,S.; Tsuboi,S.; Tokumiya,T.; Imura,H.; Tachibana,S. (Third Div., Dep. Med., Kobe Univ. Sch. Med., Kobe-shi 650, Japan) *Endocrinol. Jap., 23(4), 359-363 (1976)* En;en.

2471-U2 Urinary 5α-androstane-3α,17β-diol radioimmunoassay: a new clinical evaluation. Wright,F.; Mowszowicz,I.; *Mauvais-Jarvis,P. (Serv. Endocrinol. et Gynecol. Med., Hop. Necker, 149, rue de Sevres, 75730 Paris Cedex 15, France) *J. Clin. Endocrinol. Metab., 47(4), 850-854 (1978)* En;en.

2472-U2 Automated approach to radioimmunoassays of somatotropin (human growth hormone) and insulin. Green,G.; Dyce,D.; Gimovsky,A.; Lo,D.H. (Corp. Res. Lab., Union Carbide Corp., Tarrytown, NY 10591, USA) *Clin. Chem., 22(9), 1510-1515 (1976)* En;en.

2473-U2 A practical model for steroid hormone radioimmunoassays. Seaton,B.; Lusty,J.; Watson,J. (Wellcome Inst. Comp. Physiol., Zool. Soc. Lond., Regent's Park, London NW1 4RY, UK) *J. Steroid Biochem., 7(6-7), 511-516 (1976)* En;en.

2474-U2 A solid-phase radioimmunoassay for the detection of antibodies to ribosomes. Cavanagh,D. (Dep. Immunol., Middlesex Hosp. Med. Sch., 40-50 Tottenham St., London W1P 9PG, UK) *Anal. Biochem., 79(1-2), 217-225 (1977)* En;en.

2475-U2 Determination of cortisol in human plasma by radioimmunoassay. Use of the ^{125}I-labelled radioligand. Read,G.F.; Fahmy,D.R.; Walker,R.F. (Tenovus Inst. Cancer Res., Welsh Natl. Sch. Med., Heath Park, Cardiff, CF4 4XX, UK) *Ann. Clin. Biochem., 14(6), 343-349 (1977)* En;en.

2476-U2 A simplified solid-phase radioimmunoassay for carcinoembryonic antigen. Kim,Y.D.; Tomita,J.T.; Schenck,J.R. (Immunol. Lab. D-90D, Abbott Lab., North Chicago, IL 60064, USA) *J. Immunol. Methods, 19(4), 309-316 (1978)* En;en.

2477-U2 A double antibody radioimmunoassay for mouse hemoglobins: use of polyethylene glycol in conjunction with the second antibody. Ansari,A.A.; Bahuguna,L.M.; Malling,H.V. (Lab. Biochem. Genet., Natl. Inst. Environ. Health Sci., Research Triangle Park, NC 27709, USA) *J. Immunol. Methods, 26(3), 203-211 (1979)* En;en.

2478-U2 Development of a highly sensitive radioimmunoassay for digoxin and its application in pediatric practice. Greenwood,H.; Howard,M.; Landon,J.; Fraser,B.; Shinebourne,E. (Syva Res. Inst., 3181 Porter Drive, Palo Alto, CA 94304, USA) *Eur. J. Cardiol., 5(5), 413-424 (1977)* En;en.

2479-U2 A solid phase radioimmunoassay method for ferritin in serum using ^{125}I-labelled ferritin. Wide,L.; Birgegard,G. (Dep. Clin. Chem., University Hosp., S-750 14 Uppsala, Sweden) *Upsala J. Med. Sci., 82(1), 15-19 (1977)* En;en.

2480-U2 Radioimmunoassay of bile acids: development, validation, and preliminary application of an assay for conjugates of chenodeoxycholic acid. Schalm,S.W.; van Berge-Henegouwen,G.P.; *Hofmann,A.F.; Cowen,A.E.; Turcotte,J. (Div. Gastroenterol., Dep. Med., Univ. California, University Hosp., 225 West Dickinson St., San Diego, CA 92103, USA) *Gastroenterology, 73(2), 285-290 (1977)* En;en.

2481-U2 An ^{125}I-labeled cortisol radioimmunoassay in which serum binding proteins are enzymatically denatured. Hasler,M.J.; Painter,K.; Niswender,G.D. (Micromedic Diagn., Inc., 1800 E. Lincoln, POB 464, Fort Collins, Co 80521, USA) *Clin. Chem., 22(11), 1850-1854 (1976)* En;en.

2482-U2 Determination of drugs of abuse in body fluids by radioimmunoassay. Castro,A.; Mittleman,R. (Dep. Pathol., Hormone Res. Lab., Univ. Miami Sch. Med., Miami, FL 33152, USA) *Clin. Biochem., 11(3), 103-105 (1978)* En;en.

2483-U2 Improved micro-radioimmunoassay of digoxin in serum, with use of ^{125}I-labeled digoxin. Calesnick,B.; Dinan,A. (Div. Hum. Pharmacol., Hahnemann Med. Coll. and Hosp. Philadelphia, 230 N. Broad St., Philadelphia, PA 19102, USA) *Clin. Chem., 22(6), 903-905 (1976)* En;en.

2484-U2 Assay of estriol-16α-(β-D-glucuronide) in pregnancy urine with a specific antiserum. DiPietro,D.L. (Dep. Obstet. and Gynecol., Vanderbilt Univ. Sch. Med., Nashville, TN 37232, USA) *Am. J. Obstet. Gynecol., 125(6), 841-845 (1976)* En;en.

2485-U2 A computerized semi-automated radioimmunoassay for plasma testosterone. Bonsall,R.W.; Baumgardner,D.G.; Michael,R.P. (Dep. Psychiatry, Emory Univ. Sch. Med., Atlanta, GA 30322, USA) *J. Steroid Biochem., 7(10), 853-858 (1976)* En;en.

2486-U2 Radioimmunoassay for serum cortisol with ^{125}I-labeled ligand: comparison of three methods. Kumar,M.S.; Safa,A.M.; Deodhar,S.D. (Dep. Immunopathol., Cleveland Clinic Found., 9500 Euclid Avenue, Cleveland, OH 44106, USA) *Clin. Chem., 22(11), 1845-1849 (1976)* En;en.

2487-U2 The radioimmunoassay of 3,3',5'-triiodothyronine (reverse T3) in unextracted human serum. Ratcliffe,W.A.; Marshall,J.; Ratcliffe,J.G. (Radioimmunoassay Unit, Stobhill Hosp., Glasgow G21 3UW, UK) *Clin. Endocrinol., 5(6), 631-641 (1976)* En;en.

2488-U2 Rapid radioimmunoassay of progesterone in unextracted bovine plasma. Sugden,E.A. (Anim. Pathol. Div., Health Anim. Branch, Anim. Dis. Res. Inst. (E), POB 11300 Postal Station H, Ottawa, Ont. K2H 8P9, Canada) *Can. J. Comp. Med., 42(2), 229-234 (1978)* En;en,fr.

2489-U2 A rapid radioimmunoassay for human α-fetoprotein. Blank-Liss,W.E.; Lai,P.C.W.; *Hay,D.M.;

Lorscheider,F.L. (Div. Obstet. and Gynaecol., Fac. Med., Univ. Calgary, Calgary, Alta. T2N 1N4, Canada) *Clin. Chim. Acta, 86(1), 67-72 (1978)* En;en.

2490-U2 **A simple radioimmunoassay for serum β-oestrial in pregnancy.** Nicklin,M.G.; Hewitt,J.V.; Watson,D. (Clin. Chem. Dep., Area Lab., King Edward VII Hosp., Windsor, Berks., UK) *J. Clin. Pathol., 29(6), 561-564 (1976)* En.

2491-U2 **A simple and rapid thyroxine radio-immunoassay (T_4-RIA) in unextracted human serum; a comparison of T_4-RIA and T_4 displacement assay, T_4(D), in normal and pathologic sera.** Premachandra,B.N.; Ibrahim,I.I. (Immunol-endocrinol. Res. Dep., Veterans Adm. Hosp., Jefferson Barracks, St. Louis, MO 62125, USA) *Clin. Chim. Acta, 70(1), 43-60 (1976)* En;en.

2492-U2 **Radioimmunoassay of 16α-hydroxydehydroepiandrosterone and its sulfate.** Furuya,K.; Yoshida,T.; Takagi,S.; Kanbegawa,A.; Yamashita,H.; Kurosawa,Y.; Naito,A. (Dep. Obstet. and Gynecol., Nihon Univ., Sch. Med., Tokyo, Japan) *Steroids, 27(6), 797-812 (1976)* En;en.

2493-U2 **Radioimmunoassay of angiotensin II and its metabolites. Comparison of two different extraction procedures.** Spech,H.J.; Wernze,H.; Weiss,H. (Med. Universitatsklin. Wurzburg, Wurzburg, GFR) *Acta Endocrinol., 82(Suppl. 202), 67-69 (1976)* En.

2494-U2 **[A five-minute incubation for the radioimmunoassay of human chorionic somatomammotrophin].** Cesselin,F.; Antreassian,J.; Schaffhauser,O.; Lagoguey,M.; Desgrez,P. (Serv. Biochim. Med., Fac. Med. Pitie-Salpetrie, 91, blvd. Hopital, 75634 Paris Cedex 13, France) *Ann. Biol. Clin., 34(1), 41-45 (1976)* Fr;en,fr.

2495-U2 **Silica gel radioimmunoassay for myelin basic protein.** Hsiung,H.M.; Wu,J.; McPherson,T.A. (Dep. Med. Cross Cancer Inst., 11560 University Ave., Edmonton, Alta. T6G 1Z2, Canada) *Clin. Biochem., 11(2), 54-56 (1978)* En;en.

2496-U2 **A simple method for the radioimmunologic determination of sex steroids in plasma.** Ittrich,G. (Bereich Med. (Charite) Humboldt-Univ. Berlin, Frauenklin., Abt. Gynakol. Labordiagnostik, Tucholskystr. 2, DDR 104 Berlin, GDR) *J. Clin. Chem. Clin. Biochem., 15(11), 629-634 (1977)* De;de,en.

2497-U2 **A simplified radioimmunoassay for digoxin determination using a 125-J-labelled, solid-phase kit.** Doering,W.; Blumel,E. (II. Med. Abt., Stadt. Krankenhaus Schwabing, Kolner Platz 1, D-8000 Munchen 40, GFR) *Klin. Wochenschr., 56(10), 497-502 (1978)* De;de,en.

2498-U2 **A double-antibody radioimmunoassay for serum progesterone using progesterone-3-(O-carboxymethyl)oximino-[^{125}I]-iodohistamine as radioligand.** Scott,J.Z.; Stanczyk,F.Z.; Goebelsmann,U.; Mishell,D.R.,Jr. (Dep. Obstet. and Gynecol., Univ. Southern California Sch. Med., Los Angeles, CA 90033, USA) *Steroids, 31(3), 393-405 (1978)* En;en.

2499-U2 **A 24 hour radioimmunoassay for human prolactin.** Gwee,H.M.; Mashiter,K. (Endocrine Unit, Dep. Med., R. Postgrad. Med. Sch., Du Cane Road, London W12 0HS, UK) *J. Endocrinol., 77(3), 423-424 (1978)* En.

2500-U2 **A radioimmunoassay for the estimation of serum dehydroepiandrosterone sulphate in normal and pathological sera.** Smith,M.R.; *Rudd,B.T.; Shirley,A.; Rayner,P.H.W.; Williams,J.W.; Duignan,N.M.; Bertrand,P.V. (Birmingham and Midland Hosp. Women, Showell Green Lane, Birmingham B11 4HL, UK) *Clin. Chim. Acta, 65(1), 5-13 (1975)* En;en.

2501-U2 **A rapid, sensitive and specific radioimmunoassay for human chorionic gonadotrophin.** Kardana,A.; Bagshawe,K.D. (Dep. Med. Oncol., Charing Cross Hosp., Fulham, London, UK) *J. Immunol. Methods, 9(3/4), 297-305 (1976)* En;en.

2502-U2 **[Modification of a double antibody radioimmunoassay for human chorionic somatomammotropin (HCS) suitable for the highest concentration occurring in late pregnancy].** Glockner,E.; Kunkel,S.; Herre,H.-D.; Oertel,S. (Zentralinst. Diabetes, 'Gerhardt Katsch', DDR-2201 Karlsburg, GDR) *Zentralbl. Gynakol., 97(22), 1357-1363 (1975)* En;en.

2503-U2 **Parallel radioimmunoassay for plasma cortisol and 11-deoxycortisol.** Kao,M.; Voina,S.; Nichols,A.; Horton,R. (Steroid Dep., Nichols Inst. Endocrinol., 1300 S. Beacon St., San Pedro, CA 90731, USA) *Clin. Chem., 21(11), 1644-1647 (1975)* En;en.

2504-U2 **Semi-automatic solid-phase double-antibody radioimmunoassay for β_2-microglobulin.** Sevier,E.D.; Reisfeld,R.A. (Dep. Mol. Immunol., Scripps Clin., 476 Prospect Street, La Jolla, CA 92037, USA) *Immunochemistry, 13(1), 35-37 (1976)* En;en.

2505-U2 **A solid-phase radioimmunoassay for human β_2-microglobulin.** Taniguchi,N.; Tanaka,M.; Kobayashi,K.; Matsuda,I.; Ohno,H.; Sato,T.; Takakuwa,E. (Dep. Hyg. and Prevent. Med., Hokkaido Univ. Sch. Med., Sapporo 060, Japan) *Clin. Chim. Acta, 69(3), 471-477 (1976)* En;en.

2506-U2 **Measurement of ferritin in serum by radioimmunoassay.** Barnett,M.D.; Gordon,Y.B.; Amess,J.A.L.; Mollin,D.L. (Dep. Haematol., Med. Coll. St. Bartholomew's Hosp., London EC1A 7BE, UK) *J. Clin. Pathol., 31(8), 742-748 (1978)* En;en.

2507-U2 **Quantitation of human serum antithyroglobulin antibodies by a two-site radioimmunoassay.** Leonard,J.P.; Taymans,F.; Beckers,C. (Cent. Med. Nuclearire, Univ. Louvain, Med. Sch., UCL 54.30, B-1200 Brussels, Belgium) *Clin. Chim. Acta, 87(1), 1-9 (1978)* En;en.

2508-U2 **A simple and economical method or the radioimmunoassay of cortisol in serum.** Morris,R. (Dep. Clin. Endocrinol., Women's Hosp., Showell Green Lane, Sparkhill, Birmingham, B11 4HL, UK) *Ann. Clin. Biochem., 15(Part 3),*

178-183 (1978) En;en.

2509-U2 Fibrinogen ^{131}I scans for detection of experimental abscesses. Sham,R.; Silver,L.; Deysine,M. (Dep. Nucl. Med., Queens Hosp. Cent., New Hyde Park, NY 11040, USA) *Surg. Forum, 27, 20-22 (1976)* En.

2510-U2 Polyethylene glycol in radioimmunoassay of TSH. Kadival,G.V.; Samuel,A.M. (Radiat. Med. Cent., Bhabha At. Res. Cent., Tata Mem. Hosp., Parel, Bombay, India) *Indian J. Med. Res., 64(10), 1537-1542 (1976)* En;en.

2511-U2 An improved double antibody radioimmunoassay method for the measurement of ragweed antigen E. Schable,C.A.; Northey,W.T. (Phoenix Lab. Div., Bureau Epidemiol., Cent. Dis. Cont., Public Health Serv., US Dep. Health, Educat., and Welfare, 4402 North Seventh St., Phoenix, AZ 85014, USA) *Ann. Allergy, 38(6), 413-415 (1977)* En;en.

2512-U2 Radioimmunoassay for the determination of digoxin and related compounds in *Digitalis lanata*. Weiler,E.W.; Zenk,M.H. (Lehrstuhl Pflanzenphysiol., Ruhr-Univ. Bochum D-4630 Bochum, GFR) *Phytochemistry., 15(10), 1537-1545 (1976)* En;en.

See: 808, 1506, 1931, 2116, 2570, 2580, 2611, 2630, 3249, 3254, 3300, 3302, 3243, 3247, 3252, 3255, 3279, 3301, 3253, 3263, 3256, 3264

Enzyme-linked immunoassays

2513-U2 Immunoperoxidase staining for detection of Colorado tick fever virus, and a study of congenital infection in the mouse. Desmond,E.P.; Schmidt,N.J.; Lennette,E.H. (Microbiol. Lab., Dep. Public Health, 101 Grove St., San Francisco, CA 94102, USA) *Am. J. Trop. Med. Hyg., 28(4), 729-732 (1979)* En;en.

2514-U2 The immunoperoxidase technique for rapid human cytomegalovirus identification. Gerna,G.; Vasquez,A.; McCloud,C.J.; Chambers,R.W. (Viral Diagn. Serv., Dep. Pathol., Georgetown Univ. Med. Cent., Washington, DC 20007, USA) *Arch. Virol., 50(4), 311-321 (1976)* En;en.

2515-U2 Immunoperoxidase technique for detection of antibodies to human cytomegalovirus. Gerna,G.; McCloud,C.J.; Chambers,R.W. (Dep. Pathol., Georgetown Univ. Med. Cent., Washington, DC 20007, USA) *J. Clin. Microbiol., 3(3), 364-372 (1976)* En;en.

2516-U2 Rapid detection of human cytomegalovirus and *Herpesvirus hominis* IgM antibody by the immunoperoxidase technique. Gerna,G.; Chambers,R.W. (Ist. Mal. Infett., Univ. Pavia, I 27100 Pavia, Italy) *Intervirology, 8(5), 257-271 (1977)* En;en.

2517-U2 An application of PAP (peroxidase-anti-peroxidase) staining technique for the rapid titration of dengue virus type 4 infectivity. Okuno,Y.; Sassao,F.; Fukunaga,T.; Fukai,K. (Dep. Prev. Med., Res. Inst. Microbial Dis., Osaka Univ., Yamada-Kami, Suita, Osaka, Japan) *Biken J., 20(1), 29-33 (1977)* En.

2518-U2 Use of indirect immunoperoxidase test for detection of porcine enteroviral antigens in infected PK15 cell cultures. Sulochana,S.; Derbyshire,J.B. (Dep. Vet. Microbiol. and Immunol., Univ. Guelph, Guelph, Ont., Canada) *Kerala J. Vet. Sci., 9(1), 111-119 (1978)* En;en,tamil.

2519-U2 An immunoperoxidase technique for demonstrating membrane localized HBsAg in paraffin sections of liver biopsies. Busachi,C.A.; Ray,M.B.; Desmet,V.J. (Lab. Histochem. en Cytochem., Dep. Med. Navorsing, Acad. Ziekenhuis Sint Rafael, B-3000 Leuven, Belgium) *J. Immunol. Methods, 19(1), 95-99 (1978)* En;en.

2520-U2 Techniques for typing *Herpesvirus hominis* antibody: a comparison of inhibition of peroxidase-labelled antibody staining with inhibition of indirect haemagglutination and with microneutralisation. Gerna,G.; *Chambers,R.W. (Viral Diagnostic Serv., Dep. Pathol., Georgetown Univ., Med. Cent., Washington, DC 20007, USA) *J. Med. Microbiol., 10(3), 309-316 (1977)* En;en.

2521-U2 Use of immunoperoxidase for rapid diagnosis of mucocutaneous herpes simplex virus infection. Benjamin,D.R. (Dep. Lab., Child. Orthopedic Hosp., Seattle, WA 98195, USA) *J. Clin. Microbiol., 6(6), 571-573 (1977)* En;en.

2522-U2 Peroxidase-labeled antibody technique for rapid detection of mouse hepatitis virus in cases of natural outbreaks. Carthew,P.(Med. Res. Counc. Lab. Anim. Cent., Woodmansterne Rd., Carshalton, Surrey SM5 4EF, UK) *J. Infect. Dis., 138(3), 410-412 (1978)* En;en.

2523-U2 Detection of viral proteins in mouse mammary tumors by immunoperoxidase staining of paraffin sections. Keydar,I.; Mesa-Tejada,R.; Ramanarayanan,M.; Ohno,T.; Fenoglio,C.; Hu,R.; Spieglman,S. (Inst. Cancer Res., Coll. Physicians and Surg., Columbia Univ., 701 West 168th St., New York, NY 10032, USA) *Proc. Natl. Acad. Sci. USA, 75(3), 1524-1528 (1978)* En;en.

2524-U2 Use of immunoenzyme technique for the detection of Aujesky's disease virus in cell culture. Roszkowski,J.; Bartoszcze,M.; Zadura,J.; Swiatek,Z. (Vet. Res. Inst., 24-100 Pulawy, Poland) *Vet. Rec., 102(21), 462-463 (1978)* En. [Immunoperoxidase test].

2525-U2 Rubella antibody assay by the immunoperoxidase technique: comparison with the hemagglutination inhibition test for determination of immune status. Gerna,G.; Chambers,R.W. (Viral Diagn. Serv., Dep. Pathol., Georgetown Univ., Washington, DC 20007, USA) *J. Infect. Dis., 133(4), 469-472 (1976)* En;en.

2526-U2 [Use of immunoperoxidase technique for studying Teschen virus]. Metianu,T.; Virat,J.; Cuong,T. (Inst. Pasteur, 25, rue du Dr-Roux, 75015 Paris, France) *Rec. Med. Vet., 151(8-9), 491-497 (1975)* Fr;en,es,fr.

2527-U2 Immunoperoxidase localization of *Treponema pallidum*. Its use in formaldehyde-fixed and paraffin-

embedded tissue sections. Beckett,J.H.; Bigbee,J.W. (Dep. Dermatol., R-132, Stanford Univ. Med. Cent., Stanford, CA 94305, USA) *Arch. Pathol. Lab. Med., 103(3), 135-138 (1979)* En;en.

2528-U2 [Significance of indirect immunoperidoxase reaction for amoebiasis serodiagnosis]. Ahmed,L.; Barbier,M.; Dodin,A. (Fac. Med. Caire (Serv. Med. Trop.), Cairo, Egypt) *Bull. Soc. Pathol. Exot., 70(2), 155-160 (1977)* Fr;en.

2529-U2 Non-equilibrium enzyme immunoassay of gentamicin. Matiasson,B.; Svensson,K.; Borrebaeck,C.; Jonsson,S.; Kronvall,G. (Biochem. 2, Chem. Cent., Univ. Lund, S-220 07 Lund 7, Sweden) *Clin. Chem., 24(10), 1770-1773 (1978)* En;en.

2530-U2 Practical colorimeter for direct measurement of microplates in enzyme immunoassay systems. Clem,T.R.; Yolken,R.H. (Biomed. Eng. and Instrument. Branch, Div. Res. Serv., Natl. Inst. Health, Bethesda, MD 20014, USA) *J. Clin. Microbiol., 7(1), 55-58 (1978)* En;en.

2531-U2 A new technique of heterogeneous enzyme linked immunosorbent assay, stick-ELISA. I. Description of the technique. Felgner,P. (Tropeninst., Benhard-Nocht-Str. 74, D-2000 Hamburg 4, GFR) *Zentralbl. Bakteriol. Parasitenkd. Infektionskr. Hyg., I Abt. A., 240(1), 112-117 (1978)* En;de,en.

2532-U2 ELISA - simple, sure and speedy diagnosis. Newell,J. (Address not stated) *Curr. Med. Pract., 21(2), 81-82 (1977)* En;en.

2533-U2 Rapid micromethod of screening for antibodies to disease agents using the indirect enzyme-labeled antibody test. Saunders,G.C.; Clinard,E.H. (Los Alamos Sci. Lab., Univ. California, Los Alamos, NM 87545, USA) *J. Clin. Microbiol., 3(6), 604-608 (1976)* En;en.

2534-U2 ELISA - simple, sure and speedy diagnosis. Newell,J. (Address not stated) *Curr. Med. Pract., 21(2), 81-82 (1977)* En;en.

2535-U2 Enzyme immunoassay of serum thyroxine with the AutoChemist multichannel analyzer. Galen,R.S.; Forman,D. (MetPath, Inc., 60 Commerce Way, Hackensack, NJ 07606, USA) *Clin. Chem., 23(1), 119-121 (1977)* En;en.

2536-U2 Automated computer analysis for enzyme-multiplied immunological techniques. Rodbard,D.; McClean,S.W. (Reprod. Res. Branch, Natl. Inst. Child Health and Human Dev., NIH, Bethesda, MD 20014, USA) *Clin. Chem., 23(1), 112-115 (1977)* En;en.

2537-U2 Direct measurement of microplates and its application to enzyme-linked immunosorbent assay. Ruitenberg,E.J.; Brosi,B.J.M.; Steerenberg,P.A. (Natl. Inst. Public Health, Bilthoven, Netherlands) *J. Clin. Microbiol., 3(5), 541-542 (1976)* En;en.

2538-U2 Mechanisation of the enzyme-linked immunosorbent assay (ELISA) for large scale screening of sera. Ruitenberg,E.J.; van Amstel,J.A.; Brosi,B.J.M.; Steerenberg,P.A. (Natl. Inst. Public Health, POB 1, Bilthoven, Netherlands) *J. Immunol. Methods, 16(4), 351-359 (1977)* En;en.

2539-U2 Direct measurement of microplates and its application to enzyme-linked immunosorbent assay. Ruitenberg,E.J.; Brosi,B.J.M.; Steerenberg,P.A. (Natl. Inst. Public Health, Bilthoven, Netherlands) *J. Clin. Microbiol., 3(5), 541-542 (1976)* En;en.

2540-U2 Enzyme immunoassays with special reference to ELISA techniques. Voller,A.; Bartlett,A.; Bidwell,D.E. (Dep. Clin. Trop. Med., London Sch. Hyg. and Trop. Med., Keppel Street, London WC1, UK) *J. Clin. Pathol., 31(6), 507-520 (1978)* En;en.

2541-U2 Evaluation of short reaction times and some characteristics of the enzyme-conjugation in enzyme-linked immunosorbent assay (ELISA). Korpraditskul,P.; Casper,R.; Lesemann,D.E. (Biol. Bundesanstalt Land- und Forstwirtschaft. Inst. Viruskrankheiten Pflanzen, Braunschweig, GFR) *Phytopathol. Z., 96(3), 281-285 (1979)* En;de,en.

2542-U2 Some characteristics of a new multiple channel photometer for through-the-plate reading of microplates to be used in enzyme-linked immunosorbent assay. Ruitenberg,E.J.; Sekhuis,V.M.; Brosi,B.J.M. (Natl. Inst. Public Health, Bilthoven, Netherlands) *J. Clin. Microbiol., 11(2), 132-134 (1980)* En;en.

2543-U2 Use of rabbit antibody IgG-loaded silicone pieces for the sandwich enzymoimmunoassay of macromolecular antigens. Kato,K.; Hamaguchi,Y.; Okawa,S.; Ishikawa,E.; Kobayashi,K.; Katunuma,N. (Dep. Biochem., Med. Coll. Miyazaki, Kiyotake, Miyazaki 889-16, Japan) *J. Biochem., 81(5), 1557-1566 (1977)* En;en.

2544-U2 Magnetic solid phase enzyme-immunoassay. Guesdon,J.-L.; Avrameas,S. (Unite Immunocytochim., Dep. Biol. Mol., Inst. Pasteur, 75015 Paris, France) *Immunochemistry, 14(6), 443-447 (1977)* En;en.

2545-U2 Use of rabbit antibody IgG bound onto plain and aminoalkylsilyl glass surface for the enzyme-linked sandwich immunoassay. Kato,K.; Hamaguchi,Y.; Okawa,S.; Ishikawa,E.; Kobayashi,K.; Katunuma,N. (Dep. Biochem., Med. Coll. Miyazaki, Kiyotake, Miyazaki-gun, Miyazaki 889-16, Japan) *J. Biochem., 82(1), 261-266 (1977)* En;en.

2546-U2 A new technique of heterogeneous enzyme-linked immunosorbent assay, stick-ELISA. I. Description of the technique. Felgner,P. (Tropeninst., Bernhard-Noecht-Str. 74, D-2000 Hamburg 4, GFR) *Zentralbl. Bakteriol. Parasitenkd. Infektionskr. Hyg., I. Abt. A., 240(1), 112-117 (1978)* En;de,en.

2547-U2 Automation of enzyme-linked immunosorbent assay (ELISA). Carlier,Y.; Bout,D.; Capron,A. (Cent. Immunol. et Biol. Parasit., Inst. Pasteur, Lille, France) *J. Immunol. Methods, 31(3-4), 237-246 (1979)* En;en.

2548-U2 A cheap ELISA-reader combined with the calculator/printer TI 59 of Texas Instruments. Fey,H.; Gottstein,B. (Postfach 2735, CH-3001 Bern, Switzerland) *Schweiz. Arch. Tierheilkd., 121(8), 387-394 (1979)* De;de,en,fr,it.

2549-U2 An 'antibody electrode', preliminary report on a new approach in enzyme immunoassay. Boitieux,J.-L.; *Desmet,G.; Thomas,D. (Lab. Biochim. Pathol., Fac. Med., 80036-Amiens Cedex, France) *Clin. Chem., 25(2), 318-321 (1979)* En;en.

2550-U2 Diffusion in gel-enzyme linked immunosorbent assay (DIG-ELISA): a simple method for quantitation of class-specific antibodies. Elwing,H.; Nygren,H. (Inst. Med. Microbiol., Guldhedsgagtan 10, S-413 46 Goteborg, Sweden) *J. Immunol. Methods, 31(1-2), 101-107 (1979)* En;en.

2551-U2 Enzyme-linked immunosorbent assay (ELISA). Duvivier,J. (Lab. Chim. Med., 1, rue des Bonnes-Villes, B 4020 Liege, Belgium) *Ann. Biol. Clin., 36(4), 389-391 (1978)* Fr;en,fr.

2552-U2 Enzyme immunoassays in diagnostic medicine. Theory and practice. Voller,A.; Bidwell,D.E.; Bartlett,A. (Dep. Clin. Trop. Med., Lond. Sch. Hyg. and Trop. Med., Keppel St., London, WC1, UK) *Bull. WHO, 53(1), 55-65 (1976)* En;en,fr.

2553-U2 Enzyme-linked sandwich immunoassay of macromolecular antigens using the rabbit antibody-coupled glass rod as a solid phase. Hamaguchi,Y.; Kato,K.; Fukui,H.; Shirakawa,I.; Okawa,S.; Ishikawa,E.; Kobayashi,K.; Katunuma,N. (Dep. Biochem., Med. Coll. Miyaaki, Miyazaki, Japan) *Eur. J. Biochem., 71(2), 459-467 (1976)* En;en.

2554-U2 [Enzyme-linked immunosorbent assay (ELISA)]. Anon (Div. Paludisme et autres mal. paras., O.M.S., 1211 Geneve 27, Switzerland) *Bull. WHO, 55(5), 557-568 (1977)* Fr.

2555-U2 Magnetic enzyme immunoassay of anti-grass pollen specific IgE in human sera. Guesdon,J.L.; David,B.; Lapeyre,J. (Unite Immunocytochim., Dep. Biol. Mol., 25, Rue du Docteur Roux, 75724 Paris Cedex 15, France) *Clin. Exp. Immunol., 33(3), 430-436 (1978)* En;en.

2556-U2 A new technique of heterogeneous enzyme-linked immunosorbent assay, stick-ELISA. I. Description of the technique. Felgner,P. (Tropeninst., Bernhard-Nocht-Str. 74, D-2000 Hamburg 4, GFR) *Zentralbl. Bakteriol. Parasitenkd. Infektionskr. Hyg., I Abt. A., 240(1), 112-117 (1978)* En;de,en.

2557-U2 A new technique of heterogeneous enzyme-linked immunosorbent assay, stick-ELISA. II. Application of stick-ELISA to antigens of various infective agents. Felgner,P. (Tropeninst., Bernhard-Nocht-Str. 74, D-2000 Hamburg 4, GFR) *Zentralbl. Bakteriol. Parsitenkd. Infektionskr. Hyg., I Abt. A., 240(1), 118-122 (1978)* En;de,en.

2558-U2 An enzyme-linked immunosorbent assay for the detection of canine antibodies to canine adenoviruses. Noon,K.F.; Rogul,M.; Binn,L.N.; Keefe,T.J.; Marchwicki,R.H.; Thomas,R.; De Cicco,B.T. (Dep. Clin. Invest. and Res., Div. Vet. Med., Walter Reed Army Inst. Res., Washington, DC 20012, USA) *Lab. Anim. Sci., 29(5), 603-609 (1979)* En;en.

2559-U2 Rapid detection and titer evaluation of viruses in pepper by enzyme-linked immunosorbent assay. Marco,S.;Cohen,S. (Dep. Virol., Volcani Cent., ARO, Bet Dagan, Israel) *Phytopathology, 69(12), 1259-1262 (1979)* En;en.

2560-U2 An enzyme-linked immunosorbent assay for detecting avian leukosis-sarcoma viruses. Smith,E.J.; Fadly,A.; Okazaki,W. (US Dep. Agric., Sci. and Educ. Adm., Agric. Res. Reg. Poult. Res. Lab., 3606 East Mount Hope Road, East Lansing, MI 48823, USA) *Avian Dis., 23(3), 698-707 (1979)* En;en.

2561-U2 Detection of avian oncovirus group-specific antigens by the enzyme-linked immunosorbent assay. Clark,D.P.; Dougherty,R.M. (Dep. Microbiol., State Univ. New York, Upstate Med. Cent., 766 Irving Ave., Syracuse, NY 13210, USA) *J. Gen. Virol., 47(2), 283-291 (1980)* En;en.

2562-U2 Detection of barley yellow dwarf virus by enzyme-linked immunosorbent assay. Lister,R.M.; Rochow,W.F. (Dep. Bot. and Plant Pathol., Purdue Univ., West Lafayette, IN 47907, USA) *Phytopathology, 69(6), 649-654 (1979)* En;en.

2563-U2 Comparative diagnosis of barley yellow dwarf by serological and aphid transmission tests. Rochow,W.F. (Agric. Res., Sci. and Educ. Adm., US Dep. Agric., Ithaca, NY 14853, USA) *Plant Dis. Rep., 63(5), 426-430 (1979)* En;en.

2564-U2 Serologic detection of bean common mosaic and lettuce mosaic viruses in seed. Jafarpour,B.; Shepherd,R.J.; Grogan,R.G. (Dep. Plant Pathol., Univ. California, Davis, CA 95616, USA) *Phytopathology, 69(10), 1125-1129 (1979)* En;en.

2565-U2 Assay for the detection of antibodies against bovine leucosis virus. Behrens,F.; Ziegelmaier,R.; Toth,T.; v.Keyserlingk,M.; Forschner,E. (Behringwerke AG, Postfach 1140, D-3550 Marburg 1, GFR) *Berl. Munch. Tierarztl. Wochenschr., 92(22), 429-432 (1979)* De;de,en.

2566-U2 Detection of bovine leucosis virus infection in cattle and sheep by the enzyme-linked immunosorbent assay (ELISA). Ferencz,G.; Balla,E.; Horvath,Z.; Kassai,T. (Univ. Vet. Sci. Budapest, Belgyogyaszati Tanszeke and Klin., H-1078 Budapest 7, Hungary) *Magy. Allatorv. Lapja, 34(12), 799-801 (1979)* Hu;de,en,hu,ru.

2567-U2 The use of enzyme-linked immunosorbent assay for detection of citrus tristeza virus. Bar-Joseph,M.; Garnsey,S.M.; Gonsalves,D.; Moscovitz,M.; Purcifull,D.E.; Clark,M.F.; Loebenstein,G. (Virus Lab., Agric. Res. Org., Volcani Cent., Bet Dagan, Israel) *Phytopathology, 69(2), 190-*

194 (1979) En;en.

2568-U2 Serological detection and characterization of natural infections by cucumber mosaic virus (CMV). Devergne,J.C.; Cardin,L.; Quiot,J.B. (Stn. Bot. et Pathol. Veg., Cent. Rech. Agron. Antibes, INRA, 06602 Antibes Cedex, France) *Ann. Phytopathol., 10(2), 233-246 (1978)* Fr;en,fr.

2569-U2 Rapid detection of IgG and IgM antibodies for cytomegalovirus by the enzyme linked immunosorbent assay (ELISA). Cappel,R.; de Cuyper,F.; de Braekeleer,J. (Dep. Virol., Inst. Pasteur du Brabant, Brussels, Belgium) *Arch. Virol., 58(3), 253-258 (1978)* En;en.

2570-U2 Application of an immunoenzymatic method and an immunoautoradiographic method for a mass survey of nasopharyngeal carcinoma. Zeng,Y.; Liu,Y.; Liu,C.; Chen,S.; Wei,J.; Zhu,J.; Zai,H. (Inst. Virol., Chinese Acad. Med. Sci., Beijing, China) *Intervirology, 13(3), 162-168 (1980)* En;en.

2571-U2 Specific detection of IgM-antibodies by ELISA, applied in hepatitis-A. Duermeyer,W.; van der Veen,J. (Organon Sci. Dev. Group, POB 20, Oss, Netherlands) *Lancet, 2(8091), 684-685 (1978)* En.

2572-U2 Solid-phase enzyme-linked immunosorbent assay for detection of hepatitis A virus. Locarnini,S.A.; Garland,S.M.; Lehmann,N.I.; Pringle,R.C.; Gust,I.D. (Virus Lab., Fairfield Hosp. Commun. Dis., Fairfield, 3078, Australia) *J. Clin. Microbiol., 8(3), 277-282 (1978)* En;en.

2573-U2 ELISA in hepatitis A. Duermeyer,W.; van der Veen,J.; Koster,B. (Organon Scientific Development Group, PO Box 20, Oss, Netherlands) *Lancet, 1(8068), 823-824 (1978)* En.

2574-U2 Detection of anti-HBs antibody by means of an indirect micro-ELISA method. Spano,C.; Patti,S.; Palazzo,U. (Bacteriol. and Virol. Serv., Osp. 'V. Cervello', Via Trabucco, 180 Palermo, Italy) *Bull. WHO, 56(4), 653-654 (1978)* En;en.

2575-U2 Enzyme-linked immunoassay (ELISA): its practical application to the diagnosis of hepatitis B. Vandervelde,E.M. (Virus Ref. Lab., Cent. Public Health Lab., Colindale Ave., Colindale, London NW9, UK) *J. Med. Virol., 3(1), 17-18 (1978)* En.

2576-U2 Enzyme-immunoassay for the detection of hepatitis B surface antigen. Zanetti,A.R.; Ferroni,P.; Quartodipalo,M.E. (Inst. Virol., Univ. Milano, Via Pascal, 38-20133 Milano, Italy) *Boll. Ist. Sieroter. Milan, 56(4), 397-398 (1977)* En.

2577-U2 Demonstration of antibodies against HBs antigen with the modified MICROELISA-hepanostika-test. Lange,W.; Kohler,H. (Abt. Virol., Robert-Koch-Inst. Bundesgesundheitsamtes, 1000 Berlin 65, Nordufer 20, GFR) *Dtsch. Med. Wochenschr., 103(47), 1873-1877 (1978)* De;de,en.

2578-U2 Selective detection of IgM-antibody against core antigen of the hepatitis B virus by a modified enzyme immune assay. Gerlich,W.H.; Luer,W. (Hyg.-Inst., Univ. Gottingen (Lehrstuhl Med. Mikrobiol.) and Natl. Ref. Lab. Viral Hepatitis, Kreuzbergring 57, 34 Gottingen, GFR) *J. Med. Virol., 4(3), 227-238 (1979)* En;en.

2579-U2 Enzyme immunoassay for hepatitis B and its comparison to other methods. Caldwell,C.W.; *Barrett,J.T. (Dep. Microbiol., Univ. Missouri, M-264 Med. Sci. Build., Columbia, MO 65201, USA) *Clin. Chim. Acta, 81(3), 305-309 (1977)* En;en.

2580-U2 [Enzyme-immunoassay (EIA): comparative evaluation of the method for the study of HB_sAg]. Taliani,G.; Muglia,M.; Franco,E.; Nigro,A. (Clin. Malattie Infettive, Univ. Roma, Roma, Italy) *Boll. Soc. Ital. Biol. Sper., 53(20), 1752-1754 (1977)* It.

2581-U2 Specificity control in solid-phase enzyme immunoassay for HB_sAg by one-step in situ blocking with human anti-HB_s. Kacaki,J.; Wolters,G.; Kuijpers,L.; Schuurs,A. (Gemeente Ziekenhuis, Streeklab. Bacteriol., Wagnerlaan 55, Arnhem, Netherlands) *J. Clin. Pathol., 30(9) 894-898 (1977)* En.

2582-U2 Serotyping herpes simplex virus isolates by enzyme-linked immunosorbent assays. Mills,K.W.; *Gerlach,E.H.; Bell,J.W.; Farkas,M.E.; Taylor,R.J. (Microbiol. Div., Dep. Lab., St. Francis Hosp., Wichita, KS 67214, USA) *J. Clin. Microbiol., 7(1), 73-76 (1978)* En;en.

2583-U2 Enzyme immunoassay for measurement of antibodies to herpes simplex virus infection: comparison with complement fixation, immunofluorescent-antibody, and neutralization techniques. Denoyel,G.A.; Gaspar,A.; Nouyrigat,C. (Serv. Virus, Inst. Pasteur Lyon, 69365 Lyons Cedex 2, France) *J. Clin. Microbiol., 11(2), 114-119 (1980)* En;en.

2584-U2 Development and evaluation of an enzyme-labeled antibody test for the rapid detection of hog cholera antibodies. Saunders,G.C. (Agric. Biosci., Group H-6, MS 544, Los Alamos Sci. Lab., PO Box 1663, Los Alamos, NM 87545, USA) *Am. J. Vet. Res., 38(1), 21-25 (1977)* En;en.

2585-U2 The detection of three viruses of hop (*Humulus lupulus*) by enzyme-linked immunosorbent assay (ELISA). Thresh,J.M.; Adams,A.N.; Barbara,D.J.; Clark,M.F. (East Malling Res. Stn., Maidstone, Kent, ME19 6BJ, UK) *Ann. Appl. Biol., 87(1), 57-65 (1977)* En;en.

2586-U2 A microplate immunoenzyme assay for antiinfluenza antibodies. Lambre,C.; Kasturi,K.N. (Unite Ecol. Virale, Inst. Pasteur, 25, Rue du Docteur Roux, 75724 Paris Cedex 15, France) *J. Immunol. Methods, 26(1), 61-67 (1979)* En;en.

2587-U2 Enzyme-linked immunosorbent assay for detection of antibodies to influenza A and B and parainfluenza type 1 in sera of patients. Bishai,F.R.; Galli,R. (Lab. Serv. Branch, Ontario Minist. Health, Toronto, Ont., Canada) *J. Clin. Microbiol., 8(6), 648-656 (1978)* En;en.

2588-U2 Detection of maize dwarf mosaic virus with enzyme-linked immunosorbent assay (ELISA). Sum,I.; Nemeth,M.; Pacsa,A.S. (Plant Protection and Agrochem. Cent., Minist. Agric. and Food XI Budaorsi ut 141-145, H-1502 Budapest, Hungary) *Phytopathol. Z., 95(3), 274-278 (1979)* En;de,en.

2589-U2 Evaluation of enzyme-linked immunosorbent assay (ELISA) for mumps virus antibodies. Leinikki,P.O.; Shekarchi,I.; Tzan,N.; Madden,D.L.; Sever,J.L. (Infect. Dis. Branch, Natl. Inst. Neurol. and Commun. Disorders and Stroke, Bethesda, MD 20014, USA) *Proc. Soc. Exp. Biol. Med., 160(3), 363-367 (1979)* En;en.

2590-U2 Enzyme immunoassay for feline oncornavirus-associated cell membrane antigen (FOCMA) and detection of FOCMA in cell extract by enzyme immunoassay inhibition test. Log,T.; Chang,K.S.S. (Viral Oncog. Sect., Lab. Cell Biol., Natl. Cancer Inst., Bethesda, MD 20014, USA) *J. Immunol. Methods, 26(3), 291-303 (1979)* En;en.

2591-U2 Detection of bacteriophages in whey by an enzyme-linked immunosorbent assay (ELISA). Lembke,J.; Teuber,M. (Inst. Mikrobiol., Fed. Dairy Res. Cent., Kiel, GFR) *Milchwissenschaft, 34(8), 457-458 (1979)* En;de,en.

2592-U2 An enzyme-linked immunosorbent assay (ELISA) for PBS Z1, a defective phage of *Bacillus subtilis*. Steensma,H.Y.; Duermeyer,W. (Lab. Microbiol., Delft Univ. Technol., Delft, Netherlands) *J. Gen. Virol., 44(3), 741-746 (1979)* En;en.

2593-U2 Rapid detection and serotyping of prunus necrotic ringspot virus in perennial crops by enzyme linked immunosorbent assay. Barbara,D.J.; Clark,M.F.; Thresh,J.M.; Casper,R. (East Malling Res. Stn., Maidstone, Kent, ME19 6BJ, UK) *Ann. Appl. Biol., 90(3), 395-399 (1978)* En;en.

2594-U2 Application of an enzyme linked immunosorbent assay for diagnosis of Aujeszky's disease in swine. Moutou,F.; Toma,B.; Fortier,B. (Serv. Mal. Contagieuses, Ecole Vet., 94701 Maisons-Alfort, France) *Vet. Rec., 103(12), 264 (1978)* En.

2595-U2 Rapid detection of rabies antibodies by immunoenzymatic assay. Atanasiu,P.; Savy,V.; Perrin,P. (Unite Rage Rech. Rhabdovirus, Inst. Pasteur, 75724 Paris Cedex 15, France) *Ann. Microbiol., 128A(4), 489-498 (1977)* Fr;en,fr.

2596-U2 Rapid immunoenzymatic technique for titration of rabies antibodies IgG and IgM results. Atanasiu,P.; Savy,V.; Gibert,C. (Serv. Rage, Inst. Pasteur, Paris, France) *Med. Microbiol. Immunol., 166(1-4), 201-208 (1978)* En;en.

2597-U2 Rapid immunoenzymatic technique for titration of rabies antibodies IgG and IgM. Savy,V.; Atanasiu,P. (Unite Rage Recherche et Rhabdovirus, Inst. Pasteur, 25 rue du Dr Roux, 75015 Paris, France) *Dev. Biol. Stand., 40, 247-253 (1978)* En;en.

2598-U2 Enzyme immunoassay for demonstration of rabies-virus antibodies after immunisation. Thraenhart,O.; Kuwert,E.K. (Inst. Med. Virol. and Immunol., Univ. Essen, University Hosp., 4300 Essen 1, GFR) *Lancet, 2(8034), 399-400 (1977)* En.

2599-U2 Enzyme-linked immunosorbent assay (ELISA) for detection of human reovirus-like agent of infantile gastroenteritis. Yolken,R.H.; Kim,H.W.; Clem,T.; Wyatt,R.G.; Kalica,A.R.; Chanock,R.M.; Kapikian,A.Z. (Epidemiol. Sect., Lab. Infect. Dis., Natl. Inst. Allergy and Infect. Dis., Bethesda, MD 20014, USA) *Lancet, 2(8032), 263-266 (1977)* En;en.

2600-U2 Adaptation of enzyme-linked immunosorbent assay to the avian system. Slaght,S.S.; *Yang,T.-J.T.; van der Heide,L. (Dep. Pathobiol., Univ. Connecticut, Storrs, CT 06268, USA) *J. Clin. Microbiol., 10(5), 698-702 (1979)* En;en.

2601-U2 Comparison of an enzyme-linked immunosorbent assay for quantitiation of rotavirus antibodies with complement fixation in an epidemiological survey. Ghose,L.H.; Schnagl,R.D.; Holmes,I.H. (Microbiol. Dep., Univ. Melbourne, Parkville, Vic., Australia) *J. Clin. Microbiol., 8(3), 268-276 (1978)* En;en.

2602-U2 Rapid diagnosis of rotavirus gastroenteritis in children by immuno-enzymological assay. Peigne,H.; Beytout-Monghal,M.; Beytout-Mamouret,A.; Laluque,J.B.; Laveran,H.; Beytout,D.; Bourges,M. (Lab. Bacteriol.-Virol., Fac. Med., 28, Place Henri Dunant, 63001 Clermont-Ferrand Cedex, France) *Union Med. Can., 108(2), 146-149 (1979)* Fr;en,fr.

2603-U2 Enzyme-linked fluorescence assay: ultrasensitive solid-phase assay for detection of human rotavirus. Yolken,R.H.; Stopa,P.J. (Dep. Pediatr., Eudowood Div. Infect. Dis., Johns Hopkins Univ. Sch. Med., Baltimore, MD 21205, USA) *J. Clin. Microbiol., 317-321 (1979)* En;en.

2604-U2 Rapid diagnosis of rotavirus infection: comparison of electron microscopy and enzyme linked immunosorbent assay (ELISA). Robertson,D.M.; Harrison,M.; Hosking,C.S.; Adams,L.C.; *Bishop,R.F. (Dep. Gastroenterol., R. Child. Hosp., Flemington Road, Parkville, Vic. 3052, Australia) *Aust. Paediatr. J., 15(4), 229-232 (1979)* En;en.

2605-U2 Enyme-linked immunosorbent assay for diagnosis of rotavirus infections in calves. Ellens,D.J.; de Leeuw,P.W. (Cent. Vet. Inst., Virol. Dep., Lelystad, Netherlands) *J. Clin. Microbiol., 6(5), 530-532 (1977)* En;en.

2606-U2 Enzyme-linked immunosorbent assay determination of specific rubella antibody levels in micrograms of immunoglobulin G per milliliter of serum in clinical samples. Leinikki,P.O.; Shekarchi,I.; Dorsett,P.; Sever,J.L. (Inst. Biomed. Sci., Univ. Tampere, Tampere, Finland) *J. Clin. Microbiol., 8(4), 419-423 (1978)* En;en.

2607-U2 Solid-phase enzyme immunoassay for specific immunoglobulin G and immunoglobulin M antibodies of

rubella virus. Prevot,J.; Guesdon,J.-L. (Unite Virol. Med. et des Vaccins Viraux, Dep. Virol., Inst. Pasteur, 75724 Paris Cedex 15, France) *Ann. Microbiol., 128B(4), 531-540 (1977)* Fr;en,fr.

2608-U2 Diagnosis of postnatal rubella by the enzyme-linked immunosorbent assay for rubella IgM and IgG antibodies. Vejtorp,M.; Fanoe,E.; Leerhoy,J. (Rubella Dep., Statens Seruminst., Amager Blvd. 80, DK-2300 Copenhagen S, Denmark) *Acta Pathol. Microbiol. Scand. Ser. B, 87(3), 155-160 (1979)* En;en.

2609-U2 Enzyme-linked immunosorbent assay for determination of rubella IgG antibodies. Vejtorp.M. (Rubella Dep. Statens Seruminst., Amager Boulevard 80, DK-2300 Copenhagen S, Denmark) *Acta Pathol. Microbiol. Scand. Ser. B, 86(6), 387-392 (1978)* En;en.

2610-U2 A solid-phase enzyme-linked immunosorbent assay (ELISA) for detection of antibody to rubella virus. Garland,S.M.; Locarnini,S.A.; Gust,I.D. (Virus Lab., Fairfield Hosp. Communic. Dis., Yarra Bend Road, Fairfield, Vic. 3078, Australia) *Pathology, 11(3), 393-399 (1979)* En;en.

2611-U2 A comparison of labelled antibody methods for the detection of virus antigens in cell monolayers. Oram,J.D.; Crooks,A.J. (Microbiol. Res. Est., Porton Down, Salisbury, Wilts. SP4 0JG, UK) *J. Immunol. Methods, 25(4), 297-310 (1979)* En;en.

2612-U2 Use of the ELISA technique for the detection of soybean mosaic virus in soybean seeds. Bossennec,J.M.; Maury,Y. (Stn. Pathol. Veg., Cent. Natl. Rech. Agron., INRA, route de St-Cyr, 78000 Versailles, France) *Ann. Phytopathol., 10(2), 263-268 (1978)* En;en,fr.

2613-U2 Rapid diagnosis of tick-borne encephalitis by means of enzyme linked immunosorbent assay. Hofmann,H.; Frisch-Niggemeyer,W.; Heinz,F. (Inst. Virol., Univ. Vienna, Kinderspitalgasse 15, A-1095 Vienna, Austria) *J. Gen. Virol., 42(3), 505-511 (1979)* En;en.

2614-U2 Application of the enzyme-linked immunosorbent assay for detecting viruses in soybean seed and plants. Lister,R.M. (Dep. Bot. and Plant Pathol., Purdue Univ., West Lafayette, IN 47907, USA) *Phytopathology, 68(10), 1393-1400 (1978)* En;en.

2615-U2 Determination of virus-specific IgM antibodies by using ELISA: elimination of false-positive results with protein A-Sepharose absorption and subsequent IgM antibody assay. Leinikki,P.O.; Shekarchi,I.; Dorsett,P.; Sever,J.L. (Infect. Dis. Branch, Natl. Inst. Neurol. and Commun. Disorders and Stroke, NIH, Bethesda, MD 20014, USA) *J. Lab. Clin. Med., 92(6), 849-857 (1978)* En;en.

2616-U2 The application of immunoenzymatic method ELISA in virological diagnosis. Bartoszcze,M.; Roszkowski,J.; Brodzicki,S. (ul. Krancowa 1/19, 24-100 Pulawy, Poland) *Med. Weter., 34(10), 582-584 (1978)* Pl. [review]

2617-U2 Rapid micromethod of screening for antibodies to disease against using the indirect enzyme-labeled antibody test. Saunders,G.C.; Clinard,E.H. (Los Alamos Sci. Lab. (Univ. California), Los Alamos, NM 87545, USA) *J. Clin. Microbiol., 3(6), 604-608 (1976)* En;en.

2618-U2 New enzyme immunoassay for measurement of influenza A/Vatoria/3/75 virus in nasal washes. Berg,R.A.; Rennard,S.I.; Murphy,B.R.; Yoken,R.H.; Dolin,R.; Straus,S.E. (Lab. Clin. Invest., Natl. Inst. Allergy and Infat. Dis., NIH, Build. 10, Rm. 11N-113, Bethesda, MD 20205, USA) *Lancet, 1(8173), 851-853 (1980)* En;en.

2619-U2 An enzyme-labelled immunosorbent assay for *Brucella abortus* antibodies. Magee,J.T. (Bacteriol. Dep., Hallamshire Hosp., Glossop Road, Sheffield S10 2JF, UK) *J. Med. Microbiol., 13(1), 167-172 (1980)* En;en.

2620-U2 Single-antigen immunofluorescence test for chlamydial antibodies. Saikku,P.; *Paavonen,J. (Dep. Virol., Univ. Helsinki, Finland) *J. Clin. Microbiol., 8(2), 119-122 (1978)* En;en.

2621-U2 [Measurement of human tetanus antitoxin by enzyme-linked immunosorbent assay (ELISA).] Stiffler-Rosenberg,G.; Fey,H. (Vet.-Bakteriol. Inst., Univ. Bern, Langgasstr. 122, CH-3012 Bern, Switzerland) *Schweiz. Med. Wochenschr., 107(31), 1101-1104 (1977)* De;de,en.

2622-U2 [Determination of tetanus antitoxin in the horse with an enzyme linked immuno sorbent assay (ELISA)]. Fey,H.; Stiffler-Rosenberg,G. (Vet.-Bakteriol. Inst., Univ. Bern, Postfach 2735, CH-3001 Bern, Switzerland) *Schweiz. Arch. Tierheilkd., 119(11), 437-446 (1977)* De;de,en,fr,it.

2623-U2 Enzyme-linked immunosorbent assay for detection of *Escherichia coli* heat-labile enterotoxin. Yolken,R.H.; Greenberg,H.B.; Merson,M.H.; Sack,R.B.; Kapikian,A.Z. (Lab. Infect. Dis., Natl. Inst. Allergy and Infect. Dis., NIH, Bethesda, MD 20014, USA) *J. Clin. Microbiol., 6(5), 439-444 (1977)* En;en.

2624-U2 Detection of the K99 antigen of *Escherichia coli* in calf faeces by enzyme-linked immunosorbent assay (ELISA). Ellens,D.J.; de Leeuw,P.W.; Rozemond,H. (Cent. Vet. Inst., Virol. Dep., 8221 RA Lelystad, Netherlands) *Vet. Q., 1(4), 169-175 (1979)* En;en.

2625-U2 Enzyme-linked immunosorbent assay for detection and quantitation of capsular antigen of *Haemophilus influenzae* type b. Crosson,F.J.,Jr.; Winkelstein,J.A.; Moxon,E.R. (Dep. Pediatr., Kaiser Found. Hosp., Hayward, CA 94545, USA) *Infect. Immun., 22(2), 617-619 (1978)* En;en.

2626-U2 Indirect sandwich enzyme-linked immunosorbent assay for rapid detection of *Haemophilus influenzae* type b infection. Drow,D.L.; Maki,D.G.; Manning,D.D. (Dep. Pathol., Texas Tech. Univ. Med. Sch. at El Paso, El Paso, TX 79905, USA) *J. Clin. Microbiol., 10(4), 442-450 (1979)* En;en.

2627-U2 Rapid, simple enzyme immunoassay for gentamicin. Wills,P.J.; *Wise,R. (Dep. Med. Microbiol., Dudley Road Hosp., Birmingham, UK) *Antimicrob. Agents Chemother., 16(1), 40-42 (1979)* En;en.

2628-U2 The detection of antibodies to *Mycobacterium tuberculosis* by microplate enzyme-linked immunosorbent assay (ELISA). Nassau,E.; Parsons,E.R.; Johnson,G.D. (Dep. Pathol., Harefield Hosp., Harefield, Middx., UK) *Tubercle, 57(1), 67-70 (1976)* En;en,es,fr.

2629-U2 Sensitive enzyme-linked immunosorbent assay for detection of antibodies against typhus rickettsiae, *Rickettsia prowazekii* and *Rickettsia typhi*. Halle,S.; Dasch,G.A.; Weiss,E. (Dep. Microbiol., Naval Med. Res. Inst., Bethesda, MD 20014, USA) *J. Clin. Microbiol., 6(2), 101-110 (1977)* En;en.

2630-U2 Enzyme immunoassay and radioimmunoprecipitation test for the detection of antibodies to *Rochalimaea (Rickettsia) quintana*. Herrmann,J.E.; Hollingdale,M.R.; Collins,.F.; Vinson,J.W. (Dep. Microbiol., Harvard Sch. Public Health, Boston, MA 02115, USA *Proc. Soc. Exp. Biol. Med., 154(2), 285-288 (1977)* En;en.

2631-U2 Use of enzyme-labeled antibodies to detect *Salmonella* in foods. Krysinski,E.P.; *Heimsch,R.C. (Dep. Bacteriol. and Biochem., Univ. Idaho, Moscow, ID 83843, USA) *Appl. Environ. Microbiol., 33(4), 947-954 (1977)* En;en.

2632-U2 Parameters affecting the enzyme-linked immunosorbent assay of immunoglobulin G antibody to a rough mutant of *Salmonella minnesota*. Bruins,S.C.; Ingwer,I.; Zeckel,M.L.; White,A.C. (Dep. Intern. Med., Indiana Univ. Sch. Med., Indianapolis, IN 46202, USA) *Infect. Immun., 21(3), 721-728 (1978)* En;en.

2633-U2 Double-antibody solid-phase enzyme immunoassay for the detection of staphylococcal enterotoxin A. Saunders,G.C.; Bartlett,M.L. (Los Alamos Sci. Lab., Univ. California, Los Alamos, NM 87545, USA) *Appl. Environ. Microbiol., 34(5), 518-522 (1977)* En;en.

2634-U2 Simple assay for staphylococcal enterotoxins A, B, and C: modification of enzyme-linked immunosorbent assay. Stiffler-Rosenberg,G.; Fey,H. (Vet. Bacteriol. Inst., Univ. Berne, Berne, Switzerland) *J. Clin. Microbiol., 8(5), 473-479 (1978)* En;en.

2635-U2 Detection of staphylococcal enterotoxin B by means of the ELISA test. Simon,E.; Terplan,G. (Lehrstuhl Hyg. und Technol. Milch, Veterinarstr. 13, 8000 Munchen 22, GFR) *Zentralbl. Veterinarmed. B, 24(10), 842-844 (1977)* De;de,en.

2636-U2 Syphilis serology: ELISA, a third generation test. Sacchi,G.; Alessi,E.; Valcurone,G.; Buzzetti,C.; Tassi,G.C. (Ist. Sieroter. Milanese S. Belfanti, Via Darwin 22, 20143 Milano, Italy) *Boll. Ist. Sieroter. Milan., 57(4), 490-515 (1978)* It;en,it.

2637-U2 Microtiter enzyme-linked immunosorbent assay for immunoglobulin G cholera antitoxin in humans: method and correlation with rabbit skin vascular permeability factor technique. Young,C.R.; *Levine,M.M.; Craig,J.P.; Robins-Browne,R. (Cent. Vaccine Dev., Div. Infect. Dis., Univ. Maryland Sch. Med., Baltimore, MD 21201, USA) *Infect. Immun., 27(2), 492-496 (1980)* En;en.

2638-U2 ELISA methodology for polysaccharide antigens: protein coupling of polysaccharides for adsorption to plastic tubes. Gray,B.M. (Dep. Pediatr., Univ. Alabama in Birmingham, Sch. Med., University Station, Birmingham, AL 35294, USA) *J. Immunol. Methods, 28(1-2), 187-192 (1979)* En;en.

2639-U2 Detection by immuno-enzymology (micro-E.L.I.S.A.) of antibody and circulating antigens in visceral candidosis. Desgeorges,P.T.; *Ambroise-Thomas,P. (Serv. Parasitol., Consult. Mal. Parasitaires, Centre Hosp. Univ., B.P. 217 X, F 38043 Grenoble Cedex, France) *Lyon Med., 241(10), 653-657 (1979)* Fr;en,fr.

2640-U2 Diagnosis of parasitic diseases by means of enzyme-linked immunosorbent assay (ELISA) using a modified micromethod. 1. Technical aspects. Ambroise-Thomas,P.; Desgeorges,P.T. (Fac. Med., Univ. Grenoble, France) *Bull WHO, 56(4), 609-613 (1978)* Fr;en,fr.

2641-U2 A comparison of enzyme-linked immunosorbent assay and indirect fluorescent antibody test in the sero-diagnosis of cutaneous and visceral leishmaniasis in Iran. Edrissian,G.H.; Darabian,P. (Sch. Public Health., Univ. Teheran, P.O. Box 1310, Teheran, Iran) *Trans. R. Soc. Trop. Med. Hyg., 73(3), 289-292 (1979)* En;en.

2642-U2 Diagnosis and seroepidemiologic study of malaria by immunofluorescence, indirect hemagglutination and immunoenzymology. Ambroise-Thomas,P. (Dep. Parasitol., Univ. Grenoble, Grenoble, France) *Isr. J. Med. Sci., 14(6), 690-691 (1978)* En.

2643-U2 Immunoglobulin G and immunoglobulin M enzyme-linked immunosorbent assays and defined toxoplasmosis serological patterns. Camargo,M.E.; Ferreira,A.W.; Mineo,J.R.; Takiguti,C.K.; Nakahara,O.S. (Lab. Immunol., Inst. Med. Trop. Sao Paulo, Univ. Sao Paulo, Sao Paulo, Brazil) *Infect. Immun., 21(1), 55-58 (1978)* En;en.

2644-U2 Mechanizing the enzyme-linked immunosorbent assay (ELISA) in repressive testing for *Trichinella spiralis* infections in fattening pigs. Ruitenberg,E.J.; van Amstel,J.A.; Brosi,B.J.M.; Steerenberg,P.A. (Lab. Pathol., Rijks Inst. Volksgezondheid, Postbus 1, Bilthoven, Netherlands) *Tijdschr. Diergeneeskd., 102(17), 1021-1023 (1977)* Nl;en,nl.

2645-U2 Comparison of the diagnostic value of serum immunoglobulin levels, an enzyme-linked immunosorbent assay and a fluorescent antibody test in experimental infections with *Trypanosoma evansi* in rabbits. Luckins,A.G.; Gray,A.R.; Rae,P. (Cent. Trop. Vet. Med., Easter Bush, Roslin, Midlothian, UK) *Ann. Trop. Med. Parasitol., 72(5), 429-441 (1978)* En;en.

2646-U2 **Diagnosis of parasitic diseases by means of enzyme linked immunosorbent assay (ELISA) using a modified micromethod. I. Technical aspects.** Ambroise-Thomas,P.; Desgeorges,P.T. (Fac. Med., Univ. Grenoble, France) *Bull. WHO, 56(4), 609-613 (1978)* Fr;en,fr.

2647-U2 **The enzyme-linked immunosorbent assay (ELISA) - a new serodiagnostic method for the detection of parasitic infections.** Weiland,G. (Inst. Vgl. Tropenmed. und Parasitol., Fachber Tiermed., Univ. Munchen, Leopoldstr. 5, D-8000 Munchen 40, GFR) *Munch. Med. Wochenschr., 120(44), 1457-1460 (1978)* De;de,en.

2648-U2 **[Use of an immunoenzymatic method for the diagnosis of bilharziasis. Comparison with indirect immunofluorescence and immunoelectrophoresis techniques].** Gysin,J.; le Corroller,Y.; Pariaud,P. (Inst. Pasteur Gaudeloupe, Ponte-a-Pitre, Guadeloupe) *Med. Mal. Infect., 5(12), 560-563 (1975)* Fr;en,fr.

2649-U2 **The use of the enzyme-linked immunosorbent assay (ELISA) for the detection and quantification of specific antibody from cell cultures.** Kelly,B.S.; *Levy,J.G.; Sikora,L. (Dep. Microbiol., Univ. British Columbia, Vancouver, BC V6T 1W5, Canada) *Immunology, 37(1), 45-52 (1979)* En;en.

2650-U2 **Enzyme-immunoassay of grass-pollen and mites antibodies in rabbit antisera on magnetic polyacrylamide agarose beads.** Guesdon,J.L.; David,B.; Lapeyre,J. (Unit Immunocytochim., Dep. Biol. Mol., Inst. Pasteur, 75724 Paris Cedex 15, France) *Ann. Immunol., 128C(4-5), 799-810 (1977)* Fr;en,fr.

2651-U2 **Serodiagnosis of human cysticercosis by microplate enzyme-linked immunospecific assay (ELISA).** Arambulo,P.V.,III; Walls,K.W.; Bullock,S.; Kagan,I.G. (Dep. Vet. Microbiol., Iowa State Univ., Ames, IA 50011, USA) *Acta Trop., 35, 63-67 (1978)* En;en.

2652-U2 **A competitive enzyme-immunoassay using labelled antibody.** Halliday,M.I.; Wisdom,G.B. (Dep. Biochem., Queen's Univ., Belfast BT9 7BL, UK) *FEBS Lett., 96(2), 298-300 (1978)* En.

2653-U2 **Homogeneous enzyme immunoassay of serum digoxin with use of a bichromatic analyzer.** Bonne Eriksen,P.; Anderson,O. (Dep. Clin. Chem., Centralsygehust i Naestved, 4700 Naestved, Denmark) *Clin. Chem., 25(1), 169-171 (1979)* En;en.

2654-U2 **A sandwich method of enzyme-immunoassay. II. Quantification of rheumatoid factor.** Maiolini,R.; Ferrua,B.; Quaranta,J.F.; Pinoteau,A.; Euller,L.; Ziegler,G.; Masseyeff,R. (Lab. Immunol. et Serv. Rhumatol., Cent. Hosp.-Univ. Nice, 06031 Nice Cedex, France) *J. Immunol. Methods, 20, 25-34 (1978)* En;en.

2655-U2 **A totally automated system for enzyme immunoassay of theophylline in serum.** Castro,A.; Ibanez,J.; Voight,W.; Noto,T.; Malkus,H. (Dep. Pathol., Hormone Res. Lab., Univ. Miami, Sch. Med., Miami, FL 33152, USA) *Clin. Chem., 24(6), 944-946 (1978)* En;en.

2656-U2 **A sandwich method of enzymoimmunoassay. I. Application to rat and human alpha-fetoprotein.** Maiolini,R.; Masseyeff,R. (Lab. Immunol., Fac. Med., Ave. Vallombrose, 06000 Nice, France) *J. Immunol. Methods, 8(3), 223-234 (1975)* En;en.

2657-U2 **An enzyme immunoassay (EIA) for progesterone in horse plasma.** Seeger,K.; Thurow,H.; Haede,W.; Knapp,E. (Hoechst AG, D-6230 Frankfurt 80, GFR) *J. Immunol. Methods, 28(3-4), 211-217 (1979)* En;en.

2658-U2 **The rapid quantitation of serum alpha-fetoprotein by two-site micro enzyme immunoassay.** Macdonald,D.J.; Kelly,A.M. (Dep. Biochem., Glasgow R. Maternity Hosp., Rottenrow, Glasgow G4 0NA, UK) *Clin. Chim. Acta, 87(3), 367-372 (1978)* En;en.

2659-U2 **Factor-VIII-related antigen: measurement by enzyme immunoassay.** Barlett,A.; Dormandy,K.M.; Hawkey,C.M.; Stableforth,P.; Voller,A. (Nuffield Inst. Comp. Med., Zool. Soc. London, London NW1 4RY, UK) *Br. Med. J., 1(6016), 994-996 (1976)* En;en.

2660-U2 **Enzyme-linked immunoassay of thyrotropin.** Miyai,K.; Ishibashi,K.; Kumahara,Y. (Cent. Lab. Clin. Invest., Osaka Univ. Med. Sch., Dojima, Fukushima-ku, Osaka, Japan) *Clin. Chim. Acta, 67(3), 263-268 (1976)* En;en.

2661-U2 **An immunoenzyme assay for quantitation of human IgG antibodies to honeybee venom phospholipase A_2.** Tracy,R.P.; Alport,E.C.; Yunginger,J.W.; *Homburger,H.A. (Mayo Clin., 260 Hilton Build., Rochester, MN 55901, USA) *J. Immunol. Methods, 32(2), 185-193 (1980)* En;en.

2662-U2 **A simple enzyme immunoassay for the demonstration of rheumatoid factor.** Gripenberg,M.; Wafin,F.; Isomaki,H.; Linder,E. (Dep. Bacteriol. and Immunol., Univ. Helsinki, Haartmaninkatu 3, SF-00290 Helsinki 29, Finland) *J. Immunol. Methods, 31(1-2), 109-118 (1979)* En;en.

2663-U2 **Enzyme-linked immunosorbent assay for determination of IgM rheumatoid factor.** Vejtorp,M.; Hoier-Madsen,M.; Halberg,P. (Rubella Dep., Statens Seruminst., DK 2300 Copenhagen S, Denmark) *J. Rheumatol., 8(2), 65-70 (1979)* En;en.

2664-U2 **ELISA inhibition technique for the demonstration of sulphones in body fluids. I. Sulphones specific antibody-enzyme conjugate.** Huikeshoven,H.; de Wit,M.; Soeters,A.; Eggelte,T.A.; Landheer,J.E.; Leiker,D.L. (R. Trop. Inst., Amsterdam, Netherlands) *Lepr. Rev., 50(4), 275-281 (1979)* En;en.

2665-U2 **An in vitro immuno-enzymatic assay of tumor antigens in the mouse with β-galactosidase.** Celada,F.; Natali,P.G.; Radojkovic,J. (Lab. Biol. Cell., CNR, Roma, Italy) *J. Immunol., 117(3), 904-910 (1976)* En;en.

2666-U2 **Evaluation of the homogeneous enzyme immunoassay (EMIT) for serum digoxin determination.** Sparkes,M.D. (Misericordia Gen. Hosp., Winnipeg, Manit.,

Canada) *Can. J. Med. Technol., 40(5), 144-148 (1978)* En;en,fr.

2667-U2 A solid-phase enzyme-linked immunosorbent assay for the quantitation of human plasma α_1-acid glycoprotein. Wang,H.-P.; *Chu,C.Y.-T. (Dep. Med., Wayne State Univ., 3990 John R St., Detroit, MI 48201, USA) *Clin. Chem., 25(4), 546-549 (1979)* En;en.

2668-U2 An enzyme-linked immunoassay of testosterone. Rajikowski,K.M.; Cittanova,N.; Urios,P.; Jayle,M.F. (Lab. Chimie Biol., Fac. Med., 45, rue des Saints-Peres, 75006-Paris, France) *Steroids, 30(1), 129-137 (1977)* En;en.

2669-U2 A highly sensitive and specific enzyme-immunoassay method for oestradiol-17β. Exley,D.; Abuknesha,R. (Dep. Biochem., Queen Elizabeth Coll., Univ. London, London W8 7AH., UK) *FEBS Lett., 91(2), 162-165 (1978)* En.

See: 1104, 1105, 1273, 1395, 1398, 1750, 1958, 2008, 2009, 2028, 2042, 2190, 2267, 2388, 3262, 3274, 3301, 1044, 2040, 3249, 3254, 3264, 32 3276, 3302, 3257, 3266, 3279

General biochemical techniques

2670-U2 Methods in microbiology. Volume 9. Norris,J.R. (ed.) *Publ. by:* Academic Press Inc., (London) Ltd., 24-28 Oval Road, London NW1 7DX,UK, 13 Dec 1976. ISBN 0-12-521509-6 at £9.00 or $19.75. En.

2671-U2 Use of hydroxyapatite-coated glass beads for preclinical testing of potential antiplaque agents. Sudo,S.Z.; Schotzko,N.K.; Folke,L.E.A. (Div. Periodontol.,ch. Dent., Univ. Minnesota, Minneapolis, MN 55455, USA) *Appl. Environ. Microbiol., 32(3), 428-432 (1976)* En;en.

2672-U2 A simple method of preserving bacteria. Segerstrom,N.; Sabiston,C.B.,Jr. (710 Diana Court, Iowa City, IA 52240, USA) *J. Am. Med. Technol., 39(1), 22-23 (1977)* En;en.

2673-U2 Preservation of enzymic activity in marine plankton by low-temperature freezinç Ahmed,S.I.; Kerner,R.A.; King,F.D. (Dep. Oceanogr., Univ. Washington, Seattle, WA 98122, USA) *Mar. Chem., 4(2), 133-139 (1976)* En;en.

2674-U2 Permeabilization of microorganisms by Triton X-100. Miozzari,G.F.; Niederberger,P.; Hutter,R. (Mikrobiol. Inst., Eidg. Tech. Hochsch. Zurich, ETH-Zentrum CH-8092 Zurich, Switzerland) *Anal. Biochem., 90(1), 220-233 (1978)* En;en.

2675-U2 Detection of immune complexes. Margni,R.A.; Perdigon,G.; Manghi,M.; Hajos,S. (Dep. Microbiol. e Immunol., Fac. Farm. y Bioquim., Univ. Buenos Aires, Junin 956, 1113 Buenos Aires, Argentina) *Medicina (B. Aires), 39(2), 183-186 (1979)* En;en,es.
[immune complexes, Fc receptors, chicken erythrocytes.]

2676-U2 Direct detection of antigens in sodium dodecyl sulfate-polyacrylamide gels. Olden,K.; Yamada,K.M. (Lab. Mol. Biol., Natl. Cancer Inst., Natl. Inst. Health, Bethesda, MD 20014, USA) *Anal. Biochem., 78(2), 483-490 (1977)* En;en.

2677-U2 Membrane electrode measurement of lysozyme enzyme using living bacterial cells. D'Orazio,P.; Meyerhoff,M.E.; *Rechnitz,G.A. (Dep. Chem., Univ. Delaware, Newark, DE 19711, USA) *Anal. Chem., 50(11), 1531-1534 (1978)* En;en.

2678-U2 Analytical isoelectric focusing in polymerizable thin layers containing Sephadex. Ziegler,A.; Kohler,G. (Basel Inst. Immunol., 487 Grenzacherstr., Postfach 4005 Basel 5, Switzerland) *FEBS Lett., 64(1), 48-51 (1976)* En.

2679-U2 A simple method for the preparation of enzyme-antibody conjugates. O'Sullivan,M.J.; Gnemmi,E.; Morris,D.; Chieregatti,G.; Simmons,M.; Simmonds,A.D.; Bridges,J.W.; Marks,V. (Dep. Biochem., Div. Clin. Biochem., Univ. Surrey, Guildford, Surrey GU2 5XH, UK) *FEBS Lett., 95(2), 311-313 (1978)* En.

2680-U2 Colloidal silica aluminium modified PVP density gradient centrifugation: centrifuge tube wall cell adherence, aggregation, separation properties and comparison to BSA and Ficoll. Dettman,G.L.; Wilbur,S.M. (Dep. Radiol. Sci., Univ. California at Irvine, Coll. Med., Irvine, CA 92717, USA) *J. Immunol. Methods, 27(3), 205-217 (1979)* En;en.

2681-U2 Reversed immunosorbents: a simple method for specific antibody immobilization. Levy,D.E.; Eveleigh,J.W. (Univ. Tennessee Mem. Res. Cent., 1924 Alcoa Highway, Knoxville, TN 37920, USA) *J. Immunol. Methods, 22(1-2), 131-142 (1978)* En;en.

2682-U2 Immunocapillarymigration - a new method for immunochemical quantitation. Glad,C.; Grubb,A.O. (Dep. Clin. Chem., Univ. Lund, Malmo Gen. Hosp., S-214 01 Malmo, Sweden) *Anal. Biochem., 85(1), 180-187 (1978)* En;en.

2683-U2 Immunofixation. I. General principles and application to agarose gel electrophoresis. Ritchie,R.F.; Smith,R. (Rheum. Dis. Lab., Maine Med. Cent., Portland, ME 04102, USA) *Clin. Chem., 22(4), 497-499 (1976)* En;en.

2684-U2 A cellulose acetate immunofixation technique. Kohn,J.; Riches,P.G. (Suprareg. Prot. Ref. Unit, Putney Hop., London SW15 1HW, UK) *J. Immunol. Methods, 20, 325-331 (1978)* En;en.

2685-U2 A review of automation and rapid methods in microbiology. Newsom,S.W.B. (John Bonnett Clin. Lab., Addenbrooke's Hosp., Cambridge CB2 2QQ, UK) *Med. Lab. Sci., 35(3), 215-222 (1978)* En.

2686-U2 Isoelectric focusing in agarose. Saravis,C.A.; Zamcheck,N. (Mallory Gastrointestinal Res. Lab., Boston City Hosp., Boston, MA 02118, USA) *J. Immunol. Methods, 29(1), 91-96 (1979)* En;en.

2687-U2 Evaluation of low ionic strength in the screening and identification of antibodies. Carbonell,F.; Rafecas,F.J.; Vinas,M.T.; Montoro,J.A.; Marty,M.L. (Serv. Hematol. y Hemoter., Ciudad Sanitaria Seguridad Social 'La Fe', Valencia 9, Spain) *Sangre, 23(3), 271-277 (1978)* Es;en,es.

2688-U2 Least-squares method for restriction mapping. Schroeder,J.L.; *Blattner,F.R. (Dep. Genet., Univ. Wisconsin, Madison, WI 53706, USA) *Gene, 4(2), 167-174 (1978)* En;en.

2689-U2 Modified Farr's method and methods used for determination of the antibody heterogeneity index and of the mean association constant. Levi,M.I.; Baturova,R.S.; Apanovsky,L.N. (Cent. Control Res. Lab., Moscow Disinfect. Stn., Moscow, USSR) *Zh. Mikrobiol. Epidemiol. Immunobiol., 55(8), 117-122 (1978)* Ru;en,ru.

2690-U2 Rapid enzymologic anti-neuraminidase antibody microtest. Thraenhart,O.; Kuwert,E.K. (Inst. for Med. Virol. and Immunol., Univ. Essen, Gesamthochschule, Hufelandstr. 55, 4300 Essen 1, GFR) *Dev. Biol. Stand., 39, 475-480 (1977)* En;en.

2691-U2 An improved modification of cycle filtration leucopheresis. Kretschmer,V.; Mueller-Eckhardt,Ch. (Inst. Klin. Immunol. und Transfusionsmed., Langhausstr. 7, D-6300 Giessen, GFR) *Blut, 35(5), 415-418 (1977)* En;de,en.

2692-U2 Rapid screening procedure for detection of plasmids in streptococci. Leblanc,D.J.; Lee,L.N. (Lab. Microbiol. and Immunol., Natl. Inst. Dent. Les., Bethesda, MD 20205, USA) *J. Bacteriol., 140(3), 1112-1115 (1979)* En;en.

2693-U2 Simple method for demonstrating small plasmid deoxyribonucleic acid molecules in oral streptococci. Macrina,F.L.; Wood,P.H.; Jones,K.R. (Dep. Microbiol., Virginia Commonwealth Univ., Richmond, VA 23298, USA) *Appl. Environ. Microbiol., 39(5), 1070-1073 (1980)* En;en.

2694-U2 Detection of FP plasmids in hospital isolates of *Pseudomonas aeruginosa*. Dean,H.F.; Royle,P.; *Morgan,A.F. (Dep.Genet., Monash Univ., Clayton, Vic. 3168, Australia) *J. Bacteriol., 138(1), 249-250 (1979)* En;en.

2695-U2 Purification and mapping of specific mRNAs by hybridization-selection and cell-free translation. Ricciardi,R.P.; Miller,J.S.; Roberts,B.E. (Dep. Biol. Chem., Harvard Med. Sch., Boston, MA 02115, USA) *Proc. Natl. Acad. Sci. USA, 76(10), 4927-4931 (1979)* En;en.

2696-U2 Immunodiffusion and neutralization studies on porcine adenoviruses. Hafez,S.M.;Liess,B. (Inst. Virol., Hannover Sch. Vet. Med., Bischofscholer Damm 15, 3000 Hannover, GFR) *Vet. Microbiol., 5(2), 101-111 (1980)* En;de,en.

2697-U2 A rapid screening for the specific DNA sequence: analysis of transforming DNA segments in adenovirus-transformed cells. Fujinaga,K.; Sawada,Y.; Uemizu,Y. (Dep. Mol. Biol., Cancer Res. Inst., Sapporo Med. Coll., Minami-1-jo, Nishi-17-chome, Chuo-ku, Sapporo 060, Japan) *Gann, 70(2), 239-243 (1979)* En;en.

2698-U2 Antihemagglutinin exhaustion test in the study of the activity of antigenic preparations of arboviruses. Kokorev,V.S.; Gaidamovich,S.Ya.; Fedotova,T.T.; Filippovets,R.V. (Sci. Res. Inst. Virus Infect., Sverdlovsk, USSR) *Vopr. Virusol., No. 6, 646-651 (1979)* Ru;en.

2699-U2 The use of guanidine-HCl for the isolation of both RNA and protein from RNA tumour viruses. Pavlovec,A.; Evenson,D.P.; Hamilton,M.G. (Sloan-Kettering Inst. Cancer Res., New York, NY 10021, USA) *J. Gen. Virol., 40(1), 239-243 (1978)* En;en.

2700-U2 Early polykaryocytosis inhibition: a simple quantitative in vitro assay for the detection of bovine leukemia virus infection in cattle. Guillemain,B.; Mamoun,R.; Levy,D.; Astier,T.; Irgens,K.; Parodi,A.A. (Groupe d'Etude de la Leucose Bovine, Inst. Natl. de la Sante et de la Rech. Med., Unite de Rech. Radiobiol. Exp. et de Cancerol., U.117, 229 cours de l'Argonne, 33076 Bordeaux, France) *Eur. J. Cancer, 14(8), 811-827 (1978)* En;en.

2701-U2 Enzyme linked immunosorbent assay (ELISA) for determination of IgG antibodies to human cytomegalovirus. Sarov,J.; Andersen,P.; Andersen,H.K. (Virol. Unit, Fac. Health Sci., Soroka Med. Cent., Ben-Gurion Univ., Negev Beer-Sheeva, Israel) *Acta Pathol. Microbiol. Scand., Ser. B, 88(1), 1-9 (1980)* En;en.

2702-U2 Epstein-Barr virus-IgM antibody test in infectious mononucleosis. Nikoskelainen,J.; Klemola,E.; Leikola,J. (Div. Infect. Dis., Dep. Med., Santa Clara Valley Med. Cent., San Jose, CA, USA) *J. Infect. Dis., 134(3), 313 (1976)* En.

2703-U2 Epstein-Barr virus-IgM antibody test in infectious mononucleosis. Evans,A.S. (Dep. Epidemiol. and Public Health, Yale Univ. Sch. Med., New Haven, CT, USA) *J. Infect. Dis., 134(3), 314 (1976)* En.

2704-U2 Simple method for preparation of specific antisera against viral proteins: rabbit antisera against equine infectious anemia virus proteins p26 and p16. Barta,V.; Issel,C.J. (Dep. Vet. Sci., Sch. Vet. Med., Louisiana State Univ., Baton Rouge, LA 70803, USA) *Am. J. Vet. Res., 39(11), 1856-1857 (1978)* En;en.

2705-U2 A rapid procedure for isolation of large quantities of *Escherichia coli* DNA polymerase I utilizing a λ*polA* transducing phage. Kelley,W.S.; Stump,K.H. (Dep. Biol. Sci., Mellon Inst. Sci., Carnegie-Mellon Univ., Pittsburgh, PA 15213, USA) *J. Biol. Chem., 254(9), 3206-3210 (1979)* En;en.

2706-U2 Preparation of anti-HBs immune serum in sheep. Babes,V.T.; Lenkei,R.; Mironescu,N. ('Stefan S. Nicolau' Inst. Virol., 285 Sos. Mihai Bravu, R-74339, Bucharest, Romania) *Rev. Roum. Med., Ser. Virol., 28(3), 231 (1977)* En.

2707-U2 A rapid technique for distinguishing herpes-simplex virus type 1 from type 2 by restriction-enzyme technology. Lonsdale,D.M. (Inst. Virol., Church St., Glasgow

G11 5JR, UK) *Lancet, 1(8121), 849-852 (1979)* En;en.

2708-U2 A convenient method for isolation of the M-protein of influenza virus. Joassin,L.; Reginster,M. (Lab. Microbiol. Gen. et Med., Univ. Liege, Liege, Belgium) *FEMS Microbiol. Lett., 4(6), 315-318 (1978)* En.

2709-U2 A comparison of methods for detecting bacteriophage contamination of tissue culture sera. Mayo,J.A. (Dep. Microbiol., LSU Med. Cent., 1100 Florida Ave., New Orleans, LA 70119, USA) *In Vitro, J. Tissue Cult. Assoc., 14(5), 413-417 (1978)* En;en.

2710-U2 A simple method for the isolation of phages from *Listeria monocytogenes*. Durst,J.; Rau,E.; Kemenes,F.; Berencsi,G. (Ady E. u. 117/b, H-5000 Szolnok, Hungary) *Zentralbl. Bakterial. Mikrobiol. Hyg., I Abt. A, 246(1), 23-25 (1980)* En;de,en.

2711-U2 Automated sequencing of insoluble peptides using detergent: bacteriophage fl coat protein. Bailey,G.S.; Gillett,D.; Hill,D.F.; Petersen,G.B. (Univ. Otago, Dep. Biochem., Dunedin, New Zealand) *J. Biol. Chem., 252(7), 2218-2225 (1977)* En;en.

2712-U2 Spectral analysis on the melting fine sructure of λ DNA and T2 DNA. Akiyama,C.; Gotoh,O.; *Wada,A. (Dep Phys., Fac. Sci., Univ. Tokyo, Hongo, Tokyo, Japan) *Biopolymers, 16(2), 427-435 (1977)* En;en.

2713-U2 A fast and simple method for sequencing DNA cloned in the single-stranded bacteriophage M13. Schreier,P.H.; Cortese,R. (MRC Lab. Mol. Biol., Hills Rd., Cambridge CB2 2QH, UK) *J. Mol. Biol., 129(1), 169-172 (1979)* En;en.

2714-U2 A simple method for identifying the palindromic sequences recognized by restriction endonucleases: the nucleotide sequence of the AvaII site. Fuchs,C.; Rosenvold,E.C.; Honigman,A.; *Szybalski,W. (McArdle Lab. Cancer Res., Univ. Wisconsin, Madison, WI 53706, USA) *Gene, 4(1), 1-23 (1978)* En;en.

2715-U2 A convenient sequencing method for 5′ protein-linked RNAs. Nomoto,A.; Imura,N. (Dep. Public Health, Sch. Pharmaceut. Sci., Kitasato Univ., 9-1, Shirokane 5-chome, Minato-ku, Tokyo 108, Japan) *Nucleic Acids Res., 7(5), 1233-1246 (1979)* En;en.

2716-U2 A rational method of obtaining sera with a high virus-neutralizing antibody titre. Kravchenko,A.T.; Tsettin,E.M.; Omelchenko,T.N. (State Sci.-Res. Inst. Stand. and Control Med. Biol. Prep., Sivtsev-Vrazhek per 41, Moscow, USSR) *Zh. Mikrobiol., Epidemiol., Immunobiol., 54(1), 41-45 (1977)* Ru;en,ru.

2717-U2 Preparation of rabies fluorescein isothiocyanate-labeled immune globulin from mouse hyperimmune ascitic fluids. Tzlanabos,T.; Hebert,G.A.; White,L.A. (Cent. Dis. Control, Public Health Serv., US Dep. Health, Educ., and Welfare, Atlanta, GA 30333, USA) *J. Clin. Microbiol., 3(6), 609-612 (1976)* En;en.

2718-U2 Preparation of rabies fluorescein isothiocyanate-labeled immune globulin from mouse hyperimmune ascitic fluids. Tzianabos,T.; Hebert,G.A.; White,L.A. (Cent. for Dis. Control, Public Health Serv., US Dep. Health, Educ., and Welfare, Atlanta, GA 30333, USA) *J. Clin. Microbiol., 3(6), 609-612 (1976)* En;en.

2719-U2 An efficient method for reassembly of fusogenic Sendai virus envelopes after solubilization of intact virions with Triton X-100. Volsky,D.J.; Loyter,A. (Hebrew Univ. Jerusalem, Dep. Biol. Chem., Jerusalem, Israel) *FEBS Lett., 92(2), 190-194 (1978)* En.

2720-U2 Differential affinities of simian virus 40 large tumor antigen for DNA. Oren,M.; Winocour,E.; Prives,C. (Dep. Microbiol., State Univ. New York, Stony Brook, NY 11974, USA) *Proc. Natl. Acad. Sci. USA, 77(1), 220-224 (1980)* En;en.

2721-U2 Rapid mapping of restriction sites of a DNA: restriction of DNA in agarose gel and two-dimensional analysis of end-labeled DNA. Kovacic,R.T.; Wang,J.C. (Dep. Biochem. and Mol. Biol., Harvard Univ., Cambridge, MA 02138, USA) *Plasmid, 2(3), 394-402 (1979)* En;en.

2722-U2 A simplified procedure for the quantitative isolation of DNA and RNA transcripts synthesized in vitro. Modak,M.J. (Mem. Sloan-Kettering Cancer Cent., 1275 York Ave., New York, NY 10021, USA) *Anal. Biochem., 75(1), 340-344 (1976)* En;en.

2723-U2 Synthesis of modified nucleoside 3′,5′-bisphosphates and their incorporation into oligoribonucleotides with T4 RNA ligase. Barrio,J.R.; del Carmen G. Barrio,M.; Leonard,N.J.; England,T.E.; *Uhlenbeck,O.C. (Dep. Chem. and Biochem., Sch. Chem. Sci., Univ. Illinois, Urbana, IL 61801, USA) *Biochemistry (Wash.), 17(11), 2077-2081 (1978)* En;en.

2724-U2 Isolation of viral double-stranded RNAs using a LiCl fractionation procedure. Diaz-Ruiz,J.R.; Kaper,J.M. (Plant Virol. Lab., Plant Prot. Inst., ARS, US Dep. Agric., Beltsville, MD 20705, USA) *Prep. Biochem., 8(1), 1-17 (1978)* En;en.

2725-U2 Simple method to detect virus specific IgM antibodies in patients' serum samples after immunosorption of immunoglobulins G and A. Geisen,H.P.; Frank,R.; *Doerr,H.W.; Enders,G. (Inst. Med. Virol., Univ. Heidelberg. Im Neuenheimer Feld 324, D 6900 Heidelberg, GFR) *Med. Microbiol. Immunol., 167(2), 77-82 (1979)* En;en.

2726-U2 Simple immunochemical method for measuring DNA repair rate in u.v.-irradiated bacteria. Saenko,A.S.; Ilynia,T.P.; Podgorodnichenko,V.K.; Poverenny,A.M. (Inst. Med. Radiol., Acad. Med. Sci., Obninsk, Kaluga Region, USSR) *Immunochemistry, 13(9), 779-781 (1976)* En;en.

2727-U2 Rapid biochemical tests for characterization of the Mycoplasmatales. Bradbury,J.M. (Sub-Dep. Avian Med., Univ. Livrpool, Leahurst, Neston, Wirral L64 7TE, UK) *J. Clin. Microbiol., 5(5), 531-534 (1977)* En;en.

2728-U2 Rapid method for the detection and identification of mycolic acids in aerobic actinomycetes and related bacteria. Hecht,S.T.; Causey,W.A. (c/o Libero Ajello, Mycol. Div., Cent. Dis. Control., Atlanta, GA 30333, USA) *J. Clin. Microbiol., 4(3), 284-287 (1976)* En;en.

2729-U2 Comparative evaluation of the effectiveness of protective media in lyophilization of actinomycetes using EPR method. Kuznetsov,V.D.; Filippova,S.N.; Ruuge,E.K. (Inst. Microbiol., Acad. Sci. USSR, Moscow, USSR) *Mikrobiologiya, 47(4), 706-710 (1978)* Ru;en,ru.

2730-U2 Automatic monitoring of protease activity during fermentation. Leisola,M.; Ojamo,H.; Kauppinen,V. (Dep. Chem., Helsinki Univ. Technol., SF-02150 Espoo 15, Finland) *Enzyme Microb. Technol., 1(1), 51-52 (1979)* En;en.

2731-U2 A rapid method for constructing multiply marked strains of *Bacillus subtilis*. Piggot,P.J.; de Lencastre,H. (Div. Microbiol., Natl. Inst. Med. Res., Mill Hill, London NW7 1AA, UK) *J. Gen. Microbiol., 106(1), 191-194 (1978)* En.

2732-U2 A procedure to remove protease activities from *Bacillus subtilis* sporulating cells and their crude extracts. Nakayama,T.; Munoz,L.; Doi,R.H. (Dep. Biochem. and Biophys., Univ. California, Davis, CA 95616, USA) *Anal. Biochem., 78(1), 165-170 (1977)* En;en.

2733-U2 The results of studying the diagnostic properties of brucella and tularaemia antigenic erythrocytic diagnostic agents. Tsybin,B.P.; Taran,I.F.; Tinker,A.I. (Anti-Plague Sci. Res. Inst. Caucasus and Transcaucasia, Stavropol, USSR) *Zh. Mikrobiol. Epidemiol. Immunobiol., 52(9), 83-88 (1975)* Ru;en,ru.

2734-U2 A method for the preparation of a chlamydiae group-specific antigen on HeLa-229 cells infected with a strain of *Chlamydia trachomatis* for use in the complement fixation test. Colimon,R.; Ferchal,F.; Perol,Y. (Lab. Cent. Bacteriol.-Virol., Hop. Saint Louis, 75010 Paris, France) *Ann. Microbiol., 130B(3), 313-321 (1979)* Fr;en,fr.

2735-U2 A micro-method for detecting toxins in pseudomembranous colitis. George,R.H. (Dep. Microbiol., Birmingham Child. Hosp., Ladywood, Birmingham B16 8ET, UK) *J. Clin. Pathol., 32(3), 303-304 (1979)* En.

2736-U2 System for evaluating clostridial inhibition in cured meat products. Robach,M.C.; Ivey,F.J.; Hickey,C.S. (Monsanto Co., St. Louis, MO 63166, USA) *Appl. Environ. Microbiol., 36(1), 210-211 (1978)* En;en.

2737-U2 Pyrazine carboxylamidase activity in *Corynebacterium*. Sulea,I.T.; Pollice,M.C.; *Barksdale,L. (Dep. Microbiol., New York Univ. Sch. Med. and Med. Cent., New York, NY 10016, USA) *Int. J. Syst. Bacteriol., 30(2), 466-472 (1980)* En;en.

2738-U2 Deoxyribonuclease: detection with a three-hour test. Greenwood,J.R.; *Pickett,M.J. (Dep. Bacteriol., Univ. California, Los Angeles, CA 90024, USA) *J. Clin. Microbiol., 4(5), 453-454 (1976)* En;en.

2739-U2 Rapid test for assay of ozone sensitivity in *Escherichia coli*. Hamelin,C.; Chung,Y.S. (Dep. Sci. Biol., Univ. Montreal, CP 6128, Montreal 101, Que., Canada) *Mol. Gen. Genet., 145(2), 191-194 (1976)* En;en,fr.

2740-U2 Nitroblue tetrazolium (NBT) reduction by bacteria employed for rapid determination of antibiotic concentrations in serum. Urban,T.; Jarstrand,C. (Dep. Clin. Bacteriol., Serafimerlasarettet, 112 83 Stockholm, Sweden) *Acta Pathol. Microbiol. Scand., Ser. B, 86(3), 159-164 (1978)* En;en.

2741-U2 Rapid solid phase isolation of 20 specific tRNAs from *Escherichia coli*. Goss,D.J.; Parkhurst,L.J. (Dep. Chem. and Sch. Life Sci., Univ. Nebraska, Lincoln, NE 68588, USA) *J. Biol. Chem., 253(21), 7804-7806 (1978)* En;en.

2742-U2 Rapid isolation of highly active RNA polymerase from *Escherichia coli* and its subunits by matrix-bound heparin. Sternbach,H.; Engelhardt,R.; Lezius,A.G. (Max-Planck-Inst. Exp. Med., Abt. Chem., Gottingen, GFR) *Eur. J. Biochem., 60(1), 51-55 (1975)* En;en.

2743-U2 A simple method for the identification of altered subunits in mutant RNA polymerases of *Escherichia coli*. Sugiura,M.; Ito,N.; Suzuki,M. (Natl. Inst. Genet., Mishima 411, Japan) *Anal. Biochem., 84(1), 337-339 (1978)* En.

2744-U2 Partition of ribosomes in two-polymer aqueous phase systems. Pestka,S.; Weiss,D.; Vince,R. (Roche Inst. Mol. Biol., Nutley, NJ 07110, USA) *Anal. Biochem., 71(1), 137-142 (1976)* En;en.

2745-U2 A simple method for studying chemotaxis using sustained release of attractants from inert polymers. Langer,R.; Fefferman,M.; Gryska,P.; Bergman,K. (Rm 16-325, Massachusetts Inst. Technol., Cambridge, MA 02139, USA) *Can. J. Microbiol., 26(2), 274-278 (1980)* En;en,fr.

2746-U2 Evaluation of a DNA polymerase-deficient mutant of *E. coli* for the rapid detection of carcinogens. Fluck,E.R.; Poirier,L.A.; Ruelius,H.W. (Res. Div., Wyeth Lab., Inc., Radnor, PA 19087, USA) *Chem.-Biol. Interactions, 15(3), 219-231 (1976)* En;en.

2747-U2 A simple method for estimating the biological activity of irradiated *Escherichia coli* DNA by the use of the prophage. Petranovic,D.; Petranovic,M.; Salaj-Smic,E.; Trgovcevic,Z. (Inst. 'Rudjer Boskovic', 41001 Zagreb, Croatia, Yugoslavia) *Int. J. Radiat. Biol., 29(2), 187-190 (1976)* En.

2748-U2 The use of dextran-coated charcoal for kinetic measurements: interaction between rifampicin and DNA-dependent RNA polymerase of *Escherichia coli*. Wyss,E.; *Wehrli,W. (Biol. Res. Lab., Pharm. Div., Ciba-Geigy Ltd., CH-4002 Basle, Switzerland) *Anal. Biochem., 70(2), 547-553 (1976)* En;en.

2749-U2 A simple procedure for the preparation of large amounts of 50S ribosomal subunits from *Escherichia coli* Armstrong,J.L.; Tate,W.P. (Dep. Biochem., Univ. Otago, Dunedin, New Zealand) *Anal. Biochem., 99(2), 332-339*

(1979) En;en.

2750-U2 A rapid method for the isolation of circular DNA using an aqueous two-phase partition system. Ohlsson,R.; Hentschel,C.C.; Williams,J.G. (Natl. Inst. Med. Res., Ridgeway, Mill Hill, London NW7 1AA, UK) *Nucleic Acids Res., 5(2), 583-590 (1978)* En;en.

2751-U2 A rapid test for beta-lactamase production by *Haemophilus influenzae*. Slack,M.P.E.; Wheldon,D.B.; Turk,D.C. (Dep. Bacteriol., Radcliffe Infirm., Oxford OX2 6HE, UK) *Lancet, 2(8044), 906 (1977)* En.

2752-U2 Detection of beta lactamase activity of *Haemophilus influenzae*. McGhie,D.; Clarke,P.D.; Johnson,T.; Hutchison,J.G.P. (Reg. Public Health Lab., East Birmingham Hosp., Birmingham, UK) *J. Clin. Pathol., 30(6), 585-587 (1977)* En.

2753-U2 Procedure for the simultaneous large-scale isolation of pullulanase and 1,4-α-glucan phosphorylase from *Klebsiella pneumoniae* involving liquid-liquid separations. Hustedt,H.; Kroner,K.H.; Stach,W.; Kula,M.R. (Ges. Biotechnol. Forsch. mbH, Mascheroder Weg 1, D-3300 Branschweig-Stockheim, GFR) *Biotechnol. Bioeng., 20(12), 1989-2005 (1978)* En;en.

2754-U2 Hot-loop test for the determination of carbon dioxide production from glucose by lactic acid bacteria. Sperber,W.H.; Swan,J. (Pillsbury Co., Res. and Dev. Lab., Minneapolis, MN 55414, USA) *Appl. Environ. Microbiol., 31(6), 990-991 (1976)* En;en.

2755-U2 Drop method for determination of micrococcal and staphylococcal sensitivity to egg lysozyme. Safonova,T.B.; Givental,N.I.; Afanasieva,T.I.; Solovyova,V.E.; Sobolev,V.R. (Cent. Inst. Adv. Med. Training, Moscow, USSR) *Antibiotiki, 23(10), 886-888 (1978)* Ru;en.

2756-U2 A rapid and simple procedure for the preparation of the two bacterial cell wall peptidoglycan nucleotide precursors labeled in their amino sugars. Mirelman,D. (Dep. Biophys., Weizmann Inst. Sci., Rehovot, Israel) *Anal. Biochem., 70(2), 424-429 (1976)* En;en.

2757-U2 A relatively rapid procedure for the preparation of lysis-susceptible forms of *Mycobacterium smegmatis*. Winder,F.G.; MacNaughton,A.W. (Dep. Biochem., Trinity Coll., Dublin 2, Eire) *J. Gen. Microbiol., 109(1), 177-180 (1978)* En.

2758-U2 Two methods of large-scale extraction of an antibiotic produced by *Myxococcus coralloides*. Arias,J.M.; Almendral,J.M.; Montoya,E. (Dep. Microbiol., Fac. Sci., Univ. Granada, Granada, Spain) *Microbios, 25(99), 19-23 (1979)* En;en.

2759-U2 [Results of a simple test system for the determination of β-lactamase (penicillinase) in *Neisseria gonorrhoea*]. Muller,G. (Univ.-Hautklin. (Charite), DDR-104 Berlin, Schumannstr. 20/21, GDR) *Dermatol. Monatsschr., 164(9), 649-651 (1978)* De;de,en.

2760-U2 Simple method for the isolation of astaxanthin from the basidiomycetous yeast *Phaffia rhodozyma*. Johnson,F.A.; Villa,T.G.; Lewis,M.J.; *Phaff,H.J. (Dep. Food Sci. and Technol., Univ. California, Davis, CA 95616, USA) *Appl. Environ. Microbiol., 35(6), 1155-1159 (1978)* En;en.

2761-U2 Bacterial membrane electrode for L-cysteine. Jensen,M.A.; Rechnitz,G.A. (Dep. Chem., Univ. Delaware, Newark, DE 19711, USA) *Anal. Chim. Acta, 101(1), 125-130 (1978)* En;en.

2762-U2 Method for testing degree of infectivity of *Rhizobium meliloti* strains. Olivares,J.; Casadesus,J.; Bedmar,E.J. (Dep. Microbiol., Estacion Exp. Zaidin CSIC, Granada, Spain) *Appl. Environ. Microbiol., 39(5), 967-970 (1980)* En;en.

2763-U2 Preparation of cytochrome c_2 from *Rhodospirillum rubrum*. Sponholtz,D.K.; Brautigan,D.L.; Loach,P.A.; Margoliash,E. (Dep. Biochem. and Mol. Biol., Northwestern Univ., Evanston, IL 60201, USA) *Anal. Biochem., 72(1/2), 255-260 (1976)* En;en.

2764-U2 A novel method for the production of *Salmonella* flagellar antigen. II. Further purification for the preparation of H antisera. Fey,H. (Vet. Bacteriol. Inst., Univ. Bern, Langgass Str. 122, CH-3012 Berne, Switzerland) *Zentralbl. Bakteriol. Parasitenkd. Infektionskr. Hyg., I Abt. A, 245(1-2), 55-66 (1979)* En;de,en.

2765-U2 Accurate automatic determination of acid phosphatase substrate saturation curves. Cinci,A. (Ist. Ricerche Biomed. 'A Marxer', RBM, Casella Postale 86, 10015 Ivrea, Italy) *Anal. Biochem., 89(2), 372-384 (1978)* En;en.

2766-U2 Rapid qualitative method for detecting staphylococcal nuclease in foods. Koupal,A.; *Deibel,R.H. (Dep. Bacteriol., Coll. Agric. and Life Sci., Univ. Wisconsin, Madison, WI 53706, USA) *Appl. Environ. Microbiol., 35(6), 1193-1197 (1978)* En;en.

2767-U2 A rapid paper-strip method for the detection of penicillinase production by penicillin resistant strains of *Staphylococcus aureus*. Wheldon,D.B.; Slack,M.P.E. (Dep. Bacteriol., Reg. Public Health Lab., Radcliffe Infirm., Oxford, UK) *J. Clin. Pathol., 31(4), 388-389 (1978)* En.

2768-U2 Rapid diagnosis using surface analysis by bacterial adherence. Huang,A.S.; Okorie,T. (Dep. Microbiol. and Mol. Genet., Harvard Med. Sch., Boston, MA 02115, USA) *Lancet, 2(8100), 1146 (1978)* En.

2769-U2 Rapid isolation of antigens from cells with a staphylococcal protein A-antibody adsorbent: parameters of the interaction of antibody-antigen complexes with protein A. Kessler.S.W. (Dep. Microbiol. and Immunol., UCLA Sch. Med., Los Angeles, CA 90024, USA) *J. Immunol., 115(6), 1617-1624 (1975)* En;en.

2770-U2 A simple method of detecting staphylococcal hemolysins. Skalka,B.; Smola,J.; Pillich,J. (High Sch. Vet. Med., Palackeho 1 Brno, Czechoslovakia) *Zentralbl.*

Bakteriol. Parasitenkd. Infektionskr. Hyg., I Abt. A, 245(3), 283-286 (1979) En;de,en.

2771-U2 Isolation of DNA from streptomycetes. Petrov,P.A. (Dep. Phys. and Biophys., Med.-Biol. Inst., Sofia, Bulgaria) *Dokl. Bolg. Akad. Nauk, 31(2), 233-236 (1978)* En.

2772-U2 Determination of total and free cholesterol by using cholesterol oxidase from *Streptomyces*. Otani,T.; Ishimaru,K.; Nakamura,S.; Kamei,T.; Suzuki,H. (Inst. Pharm. Sci., Hiroshima Univ., Sch. Med., 1-2-3 Kasumi, Hiroshima 734, Japan) *Chem. Pharm. Bull., 25(6), 1452-1455 (1977)* En;en.

2773-U2 Application of a new method for detecting streptococcal nicotinamide adenine dinucleotide glycohydrolase to various M types of *Streptococcus pyogenes*. Lutticken,R.; Lutticken,D.; Johnson,D.R.; Wannamaker,L.W. (Inst. Hyg., Univ. Koln, 5 Koln 41, GFR) *J. Clin. Microbiol., 3(5), 533-536 (1976)* En;en.

2774-U2 A simple method for the rapid and economical immobilization of glucose isomerase. Bhatt,R.R.; Joshi,S.; Kothari,R.M. (Cadila Lab., PO Box 9004 Maninagar, Ahmedabad 380008, India) *Enzyme Microb. Technol., 1(2), 113-116 (1979)* En;en.

2775-U2 Evaluation of the rapid hippurate hydrolysis test with enterococcal group D streptococci. Lee,M.R.; *Ederer,G.M. (Dep. Lab. Med. and Pathol., Univ. Minnesota, Minneapolis, MN 55455, USA) *J. Clin. Microbiol., 5(3), 290-292 (1977)* En;en.

2776-U2 Detection of isoelectric focused antibody by autoradiography and hemolysis of antigen-coated erythrocytes. A comparison of methods. Briles,D.E.; Davie,J.M. (Dep. Pathol., Washington Univ. Sch. Med., St. Louis, MO 63110, USA) *J. Immunol. Methods, 8(4), 363-372 (1975)* En;en.

2777-U2 A rapid procedure for the isolation of endonucleases from two thermophilic bacteria. Runswick,M.J.; Harris,J.I. (deceased) (MRC Lab. Mol. Biol., Hills Road, Cambridge CB2 2QH, UK) *FEBS Lett., 94(2), 380-382 (1978)* En.

2778-U2 A rapid method for the determination of cellular protein in the presence of elemental sulfur. Proteau,G.; Silver,M. (Dep. Biochim., Fac. Sci. et Genie, Univ. Laval, Que. G1K 7P4, Canada) *Can. J. Microbiol., 23(10), 1492-1494 (1977)* En;en,fr.

2779-U2 Rapid test for diagnosis of syphilis by means of the surface fixation method using as antigens VDRL and *Treponema pallidum* (strain Nichols). Ruiz-Castaneda,M.; Naranjo,Y. (La. Desarrollo, Hosp. Infantil Mexico, S.S.A. Dr. Marquez Num. 162, Mexico 7, DF, Mexicol) *Arch. Invest. Med., 9(4), 559-564 (1978)* En/Es;en,es.

2780-U2 Rapid test for diagnosis of syphilis by means of the surface fixation method using as antigens VDRL and *Treponema pallidum* (strain Nichils). Ruiz-Castaneda,M.; Naranjo,Y. (Lab. Desarrollo, Hosp. Infantil Mexico, SSA Dr. Marquez Num. 162, Mexico 7, DF, Mexico) *Arch. Invest. Med., 9(4), 559-564 (1978)* En/Es;en,es.

2781-U2 Partial characterization of a beta-lactamase from *Vibrio parahaemolyticus* by a new automated microiodometric technique. DeBell,R.M.; Hickey,T.M.; Uddin,D.E. (Dep. Microbiol., Naval Med. Res. Inst., Bethesda, MD 20014, USA) *Antimicrob. Agents Chemother., 13(2), 165-169 (1978)* En;en.

2782-U2 [Determination of lethal doses of microorganisms]. Baracho,I.R.; Piedrabuena,A.E. (Dep. Gen. e Evolucao, Univ. Estadual de Campinas, Campinas, SP, Brazil) *Cienc. Cult., 29(3), 314-316 (1977)* Pt;en,pt.

2783-U2 A physiological method for the quantitative measurement of microbial biomass in soils. Anderson,J.P.E.; Domsch,K.H. (Inst. Bodenbiol., FAL, Bundesallee 50, D-3300 Braunschweig, GFR) *Soil Biol. Biochem., 10(3), 215-221 (1978)* En;en.

2784-U2 Simple and sensitive procedure for screening collagenolytic bacteria and the isolation of collagenase mutants. Robbertse,P.J.; Woods,D.R.; Reay,A.H.; Robb,F.T. (Dep. Genet., Univ. Pretoria, Pretoria 0002, South Africa) *J. Gen. Microbiol., 106(2), 373-376 (1978)* En.

2785-U2 A rapid sensitive method for the measurement of guanine ribonucleotides in bacterial and environmental extracts. Karl,D.M. (Dep. Oceanogr., Univ. Hawaii, Honolulu, HI 96822, USA) *Anal. Biochem., 89(2), 581-595 (1978)* En;en.

2786-U2 Improved techniques for the preparation of bacterial lipopolysaccharides. Johnson,K.G.; Perry,M.B. (Div. Biol. Sci., Natl. Res. Council Canada, Ottawa, K1A 0R6, Canada) *Can. J. Microbiol., 22(1), 29-34 (1976)* En;en,fr.

2787-U2 A simple screening test for determining the β-lactamase activity of bacteria. Ullmann,U. (Dep. Microbiol., Hyg.-Inst., Eberhard-Karls-Univ. Tubingen, Silcherstr. 7, D-7400 Tubingen 1, GFR) *Microbios Lett., 3(9), 35-39 (1976)* En;en.

2788-U2 Diffusion coefficient of oxygen in microbial aggregates. Nglan,K.F.; Lin,S.H. (Dep. Chem. Eng., Univ. Melbourne, Parkville, Vic. 3052, Australia) *Biotechnol. Bioeng., 18(11), 1623-1627 (1976)* En.

2789-U2 Simple and sensitive method for determination of asparaginase activity in biological materials. Magarlamov,A.G.; Zaikin,A.A.; Beljaeva,L.V. (Kiev Sci. Res. Inst., Haematol. and Blood Transfus., Kiev, USSR) *Mikrobiol. Zh., 40(5), 645-648 (1978)* Uk;en.

2790-U2 A simple, quantitative method for measuring chemotaxis and motility in bacteria. Armitage,J.P.; Josey,D.P.; Smith,D.G. (Dep. Bot. and Microbiol., University Coll. London, Gower St., London WC1E 6BT, UK) *J. Gen. Microbiol., 102(1), 199-202 (1977)* En.

2791-U2 Entrapped glucose isomerase for high fructose syrup production. Linko,Y.-Y.; Pohjola,L.; Linko,P. (Address not stated) *Process Biochem., 12(6), 14-16, 32 (1977)* En.

2792-U2 Rapid characterization of mixed microbial populations in ruminal contents, cecal contents and in feces by a semi-quantitative assay of some hydrolytic enzymes (API ZYME). Lankhorst,A.; Counotte,G.H.M.; Koopman,J.P.; Prins,R.A. (Lab. Anim. Nutr., Vet. Fac., State Univ. Utrecht, Utrecht, Netherlands) *Z. Tierphysiol. Tierernahr. Futtermittelkd., 41(3), 162-171 (1979)* En;de,en.

2793-U2 New methods to assess bacterial injury in water. Zaske,S.K.; Dockins,W.S.; Schillinger,J.E.; *McFeters,G.A. (Dep. Microbiol., Montana State Univ., Bozeman, MT 59717, USA) *Appl. Environ. Microbiol., 39(3), 656-658 (1980)* En;en.

2794-U2 Rapid procedure for the approximate determination of the deoxyribonucleic acid base composition of micrococci, staphylococci, and other bacteria. Meyer,S.A.; *Schleifer,K.H. (Lehrstuhl Mikrobiol., Tech. Univ. Munchen, 8 Munchen 2, Arcisstr. 21, GFR) *Int. J. Syst. Bacteriol., 25(4), 383-385 (1975)* En;en.

2795-U2 A method for the demonstration of extracellular hydrolysis of poly-β-hydroxybutyrate. Malik,K.A.; Claus,D. (Dtsch. Sammlung Mikroorg., Gesellschaft Strahlen- und Umweltforsch. mbh., Gottingen, GFR) *J. Appl. Bacteriol., 45(1), 143-146 (1978)* En;en.

2796-U2 A novel and simple method for the preparation of adenine arabinoside by bacterial transglycosylation reaction. Utagawa,T.; Morisawa,H.; Miyoshi,T.; Yoshinaga,F.; Yamazaki,A.; Mitsugi,K. (Central Res. Lab., Ajinomoto Co., 1-1 Suzuki-cho, Kawasaki-ku, Kawasaki 210, Japan) *FEBS Lett., 109(2), 261-263 (1980)* En.

2797-U2 A rapid colony test for thiaminase activity. Edwin,E.E.; Shreeve,J.E.; Jackman,R. (Cent. Vet. Lab., Mhist. AgEdwin,E.E.; Shreeve,J.E.; Jackman,R. (Cent. Vet. Lab., Minist. Agric., Fish. and Food, New Haw, Weybridge, Surrey KT15 3NB, UK) *J. Appl. Bacteriol., 44(2), 305-312 (1978)* En;en.

2798-U2 Characterization of anaerobic gram-negative bacilli by using rapid slide tests for β-lactamase production. Bourgault,A.-M.; *Rosenblatt,J.E. (Mayo Clin. and Mayo Found., Rochester, MN 55901, USA) *J. Clin. Microbiol., 9(6), 654-656 (1979)* En;en.

2799-U2 A rapid method for the base ratio determination of bacterial DNA. Cashion,P.; Holder-Franklin,M.A.; McCully,J.; Franklin,M. (Microbiol. Res. Lab., Dep. Biol., Univ. New Brunswick, Fredericton, NB, E3B 5A3, Canada) *Anal. Biochem., 81(2), 461-466 (1977)* En;en.

2800-U2 Utilization of promoter and terminator sites on bacteriophage T7 DNA by RNA polymerases from a variety of bacterial orders. Wiggs,J.L.; Bush,J.W.; Chamberlin,M.J. (Dep. Biochem., Univ. California, Berkeley, CA 94720, USA) *Cell, 16(1), 97-109 (1979)* En;en.

2801-U2 Application of photo-crosslinkable resin to immobilization of an enzyme. Fukui,S.; Tanaka,A.; Iida,T.; Hasegawa,E. (Lab-Ind., Biochem., Dep. Ind. Chem., fac. Eng., Kyoto Univ., Sakyo-ku, Kyoto, Japan) *FEBS Lett., 66(2), 179-182 (1976)* En.

2802-U2 A rapid method for extracting DNA from agarose gels. Finkelstein,M.; *Rownd,R.H. (Lab. Mol. Biol. and Dep. Biochem., Univ. Wisconsin, Madison, WI 53706, USA) *Plasmid, 1(4), 557-562 (1978)* En;en.

2803-U2 A rapid micromethod for the detection of indole production from tryptophan. Borriello,S.P.; Cohen,B.A. (Bact. Metab. Res. Lab., Cent. Public Health Lab., Colindale Ave., London NW9 5HT, UK) *Microbios Lett., 5(17), 7-11 (1977)* En;en.

2804-U2 Detection of bacterial phosphatase activity by means of an original and simple test. Satta,G.; Grazi,G.; Varaldo,P.E.; Fontana,R. (Ist. Microbiol., Univ. Genova, Viale Benedetto XV, 10, 16132 Genova, Italy) *J. Clin. Pathol., 32(4), 391-395 (1979)* En;en.

2805-U2 A rapid slide test for penicillinase. Rosenblatt,J.F.; Neumann,A.M. (Dep. Lab. Med., Sect. Clin. Microbiol., Mayo Clinic and Mayo Found., Rochester, MN 55901, USA) *Am. J. Clin. Pathol., 69(3), 351-354 (1978)* En;en.

2806-U2 Surface fixation as a rapid test for detection of penicillin antibodies. Ruiz-Castaneda,M.; Symes,I. (Hosp. Infantil Mexico, Calle Dr. Marquez 162, Mexico 7, DF) *Arc. Invest. Med., 8(2), 85-90 (1977)* En;en,es.

2807-U2 Detection of bacterial phosphatase activity by means of an original and simple test. Satta,G.; Grazi,G.; Varaldo,P.E.; Fontana,R. (Ist. Microbiol., Univ. Genova, Viale Benedetto XV, 10, 16132 Genova, Italy) *J. Clin. Pathol., 32(4), 391-395 (1979)* En;en.

2808-U2 Distinction between β-lactamases of immunotypes 1 or 2 using a new β-lactam antibiotic. Philippon,A.; Paul,G.; Labia,R.; Nevot,P. (Serv. Bacteriol., CHU Cochin, 75014 Paris, France) *Ann. Microbiol., 127A(4), 487-491 (1976)* Fr;en,fr.

2809-U2 Method of rapid decompression for disintegration of microbial cell walls. Rakitin,V.Y.; Monosov,E.Z.; Prokofiev,N.V.; Grigopian,A.N. (All-Union Res. Inst. Synthesis of Proteins and Protein Cpds., Moscow, USSR) *Prikl. Biokhim-Mikrobiol., 12(2), 278-282 (1976)* Ru;en,ru.

2810-U2 Aquatic microflora enumeration by means of adenylic nucleotides dosage. Champiat,D.; Larpent,J.P. (Lab. Microbiol., 4 rue Ledru-63000-Clermont Ferrand, France) *Hydrobiologia, 58(1), 37-42 (1978)* En;en.

2811-U2 The application of miniaturized methods for the characterization of various organisms isolated from the animal gut. Jayne-Williams,D.J. (Natl. Inst. Res. in Dairying, Univ. Reading, Shinfield, Nr. Reading, Berks., UK) *J. Appl. Bacterial., 40(2), 189-200 (1976)* En;en.

2812-U2 [The rapid series determination of proteins with bromophenol blue in whole microorganisms and other cell suspensions]. Hesse,G.; Lindner,R. (Zentralinst. Mikrobiol. und Exp. Ther., DDR-69 Jena, Beuthenbergerstr. 11, GDR) *Z. Allg. Mikrobiol. Morphol. Physiol. Okol. Mikroorg., 15(7), 559-561 (1975)* De.

2813-U2 A review of automation and rapid methods in microbiology. Newsom,S.W.B. (John Bonnett Clin. Lab., Addenbrooke's Hosp., Cambridge CB2 2QQ, UK) *Med. Lab. Sci., 35(3), 215-222 (1978)* En.

2814-U2 Simple qualitative methods for detection of β-lactamase-producing microbial strains. Givental,N.I.; Bogdanova,L.F. (Inst. Adv. Med. Training, Moscow, USSR) *Antibiotiki, 23(11), 975-981 (1979)* Ru;en.

2815-U2 Attempt to improve and standardize cutaneous microorganisms sampling using electric apparatus. Fleurette,J.; Transy,M.J. (Lab. Bacteriol., Fac. Med. Alexis-Carrel, rue Guillaume-Paradin, 69372 Lyon Cedex 2, France) *Rev. Inst. Pasteur Lyon, 11(3), 493-501 (1978)* Fr;en,fr.

2816-U2 A new method for the quantitative determination of microorganisms on human skin. Staal,E.M.; Noordzij,A.C. (Intradal Res. Lab., Brabantsestr. 17, Amersfoort, Netherlands) *J. Cosmet. Chem., 29(10), 607-615 (1978)* En;en.

2817-U2 A simplified procedure for the examination of drinking water for bacteria of public health significance: the differential hydrobacteriogramme. Mossel,D.A.A.; van Ekeren,A.J.W.M.; Eelderink,I. (Chair Med. Food Microbiol., Dep. Sci. Food of Anim. Origin, Fac. Vet. Med., Univ. Utrecht, Utrecht, Netherlands) *Zentralbl. Bakteriol. Parasitenkd. Infektionskr. Hyg., I Abt. B, 165(5-6), 498-516 (1977)* En;de,en.

2818-U2 [Hand disinfection in surgery with a solution of performic acid]. Szechy,M.; Csete,A.; Vitez,I. (H-2800 Tatabanya, Semmelweis u. 2, Hungary) *Zentralbl. Chir., 102(19), 1191-1193 (1977)* De;de,en.

2819-U2 Rapid biological methods for continuous water quality monitoring. Kingsbury,R.W.S.M.; Rees,C.P. (Atkins Res. and Dev., Epsom, Surrey, UK) *Effluent Water Treat. J., 18(7), 319-321, 323-325, 328-331 (1978)* En.

2820-U2 Use of baker's yeast to trace microbial movement in ground water. Wood,W.W.; Ehrlich,G.G. (U.S. Geol. Surv., Reston, VA 22092, USA) *Ground Water, 16(6), 398-403 (1978)* En;en.

2821-U2 Simple method for carbon determination in microbiological experiments. Novak,B.(Drnovska 507, Praha 6 - Ruzyne, Czechoslovakia) *Zentralbl. Bakteriol. Parasitenkd. Infektionskr. Hyg., II, 131(7), 588-591 (1976)* En;de,en.

2822-U2 Large-scale disintegration of microorganisms by freeze-pressing. Magnusson,K.-E.; Edebo,L. (Dep. Med. Microbiol., Linkoping Univ., S-581 85 Linkoping, Sweden) *Biotechnol. Bioeng., 18(7), 975-986 (1976)* En;en.

2823-U2 Adsorption of amyloglucosidase on inorganic carriers. Solomon,B.; Levin,Y. (Dep. Biophys., Weizmann Inst. Sci., Rehovot, Israel) *Biotechnol. Bioeng., 17(9), 1323-1333 (1975)* En;en.

2824-U2 The use of immobilized enzyme-membrane sandwich reactors in automated analysis. Campbell,J.; Chawla,A.S.; Chang,T.M.S. (Physiol. Dep., Artif. Organs Res. Unit, McGill Univ., Montreal, Que., Canada) *Anal. Biochem., 83(1), 330-335 (1977)* En;en.

2825-U2 Sedimentation of *Beauveria bassiana* (Bals) Vuill conidia by aluminium sulphate. Globa,L.I.; Rotmistrov,M.N.; Jakovnko,V.V.; Bevzenko,T.M. (Inst. Chem., Acad. Sci., Ukr.SSR, Kiev, USSR) *Mikrobiol. Zh., 39(1), 111-114 (1977)* Uk;en,uk.

2826-U2 A micro method for the estimation of killing and phagocytosis of *Candida albicans* by human leucocytes. Wood,S.M.; White,A.G. (Immunol. Lab. South East Scotland Region Transfus. Serv., R. Infirm., Edinburgh, UK) *J. Immunol. Methods, 20, 43-52 (1978)* En;en.

2827-U2 A convenient method for permeabilizing the fungus *Cephalosporium acremonium*. Felix,H.; Nuesch,J.; Wehrli,W. (Biol. Res. Lab., Pharm. Div., Ciba-Geigy Ltd., CH-4002 Basel, Switzerland) *Anal. Biochem., 103(1), 81-86 (1980)* En;en.

2828-U2 Development of the agar disk method for the rapid selection of cephalosporin producers with improved yields. Trilli,A.; Michelini,V.; Mantovani,V.; *Pirt,S.J. (Microbiol. Dep., Queen Elizabeth Coll., Campden Hill, London W8, UK) *Antimicrob. Agents Chemother., 13(1), 7-13 (1978)* En;en.

2829-U2 A seedling-box test for evaluating alfalfa for resistance to anthracnose. Morrison,R.H. (Res. Serv. Dep., Northrup, King and Co., 13410 Research Road, Eden Prairie, MN 55343, USA) *Plant Dis. Rep., 61(1), 35-37 (1977)* En;en.

2830-U2 A rapid chitin synthase preparation for the assay of potential fungicides and insecticides. Adams,D.J.; Gooday,G.W. (Dep. Microbiol., Univ. Aberdeen, Marischal Coll., Aberdeen AB9 1AS, UK) *Biotechnol. Lett., 2(2), 75-78 (1980)* En;en.

2831-U2 Identification of concanavalin A receptors and galactose-binding proteins in purified plasma membranes of *Dictyostelium discoideum*. West,C.M.; McMahon,D. (Div. Biol., California Inst. Technol., Pasadena, CA 91125, USA) *J. Cell Biol., 74(1), 264-273 (1977)* En;en.

2832-U2 Methods in coccidiosis research: separation of oocysts from faeces. Ryley,J.F.; Meade,R.; Hazelhurst,J.; Robinson,T.E. (Imperial Chem. Ind. Ltd., Pharm. Div., Alderley Park, Macclesfield, Cheshire, UK) *Parasitology, 73(3), 311-326 (1976)* En;en.

2833-U2 Method of rapid comparative-quantitative investigations of polygalacturonase activity of *Fusarium oxysporum vasinfectum* (Robin) Berkh strains.

Morozova,A.V. (Address not stated) *Izv. Akad. Nauk Turk. SSR, Ser. Biol., No. 5, 3-8 (1977)* Ru;en,ru.

2834-U2 [Flow cytometric determination of the DNA-content of *Nectria coccinea* ex Fr. as a response to fungicidal action]. Hutter,K.-J.; Gortz,T.; Oldiges,H.; Eipel,H.E. (Fraunhofer-Geselisch., Inst. Aerobiol. Grafschaft, D-5948 Schmallenberg/Hochsauerland, GFR) *Chemosphere, 7(1), 51-58 (1978)* De;en.

2835-U2 [A simple method for the isolation of L-saccharopin from *Neurospora crassa*] Hermann,P.; Meinel,K. (DDR-402 Halle Saale, Hollystr. 2, GDR) *Pharmazie, 31(2), 132-133 (1976)* De.

2836-U2 Large scale isolation of ribosomal DNA from giant surface cultures of *Physarum polycephalum*. Affolter,H.U.; Behrens,K.; Seebeck,T.; Braun,R. (Inst. Gen. Microbiol., Univ. Berne, Altenbergrain 21, CH-3013 Berne, Switzerland) *FEBS Lett., 107(2), 340-342 (1979)* En.

2837-U2 A simple method for the isolation of actin from myxomycete plasmodia. Hatan,S.; Owaribe,K. (Inst. Ml. Biol., Fac. Sci., Nagoya Univ., Chikusa-ku, Nagoya, Aichi 464, Japan) *J. Biochem., 82(1), 201-205 (1977)* En;en.

2838-U2 Oscillating contractions in protoplasmic strands of *Physarum*: infrared reflexion as a non-invasive registration technique. Samans,K.E.; Gotz von Olenhusen,K.; Wohlfarth-Bottermann,K.E. (Inst. Cytol., Univ. Bonn, D-5300 Bonn 1, Ulrich-Haberland-Str. 61a, GFR) *Cell Biol. Int. Rep., 2(3), 271-278 (1978)* En;en.

2839-U2 A simplified, non-computerized program for forecasting potato late blight. MacHardy,W.E.(Dep. Bot. and Plant Pathol., Univ. New Hampshire, Durham, NH 03824, USA) *Plant Dis. Rep., 63(1), 21-25 (1979)* En;en.

2840-U2 A new method of testing viability of chlamydospores of *Protomyces macrosporus* Ung. Gupta,R.N. (Dep. Bot., Gov. Sci. Coll., Gwalior 474004, India) *Acta Bot. Indica, 6(2), 209-210 (1978)* En;en.

2841-U2 A fast method for quantitative determination of pullulan in the cultural broth of *Pullularia pullulans*. Imshenetsky,A.A.; Kondratyeva,T.F. (Inst. Microbiol., Acad. Sci. USSR, Moscow, USSR) *Mikrobiologiya, 47(3), 566-568 (1978)* Ru;en,ru.

2842-U2 Rapid laboratory screening of sugar beet cultivars for resistance to *Rhizoctonia solani*. Campbell,C.L.; Altman,J. (Dep. Bot. and Plant Pathol., Colorado State Univ., Fort Collins, CO 80523, USA) *Phytopathology, 66(11), 1373-1374 (1976)* En;en.

2843-U2 Continuous fermentation by immobilized brewers yeast. White,F.H.; Portino,A.D. (Bass Production Ltd., High St., Burton-on-Trent, Staffs., UK) *J. Inst. Brew., 8(4), 228-230 (1978)* En;en.

2844-U2 Isolation of DNA from yeasts grown in an osmotically balanced medium in the presence of 2-deoxy-D-glucose. Larionov,V.L.; Gause,G.G.,Jr.; Neyfakh,S.A. (Inst. Exp. Med., USSR Acad. Med. Sci., Leningrad, USSR) *Mol. Biol., 11(1), 242-247 (1977)* Ru;en,ru.

2845-U2 Measurement in vitro of the esterification of yeast sterols. Parks,L.W.: Stromberg,V.K. (Dep. Microbiol., Oregon State Univ., Corvallis, OR 97331, USA) *Lipids, 13(1), 29-33 (1978)* En;en.

2846-U2 A simplified homogenization method to demonstrate sarcosporidia (Miescher's tubes) in slaughtered cattle. Hinaidy,H.K. (Linke Bahngasse 11, A-1030 Wien, Austria) *Wien. Tierarztl. Monatsschr., 67(2), 54-55 (1980)* De;de,en.

2847-U2 A technique for the assessment for resistance of the potato varieties to *Synchytrium endobioticum*. Phadtare,S.G.; Cammack,R.H.; Sharma,R. (Wart Testing Stn., Darjeeling, (WB), India) *J. Indian Potato Assoc., 2(1), 2-6 (1975)* En;en,hindi.

2848-U2 [A rapid method for determination of the pathogeniety of *Thielaviopsis basicola* (Berk et Br.) Ferr]. Karimov,Kh.M. (Cent. Asia Plant Prot. Res. Inst., Tashkent, USSR) *Mikol. Fitopatol., 11(5) 445 (1977)* Ru.

2849-U2 Yeast protoplasts from stationary and starved cells: preparation, ultrastructure and vacuolar development. Schwente,J.; Magana-Schwente,N.; Laporte,J. (Lab. Enzymol., CNRS, 91190 Gif-sur-Yvette, France) *Ann. Microbiol., 128(A)(1), 3-16 (1977)* Fr;en,fr.

2850-U2 Modification of a rapid method for determining carbohydrate assimilation patterns of yeasts. Marraro,R.V.; Rodgers,E.M.; O'Brien,P.N. (12511 El Domingo, San Antonio, TX 78233, USA) *J. Am. Med. Technol., 39(3), 124-126 (1977)* En;en.

2851-U2 A novel method for the rapid preparation of coupled yeast mitochondria. Pena,A.; Pina,M.Z.; Escamilla,E.; Pina,E. (Inst. Biol., Apartado Postal 70-600, Mexico 20, D.F., Mexico) *FEBS Lett., 80(1), 209-213 (1977)* En.

2852-U2 Permeabilization of yeast for enzyme assays: an extremely simple method for small samples. Mowshowitz,D.B. (Dep. Biol. Sci., Columbia Univ., Broadway and 116th St., New York, NY 10027, USA) *Anal. Biochem., 70(1), 94-99 (1976)* En;en.

2853-U2 Some properties of yeast mitochondria prepared by a rapid mechanical procedure. Chambon,H.; Labbe,P. (Lab. Biochem. Porphyrines (ER 186), Univ. Pariws VII, Tour 43, 2 place Jussieu, 75005 Paris, France) *Biochimie, 58(7), 837-842 (1976)* En;en.

2854-U2 Rapid method for determining nitrate utilization by yeasts. Hopkins,J.M.; *Land,G.A. (Granville C. Morton Cancer and Res. Hosp., Div. Wadley Inst. Mol. Med., Dallas, TX 75235, USA) *J. Clin. Microbiol., 5(4), 497-500 (1977)* En;en.

2855-U2 A rapid screening method for the aflatoxins and ochratoxin A. Holaday,C.E. (Natl. Peanut Res. Lab., PO Box

637, Forrester Drive, Dawson, GA 31742, USA) *J. Am. Oil Chem. Soc., 53(9), 603-605 (1976)* En;en.

2856-U2 Rapid chemical confirmation method for aflatoxins B_1 and G_1 by direct acetylation on a thin layer plate before chromatography. Cauderay,P. (Tech. Assist. Co. Ltd., Control Lab. Nestle Prod., La Tour-de-Petilz, Switzerland) *J. Assoc. Off. Anal. Chem., 62(1), 197 (1979)* En;en.

2857-U2 Simple procedure for disruption of fungal spores. van Etten,J.L.; Freer,S.N. (Dep. Plant Pathol., Univ. Nebraska, Lincoln, NE 68583, USA) *Appl. Environ. Microbiol., 35(3), 622-623 (1978)* En;en.

2858-U2 A simple method for the large-scale preparation of mitochondria from microrganisms. Lang,B.; Burger,G.; Doxiadis,I.; Thomas,D.Y.; Bandlow,W.; Kaudewitz,F. (Genet. Inst. Univ. Munchen, Marla-Ward-Str. 1a, D-8000 Munchen19, GFR) *Anal. Biochem., 77(1), 110-121 (1977)* En;en.

2859-U2 Cryofracturing as a technique for the study of fungal structures in the scanning electron microscope. O'Donnell,K.L.; Hooper,G.R. (NRRL, ARS, USDA, 1815 N. Univ., Peoria, IL 61604, USA) *Mycologia, 69(2), 309-320 (1977)* En;en.

2860-U2 Development of an improved technique for rapid screening of wood preservative fungicides. I. Selection of test substrate. Bravery,A.F.; Carey,J.K. (Build. Res. Establ., Princes Risborough Lab., Princes Risborough, Aylesbury, Bucks, HP17 9PX, UK) *Int. Biodeterior. Bull., 13(1), 18-24 (1977)* En;de,en,es,fr.

2861-U2 Confirmatory test for aflatoxin M_1 on a thin layer plate. Van Egmond,H.P.; Paulsch,W.E.; Schuller,P.L. (Natl. Inst. Public Health, Lab. Chem. Anal. Foodstuffs, PO Box 1, Bilthoven, Netherlands) *J. Assoc. Off. Anal. Chem., 61(4), 809-812 (1978)* En;en.

2862-U2 Indirect method of determining mass of epilithic lichen. Byazrov,L.G.; Starostina,I.E. (Sov.-Mong. Complex Biol. Exped., Moscow, USSR) *Bot. Zh., 62(2), 222-225 (1977)* Ru;ru.

2863-U2 A screening method for the estimation of filter paper activity. Wabnegg,F.; Messner,K.; Rohr,M. (Inst. Biochem. Technol. und Mikrobiol., Tech. Univ. Wien, Getreidemarkt 9, A-1060 Wien, Austria) *J. Gen. Microbiol., 117(1), 267-269 (1980)* En;en.

2864-U2 A simple method for determining efficacy and weatherability of fungicides on foliage. Ko,W.H.; Lin,H.-H.; Kunimoto,R.K. (Dep. Plant Pathol., Univ. Hawaii, Beaumont Agric. Res. Cent., Hilo, HI 96720, USA) *Phytopathology, 65(9), 1023-1025 (1975)* En;en.

2865-U2 Improved method for preparation of microcrystalline chlorophyll a with *Anacystis nidulans* as a source. Kis,P. (Inst. Phys. Chem., Univ. Vienna, Wahringerstr. 42, Vienna, Austria) *Experientia, 34(10), 1289-1290 (1978)* En;en.

2866-U2 Sinking movements of phytoplankton indicated by a simple trapping method. I. A fragilaria population. Reynolds,C.S. (Freshwater Biol. Assoc., Ferry House, Ambleside, Cumbria LA22 0LP, UK) *Br. Phycol. J., 11(3), 279-291 (1976)* En;en.

2867-U2 An indirect method for the rapid estimation of carotenoid contents in *Phaeodactylum tricornutum*: possible application to other marine algae. Carreto,J.I.; Catoggio,J.A. (Inst. Biol. Mar., Casilla de Correo 175, Mar del Plata, Argentina) *Mar. Biol., 40(2), 109-116 (1977)* En;en.

2868-U2 [A method for the rapid determination of the nucleic acid content of algal biomass]. Paoletti,C.; Materassi,R.; Balloni,W. (Cent. Studio dei Microrg. Autrofi, CNR, Ist. Microbiol. Agraria e Tecnica, Univ. Firenze, Florence, Italy) *Ann. Microbiol. Enzimol., 23(4-6), 95-108 (1973)* It;en,it.

2869-U2 Determination of halogens in marine algae by use of an ion-selective electrode. Whyte,J.N.C.; Englar,J.R. (Fish. and Mar. Serv., Dep. Environ., Vancouver Lab., 6640 N.W. Marine Drive. Vancouver, BC V6T 1X2, Canada) *Analyst, 101(1207), 815-819 (1976)* En;en.

2870-U2 A simplified method for the simultaneous extraction of phytoplanktonic chlorophyll and fecal sterol from water. Wun,C.K.; Rho,J.; Walker,R.W.; Litsky,W. (Dep. Environ. Sci., Univ. Massachusetts, Amherst, MA 01003, USA) *Water Air Soil Pollut., 11(2), 173-178 (1979)* En;en.

2871-U2 Rapid screening for copper tolerance in ship-fouling algae. Goodman,C.; Newall,M.; Russell,G. (Hartley Bot. Lab., Univ. Liverpool, Liverpool L69 3BX, UK) *Int. Biodeterior. Bull., 12(3), 81-83 (1976)* En;de,en,es,fr.

2872-U2 A new method of removal of algal scum. Kumar,D. (Cent. Inland Fish. Res. Inst., Barrackpore, West Bengal, India) *Sci. Cult., 43(8), 364-365 (1977)* En.

2873-U2 Direct determination of cadmium in unicellular green algae by flameless atomic absorption. Meisch,H.-U.; Reinle,W. (Fachber. 15.2, Biochem., Univ. Saarlandes, D-6600 Saarbrucken 11, GFR) *Mikrochim. Acta, 1(5-6), 505-510 (1977)* De;de,en.

2874-U2 Silicon content of five marine plankton diatom species measured with a rapid filter method. Paasche,E. (Dep. Mar. Biol. and Limnol., Univ. Oslo, PO Box 1069, Blindern, Oslow 3, Norway) *Limnol. Oceanogr., 25(3), 474-480 (1980)* En;en.

2875-U2 A modified india-ink immunoreaction for the detection of encephalitozoonosis. Kellett,B.S.; Bywater,J.E.C. (Basel Inst. Immunol., 487 Grenzacherstr., CH 4005, Basel 5, Switzerland) *Lab. Anim., 12(2), 59-60 (1978)* En;de,en.

2876-U2 The preparation and characterization of *Gonyaulax* spheroplasts. Adamich,M.; Sweeney,B.M. (Dep. Biol. Sci., Univ. California, Santa Barbara, CA 93106, USA) *Planta, 130(1), 1-6 (1976)* En;en.

2877-U2 A method for production of rabbit antisera with high titres of antibody to the causative agent of cutaneous leishmaniasis. Dobrzhanskaya,R.S. (Turkmenian Dermatol. Inst., Ashkhabad, USSR) *Vestn. Dermatol., Venerol., No. 5, 80-84 (1978)* Ru;en.

2878-U2 A simple diphasic medium lacking whole blood for culturing *Leishmania* spp. Aljeboori,T.I. (Dep. Med. Protozool., London Sch. Hyg. and Trop. Med., Keppel St., London WC1E 7HT, UK) *Trans. R. Soc. Trop. Med. Hyg., 73(1), 117 (1979)* En.

2879-U2 A simplified method for in vitro production of schizonts of primate malarias useful as antigen in serologic tests. Sulzer,A.J.; Latorre,C.R. (Parasitol. Div., Bur. Lab., Cent. Dis. Control, Public Health Serv., US Dep. Health, Educ. and Welfare, Atlanta, GA 30333, USA) *Trans. R. Soc. Trop. Med. Hyg., 71(6), 553 (1977)* En.

2880-U2 Axenic cultures of *Tetrahymena pyriformis* as toxicological tools. Moravcova,V. (Water Dev. and Construct., Prague, Czechoslovakia) *Acta Hydrochim. Hydrobiol., 4(1), 83-94 (1976)* En;en.

2881-U2 A rapid high-yield isolation method for nuclear envelope ghosts. Giese,G.; *Wunderlich,F. (Lehrstuhl Zellbiol., Inst. Biol. II, Univ. Freiburg D-7800 Freiburg, GFR) *Anal. Biochem., 100(2), 282-288 (1979)* En;en.

2882-U2 In vitro maintenance of *Toxocara canis* larvae and a simple method for the production of *Toxocara* ES antigen for use in serodiagnostic tests for visceral larva migrans. de Savigny,D.H. (Dep. Vet. Microbiol. and Immunol., Univ. Guelph, Guelph, Ont., Canada) *J. Parasitol., 61(4), 781-782 (1975)* En.

2883-U2 A qualitative method for detection of nematode attracting substances and proof of production of three different attractants by the fungus *Monacrosporium rutgeriensis*. Balan,J.; Krizkova,L.; Nemec,P.; Kotozsvary,A. (Dep. Biol. Properties Low Mol Wt Substances, Inst. Mol. Biol., Slovak Acad. Sci., 885 34 Bratislava, Janska 1, Czechoslovakia) *Nematologica, 22(3), 306-311 (1976)* En;de,en.

2884-U2 A simple efficient production of neoantigenic antisera against fibrinolytic degradation products: radioimmunoassay of fragment E. Chen,J.P.; Shurley,H.M. (Dep. Human Biol. Chem. and Genet. Univ. Texas Med. Branch, Galveston, TX 77550, USA) *Thromb. Res., 7(3), 425-434 (1975)* En;en.

2885-U2 Direct demonstration and quantitation of the first complement component in human serum. Ziccardi,R.J.; Cooper,N.R. (Dep. Mol. Immunol., Scripps Clinic and Res. Found., La Jolla, CA 92037, USA) *Science (Wash.), 199(4333), 1080-1082 (1978)* En;en.

2886-U2 An efficient method to produce specific anti-actin. Jockusch,B.M.; Kelley,K.H.; Meyer,R.K.; Burger,M.M. (European Mol. Biol. Lab., D-6900 Heidelberg, GFR) *Histochemistry, 55(3), 177-184 (1978)* En;en.

2887-U2 An improved method for detection and titration of antibodies cytophilic for macrophages. Vasquez,R.; Herbert,W.J. (Natl. Inst. Med. Res., The Ridgeway, Mill Hill, London NW7 1AA, UK) *J. Immunol. Methods, 20, 77-85 (1978)* En;en.

2888-U2 A simplified method for the in vitro induction of IgG antibody in collagen coated dishes. Kemshead,J.T.; Askonas,B.A. (Natl. Inst. Med. Res., The Ridgeway, Mill Hill, London W7N 1AA, UK) *Immunology, 34(6), 1071-1076 (1978)* En;en.

2889-U2 Use of modified automated polybrene test for screening IgG sensitized RBC in routine laboratory. Boudart,D.; Guimbretiere,J. (Cent. Transfus. Sang., BP 349, 44011 Nantes, Cedex, France) *Pathol. Biol., 25(10), 755-757 (1977)* En;en,fr.

2890-U2 A simple microtitration technique for quantitating total haemolytic complement levels. Ellis,S.T.; Clancy,R.L. (Dep. Clin. Immunol., R. Prince Alfred Hosp., Camperdown, NSW 2050, Australia) *J. Immunol. Methods, 20, 311-316 (1978)* En;en.

2891-U2 Rapid method for preparing bone marrow cells from small laboratory animals. Oliver,J.P.; Goldstein,A.L. (Div. Biochem., Univ. Texas Med. Branch, Galveston, TX 77550, USA) *J. Immunol. Methods, 19(2-3), 289-292 (1978)* En;en.

2892-U2 Isolation procedure and some properties of myeloperoxidase from human leucocytes. Bakkenist,A.R.J.; Wever,R.; Vulsma,T.; Plat,H.; Van Gelder,B.F. (Lab. Biochem., BCP Jansen Inst., Univ. Amsterdam, Plantage Muidergracht 12, Amsterdam, Netherlands) *Biochim. Biophys. Acta, 524(1), 45-54 (1978)* En;en.

2893-U2 [Automated measurement of T lymphocyte transformation by the lymphoblastic megarosette method]. Abuaf,N.; Leynadier,F.; Luce,H.; Parent,C.; Dry,J. (Cent. Allergie, Hop. Rothschild, 43, bivd. de Picpus, 75012 Paris, France) *Rev. Fr. Allergol., 17(5), 275-278 (1977)* Fr;en,fr.

2894-U2 A simple method for production of specific xenoantisera to human histocompatibility (HLA-A, -B, -C,) antigens. Wilson,B.S.; Pellegrino,M.A.; Reisfeld,R.A.; Ferrone,S. (Dep. Mol. Immunol., Scripps Clin. and Res. Found., La Jolla, CA 92037, USA) *Transplant. Proc., 10(4), 741-743 (1978)* En.

2895-U2 A simple and reproducible method to evaluate granulocyte adherence. Lorente,F.; Fontan,G.; Garcia Rodriguez,M.C.; Ojeda,J.A. (Serv. Immunoalerg., Clin. Infantil 'La Paz', Madrid-34, Spain) *J. Immunol. Methods, 19(1), 47-51 (1978)* En;en.

2896-U2 A simple technique for separating and fixing human polynuclear basophils. Leynadier,F.; Luce,H.; Dry,J. (Serv. Med. Intern., Hop. Rothschild, 43, blvd de Picpus, 75571 Paris Cedex 12, France) *Rev. Fr. Allergol., 17(4), 215-218 (1977)* Fr;en,fr.

2897-U2 A simple procedure for the isolation of germinal centres from chicken spleen. Smithyman,A.M. (Dep. Bacteriol. Immunol., Glasgow Univ., Glasgow, UK) *Dev. Comp. Immunol., 1(3), 263-270 (1977)* En.

2898-U2 Quantitation of J chain in human biological fluids by a simple immunochemical procedure. Grubb,A.O. (Dep. Clin. Chem., Univ. Lund, Malmo Gen. Hosp., Malmo, Sweden) *Acta Med. Scand., 204(6), 453-465 (1978)* En;en.

2899-U2 A simple method for immediate determination of γ globulin in the serum of cattle and swine. Segal,L. (Sekt. Tierprodukt. und Veterinarmed., Humboldt-Univ. Berlin, Berlin, GDR) *Monatsh. Veterinarmed., 33(24), 953-956 (1978)* De;de,en,ru.

2900-U2 Rapid quantitative surface immunofixation of proteins in polyacrylamide gels. Kahn,S.N.; Thompson,E.J. (Dep. Chem. Pathol., Inst. Neurol., Natl. Hosp., Queen Square, London WC1N 3BG, UK) *Clin. Chim. Acta, 89(2), 253-265 (1978)* En;en.

2901-U2 Detection of circulating immune complexes by three techniques using polyethylene glycol. Digeon,M.; *Bach,J.-F. (INSERM U 25, Hop. Necker, 161 rue de Sevres, F 75730 Paris Cedex 15, France) *Nouv. Presse Med., 6(43), 4031-4037 (1977)* Fr;en,fr.

2902-U2 Quantitation of potential T-lymphocyte function in rats. Miller,T.E.; Creaghe,E. (Dep. Med., Univ. Auckland Sch. Med., Auckland 3, New Zealand) *Infect. Immun., 12(4), 722-727 (1975)* En;en.

2903-U2 Quantification of in vitro antibody secretion by immune spleen cells. van Dijk,H.; Bloksma,N. (Dep. Immunol., Lab. Microbiol., State Univ. Utrecht, Catharijnesingel 59, Utrecht, Netherlands) *J. Immunol. Methods, 14(3,4), 325-331 (1977)* En;en.

2904-U2 New method for obtaining IgA-specific protease. Higerd,T.B.; Virella,G.; Cardenas,R.; Koistinen,J.; Fett,J.W. (Dep. Basic and Clin. Immunol. and Microbiol., Med. Univ. South Carolina, Charleston, SC 29401, USA) *J. Immunol. Methods, 18(3-4), 245-249 (1977)* En;en.

2905-U2 A rapid measure of primary antigen binding capacity of antiserum. Caputo,R.A.; Barnhart,D.D.; Treick,R.W. (Dep. Microbiol., Miami Univ., Oxford, OH 45056, USA) *Microchem. J., 21(1), 85-91 (1976)* En;en.

2906-U2 Improvements of the electrophoretic mobility test to measure lymphocyte sensitization. Zwergel,T.; Nitzschke,U.; Franke,F.; Lampert,F. (Kinderpoliklin., Zent. Kinderheilkunde, Justus-Liebig-Univ., Giessen, GFR) *Microsc. Acta, 80(5), 383-390 (1978)* En;en.

2907-U2 Simple immunofixation modification: a practical way to identify monoclonal protein fractions. Irjala,K.; Rajamaki,A. (Dep. Hematol., Turku Univ. Cent. Hosp., 20520 Turku 52, Finland) *Scand. J. Lab. Clin. Invest., 39(3), 277-278 (1979)* En;en.

2908-U2 A test for antigen-antibody complexes in human sera using IgM of rabbit antisera to human immunoglobulins. Levinsky,R.J.; Soothill,J.F. (Dep. Immunol., Inst. Child Health, London WC1, UK) *Clin. Exp. Immunol., 29(3), 428-435 (1977)* En;en.

2909-U2 Alveolar macrophages. I. A simple technique for the preparation of high numbers of viable alveolar macrophages from small laboratory animals. Holt,P.G. (Child. Med. Res. Found., Princess Margaret Hosp., Subiaco, WA, Australia) *J. Immunol. Methods, 27(2), 189-198 (1979)* En;en.

2910-U2 [Examination of the immunocompetent cell composition of contact dermatitis. Methods for liberating infiltrative dermal cells and measuring antigen-dependent chromatin birefringence]. Kiraly,K.; Balo-Benga,J.M.; Molnar,L. (Natl. Res. Inst. Dermatol. and Venerol., Budapest VIII. Maria utca 41, Hungary) *Allerg. Immunol. (Leipzig), 25(1), 32-44 (1979)* En;de,en.

2911-U2 Evaluation of intercellular adhesion with a very simple technique. Bongrand,P.; Capo,C.; Benoliel,A.M.; Depieds,R. (Lab. Immunol., Hop. Sainte Marguerite, B.P. 29, 13274 Marseilles, France) *J. Immunol. Methods, 28(1-2), 133-141 (1979)* En;en.

2912-U2 A simpler method for combined detection of human peripheral lymphocyte subpopulations. Lee,Y.; Yoshizawa,Y.; Carr,R.G.; Yokoyama,M.M. (Clin. Immunol. and Serol., Hosp. Clin. Lab., Univ. Illinois at the Med. Cent., POB 6998, Chicago, IL 60680, USA) *Int. Arch. Allergy Appl. Immunol., 59(1), 114-116 (1979)* En;en.

2913-U2 A methodological study of E-rosette formation using AET-treated sheep red blood cells. Madsen,M.; Johnsen,H.E. (Tissue Typing Lab., Blood Bank and Blood Grouping Lab., Aarhus Kommunehospital, DK-8000 Aarhus C, Denmark) *J. Immunol. Methods, 27(1), 61-74 (1979)* En;en.

2914-U2 An attempt to separate mononuclear cells fused with human red blood cell-ghosts from a cell mixture treated with HVJ (Sendai virus) using a fluorescence activated cell sorter (FACS II). Mekada,E.; Yamaizumi,M.; Okada,Y. (Dep. Anim. Virol., Res. Inst. for Microb. Dis., Osaka Univ., Suita, Osaka 565, Japan) *J. Histochem. Cytochem., 26(1), 62-67 (1978)* En;en.

2915-U2 A simple estimation of the immunoglobulin content of ewe colostrum. Harker,D.B. (ICI Pharmaceut. Div., Alderley Park, Macclesfield, Cheshire, UK) *Vet. Rec., 103(1), 8-9 (1978)* En;en.

2916-U2 A simple method for the preparation of the immunoglobulins to hydroxyindole O-methyltransferase. Kuwano,R.; Takahashi,Y. (Dep. Neuropharmacol., Brain Res. Inst., Niigata Univ., Asahimachi 1, Niigata 951, Japan) *J. Neurochem., 31(4), 809-814 (1978)* En;en.

2917-U2 Automated immunochemical method for determination of urinary protein of plasma origin. Killingsworth,L.M.; Britain,C.E.; Woodward,L.L. (Clin.

Chem. Lab., N.C. Mem. Hosp., Chapel Hill, NC 27514, USA) *Clin. Chem., 21(10), 1465-1468 (1975)* En;en.

2918-U2 Microtechnique for simultaneous determination of immobilizing and cytotoxic sperm antibodies. Methodological and clinical studies. Husted,S.; *Hjort,T. (Inst. Med. Microbiol., Batholin Bygningen, Univ. Aarhus, 8000 Aarhus C, Denmark) *Clin. Exp. Immunol., 22(2), 256-264 (1975)* En;en.

2919-U2 An attempt at the simplification of the rosette-inhibition test for routine use in patients under immunosuppressive therapy. Nowaczyk,M.; Laudanski,Z.; Skopinska,E. (Inst. Transplantol. AM, ul. Nowogrodzka 59, 02-006 Warszawa, Poland) *Ann. Med. Sect. Pol. Acad. Sci., 21(3), 159-167 (1976)* En.

2920-U2 C1q deviation test for the detection of immune complexes, aggregates of IgG, and bacterial products in human serum. Sobel,A.T.; Bokisch,V.A.; Muller-Eberhard,H.J. (Serv. Nephrol., Hop. Henri-Mondor, 94010-Creteil, France) *J. Exp. Med., 142(1), 139-150 (1975)* En;en.

2921-U2 Platelet migration inhibition test for platelet antibodies. Parameters and simplified methodology. Maples,J.A.; *Yokoyama,M.M.; Inboriboon,P.; Inboriboon,A.; Chao,W. (Hosp. Clin. Lab., Univ. Illinois Med. Cent., POB 6998, Chicago, IL 60612, USA) *Int. Arch. Allergy Appl. Immunol., 54(4), 374-377 (1977)* En;en.

2922-U2 Improved B cell typing for HLA-DR using nylon wool column enriched B lymphocyte preparations. Lowry,R.; Goguen,J.; Carpenter,C.B.; Strom,T.B.; Garovoy,M.R. (Tissue Typing Lab., Renal Div., Dep. Med., Peter Bent Brigham Hosp., Boston, MA 02115, USA) *Tissue Antigens, 14(4), 325-330 (1979)* En;en.

2923-U2 A rapid method for the preparation of high potency auto and alloantibody eluates. Jenkins,D.E.,Jr.; Moore,W.H. (321-22nd Ave., N., Nashville, TN 37203, USA) *Transfusion, 17(2), 110-114 (1977)* En;en.

2924-U2 The use of nuclear monolayers in the study of influenza virus-infected cells. Hudson,J.B.; Dimmock,N.J. (Dep. Microbiol., Univ. British Columbia, Vancouver VG1 1W5, Canada) *FEMS Microbiol. Lett., 1(6), 325-327 (1977)* En.

2925-U2 [A micromethod for lymphocyte treatment with neuraminidase. Its use for detection of lymphocytotoxic antibodies. de Mouzon,A.; Ohayon,E.; Ducos,J. (Cent. Transfus., CHU Purpan, F 31052 Toulouse Cedex, France) *Rev. Fr. Transfus. Immuno-Hematol., 20(1), 23-26 (1977)* Fr.

2926-U2 A simplified method for the production of heterologous antiserum to factor VIII related antigen. Yang,H.C.; Duffy,C.; Levine,H. (Hemophilia Cent. and Dep. Med., Mem. Hosp., Worcester, MA 06105, USA) *Thromb. Res., 11(4), 463-470 (1977)* En;en.

2927-U2 Automatic blood film preparation by rheologically conrolled spinning. Megla,G.K. (Corning Glass GmbH, 6200 Wiesbaden-Biebrich, Hagenauerstr. 47, GFR) *Am. J. Med. Technol., 42(9), 336-342 (1976)* En;en.

2928-U2 A simple and convenient method for producing anti-activated lymphocyte alloantisera which does not require prior absorption. Kerbel,R.S.; Twiddy,R.R. (Natl. Cancer Inst., Canada Res. Group, Div. Cancer Res., Dep. Pathol., Queen's Univ., Kingston, Ont., Canada) *J. Immunol. Methods, 21(1-2), 11-22 (1978)* En;en.

2929-U2 Microtechnique for studies on the role of monocytes in the stimulation of lymphocytes. Tarnvik,A. (Dep. Clin. Bacteriol., Univ. Umea, Umea, Sweden) *Infect. Immun., 24(3), 589-592 (1979)* En;en.

2930-U2 A simple method for screening human T and B lymphocyte alloantibodies. Hsia,S.; Ward,F.E.; Amos,D.B. (Div. Immunol., Dep. Microbiol. and Immunol., Duke Univ. Med. Cent., Durham, NC 27710, USA) *J. Immunol. Methods, 12(3,4), 337-343 (1976)* En;en.

2931-U2 Latex phagocytosis by polymorphonuclear leukocytes. In vitro and in vivo studies with a simple screening test. Wehinger,H.; Hofacker,M. (Univ.-Kinderklin., Mathildenstr. 1, D-7800 Freiburg i. Br., GFR) *Eur. J. Pediatr., 123(2), 125-132 (1976)* En;en.

2932-U2 A simple method to remove hemoglobin from antiserum. Sagisaka,K.; Iwasa,M. (Dep. Legal Med., Gifu Univ. Sch. Med., Gifu, Japan) *Tohoku J. Exp. Med., 120(1), 97-98 (1976)* En;en.

2933-U2 Evaluation of rosette formation on smears. Stein,G. (II Univ.-Frauenklin., Lindwurmst. 2a, 8 Munchen 2, GFR) *J. Immunol. Methods, 14(3,4), 371-380 (1977)* En;en.

2934-U2 Isolation of platelet microtubule protein by an immunosorptive method. Ikeda,Y.; Steiner,M. (Div. Hematol. Res., Memorial Hosp., Pawtucket, RI 02860, USA) *J. Biol. Chem., 251(19), 6135-6141 (1976)* En;en.

2935-U2 A simple method for the determination of complement receptor-bearing mononuclear cells. Gelfand,J.A.; Fauci,A.S.; Green,I.; Frank,M.M. (Lab. Clin. Invest., Natl. Inst. Allergy and Infect. Dis., Bethesda, MD 20014, USA) *J. Immunol., 116(3), 595-599 (1976)* En;en.

2936-U2 An economic and simplified migration inhibition test in chickens. Scheu,M.; Fiedler,H. (Inst. Physiol., Physiol. Chem. und Ernahrungsphysiol., Fach. Tiermed., Univ. Munchen, Veterinarstr. 13, D-8000 Munchen 22, GFR) *Zentralbl. Veterinarmed., B, 26(10), 843-844 (1979)* En;de,en.

2937-U2 A rapid immunological procedure for the isolation of hormonally sensitive rat fat-cell plasma membrane. Luzio,J.P.; Newby,A.C.; Hales,C.N. (Dep. Med. Biochem., Welsh Natl. Sch. Med., Heath Park, Cardiff CF4 4XN, Wales, UK) *Biochem. J., 154(1), 11-21 (1976)* En;en.

2938-U2 A simple tube technique for the detection of antinuclear factors. Mann,R. (Westman Regional Lab. Serv. Inc., Brandon, Manit., Canada) *Can. J. Med. Technol., 38(3), B121-B124 (1976)* En;en.

2939-U2 Direct observation of immunoglobulins on the erythrocyte membrane. Stolinski,C.; Romano,E.L. (Dep. Biophys., St. Mary's Hosp. Med. Sch., London W2 1PG, UK) *Vox Sang., 30(6), 420-429 (1976)* En;en.

2940-U2 A simple test for immunogenicity of colloidal infusion solutions - the draining lymph node activation. Korcakova,L.; Paluska,E.; Haskova,V.; Kopecek,J. (Inst. Clin. and Exp. Med., Budejovicka 800, 146 22 Praha 4, Czechoslovakia) *Z. Immunitatsforsch. Immunobiol., 151(3), 219-223 (1976)* En;en.

2941-U2 A simplified procedure for in vitro immunization of dispersed spleen cell cultures. Kamo,I.; Pan,S.; Friedman,H. (Dep. Microbiol., Albert Einstein Med. Cent., Philadelphia, PA 19141, USA) *J. Immunol. Methods, 11(1), 55-62 (1976)* En;en.

2942-U2 A simple perfusion system for the study of histamine release from rat peritoneal mast cells. Niederhauser,U.; Whittle,B.J.R. (Dep. Pharmacol., Inst. Basic Med. Sci., R. Coll. Surgeons of England, Lincoln's Inn Fields, London, WC2A 3PN, UK) *Br. J. Pharmacol., 56(3), 391-392 (1976)* En.

2943-U2 A comparison of the kinetics of the macrophage electrophoretic mobility (MEM) and the tanned sheep erythrocyte electrophoretic mobility (TEEM) tests. Shenton,B.K.; Jenssen,H.L.; Werner,H.; Field,E.J. (Dep. Surg., R. Victoria Infirm., Newcastle Upon Tyne NE1 4LP, UK) *J. Immunol. Methods, 14(2), 123-139 (1977)* En;en.

2944-U2 Evaluation of antigen-induced buffy coat leucocyte aggregation as a simple test of allergic reactivity. Ford,P.M.; Ford,S.E.; Gibson,J. (Dep. Med., Queen's Univ., Kingston, Ont., Canada) *J. Int. Arch. Allergy Appl. Immunol., 53(1), 56-61 (1977)* En;en.

2945-U2 Direct immunofixation after isoelectric focusing. An improved method for identification of cerebrospinal fluid and serum proteins. Stibler,H. (Dep. Neurol., Karolinska Hosp., 104 01 Stockholm, Sweden) *J. Neurol. Sci., 42(2), 275-281 (1979)* En;en.

2946-U2 Particle-labeled antibodies. I. Anti T-cell antibodies attached to plastic beads by poly-L-lysine. Gabrilovac,J.; Pachmann,K.; Rodt,H.; Jager,G.; Thierfelder,S. (Inst. Hamatol., GSF, Munich, GFR) *J. Immunol. Methods, 30(2), 161-170 (1979)* En;en.

2947-U2 Ingestion of dyed-opsonised yeasts as a simple way of detecting phagocytes in lymphocyte preparations. Cytophilic binding of immunoglobulins by ingesting cells. Shaala,A.Y.; Dhaliwal,H.S.; Bishop,S.; Ling,N.R. (Dep. Immunol., Univ. Birmingham, Birmingham B15 2TJ, UK) *J. Immunol. Methods, 27(2), 175-187 (1979)* En;en.

2948-U2 A simple cytochemical method for distinguishing EAC rosettes formed by lymphocytes and monocytes. Mullink,H.; von Blomberg,M.; Wilders,M.M.; Drexhage,H.A.; Alons,C.L. (Dep. Pathol., Free Univ. Hosp., De Boelelaan 1117, Amsterdam, Netherlands) *J. Immunol. Methods, 29(2), 133-137 (1979)* En;en.

2949-U2 A rapid method for the isolation of human peripheral null lymphocytes. Ozer,H.; Strelkauskas,A.J.; Callery,R.T.; Schlossman,S.F. (Div. Tumor Immunol., Sidney Farber Cancer Inst., Harvard Med. Sch., Boston, MA 02115, USA) *Cell. Immunol., 45(2), 334-343 (1979)* En;en.

2950-U2 Glucocorticoid receptor in polymorphonuclear leukocytes: a simple method for leukocyte glucocorticoid receptor characterization. Murakami,T.; Brandon,D.; Rodbard,D.; Loriaux,D.L.; Lipsett,M.B. (Endocrinol. and Reprod. Res. Branch, Natl. Inst. Child Health and Human Dev., Bethesda, MD 20014, USA) *J. Steroid Biochem., 10(5), 475-481 (1979)* En;en.

2951-U2 A semi-automatic method of measuring leucocyte movemement. Moss,V.A.; Simpson,H.K.L.; Roberts,J.A. (Inst. Physiol., Univ. Glasgow, Glasgow G12 8QQ, UK) *J. Immunol. Methods, 27(3), 293-300 (1979)* En;en.

2952-U2 A simple method of preparing anti-human IgA and its use in quantitation of serum IgA by radial immunodiffusion. Kelkar,S.S.; Warawdekar,W. (Grant Med. Coll., Bombay-400008, India) *Indian J. Med. Res., 69(Jan.), 113-116 (1979)* En;en.

2953-U2 Rapid measurement of total protein, albumin and IgG in cerebrospinal fluid. Muir,A.; Hensley,W.J. (Dep. Biochem., R. Prince Alfred Hosp., Sydney, NSW 2050, Australia) *Clin. Chim. Acta, 98(3), 277-279 (1979)* En.

2954-U2 Isolation and characterization of hepatocytes and Kupffer cells. Page,D.T.; Garvey,J.S. (Dep. Biol. Syracuse Univ., Syracuse, NY 13210, USA) *J. Immunol. Methods, 27(2), 159-173 (1979)* En;en.

2955-U2 Estimation of ABO system antibodies by an automatic technique. Michael,P.; Frankowska,K. (Dep. Serol., Inst. Hematol., 00-957 Warsaw, Poland) *Arch. Immunol. Ther. Exp., 25(4), 549-552 (1977)* En;en.

2956-U2 A rapid screening method for the detection of monospecific antibodies against hemoglobins. Ansari,A.A.; Malling,H.V. (Lab. Biochem. Genet., Natl. Inst. Environ. Health Sci., Research Triangle Park, NC 27709, USA) *J. Immunol. Methods, 24(3-4), 383-387 (1978)* En;en.

2957-U2 The use of [^{125}I]Clq subcomponent for the measurement of complement binding antibodies on cell surfaces. Shepherd,P.S.; Dean,C.J. (Chester Beatty Res. Inst., Clifton Ave., Belmont, Sutton, Surrey, UK) *J. Immunol. Methods, 25(1), 55-64 (1979)* En;en.

2958-U2 A simplified modified method of macrophage migration inhibition in mice. Suslov,A.P.; Chernousov,A.D. (Cancer Res. Cent., Acad. Med. Sci. USSR, Moscow, USSR) *Byull. Eksp. Biol. Med., 88(8), 236-238 (1979)* Ru;en,ru.

2959-U2 A rapid method for generating cytotoxic effector cells in vivo. Newcomb Hurt,S.; Berke,G.; Clark,W. (Dep. Biol., UCLA, Los Angeles, CA 90024, USA) *J. Immunol. Methods, 28(3-4), 321-329 (1979)* En;en.

2960-U2 A [^{3}H]thymidine paper strip method for stimulation of lymphocytes. I. Culture conditions for human or murine lymphocytes and reproducibility of the paper strip method. Schutt,C.; Harms,L.; Templin,B.-U.; Schulze,H.-A. (Forschungsabt. Immunol., Inst. Physiol. Chem., Wilhelm-Pieck-Univ. Rostock, 25 Rostock, GDR) *Acta Biol. Med. Ger., 37(7), 1091-1097 (1978)* De;de,en.

2961-U2 [A microtest with whole blood for the lymphocyte response in vitro]. Rottoli,P.; Rottoli,L.; Caramia,R. (Ist. Tisiol. e Malattie, Apparato Resp., Univ. Studi, Siena, Italy) *Boll. Soc. Ital. Biol. Sper., 54(8), 741-745 (1978)* It.

2962-U2 A simple method for detecting suppressor cells of the mixed lymphocyte reaction in man: application to a healthy population. Engleman,E.G. (Stanford Univ. Blood Cent., 3330 Hillview Ave., Palo Alto, CA 94304, USA) *Transplant. Proc., 10(4), 901-903 (1978)* En.

2963-U2 Description of a method for measuring sera with a high level of IgE: test through the inactivation of anti-IgE covered phages. Fresan-Orozco,C.; Contreras,M.F.; Zamacona,G.; *Ortiz-Ortiz,L. (Inst. Invest. Biomed., UNAM, Apartado Postal 70228, Mexico 20, DF, Mexico) *Arch. Invest. Med., 9(2), 417-428 (1978)* En;Es,en,es.

2964-U2 Separation of lymphocyte sub-populations using antibodies attached to staphylococcal protein A-coated surfaces. Nash,A.A. (Dep. Exp. Pathol., Univ. Birmingham, Birmingham, UK) *J. Immunol. Methods, 12(1, 2), 149-161 (1976)* En;en.

2965-U2 A simplified micromethod for the measurement of leukocyte candidacidalactivity. Laurenti,F.; Battaglia,M.; Lendvai,D.; Midulla,M.; Rezza,E. (Univ. Roma, Catt. I Clin. Pediatr., Roma, Italy) *Ann. Sclavo Riv. Microbiol. Immunol., 18(4), 574-581 (1976)* It;en.

2966-U2 A simple technique for the inactivation of IgM antiboies using dithiothreitol. Olson,P.R.; Weiblen,B.J.; O'Leary,J.J.; Mascowitz,J.; *McCullough,J. (Blood Bank, PO Box 198, Univ. Minnesota Hosp., Minneapolis, MN 55455, USA) *Vox Sang., 30(2), 149-159 (1976)* En;en.

2967-U2 Production of adjuvant-induced ascitic fluid in the white rat. Price,J.L. (Asa Wright Nature Cent., PO Bag 10, Port of Spain, Trinidad, West Indies) *Am. J. Trop. Med. Hyg., 27(1, Part 1), 150-152 (1978)* En;en.

2968-U2 Immunofixation on cellulose acetate: an improved screening method for monoclonal immunoglobulins. Pizzolato,M.A.; Goni,F.R.; Salvarezza,R.C. (Dep. Clin. Chem., Fac. Biochem., Univ. Buenos Aires, Junin 956, 113 Buenos Aires, Argentina) *J. Immunol. Methods, 26(4), 365-368 (1979)* En;en.

2969-U2 A rapid, simple, and reliable technique for preparation of antisera against idiotypes of homogeneous immunoglobulins. Radl,J.; de Glopper,E.; de Groot,G. (Inst. Exp. Gerontol., Org. Health Res., TNO, Lange Kleiweg 151, Rijswijk, ZH, Netherlands) *Vox Sang., 35(1 and 2), 10-12 (1978)* En;en.

2970-U2 Rapid screening technique for polymeric IgM. Krolikowski,F.J.; Bell,F.T.; Gerson,B. (Lab. Pathol., New England Deaconess Hosp., Boston, MA 02215, USA) *Clin. Chem., 25(9), 1673-1674 (1979)* En.

2971-U2 A simple and efficient method for preparation of immunoelectrohoretically pure guinea pig IgM and isolation of monospecific anti-IgM antibodies. Mauch,H.; Kumel,G. (Med. Univ.-klin. (Innere Med. I), Univ. Saarlandes, D-6650 Homburg, GFR) *Res. Exp. Med., 175(3), 279-286 (1979)* En;en.

2972-U2 The ethanol fractionation of mouse serum. Measel,J.W.; Cozad,G.C. (Dep. Microbiol. and Immunol., Kirksville Coll. Osteopath. Med., Kirksville, MO 63501, USA) *J. Immunol. Methods, 13(3-4), 289-298 (1976)* En;en.

2973-U2 Isolation of carcinoembryonic antigen by an improved procedure. Pritchard,D.G.; Egan,M.L. (Dep. Microbiol., Univ. Alabama in Birmingham, University Station, Birmingham, AL 35294, USA) *Immunochemistry, 15(6), 385-387 (1978)* En;en.

2974-U2 A simple allogeneic test system for the study of antibody-dependent cellular cytotoxicity (ADCC) in man. Lang,I.; Fekete,B.; Gergely,P.; Petranyi,Gy. (II Dep. Med., Semmelweis Univ. Med. Sch., Budapest, Hungary) *Haematologica, 11(1-2), 85-91 (1977)* En;en.

2975-U2 Isolation of normal human IgA, IgM and IgG fragments by polyacrylamide beads immunoadsorbents. Sapin,C.; Massez,A.; Contet,A.; Druet,P. (Unit Rech. Path. Renale et Vasc., INSERM U 28, CNRS ERA 48, Hop. Broussais, 96, rue Didot, 75014 Paris, France) *J. Immunol. Methods, 9(1), 27-38 (1975)* En;en.

2976-U2 An evaluation of some factors affecting the detection of blood group antibodies by automated methods. Kolberg,J.; *Nordhagen,R. (Dep. Immunol., Natl. Inst. Public Health, Postu Hak, Oslo 1, Norway) *Transfusion, 15(4), 334-339 (1975)* En;en.

2977-U2 A micro-method for PHA-induced stimulation of human lymphocytes. I. Communication: technical considerations. Pees,H.; Pappas,A. (Med. Universitatsklin. und Poliklin. Innere Med. I, Univ. Saarlandes, D-665 Homburg/Saar, GFR) *Z. Immunitatsforsch. Exp. Klin. Immunol., 150(4), 309-317 (1975)* De;de,en.

2978-U2 Developments in rapid cell analysis and sorting techniques applicable to biomedical problems. Cram,L.S.; Salzman,G.C. (Biophys. and Instrum. Group, Los Alamos Sci. Lab., Univ. California, Los Alamos, NM 87545, USA) *Dev. Ind. Microbiol., 17, 141-151 (1976)* En;en.

2979-U2 An automated bichromatic measurement of serum immunoglobulins. Fu,P.C.; Zic,V. (Sect. Clin. Biochem., Dep. Pathol., UCLA Sch. Med., Harbor Gen. Hosp. Campus, Torance, CA 90509, USA) *Ann. Clin. Lab. Sci., 8(4), 323-329 (1978)* En;en.

2980-U2 A micro-rosette test for newborns. Moyer,R.P.; Dockhorn,R.J. (Med. and Professional Bldg., Suite N-6, 7501

Mission Rd., Prairie Village, KS 66208, USA) *Ann. Allergy, 35(5), 271-273 (1975)* En;en.

2981-U2 A simple and inexpensive method for maintaining a defined flora mouse colony. Sedlacek,R.S.; Mason,K.A. (Dep. Radiat. Med., Massachusetts Gen. Hosp., Boston, MA 02114, USA) *Lab. Anim. Sci., 27(5 Part 1), 667-670 (1977)* En;en.

2982-U2 Pretreatment of plastic Petri dishes with fetal calf serum. A simple method for macrophage isolation. Kumagai,K.; Itoh,K.; Hinuma,S.; Tada,M. (Dep. Microbiol., Tohoku Univ. Sch. Dent., Sendai, Japan) *J. Immunol. Methods, 29(1), 17-25 (1979)* En;en.

2983-U2 Separation of various B-cell subpopulations from mouse spleen. II. Depletion of antigen-specific B cells by rosetting with glutaraldehyde-fixed, antigen-coupled red blood cells. Walker,S.M.; Weigle,W.O. (Dep. Immunopathol., Scripps Clin. and Res. Found., 10666 North Torrey Pines Road, La Jolla, CA 92037, USA) *Cell. Immunol., 46(1), 170-177 (1979)* En;en.

2984-U2 Human leucocyte migration: studies with an improved skin chamber technique. Hellum,K.B.; Solberg,C.O. (Med. Dep. B, 5016, Haukeland sykehus, Bergen, Norway) *Acta Pathol. Microbiol. Scand., Ser. C, 85(6), 413-423 (1977)* En;en.

2985-U2 A simplified method for measuring basophil histamine release and blocking antibodies in hay fever patients. Basophil histamine content and cell preservation. Stahl Skov,P.; *Norn,S. (Dep. Pharmacol., Univ. Copenhagen, Juliane Maries Vej 20, DK-2100 Copenhagen 0, Denmark) *Acta Allergol., 32(3), 170-182 (1977)* En;en.

2986-U2 Anti-human Clq: rapid and simple method for preparing monospecific antisera. Yonemasu,K. (Webb-Waring Lung Inst., Univ. Colorado Sch. Med., Box B122, 4200 East Ninth Ave., Denver, CO 80220, USA) *J. Immunol. Methods, 9(1), 185-194 (1975)* En;en.

2987-U2 A rapid micro method for the simultaneous determination of phagocytic-microbiocidal activity of human peripheral blood leukocytes in vitro. Smith,D.L.; Rommel,F. (Natl. Inst. Health, Build. 10, Room 8N 214, Bethesda, MD 20014, USA) *J. Immunol. Methods, 17(3-4), 241-247 (1977)* En;en.

2988-U2 A simplified micromethod for the measurement of leucocyte chemotaxis. Laurenti,F.; Battaglia,M.; Lendvai,D.; Midulla,M.; Rezza,E. (Univ. Roma, Cattedra I Clin. Pediatr., Roma, Italy) *Ann. Sclavo Riv. Microb. Immunol., 18(4), 582-589 (1976)* It;en.

2989-U2 Quantitation of urinary immunoglobulins by a double antibody technique. Uehling,D.; Hong,R. (Univ. Wisconsin Cent. Health Sci., Dep. Surg. (Urology), Madison, WI 53706, USA) *Invest. Urol., 15(1), 39-41 (1977)* En;en.

2990-U2 In vitro evaluation of opsonic and cellular granulocyte function by luminol-dependent chemiluminescence: utility in patients with severe neutropenia and cellular deficiency states. Stevens,P.; Winston,D.J.; Van Dyke,K. (Div. Infect. Dis., Dep. Med., UCLA Sch. Med., Los Angeles, CA 90024, USA) *Infect. Immun., 22(1), 41-51 (1978)* En;en.

2991-U2 A rapid method for isolation of mesophyll protoplasts. Nagata,T.; Ishii,S. (Dep. Pure and Appl. Sci., Univ. Tokyo, 3-8-1 Komaba, Meguro-ku, Tokyo 153, Japan) *Can. J. Bot., 57(18), 1820-1823 (1979)* En;en,fr.

2992-U2 A simple method for extracting crude sesquiterpene lactones from Compositae plants for skin tests, chemical investigations and sensitizing experiments in guinea pigs. Hausen,B.M. (University Hosp., Hamburg, GFR) *Contact Derm., 3(1), 58-60 (1977)* En.

See: 207, 679

Storage and preservation techniques

2993-U2 Prolonged preservation of cell lines at 36 degrees. Farrohi,K.; Mohammadzadeh-Kiai,F. (Dep. Microbiol. et Immunol., Fac. Med., Univ. Teheran, Iran) *Bull. Soc. Pathol. Exot., 68(6), 603-608 (1975)* Fr;en,fr.

2994-U2 Freeze-drying apparatus and its application to adenovirus. Hosoi,J.; Iwata,H.; Okutomi,S. (JFOL, Akishima, Tokyo, 196 Japan) *J. Electron Microsc., 28(1), 49-50 (1979)* En.

2995-U2 The long-term preservation of potato virus Y and water-melon mosaic virus in liquid nitrogen in comparison to other preservation methods. de Wijs,J.J.; Suda-Bachmann,F. (Ciba-Giegy Ltd., Agrochem. Div., 4002 Basle, Switzerland) *Neth. J. Plant Pathol., 85(1), 23-29 (1979)* En;en,nl.

2996-U2 Preservation of mycoplasmas on anhydrous silica gel. Thorns,C.J. (Cent. Vet. Lab., New Haw, Weybridge, Surrey, UK) *J. Applied. Bacteriol., 47(1), 183-186 (1979)* En;en.

2997-U2 A simple method for preservation of inoculum of actinomycetes. Wellington,E.M.H.; Williams,S.T. (Bot. Dep., Liverpool Univ., Liverpool L69 3BX, UK) *Biol. Actinomycetes Relat. Org., 12(4), 48-52 (1977)* En.

2998-U2 Preservation of actinomycete inoculum in frozen glycerol. Wellington,E.M.H.; Williams,S.T. (Dep. Bot., Univ. Liverpool, PO Box 147, Liverpool L69 3BX, UK) *Microbios Lett., 6(23-24), 151-157 (1977)* En;en.

2999-U2 Simple bacterial preservation medium and its application to proficiency testing in water bacteriology. Brodsky,M.H.; Ciebin,B.W.; Schiemann,D.A. (Ontario Minist. Health. Lab. Serv. Branch, Environ. Bacteriol., Toronto, Ont. M5W 1R5, Canada) *Appl. Environ. Microbiol., 35(3), 487-491 (1978)* En;en.

3000-U2 A simple method for long-term preservation of stock cultures of lactic acid bacteria. Juven,B.J. (Div. Food Technol., Agric. Res. Organization, Volcani Cent., PO Box 6,

Bet Dagan 50-200, Israel) *J. Appl. Bacteriol., 47(3), 379-381 (1979)* En;en.

3001-U2 Preservation of streptococci and other bacteria by sand desiccation and filter paper techniques. Koshi,G.; Rajeshwari,K.; Philipose,L. (Dep. Microbiol., Christian Med. Coll., Vellore, India) *Indian J. Med. Res., 65(4), 500-502 (1977)* En;en.

3002-U2 Application of the soft agar stab method for preservation of *Xanthomonas citri.* Wu,W.C.; Fong,K.-T.; Tzeng,K.C. (Dep. Plant Pathol., Natl. Chung Hsing Univ., Taichung, Taiwan) *Chin. J. Microbiol., 9(3/4), 68-72 (1976)* En;ch,en.

3003-U2 Maintenance procedures for the curtailment of genetic instability: *Xanthomonas campestris* NRRL B-1459. Kidby,D.; *Sandford,P.; Herman,A.; Cadmus,M. (Northern Reg. Res. Cent., ARS, USDA, Peoria, IL 61604, USA) *Appl. Environ. Microbiol., 33(4), 840-845 (1977)* En;en.

3004-U2 Efficacy of some simple methods for long-term preservation of bacterial cultures. Suassuna,I.; Suassuna,I.R.; Ricciardi,I.D.; Formiga,L.C.D. (Dep. Microbiol. Med., Inst. Microbiol., Univ. Fed. Rio de Janeiro, 20000 Rio de Janeiro, RJ, Brazil) *Rev. Microbiol., 8(1), 16-20 (1977)* En;en,pt.

3005-U2 A simple method for maintaining fastidious organisms. Park,C.H. (Dep. Pathol., Fairfax Hosp., Falls Church, VA 22046, USA) *Am. J. Clin. Pathol., 66(5), 927-928 (1976)* En;en.

3006-U2 The use of plastic ampoules for freeze preservation of microorganisms. Simione,F.P.,Jr.; Daggett,P.-M.; McGrath,M.S.; Alexander,M.T. (ATCC, 12301 Parklawn Drive, Rockville, MD 20852, USA) *Cryobiology, 14(4), 500-502 (1977)* En.

3007-U2 Polypropylene vials for preserving fungi in liquid nitrogen. Butterfield,W.; Jong,S.C.; Alexander,M.T. (American Type Culture Coll., 12301 Parklawn Drive, Rockville, MD 20852, USA) *Mycologia, 70(5), 1122-1124 (1978)* En.

3008-U2 Application of flow microcalorimetry to analytical problems: the preparation, storage and assay of frozen inocula of *Saccharomyces cerevisiae.* Beezer,A.E.; Newell,R.D.; Tyrrell,H.J.V. (Chem. Dep., Chelsea Coll., Manresa Rd., London, SW3 6LX, UK) *J. Appl. Bacteriol., 41(2), 197-207 (1976)* En;en.

3009-U2 Cryopreservation of encysted toxoplasms in liquid nitrogen. Janitschke,K.; Jorren,H.R. (Robert Koch-Inst., D-1 Berlin 65, Nordufer 20, GFR) *Tropenmed. Parasitol., 26(3), 307-311 (1975)* De;de,en.

3010-U2 [Preservation of algal preparations for transmission electron microscopy]. Genkal,S.I.; Balonov,I.M. (Inst. Biol. Inland Waters, Borok, USSR) *Bot. Zh., 61(11), 1578-1579 (1976)* Ru;ru.

3011-U2 On making fluid mounts of plankton algae. Nygaard,G. (Freshwater Biol. Lab., DK-3400 Hillerod, Denmark) *Phycologia, 16(3), 351 (1977)* En.

3012-U2 Freeze-drying of algae: Chlorophyta and Chrysophyta. McGrath,M.S.; Daggett,P.M.; Dilworth,S. (American Type Culture Collect., 12301 Parklawn Drive, Rockville, MD 20852, USA) *J. Phycol., 14(4), 521-525 (1978)* En;en.

3013-U2 Mini-technique for the liquid nitrogen preservation of trichomonads. Propst,C. (GTE Inc., 40 Sylvan Rd., Waltham, MA 02154, USA) *Dev. Ind. Microbiol., 18, 741-744 (1977)* En;en.

3014-U2 A simplified freeze-drying technique for protozoan cells. Suzaki,T.; Shigenaka,Y.; Toyohara,A.; Otsuji,H. (Dep. Inf. and Behavior Sci., Fac. Integrated Arts and Sci., Hiroshima Univ., Hiroshima, 730 Japan) *J. Electron Microsc., 27(2), 153-156 (1978)* En.

3015-U2 A simple method for the cryopreservation of human lymphocytes at -80°C. Simon,J.D.; Flinton,L.J.; Albala,M.M. (Div. Clin. Hematol., Rhode Island Hosp., Providence, RI 02902, USA) *Transfusion, 17(1), 23-28 (1977)* En;en.

3016-U2 A method for preservation of lymphocyte rosettes in agarose. Yoshizawa,Y.; Lee,Y.; Carr,R.G.; Yokoyama,M.M. (Dep. Pathol., Abraham Lincoln Sch. Med., Univ. Illinois at the Med. Cent., Chicago, IL 60680, USA) *Immunol. Commun., 8(2), 185-191 (1979)* En;en.

3017-U2 A simple method for keeping allergenic extracts at a temperature near 4 C during in-office use for skin testing and immunotherapy. Van Metre,T.E.,Jr. (11E Chase St., Baltimore, MD 21202, USA) *J. Allergy Clin. Immunol., 59(4), 341-342 (1977)* En;en.

3018-U2 Liquid nitrogen storage of *Anaplasma marginale* complement-fixation antigen by a multiple small aliquot technique. Parker,R.; Parker,M.L.; Wilson,A.J. (Dep. Primary Ind., Anim. Health Stn., Oonoonba, Townsville, Queensl., Australia) *Res. Vet. Sci., 25(3), 401-402 (1978)* En;en.

3019-U2 A simple method for storage of bacteria at -76 C. Feltham,R.K.A.; Power,A.K.; Pell,P.A.; Sneath,P.H.A. (Dep. Microbiol., Univ. Leicester, Leicester LE1 7RH, UK) *J. Appl. Bacteriol., 44(2), 313-316 (1978)* En;en.

3020-U2 An adaptable system for timed aseptic sampling and storage of microbial cultures. Newman,P.B. (ARC, Meat Res. Inst., Langford, Bristol BS18 7DY, UK) *J. Appl. Bacteriol., 41(3), 497-504 (1976)* En;en.

3021-U2 Storage of fungal cultures in water. Boesewinkel,H.J. (Minist. Agric. and Fish., C-Plant Dis. Div., DSIR, Auckland, New Zealand) *Trans. Br. Mycol. Soc., 66(1), 183-185 (1976)* En.

3022-U2 A capillary tube method for storage of myxosporidian and microsporidian spores. Lom,J. (Parasitol. Inst., Czechoslovak Acad. Sci., 166 32 Prague 6, Czechoslovakia) *Folia Parasitol., 22(3), 275-277 (1975)*

En;en,ru.

3023-U2 MLC reactions with dog lymphocytes frozen in microtiter plates. Netzel,B.; Grosse-Wilde,H.; Mempel,W. (Inst. Hamatol., Abt. Immunol. GSF, Munchen, GFR) *Transplant. Proc., 7(3), 403-405 (1975)* En.

3024-U2 Use of serum stored in filter paper disks in complement fixation tests for adenovirus antibody. Edwards,E.A. (Biol. Sci. Div., Nav. Health Res. Cent., San Diego, CA 92152, USA) *J. Clin. Microbiol., 5(2), 253-254 (1977)* En;en.

3025-U2 Preservation and storage of pathogenic *Neisseria*. Cody,R.M. (Dep. Bot. and Microbiol., Auburn Univ., Auburn, AL 36830, USA) *Health Lab. Sci., 15(4), 206-209 (1978)* En;en.

3026-U2 Compact liquid nitrogen storage system yielding high recoveries of gram-negative anaerobes. Gilmour,M.N.; Turner,G.; Berman,R.G.; Krenzer,A.K. (Eastman Dent. Cent., Rochester, NY 14603, USA) *Appl. Environ. Microbiol., 35(1), 84-88 (1978)* En;en.

3027-U2 Studies on the preservation of the ciliate *Didinium nasutum*. McGrath,M.S.; Daggett,P.-M.; Nerad,T.A. (American Type Culture Collection, 12301 Parklawn Drive, Rockville, MD 20852, USA) *Trans. Am. Microsc. Soc., 96(4), 519-525 (1977)* En;en.

3028-U2 A simple method for the cryoperservation of lymphocytes. Retention of specific immune effector functions by frozen-stored cells. Grant,C.K. (Dep. Surg., Sch. Vet. Med., Univ. California, Davis, CA 95616, USA) *Clin. Exp. Immunol., 23(1), 166-174 (1976)* En;en.

3029-U2 Application of photo-crosslinkable resin to immobilization of an enzyme. Fukui,S.; Tanaka,A.; Iida,T.; Hasegawa,E. (Lab. Ind. Biochem., Dep. Ind. Chem., Fac. Eng., Kyoto Univ., Sakyo-ku, Kyoto, Japan) *FEBS Lett., 66(2), 179-182 (1976)* En.

3030-U2 Disinfection of anaesthesia equipment by a mechanized pasteurization method. /[presented at the Annual Meeting, Canadian Anaesthetists' Society, held at St. John's, Newfoundland, 1974]. Craig,D.B.; Cowan,S.A.; Forsyth,W.; Parker,S.E. (Dep. Anaesth., Univ. Manitoba, Winnipeg, Manit., Canada) *J., Can. Anaesth. Soc., 22(2), 219-223 (1975)* En;en,fr.

3031-U2 Preparation of enterochelin from *Escherichia coli*. Young,I.G. (Dep. Biochem., John Curtin Sch. Med. Res., Australian Natl. Univ., Canberra, PO Box 334, Canberra City, ACT, Australia) *Prep. Biochem., 6(2-3), 123-131 (1976)* En;en.

3032-U2 Rapid techniques for the examination of bacterial citrate synthases. Harford,S.; Jones,D.; *Weitzman,P.D.J. (Dep. Biochem., Sch. Biol. Sci., Adrian Build., Univ. Leicester, Leicester, UK) *J. Appl. Bacteriol., 41(3), 465-471 (1976)* En;en.

3033-U2 A simple method for the preparation of antisera specific for murine immunoglobulin heavy chains. Babcock,G.F.; Lanier,L.L.; Lynes,M.A.; Haughton,G. (Address not stated) *J. Immunol. Methods, 23(1-2), 1-6 (1978)* En;en.

Equipment (including computer applications)

3034-U2 A new polystyrene microscope tray for immunofluorescent studies. Krech,U.; Jung,M.; Pyndiah,N.; Price,P.C. (Frohbergstr. 3, CH-9000 St. Gallen, Switzerland) *Zentralbl. Bakteriol. Parasitenkd. Infektionskr. Hyg., I Abt. A, 234(1), 136-140 (1976) Immun., 14(2), 332-336 (1976)* En;en,de.

3035-U2 Design and construction of an apparatus for the growth of micro cell cultures on standard glass microscope slides and its application for screening large numbers of sera by the indirect fluorescent antibody technique. Lawman,M.J.P.; Caie,I.S. (Anim. Virus Res. Inst., Pirbright, Woking, Surrey, GU24 0NF, UK) *J. Clin. Microbiol., 2(3), 153-156 (1975)* En;en.

3036-U2 A low-cost microcomputer for the control of microbiological systems. Breame,A.J.; Spier,R.E. (Anim. Virus Res. Inst., Pirbright, Woking, Surrey GU24 0NF, UK) *Lab. Pract., 26(12), 957-960 (1977)* En.

3037-U2 Automation in microbiology and immunology. Heden,C.-G.; Illeni,T. (eds.) *Publ. by:* John Wiley and Sons Ltd., Baffins Lane, Chichester, Sussex, England. Jan. 1975. ISBN 0-471-36745-1. £25.25 or $43.00. En.

3038-U2 Simple device for washing capillary microbiological laboratory dishes. Magarlamov,A.G.; Zaikin,A.A. (Kiev Sci. Res. Inst. Haematol. and Blood Transfusions, Kiev, USSR) *Mikrobiol. Zh., 38(3), 372-373 (1976)* Uk;en.

3039-U2 Computer-assisted analysis of adenosine triphosphate data. Eerkenbrecher,C.W.; Crabtree,S.J.,Jr.; Stevenson,L.H. (Dep. Biol., Univ. South Carolina, Columbia, SC 29208, USA) *Appl. Environ. Microbiol., 32(3), 451-454 (1976)* En;en.

3040-U2 A method of flooding sensitivity agar plates using a modified Oxford automatic diluter/dispenser. Cowlishaw,W.A. (Dep. Microbiol., Gen. Hosp., Nottingham NG1 6HA, UK) *Med. Lab. Sci., 33(2), 119-124 (1976)* En;en.

3041-U2 Safe, convenient pipetting station. Songer,J.R.; Braymen,D.T.; Mathis,R.G. (Natl. Anim. Dis. Cent., ARS, Ames, IA 50010, USA) *Appl. Microbiol., 30(5), 887-888 (1975)* En;en.

3042-U2 Rapid entry port for an anaerobic glove box. Gill,V.J.; Tipton,H.W.; Gersch,S.M. (Microbiol. Serv., Dep. Clin. Pathol., NIH, Bethesda, MD 20014, USA) *J. Clin. Microbiol., 8(6), 736-739 (1978)* En;en.

3043-U2 A computerized system for clinical microbiology. Schito,G.C. (Univ. Parma, Ist. Microbiol., Parma, Italy) *Ann. Sclavo Riv. Microbiol. Immunol., 19(4), 588-609 (1977)* It;en.

3044-U2 Automated differential leucocyte counting: the present state of the art. Bentley,S.A.; Lewis,S.M. (Dep. Haematol., R. Postgrad Med. Sch., Du Cane Rd., Loneon W12 0HS, UK) *Br. J. Haematol., 35(4), 481-485 (1977)* En.

3045-U2 An all-metal block for replica plating of microbial colonies. Skodova,H.; Weisgerber,J.; Skoda,J. (Res. Inst. for Biofactors and Vet. Drugs, Pohori-Chotoun, Czechoslovakia) *Folia Microbiol., 23(11), 27-29 (1978)* En;en.

3046-U2 An epidemiological database system. McClatchie,G. (Div. Health Res. Plan., Health Comm. New South Wales, 13 Young St., NSW, Australia) *Int. J. Bio-Med. Comput., 9(1), 11-24 (1978)* En;en,fr.

3047-U2 Quantitative electronic analysis of normal and transformed BHK21 fibroblast aggregation. Whur,P.; Koppel,H.; Urquhart,C.; Williams,D.C. (Cell Biol. Unit, Marie Curie Mem. Found., Oxted, Surrey, UK) *J. Cell Sci., 23, 193-209 (1977)* En;en.

3048-U2 A simple multiple chamber apparatus for measuring chemotaxis of polymorphonuclear leukocytes utilizing centrifugation of the chambers before incubation. Swanson,M.J. (Midwest Res. Inst., 425 Volker Blvd., Kansas City, MO 64110, USA) *J. Immunol. Methods, 16(4), 385-390 (1977)* En;en.

3049-U2 Mechanized blood grouping: a hospital trial using an 8-channel grouping machine. Jenkins,G.C.; Fewell,R.G.; Lloyd,M.J.; Brown,J.; Judd,J.; Lane,R.S.; Jenkins,W.J. (London Hosp., Whitechapel, London, UK) *J. Clin. Pathol., 28(11), 860-862 (1975)* En;en.

3050-U2 Monitoring of dosing of liquids in laboratory scale fermentation. Enfors,S.-O.; Dostalek,M. (Dep. Tech. Microbiol., Chem. Cent., Univ. Lund, Lund, Sweden) *Process Biochem., 10(6), 13-15, 21 (1975)* En;en.

3051-U2 Glutamine-selective membrane electrode that uses living bacterial cells. Rechnitz,G.A.; Riechel,T.L.; Kobos,R.K.; Meyerhoff,M.E. (Dep. Chem., State Univ. New York, Buffalo, NY 14214, USA) *Science (Wash.), 199(4327), 440-441 (1978)* En;en.

3052-U2 A simplified design of capillary pedoscope for microbiological studies. Mjatlikova,K.O. (Inst. Microbiol. and Virol., Acad. Sci. Ukrainian SSR, ul. Zabolotnogo 59, Kiev, USSR) *Mikrobiol. Zh., 38(4), 513-514 (1976)* Uk;en.

3053-U2 Computer applications to fermentation processes. Weigand,W.A. (Sch. Chem. Eng., Purdue Univ., West Lafayette, IN 47907, USA) *In:* Annual reports on fermentation processes. Vol. 2. Perlman,D. (ed.). *Publ.by:* Academic Press, Inc. (London) Ltd., 24-28 Oval Rd., London NW1 7DX, UK. 1978. p.43-72 ISBN: 0-12-040302-1. En. [Review with 65 refs.]

3054-U2 A computer assisted method for the determination of restriction enzyme recognition sites. Gingeras,T.R.; Milazzo,J.P.; Roberts,R.J. (Cold Spring Harbor Lab., Cold Spring Harbor, NY 11724, USA) *Nucleic Acids Res., 5(11), 4105-4127 (1978)* En;en.

3055-U2 A new fermentor with a high oxygen transfer capacity. Minami,K.; Yamamura,M.; Shimizu,S.; Ogawa,K.; Sekine,N. (Res. Dev. Cent., Maruzen Oil Co. Ltd., PO Box 1, Satte, Saitama-ken 340-01, Japan) *J. Ferment. Technol., 56(1), 64-67 (1978)* En;en.

3056-U2 Dispenser for microbiological plating media. Quinsland,D. (Paul Masson Vineyards, PO Box 97, Saratoga, CA 95070, USA) *Am. J. Enol. Vitic., 31(1), 106-107 (1980)* En.

3057-U2 The design and operation of a chemostat in conjunction with a flow micro-calorimeter. Ackland,P.J.; Prichard,F.E.; James,A.M. (Dep. Chem., Bedford Coll. (Univ. London), Regent's Park, London NW1 4NS, UK) *Microbios Lett., 3(9), 21-24 (1976)* En;en.

3058-U2 Silicone tubing sensor for detection of methanol. Yano,T.; Kobayashi,T.; Shimizu,S. (Dep. Food Sci. and Technol., Fac. Agric., Nagoya Univ., Nagoya 464, Japan) *J. Ferment. Technol., 56(4), 421-427 (1978)* En;en.

3059-U2 A computer system for clinical microbiology. Williams,K.N.; Davidson,J.M.F.; Lynn,R.; Rice,E.; *Phillips,I. (Dep. Microbiol., St. Thomas' Hosp. Med. Sch., London SE1 7EH, UK) *J. Clin. Pathol., 31(12), 1193-1201 (1978)* En;en.

3060-U2 Modification of membrane diffusion chambers for deep-water studies. Fliermans,C.B.; Gorden,R.W. (Savannah River Lab., E.I. du Pont de Nemours and Co., Aiken, SC 29801, USA) *Appl. Environ. Microbiol., 33(1), 207-210 (1977)* En;en.

3061-U2 pH measurements of agar culture media by using a flat-bottom electrode. Gorski,T.W.; *Ritzert,R.W. (Owens-Illinois, Inc., Toledo, OH 43666, USA) *Appl. Environ. Microbiol., 34(2), 242-243 (1977)* En;en.

3062-U2 Disrupter for bacteriological and mammalian cells. Sharpe,J.E.E. (Natl. Inst. Med. Res., Mill Hill, London NW7 1AA, UK) *Lab. Pract., 25(1), 28-29 (1976)* En;en.

3063-U2 The electrolytic respirometer. II. Use in water pollution control plant laboratories. Young,J.C.; Baumann,E.R. (Dep. Civil Eng., Iowa State Univ., Ames, IA 50011, USA) *Water Res., 10(12), 1141-1149 (1976)* En;en.

3064-U2 Evaluation of an automated agar plate streaker. Tilton,R.C.; Ryan,R.W. (Univ. Connecticut Health Cent., Farmington, CT 06032, USA) *J. Clin. Microbiol., 7(3), 298-304 (1978)* En;en.

3065-U2 [Fumigation chambers for disinfestation and sterilization]. Albertazzi,G. (Colkim SNC, Via Piemonte, Ozzano Emilia (Bo), Italy) *Ind. Aliment., 14(12), 73-78 (1975)* It.

3066-U2 An agar well punching-sucking device for diffusion assays. Hallynck,Th.; Pijck,J. (Lab. Pharm. Microbiol., State Univ. Ghent, Apotheekst. 1, B-9000 Ghent, Belgium) *J. Antimicrob. Chemother., 4(1), 94-95 (1978)* En.

3067-U2 A syringe system for rapid and accurate pipetting of microlitre samples of serum. Bradwell,A.R.; Bunce,R.A. (Dep. Med., (Wolfson Res. Lab.) Queen Elizabeth Med. Cent., Edgbaston, Birmingham 15 2TH, UK) *Med. Lab. Sci., 33(3), 229-233 (1976)* En;en.

3068-U2 Design and testing of a calorimeter for microbiological uses. Fujita,T.; Nunomura,K.; Kagami,I.; Nishikawa,Y. (Inst. Appl. Microbiol., Univ. Tokyo, Tokyo, Japan) *J. Gen. Appl. Microbiol., 22(1), 43-50 (1976)* En;en.

3069-U2 A device for mass isolation of virus clones from plaques. Davydova,S.N.; Krasnobaev,E.A.; Filin,Yu.I. (All Union Sci.-Res. Foot-and-Mouth Disease Inst., Minist. Agric. USSR, Vladimir, USSR) *Vopr. Virusol., No. 3, 376-377 (1978)* Ru;en,ru.

3070-U2 A multiple sampling device for the mass screening of serum samples for hepatitis B surface antigen. Cameron,C.H.; Barbara,J.A.J. (Dep. Virol., Middlesex Hosp. Med. Sch., London W1P 7LD, UK) *J. Clin. Pathol., 31(3), 288-291 (1978)* En.

3071-U2 An automatic bacteriophage distributor. Lambert,N.G.; Jette,L.P.; Kasatiya,S.S.; Caprioli,T.; Couture,M. (Min. Affaires Soc., 20045, chemin Ste-Marie ouest, Ste-Anne-de-Bellevue, Que. H9X 3L2, Canada) *Ann. Microbiol., 130B(1), 79-84 (1979)* Fr;en,fr.

3072-U2 Production of large amounts of 35S RNA and complementary DNA from avian RNA tumor viruses. Smith,R.E.; Nebes,S.; Leis,J. (Dep. Microbiol., Duke Univ. Med. Cent., Durham, NC 27710, USA) *Anal. Biochem., 77(1), 226-234 (1977)* En;en.

3073-U2 A simple low-cost cabinet for heat therapy of virus-infected plants. Smee,L.; Ikin,R. (Plant Quarantine Res. Stn., Australian Dep. Health, PO Box 100, Woden ACT 2606, Australia) *APPS Newsl., 4(2), 15 (1975)* En.

3074-U2 Computer evaluation of antiviral activities of some thiosemicarbazones in experiments in vivo. Borysiewicz,J.; Tadeusiewicz,R. (Dep. Virol., Inst. Microbiol., Med. Acad., 31-121 Krakow, Poland) *Acta Virol., 20(5), 402-410 (1976)* En;en.

3075-U2 Virometer: an optical instrument for visual observation, measurement and classification of free viruses. Hirschfield,T.; Block,M.J.; Mueller,W. (Block Engineering, Inc., Cambridge, MA 02139, USA) *J. Histochem. Cytochem., 25(7), 719-723 (1977)* En;en.

3076-U2 Continuous aerosol therapy system using a modified collison nebulizer. Young,H.W.; Dominik,J.W.; Walker,J.S.; Larson,E.W. (US Army Med. Res. Inst. Infect. Dis., Frederick, MD 21701, USA) *J. Clin. Microbiol., 5(2), 131-136 (1977)* En;en.

3077-U2 A low-cost microcomputer for the control of microbiological systems. Breame,A.J.; Spier,R.E. (Anim. Virus Res. Inst., Pirbright, Woking, Surrey GU24 0NF, UK) *Lab. Pract., 26(12), 957-960 (1977)* En.

3078-U2 Instrument measuring the relation between the temperature and death-rate of microorganisms. Savel,J.; Prokopova,M. (Jihoceske pivovary, n.p., Ceske Budejovice, Czechoslovakia) *Kvasny Prumysl, 22(2), 25-28 (1976)* Cs;cs,de,en,ru.

3079-U2 Mechanizing microbiology. Sharpe,A.N.; Clark,D.S. (eds.) (Bur. Microb. Hazards, Ottawa, Ont., Canada) *Publ. by:* Charles B. Thomas; 301-327 E. Lawrence Ave., Springfield, IL 62717, USA 1978 352 pp. at $29.75 En.

3080-U2 Enumeration of bacterial colonies in air with an electronic colony counter. Gartner,E.; Muller,W.; Farmanara,F. (Univ. Hohenheim, Abt. Tierhyg., 7 Stuttgart 70, Postfach 106, GFR) *Zentralbl. Veterinarmed. B, 22(4), 326-334 (1975)* De;de,en,fr.

3081-U2 Use of platinum electrodes for the electrochemical detection of bacteria. Wilkins,J.R. (NASA, Langley Res. Cent., Hampton, VA 23665, USA) *Appl. Environ. Microbiol., 36(5), 683-687 (1978)* En;en.

3082-U2 Recovery of *Clostridium perfringens, Staphylococcus aureus* and molds from foods by the stomacher: effect of fat content, surfactant concentration, and blending time. Sharpe,A.N.; Harshman,G.C. (Bur. Microb. Hazards, Health Protection Branch, Food Directorate, Tunney's Pasture, Ottawa, Ont., Canada) *Can. Inst. Food Sci. Technol., 9(1), 30-34 (1976)* En;en,fr.

3083-U2 Evaluation of the Enteric Analyzer for identification of Enterobacteriaceae. Shayegani,M.; Hubbard,M.E.; Hiscott,T.; McGlynn,D.M.; Yewdall,R.C. (Div. Lab. and Res., New York State Dep. Health, Albany, NY 12201, USA) *J. Clin. Microbiol., 2(3), 186-192 (1975)* En;en.

3084-U2 A computer method for predicting the sequence of tRNA from its enzymatic digestion products based on its secondary structure. Jagadeeswaren,P.; Cherayil,J.D.; Sasisekharan,V.; Pattabiraman,N. (Dep. Biochem., Indian Inst. Sci., Bangalore 560 062, India) *Curr. Sci., 46(20), 691-694 (1977)* En.

3085-U2 A device for the incubation of *Fusarium*-inoculated tulip bulbs in a constant air stream. Bergman,B.H.H. (Bulb Res. Cent., Lisse, Netherlands) *Neth. J. Plant Pathol., 91(4), 154-156 (1975)* En;nl.

3086-U2 Rapid diagnosis of red spot disease in cultured eels by direct immunofluorescence (IF). Horiuchi,M.; Kohga,N. (Nihon Nosan Kogyo Co., Ltd., Res. Cent., Midorsku, Yokohama 226, Japan) *Bull. Jap. Soc. Sci. Fish., 45(7), 835-840 (1979)* Ja;en,ja.

3087-U2 [Use of computers and mathematical methods in epidemiological studies. IV. Methods of diagnosis of the epidemic process in dysentery by means of a computer (image

recognition theory)]. Leontieva,L.G.; Romanovsky,G.V.; Krivenko,O.V. (Inst. Biophys., Moscow, USSR) *Zh. Mikrobiol. Epidemiol. Immunobiol., 53(3), 127-132 (1976)* Ru;en,ru.

3088-U2 Apparatus for the micromanipulation of small bacteria. Isaac,L.; *Ware,G.C.; Leonard,P.G. (Dep. Bacteriol., Med. Sch. Workshops, Univ. Bristol, Bristol BS8 1TD, UK) *Lab. Pract., 24(11), 744-746 (1975)* En.

3089-U2 A model of automatic identification of streptomycetes. Gyllenberg,H.G.; Niemela,T.K.; Niemi,J.S. (Dep. Microbiol., Univ. Helsinki, SF-00710, Helsinki 71, Finland) *Post. Hig. I Med. Dosw., 29(3), 357-383 (1975)* En;en.

3090-U2 Fatty acid fingerprints of some chemostat-grown streptococci with computerized data analysis. Drucker,D.B.; Griffith,C.J.; Melville,T.H. (Dep. Bacteriol. and Virol., Univ. Manchester, Stopford Build., Oxford Road, Manchester M13 9PT, UK) *Microbios Lett., 1(1), 31-34 (1976)* En;en.

3091-U2 An apparatus for the continuous culture of micr-organisms on solid surfaces with special reference to dental plaque. Dibdin,G.H.; Shellis,R.P.; Wilson,C.M. (MRC Dent. Unit, Dent. Hosp., Lower Maudlin St., Bristol BS1 2LY, UK) *J. Appl. Bacteriol., 40(3), 261-268 (1976)* En;en.

3092-U2 Computer-aided prediction of gangrenous and perforating appendicitis. Graham,D.F. (Dep. Surg., Bangour Gen. Hosp., West Lothian EH52 6LR, UK) *Br. Med. J., 2(6099), 1375-1377 (1977)* En;en.

3093-U2 Anaerobic specimen transport device. Wilkins,T.D.; Jimenez-Ulate,F. (Anaerobe Lab., Virginia Polytech. Inst. and State Univ., Blacksburg, VA 24060, USA) *J. Clin. Microbiol., 2(5), 441-447 (1975)* En;en.

3094-U2 Automated screening for bacteriuria. Human,R.P.; Rowe,G.D. (Hosp. Microbiol. and Public Health Lab., Plymouth Gen. Hosp., Plymouth, Devon, UK) *Med. Lab. Sci., 35(3), 223-226 (1978)* En;en.

3095-U2 An improved catalyst sachet for anaerobic jars. Baldwin,A.W.R. (Dep. Microbiol., North Middlesex Hosp., Edmonton, London N18 1QX, UK) *Med. Lab. Technol., 32(4), 329-330 (1975)* En.

3096-U2 Note on a cabinet and ancillary equipment for isolation of anaerobic bacteria. Deacon,A.G.; Loutit,M. (Microbiol. Dep., Univ. Otago, Dunedin, New Zealand) *N. Z. J. Sci., 18(2), 205-207 (1975)* En;en.

3097-U2 Automatic disinfection of fiberendoscopes. Baas,E.U. (I Med. Klin. und Poliklin., Univ. Mainz, Langenbeckstr. 1, D-6500 Mainz, GFR) *Zentralbl. Bakteriol. Parasitenkd. Infektionskr. Hyg., I Abt. B, 165(5-6), 458-463 (1977)* De;de,en.

3098-U2 [A method for uniform inoculation of meat samples with bacteria]. Bomar,M.R.; Hajek,M.Z. (Bundesforschungsanstalt Ernahrung, Engesserstr. 20, 7500 Karlsruhe 1, GFR) *Fleischwirtschaft, 56(8), 1155-1157 (1976)* De;de,en.

3099-U2 Automation of water bacteriological analysis: running test of an experimental prototype. Trinel,P.A.; Hanoune,N.; *Leclerc,H. (Unite INSERM 146, Inst. Pasteur de Lille, Domaine du Certia, 59650 Villeneuve d'Ascq, France) *Appl. Environ. Microbiol., 39(5), 976-982 (1980)* En;en.

3100-U2 Chamber for bacterial chemotaxis experiments. Palleroni,N.J. (Dep. Chem. Res., Hoffmann-La Roche Inc., Nutley, NJ 07110, USA) *Appl. Environ. Mirobiol., 32(5), 729-730 (1976)* En;en.

3101-U2 Evaluation of a simple device for bacteriological sampling of respirator-generated aerosols. Ryan,K.J.; Mihalyi,S.F. (Clin. Microbiol. Lab., Dep. Pathol., Arizona Med. Cent., Univ. Arizona, Tucson, AZ 85724, USA) *J. Clin. Microbiol., 5(2), 178-183 (1977)* En;en.

3102-U2 A computer-assisted bacteriology reporting and information system. Mitchison,D.A.; Darrell,J.H.; Mitchison,R. (Dep. Bacteriol., R. Postgrad. Med. Sch., Hammersmith Hosp., London W12, UK) *J. Clin. Pathol., 31(7), 673-680 (1978)* En;en.

3103-U2 Preprototype of an automated microbial detection and identification system: a developmental investigation. Sonnenwirth,A.C. (Dep. Pathol. and Lab. Med., Jewish Hosp. St. Louis, St. Louis, MO 63110, USA) *J. Clin. Microbiol., 6(4), 400-405 (1977)* En;en.

3104-U2 Small-scale chemostat for the growth of mesophilic and thermophilic microorganisms. Gilbert,P.; Stuart,A. (Dep. Pharm., Univ. Aston, Birmingham, B4 7ET, UK) *Lab. Pract., 26(8), 627-628 (1977)* En.

3105-U2 Automated microbiological detection/identification system. Aldridge,C.; *Jones,P.W.; Gibson,S.; Lanham,J.; Meyer,M.; Vannest,R.; Charles,R. (McDonnell Douglas Astronautics Co.-East, St. Louis, MO 63166, USA) *J. Clin. Microbiol., 6(4), 406-413 (1977)* En;en.

3106-U2 Biological identification with computers. Pankhurst,R.J. (ed.) *Publ. by:* Academic Press Inc., (London) Ltd., 24-28 Oval Road, London NW1 7DX, UK. 12th November 1975. ISBN: 0-12-544850-3. at £11.00 or US $28.50.

3107-U2 Generalized indicator plate for genetic, metabolic, and taxonomic studies with microorganisms. Bochner,B.R.; Savageau,M.A. (Dep. Biochem., Univ. California, Berkeley, CA 94720, USA) *Appl. Environ. Microbiol., 33(2), 434-444 (1977)* En;en.

3108-U2 Application of microcomputers in the study of microbial processes. Hampel,W.A. (Inst. Biochem. Technol. and Microbiol., Univ. Technol., Vienna, A-1060 Wien, Austria) *In:* Advances in biochemical engineering. Vol. 13. Mass transfer and process control. Ghose,T.K. et al. (eds.) *Publ. by:* Springer-Verlag KG, Heidelberger Platz 3, Postfach, D-1000 Berlin 33, GFR. 1979 p. 1-33 ISBN 3-540-09468-7 En;en.

3109-U2 Computerization of a hospital clinical microbiology laboratory. Jorgensen,J.H.; Holmes,P.; Williams,W.L.; Harris,J.L. (Dep. Pathol., Univ. Texas Health Sci. Cent., 7703 Floyd Curl Drive, San Antonio, TX 78284, USA) *Am. J. Clin. Pathol., 69(6), 605-614 (1978)* En;en.

3110-U2 The tubular loop fermentor: oxygen transfer, growth kinetics, and design. Ziegler,H.; Meister,D.; Dunn,I.J.; Blanch,H.W.; Russell,T.W.F. (Chem. Eng. Lab., Swiss Fed. Inst. Technol., Zurich, Switzerland) *Biotechnol. Bioeng., 19(4), 507-525 (1977)* En;en.

3111-U2 Computer-aided material balancing for prediction of fermentation parameters. Cooney,C.L.; Wang,H.Y.; Wang,D.I.C. (Dep. Nutr. and Food Sci., Massachusetts Inst. Technol., Cambridge, MA 02139, USA) *Biotechnol. Bioeng., 19(1), 55-67 (1977)* En;en.

3112-U2 New instrument for the determination of resistance of microorganisms ta higher temperatures. Savel,J.; Prokopova,M. (Address not stated) *Kvasny Prum., 23(6), 127-128 (1977)* Cs;cs,de,en,ru.

3113-U2 The microbe thermistor. Mattiasson,B.; Larsson,P.O.; Mosbach,K. (Biochem. Div., Chem. Centre, Univ. Lund, Box 740, S-220 07 Lund 7, Sweden) *Nature, 268(5620), 519-520 (1977)* En.

3114-U2 Safety-device for spray-freezing of pathogenic microorganisms. Lickfeld,K.G.; Almert,U.; Menge,B. (Arbeitsgruppe Elektronenmikrosk., Inst. Med. Mikrobiol., Hufelandstr. 55, D-4300 Essen 1, GFR) *Microsc. Acta., 77(5), 441-444 (1976)* En;de,en.

3115-U2 Computer applications to fermentation processes. Weigand,W.A. (Sch. Chem. Eng., Purdue Univ., West Lafayette, IN 47907, USA). *In:* Annual reports on fermentation processes. Vol. 2. Perlman,D.(ed.). *Publ. by:* Academic Press, Inc. (London) Ltd., 24-28, Oval Rd, London NW1 7DX, UK,1978. p. 43-72. ISBN:0-12-040302-1. En.

3116-U2 The photobioluminometer, an instrument for the study of ecological factors affecting photosynthesis. Tchan,Y.T.; Chiou,A.C.M.; New,P.B.; Funnell,G.R. (Dep. Microbiol., Univ. Sydney, NSW 2006, Australia) *Microb. Ecol., 3(4), 327-332 (1977)* En;en.

3117-U2 A simple volumetric apparatus for measuring the rate of gas exchange in microorganisms. Takaoki,T. (Bot. Inst., Fac. Sci., Hiroshima Univ., Hiroshima 730, Japan) *Plant Cell Physiol., 19(1), 61-70 (1978)* En;en.

3118-U2 Computer analysis of nucleic acid regulatory sequences. Korn,L.J.; Queen,C.L.; Wegman,M.N. (Dep. Embryol., Carnegie Inst. Washington, 115 West University Parkway, Baltimore, MD 21210, USA) *Proc. Natl. Acad. Sci. USA, 74(10), 4401-4405 (1977)* En;en.

3119-U2 New instrument for the determination of resistance of microorganisms to higher temperatures. Savel,J.; Prokopova,M. (Jiboceske Pivovary, n.p., Ceske Budejovice, Czechoslovakia) *Kvasny Prum., 23(6), 127-128 (1977)* Cs;cs,de,en,ru.

3120-U2 [Photometer for determination of microbial biomass by the amount of ATP]. Erokhin,V.E.; Gordienko,A.P. (Inst. Biol. Southern Seas, Acad. Sci. Ukrainian SSR, Nakhimov prospekt 2, Sebastopol, USSR) *Mikrobiol. Zh., 38(4), 508-511 (1976)* Uk;en.

3121-U2 Computers and clinical microbiology perspectives and applications. Amsterdam,D. (Kingsbrook Jewish Med. Cent., Rutland Road and East 49th St., Brooklyn, NY 11203, USA) *Mt. Sinai J. Med., 44(1), 113-133 (1977)* En.

3122-U2 A low-cost microcomputer for the control of microbiological systems. Breame,A.J.; Spier,R.E. (Anim. Virus Res. Inst., Pirbright, Woking, Surrey GU24 0NF, UK) *Lab. Pract., 26(12), 957-960 (1977)* En.

3123-U2 Method of rapid decompression for disintegration of microbial cell walls. Rakitin,V.Yu.; Monosov,E.Z.; Prokofiev,N.V.; Grigorian,A.N. (All-Union Res. Inst. Synthesis of Proteins and Protein Compounds, Moscow, USSR) *Prikl. Biokhim.-Mikrobiol., 12(2), 278-282 (1976)* Ru,en,ru;

3124-U2 Specimen housing unit for cinemicrographic studies in the vertical plane. Wilkins,J.R.; Tynan,C.I.,Jr.; Boykin,E.H. (NASA, Langley Res. Cent., Hampton, VA 23665, USA) *Appl. Environ. Microbiol., 32(2), 294-297 (1976)* En;en.

3125-U2 Controlled cell disruption: a comparison of the forces required to disrupt different microorganisms. Kelemen,M.V.; Sharpe,J.E.E. (Microbiol. Sect., Dep. Pharm., Sch. Pharm., Univ. London, London, UK) *J. Cell Sci., 35, 431-441 (1979)* En;en.

3126-U2 Construction and operation of an anaerobic glove box. Nicolai-Scholten,M.-E. (Zentralinst. Versuchstiere, Hannover, GFR) *Zentralbl. Bakteriol. Parasitenkd. Infektionskr. Hyg., I Abt. A, 240(2), 235-245 (1978)* De;de,en.

3127-U2 Instrumentation in computer-aided fermentation. Swartz,J.R.; Cooney,C.L. (Dep. Nutr. and Food Sci., Massachusetts Inst. Technol., Cambridge, MA 02139, USA) *Process Biochem., 13(2), 3-4, 6-7, 24 (1978)* En.

3128-U2 Application of microcomputers in the study of microbial processes. Hampel,W.A. (Inst. Biochem. Technol. and Microbiol., Univ. Technol. Vienna, A-1060 Wien, Austria) *In:* Advances in biochemical engineering. Vol. 13. Mass transfer and process control. Ghose,T.K. et al. (eds.) *Publ. by:* Springer-Verlag KG,; Heidelberger Platz 3, Postfach, D-1000 Berlin 33, GFR 1979 p. 1-33 ISBN 3-540-09468-7. En;en.

3129-U2 Spore trapping under hot and humid conditions. Perrin,P.W. (Fish. and Environ. Canada, For. Direct., Western For. Prod. Lab., Vancouver, BC V6T 1X2, Canada) *Mycologia, 69(6), 1214-1218 (1977)* En.

3130-U2 A new 7-day spore sampler. Kramer,C.L.; Eversmeyer,M.G.; Collins,T.I. (Div. Biol., Kansas State Univ., Manhatten, KS 66506, USA) *Phytopathology, 66(1),*

60-61 (1976) En;en.

3131-U2 Computer-assisted structure manipulation. Studies in the biosynthesis of natural products. Varkony,T.H.; Smith,D.H.; Djerassi,C. (Dep. Chem., Stanford Univ., Stanford, CA 94305, USA) *Tetrahedron, 34(7), 841-853 (1978)* En;en.

3132-U2 Numerical analysis and computerized identification of the yeast genera *Candida* and *Torulopsis*. Campbell,I. (Dep. Brew. and Biol. Sci., Heriot-Watt Univ., Edinburgh EH1 1HX, UK) *J. Gen. Microbiol., 90(1), 125-132 (1975)* En;en.

3133-U2 Glycogen phosphorylase from *Dictyostelium*: a kinetic analysis by computer simulation. Kelly,P.J.; Kelleher,J.K.; Wright,B.E. (Boston Biomed. Res. Inst., Dep. Dev. Biol., 20 Staniford St., Boston, MA 02114, USA) *BioSystems, 11(1), 55-63 (1979)* En;en.

3134-U2 [Development of an apparatus for field study of fungal spore dispersion by water drops; application on *Kabatiella zeae, Septoria nodorum* and *Fusarium roseum*]. Rapilly,F.; Bonnet,A.; Aoucault,B. (Stn. Cent. Pathol., Cent. Natl. Rech. Agron., INRA, 78000 Versailles, France) *Ann. Phytopathol., 7(1), 45-50 (1975)* Fr;en,fr.

3135-U2 An apparatus for the mass collection of spores of entomopathogenic fungi. Hamalle,R.J.; Bell,J.V. (Bioenviron. Insect. Control Lab., ARS, USDA, Stoneville, MS 38776, USA) *J. Ga. Entomol. Soc., 11(3), 221-223 (1976)* En;en.

3136-U2 A new pellet soil-sampler and its use for the study of population dynamics of *Rhizoctonia solani* in soil. Henis,Y.; Ghaffar,A.; Baker,R.; Gillespie,S.L. (Dep. Plant Pathol. and Microbiol., Hebrew Univ. Jerusalem, Jerusalem, Israel) *Phytopathology, 68(3), 371-376 (1978)* En;en.

3137-U2 Breakage of yeast cells: large scale isolation of yeast mitochondria with a continuous-flow disintegrator. Deters,D.; Muller,U.; Homberger,H. (Biozent., Univ. Basel, CH-4056 Basel, Switzerland) *Anal. Biochem., 70(1), 263-267 (1976)* En;en.

3138-U2 A simple fermentor for growth of strictly anaerobic yeast in small volumes. Bieglmayer,C.; *Ruis,H. (Inst. Allg. Biochem., Univ. Wien, Wahringer Str. 38, A-1090 Wien, Austria) *Anal. Biochem., 83(1), 322-325 (1977)* En;en.

3139-U2 A cabinet designed for fungal growth studies and an example of its use in investigating the sensitivity of *Serpula lacrimans* to zinc oxychloride. Coggins,C.R.; Jennings,D.H. (Res. and Dev. Div., Rentokil Ltd., East Grinstead, Sussex, RH19 2JY, UK) *Int. Biodeterior. Bull., 11(2), 64-66 (1975)* En;de,en,es,fr.

3140-U2 Chamber for continuous microscopic observation of fungal growth. Falloon,R.E. (Agric. Bot. Dep., Univ. Coll. Wales, Aberystwyth, UK) *Trans. Br. Mycol. Soc., 68(3), 469-472 (1977)* En.

3141-U2 Large-scale preparation of yeast mitochondria exhibiting acceptor control. Labbe,P.; Chambon,H. (Lab. Biochim. Porphyrines, Univ. Paris VII, Tour 43, 2 Place Jussieu, 75005 Paris, France) *Anal. Biochem., 81(2), 416-424 (1977)* En;en.

3142-U2 A method to suck off spores aseptically from fungal cultures. Gisi,U. (Bot. Inst., Univ. Basel, Schonbeinstr. 6, CH-4056 Basel, Switzerland) *Phytopathol. Z., 84(4), 369-372 (1975)* De;en.

3143-U2 A cabinet designed for fungal growth studies and an example of its use in investigating the sensitivity of *Serpula lacrimans* to zinc oxychloride. Coggins,C.R.; Jennings,D.H. (Res. and Dev. Div., Rentokil Ltd., East Grinstead, Sussex, RH19 2JY, UK) *Int. Biodeterior. Bull., 11(2), 64-66 (1975)* En;de,en,es,fr.)

3144-U2 An easy and space-saving Petri dish method for testing fungi on *Heterodera avenae*. Juhl,M. (State Plant Pathol. Inst., Zool. Div., Lottenborgvej 2, DK 2800 Lyngby, Denmark) *Nematologica, 23(1), 122-123 (1977)* En.

3145-U2 Monte Carlo technique to simulate aflatoxin testing programs for peanuts. Whitaker,T.B.; Dickens,J.W.; Wiser,E.H. (Biol. and Agric. Eng. Dep., North Carolina Univ., Raleigh, NC 27607, USA) *J. Am. Oil Chem. Soc., 53(8), 545-547 (1976)* En;en.

3146-U2 Fermentation apparatus of the Microbiological Institute of the USSR Academy of Sciences for continuous cultivation of microorganisms. Rabotnova,I.L.; Perevezentzev,V.P.; Lirova,S.A. (Inst. Microbiol., Acad. Sci. USSR, Moscow, USSR) *Mikrobiologiya, 45(4), 729-732 (1976)* Ru;en,ru.

3147-U2 The tubular reactor as a simplified fermenter. Imrie,F.K.E.; Greenshield,R.N. (Tate and Lyle Ltd., Group Res. and Dev., Philip Lyle Memorial Res. Lab., Univ. Reading, PO Box 68, Reading, UK) *In:* Global impacts of applied microbiology. IVth International Conference, J.S. Furtado. *Publ. by:* Sociedade Brasileira de Microbiologia, Revista de Microbiologia, Sao Paulo, Brazil. Part 2, p.831-848. En;en.

3148-U2 A new 7-day spore sampler. Kramer,C.L.; Eversmeyer,M.G.; Collins,T.I. (Div. Biol., Kansas State Univ., Manhattan, KS 66506, USA) *Phytopathology, 66(1), 60-61 (1976)* En;en.

3149-U2 On-line computer optimization of chemostat productivity. Whaite,P.; Gray,P.P. (Sch. Biol. Technol., Univ. New South Wales, Kensington, NSW 2033, Australia) *Biotechnol. Bioeng., 19(4), 575-581 (1977)* En;en.

3150-U2 Critical evaluation of the pulsed-current resistance meter for detection of decay in wood. Piirto,D.D.; Wilcox,W.W. (Forest Prod. Lab., Univ. California, Berkeley, CA, USA) *For. Prod. J., 28(1), 52-57 (1978)* En;en.

3151-U2 A continuous culture apparatus for the mass production of algae. Palmer,F.E.; Ballard,K.A.; Taub,F.B. (Inst. Food Sci. and Technol., Coll. Fish., Univ. Washington, Seattle, WA 98105, USA) *Aquaculture, 6(4), 319-331 (1975)* En;en.

3152-U2 Computer method for predicting the secondary structure of single-stranded RNA. Studnicka,G.M.; Rahn,G.M.; Cummings,I.W.; Salser,W.A. (Mol. Biol. Inst., Univ. California, Los Angeles, CA 90024, USA) *Nucleic Acids Res., 5(9), 3365-3387 (1978)* En;en.

3153-U2 A computer analysis of cyanide stimulated oxygen uptake in *Chlorella protothecoides*. Clore,G.M.; Chance,E.M. (Dep. Biochem., University Coll. London, Gower St., London WC1E 6BT, UK) *FEBS Lett., 79(2), 353-356 (1977)* En.

3154-U2 Determination of photosynthetic rates for the marine algae *Fucus vesiculosus* and *Laminaria digitata*. King,R.J.; Schramm,W. (Bot. Sch., Univ. New South Wales, Kensington, NSW, Australia) *Mar. Biol., 37(3), 209-213 (1976)* En;en.

3155-U2 A pump system to collect water samples in shallow water without disturbance. Caljon,A. (Lab. Morfol., Syst. Ekol. Planten, Ledeganckstraat 35, B-9000 Gent, Belgium) *Bull. Soc. R. Bot. Belg., 111(1), 49-54 (1978)* En;en,fr.

3156-U2 A simple tide simulator for maintenance of marine (intertidal) algal and lichen cultures. Fletcher,A.; Jones,W.E. (Dep. Mar. Biol., Univ. Coll. North Wales, Menai Bridge, Anglesey LL59 5EH, UK) *Br. Phycol. J., 10(9), 263-264 (1975)* En.

3157-U2 An apparatus for measuring photosynthesis and respiration of intact large marine algae and comparison of results with those from experiments with tissue segments. Hatcher,B.G. (Dep. Biol., Dalhousie Univ., Halifax, NS B3H 4J1, Canada) *Mar. Biol., 43(4), 381-385 (1977)* En;en.

3158-U2 A computer orientated numerical coding system for algae. Whitton,B.A.; Diaz,B.M.; Holmes,N.T.H. (Dep. Bot., Univ. Durham, Durham DH1 3LE, UK) *Br. Phycol. J., 14(4), 353-360 (1979)* En;en.

3159-U2 A modified chamber for use with an oxygen electrode to allow measurement of incident illumination within the reaction suspension. Griffiths,D.J.; Thinh,L.-V.; Florian,Z. (Bot. Dep., James Cook Univ., North Queensland, Queensl. 4811, Australia) *Limnol. Oceanogr., 23(2), 368-372 (1978)* En;en.

3160-U2 Thirty liter tower-type pilot plant for the mass cultivation of light- and motion-sensitive planktonic algae. Juttner,F. (Inst. Chem. Pflanzenphysiol. Univ. Tubingen, Correnssstr. 41, D 74 Tubingen, GFR) *Biotechnol. Bioeng., 19(11), 1679-1687 (1977)* En;en.

3161-U2 Portable growth cabinet for algae and other small plants. Clayton,M.N.; Kajtar,M. (Bot. Dep., Monash Univ., Clayton, Vic. 3168, Australia) *Lab. Pract., 25(1), 26-27 (1976)* En.

3162-U2 A convenient method for the determination of carbon in marine 'net zooplankton'. Traganza,E.D.; Radney,J.C.; Graham,K.J. (Dep. Oceanogr., Naval Postgrad. Sch., Monterey, CA 93940, USA) *Mar. Chem., 4(2), 165-173 (1976)* En;en.

3163-U2 Automated irregular antibody screening on a modified 15-channel blood-grouping machine. Cotton,R.; Ray,T.C. (South West Reg. Blood Transfusion Cent., Bristol, UK) *Vox Sang., 31(6), 440-445 (1976)* En;en.

3164-U2 Computer analysis of presensitization and cross-reacting antibodies. Garovoy,M.R.; Myrberg,S.J.; Cooper,S.M.; Carpenter,C.B. (Tissue Typing Lab., Peter Bent Brigham Hosp., Boston, MA 02115, USA) *Transplant. Proc., 9(4), 1811-1813 (1977)* En.

3165-U2 The vehicle tray revisited: the use of the vehicle tray in assessing allergic contact dermatitis by a 24-hour application method. Iden,D.L.; *Schroeter,A.L. (Mayo Clin., 200 First St. SW, Rochester, MN 55901, USA) *Contact Derm., 3(3), 122-126 (1977)* En;en.

3166-U2 Instrumental sample-preparation technique for an automated immunohematology assay system. Blume,P.; Polesky,H.F.; Mathys,J.; Hanson,M.; Taylor,D.G. (Good Samaritan Hosp. and Med. Cent., Portland, OR 97210, USA) *Clin. Chem., 22(9), 1456-1458 (1976)* En;en.

3167-U2 A 48-well micro chemotaxis assembly for rapid and accurate measurement of leukocyte migration. Falk,W.; Goodwin,R.H.Jr.; Leonard,E.J. (Immunopathol. Sect., Lab. Immunobiol., Natl. Cancer Inst., NIH, Bethesda, MD 20205, USA) *J. Immunol. Methods, 33(3), 239-247 (1980)* En;en.

3168-U2 Studies on lymphocytes. XIII. Nuclear volume measurement as a rapid approach to estimate proliferative fraction. Sipe,C.R.; *Chanana,A.D.; Cronkite,E.P.; Gulliani,G.L.; Joel,D.D. (Div. Exp. Pathol., Med. Res. Cent., Brookhaven Natl. Lab., Upton, LI, NY 11973, USA) *Scand. J. Haematol., 16(3), 196-201 (1976)* En;en.

3169-U2 The demonstration and automatic quantitation of mast cells by image analyzing computer. Farnoush,A.; Nuki,K. (Dep. Oral Biol., Univ. Iowa, Iowa City, IO 52240, USA) *J. Periodont. Res., 10(5), 275-281 (1975)* En;en.

3170-U2 Microrespirometer chamber for determinations of viability in cell and organ cultures. Gabridge,M.G. (Dep. Microbiol., Sch. Basic Med. Sci., Univ Illinois, Urbana, IL 61801, USA) *J. Clin. Microbiol., 3(6), 560-565 (1976)* En;en.

3171-U2 Evaluation of the optical refractometer for lamb serum immunoglobulin estimation. Harker,D.B. (ICI Pharmaceut., Alderley House, Alderley Park, Macclesfield, Cheshire, UK) *Vet. Rec., 102(10), 213-214 (1978)* En;en.

3172-U2 A computer-based method for the quantification of leukocyte chemotaxis. Wine,A.C.; *Kellermeyer,R. (Dep. Med., Univ. Hosp. Cleveland, Cleveland, OH 44106, USA) *J. Lab. Clin. Med., 88(3), 487-490 (1976)* En;en.

3173-U2 Easily constructed soil percolation apparatus. Longden,A.R.; *Claridge,C.A. (Res. Div., Bristol Lab., Div. Bristol-Myers, Syracuse, NY 13201, USA) *Appl. Environ. Microbiol., 32(1), 188-189 (1976)* En;en.

See: 6, 88, 89, 96, 133, 136, 243, 244, 278, 325, 328, 344, 346, 378, 417, 423, 424, 428, 429, 432, 442, 448, 459, 461,

620, 717, 741, 746 761, 780, 811, 812, 838, 840, 841, 863, 914, 928, 981, 1058, 1059, 1067, 1092, 1147, 1213, 1377, 1397, 1420, 1440, 1606, 1696, 1924, 21 2186, 2213, 2222, 2223, 2485, 2535, 2536, 2548, 2815, 2927, 3006, 31 2198, 2530

Sterilization techniques

3174-U2 [Instrument sterilization with peracetic acid in a gynaecological outpatient clinic]. Flach,W. (Frauenklin. Stadtischen Klin., DDR-1115 Berlin-Buch, Wiltbergstr. 50, GDR) *Zentralbl. Gynakol., 98(6), 374-375 (1976)* De.

3175-U2 A new method of disinfection of the flexible fibrebronchoscope. Garcia de Cabo,A.; Martinez L.,P.L.; Checa P.,J.; Guerra S., F. (Hosp. Enferm. Torax, Victoria Eugenia, Sinesio Delgado 8, Madrid 29, Spain) *Thorax, 33(2), 270-272 (1978)* En;en.

3176-U2 A method for sterilizing ciliates without special equipment. Kosinski,R.J. (Dep. Biol., Texas A and M Univ., College Station, TX 77843, USA) *Bioscience, 29(5), 306-307 (1979)* En.

Purification and concentration techniques

3177-U2 Large scale production and concentration of bovine C-type virus. Guillemain,B.; Levy,D.; Lasneret,J.; Astier,T.; Boiron,M.; Parodi,A.L. (Groupe Etude Leucose Bovine, Ec. Natl. Vet., 94701 Alfort, France) *Vet. Microbiol., 1(2-3), 185-192 (1976)* En;en.

3178-U2 Comparative examination of methods for concentrating foot-and-mouth disease virus. Studen,M.; Panjevic,D.; Dobric,D. (Katedra Zaraze, Vet. Fak., 11000 Beograd, Yugoslavia) *Acta Vet. (Beograd), 26(4), 217-220 (1976)* En;en,sb.

3179-U2 A portable device for concentrating bacteriophages from large volumes of freshwater. Seeley,N.D.; Hallard,G.; Primrose,S.B (Dep. Biol. Sci., Univ. Warwick, Coventry CV4 7AL, UK) *J. Appl. Bacteriol., 47(1), 143-152 (1979)* En;en.

3180-U2 Rapid concentration of bacteriophages from aquatic habitats. Primrose,S.B.; Day,M. (Dep. Biol. Sci., Univ. Warwick, Coventry CV4 7AL, UK) *J. Appl. Bacteriol., 42(3), 417-421 (1977)* En;en.

3181-U2 A method for concentrating bacteriophages using polyethylene glycol. Krzywy,T.; Mulczyk,M. (ul. Czerska 12, 53-114 Wroclaw, Poland) *Arch. Immunol. Ther. Exp., 23(5), 725-731 (1975)* En;en.

3182-U2 Concentration of enteroviruses from large volumes of turbid estuary water. Sobsey,M.D.; Gerba,C.P.; Wallis,C.; *Melnick,J.L. (Dep. Virol. and Epidemiol., Baylor Coll. Med., Houston, TX 77030, USA) *Can. J. Microbiol., 23(6), 770-778 (1977)* En;en,fr.

3183-U2 [Apparatus for fractionation of immune ascites fluid in the concentrate.] Trofimov,N.M. (Belorussian Res. Inst. Epidemiol. and Microbiol., Minsk, USSR) *Vopr. Virusol., No. 6, 737-738 (1976)* Ru.

3184-U2 Large-scale production and concentration of human lymphoid interferon. Klein,F.; Ricketts,R.T.; Jones,W.I.; DeArmon,I.A.; Temple,M.J.; Zoon,K.C.; Bridgen,P.J. (Frederick Cancer Res. Cent., Frederick, MD 21701, USA) *Antimicrob. Agents. Chemother., 15(3), 420-427 (1979)* En;en.

3185-U2 [On the techniques of concentrating viruses from public treated water]. Drapeau,A.J.; Hoang van,H.; Dumoulin,P.P. (Ec. Polytech. Montreal, B.P. 6079, Succursale 'A', Montreal, Quebec H3C 3A7, Canada) *Ann. Biol., 15(11-12), 485-518 (1976)* Fr.

3186-U2 Preparation of crude protease powder from a Philippine strain of *Aspergillus oryzae* cultured in copra meal. Arguelles,A.L.L.; Baens-Arcega,L. (P.O. Box 774, Manila, 2801, Philippines) *Philipp J. Sci., 106(1), 11-23 (1977)* En;en.

3187-U2 Organic flocculation and hydroextraction: two new methods for phytoplankton concentration from lakes. Bitton,G.; Beaver,J.R.; Farrah,S.R. (Dep. Environ. Eng. Sci., Univ. Florida, Gainesville, FL 32611, USA) *Can. J. Fish. Aquat. Sci., 37(8), 1317-1320 (1980)* En;en,fr.

3188-U2 A simple device for concentration of parasite eggs, larvae, and protozoa. Zierdt,W.S. (Microbiol. Serv., Clin. Pathol. Dep., Clin. Cent., NIH, Bethesda, MD 20014, USA) *Am. J. Clin. Pathol., 70(1), 89-93 (1978)* En;en.

3189-U2 Antigen and antibody purification by immunoadsorption: elimination of non-biospecifically bound proteins. Zoller,M.; Matzku,S. (Inst. Nuklearmed., Deutsches Krebsforschungszent., Heidelberg, GFR) *J. Immunol. Methods, 11(3-4), 287-295 (1976)* En;en.

3190-U2 Purification and molecular characterization of two inhibitors of yeast proteinase B. Maier,K.; Muller,H.; Holzer,H. (Inst. Toxicol. und Biochem, Ges. Strahlen- und Umweltforsch., m.b.H. Munchen, D-8042 Neuherberg, GFR) *J. Biol. Chem., 254(17), 8491-8497 (1979)* En;en.

3191-U2 A rapid method for the purification of extrachromosomal DNA from eukaryotic cells. Shoyab,M.; Sen,A. (Meloy Lab., Inc., Springfield, VA 22151, USA) *J. Biol. Chem., 253(19), 6654-6656 (1978)* En;en.

3192-U2 Large-scale purification of two forms of active *lac* operator from plasmids. Kallai,O,B.; Rosenberg,J.M.; Kopla,M.L.; Takano,T.; Dickerson,R.E.; Kan,J.; Riggs,A.D. (Div. Chem. and Chem. Eng., California Inst. Technol., Pasadena, CA 91125, USA) *Biochim. Biophys. Acta, 606(1), 113-124 (1980)* En;en.

3193-U2 Purification of avian encephalomyelitis virus by ultracentrifugation in a non-linear cesium chloride gradient. Matsumoto,M.; Wangelin,J.R.; Murphy,M.L. (Sch. Vet.

Med., Oregon State Univ., Corvallis, OR 97331, USA) *Avian Dis., 22(3), 496-502 (1978)* En;en.

3194-U2 Use of polyethylene glycol and fluorocarbon for the purification of avian encephalomyelitis virus. Matsumoto,M.; Murphy,M.L. (Sch. Vet. Med., Oregon State Univ., Corvallis, OR 97331, USA) *Avian Dis., 21(2), 300-309 (1977)* En;en.

3195-U2 A simple method for rapid purification of avian sarcoma virus supercoiled DNA by selective precipitation of infected cell chromatin. Guntaka,R.V. (Dep. Microbiol., Coll. Physicians and Surgeons, Columbia Univ., New York, NY 10032, USA) *Anal. Biochem., 90(1), 256-261 (1978)* En;en.

3196-U2 Rhizomania disease of sugar beet in France. Attempts to develop a suitable purification procedure of beet necrotic yellow vein virus (BNYVV). Putz,C.; Kuszala,M. (Stn. Pathol. Veg., Cent. Rech. Agron. Colmar, INRA, 68021 Colmar, France) *Ann. Phytopathol., 10(2), 247-262 (1978)* Fr;en,fr.

3197-U2 Purification and characterization of the circular and linear forms of chrysanthemum stunt viroid. Palukaitis,P.; Symons,R.H.; (Dep. Biochem., Univ. Adelaide, Adelaide, South Australia, 5001, Australia) *J. Gen Virol., 46(2), 477-489 (1980)* En;en.

3198-U2 A simple and sensitive method for the purification and peptide mapping of proteins solubilized from densonucleosis virus with sodium dodecyl sulfate. Tijssen,P.; *Kurstak,E. (Fac. Med., Univ. Montreal, CP 6128, Succ.A, Montreal, Que., H3C 3J7, Canada) *Anal. Biochem., 99(1), 97-104 (1979)* En;en.

3199-U2 [A simple method for the separation and purification of the hepatitis B surface antigen from serum]. Desmet,G.; Boitieux,J.L. (Lab. Biochim. Pathol., Fac. Med., rue Frederic Petit, 80036 Amiens Cedex, France) *Clin. Chim. Acta, 74(1), 59-69 (1976)* Fr;en,fr.

3200-U2 Large-scale purification of hepatitis B surface antigen. Vnek,J.; Prince,A.M. (Lindsley F. Kimball Res. Inst. New York Blood Cent. New York, NY 10021, USA) *J. Clin. Microbiol., 3(6), 625-631 (1976)* En;en.

3201-U2 Purification of hepatitis type B antigen (HB Ag) from human plasma by heat coagulation. Rizzo,E.de (Tissue Culture and Control Lab., Virus Dep., Inst. Butantan, Caixa Postal 65, Sao Paulo, Brazil) *Rev. Inst. Med. Trop., Sao Paulo, 17(4), 225-229 (1975)* En;en,pt.

3202-U2 Use of trypsin for rapid and efficient purification of murine sarcoma and leukemia virus. Rucheton,M.; Blaas,D.; *Jeanteur,P. (Lab. Biochim., Cent Paul Lamarque, BP 5054, 34033 Montpellier Cedex, France) *Biochimie, 60(11-12), 1333-1337 (1978)* En;en.

3203-U2 Rapid stepwise solubilization and purification of type C retrovirus structural proteins by extraction with organic solvent. Olpin,J.L.; *Oroszlan,S. (Biol. Carcinogen. Program, Frederick Cancer Res. Cent., Frederick, MD 21701, USA) *Anal. Biochem., 103(2), 331-336 (1980)* En;en.

3204-U2 The purification of nuclease-free T4-RNA ligase. Moseman McCoy,M.I.; Lubben,T.H.; *Gumport,R.I. (Dep. Biochem., Sch. Basic Med. Sci., Univ. Illinois, Urbana, IL 61801, USA) *Biochim. Biophys. Acta, 562(1), 149-161 (1979)* En;en.

3205-U2 Simple methods to purify Sharka virus (plum pox). Kerlan,C.; Dunez,J.; Bellet,F. (Stn. Pathol. Veg., Cent. Rech. Bordaux, INRA, 33140 Pont de la Maye, France) *Ann. Phytopathol., 7(4), 287-297 (1975)* Fr;en,fr.

3206-U2 A modification of the polyethylene glycol technique for the purification of small quantities of tobacco mosaic virus. Bateman,J.G.; Chant,S.R. (Dep. Appl. Biol., Chelsea Coll. (Univ. London), London SW10, UK) *Microbios, 25(99), 33-43 (1979)* En;en.

3207-U2 Extraction of vaccinia virus and purification of vaccinia elementary body suspension with Mafron II. Topa,P.K.; Rao,K.V. (Indian Counc. Med. Res., Natl. Inst. Commun. Dis., 22 Alipore Rd., Delhi 6, India) *Indian J. Med. Res., 64(10), 1421-1431 (1976)* En;en.

3208-U2 A rapid procedure for the purification of ferredoxin from clostridia using polyethyleneimine. Schonheit,P.; Wascher,C.; Thauer,R.K. (Fachber. Biol., Mikrobiol., Philipps-Univ. Marburg, Auf den Lahnbergen, 3530 Marburg, GFR) *FEBS Lett., 89(2), 219-222 (1978)* En.

3209-U2 The purification and properties of the aspartate aminotransferase and aromatic-amino-acid aminotransferase from *Escherichia coli*. Powell,J.T.; *Morrison,J.F. (Dep. Biochem., John Curtin Sch. Med. Res., Australian Natl. Univ., PO Box 4, Canberra City, ACT 2601, Australia) *Eur. J. Biochem., 87(2), 391-400 (1978)* En;en.

3210-U2 A rapid high-yield purification procedure for the cyclic adenosine 3′,5′-monophosphate receptor protein from *Escherichia coli*. Boone,T.; Wilcox,G. (Dep. Bacteriol., Univ. California, Los Angeles, CA 90024, USA) *Biochim. Biophys. Acta, 541(4), 528-534 (1978)* En;en.

3211-U2 Rapid micromethod for the purification of *Escherichia coli* ribonucleic acid polymerase and the preparation of bacterial extracts active in ribonucleic acid synthesis. Gross,C.; Engbaek,F.; Flammang,T.; *Burgess,R. (McArdle Lab. Cancer Res., Univ. Wisconsin, Madison, WI 53706, USA) *J. Bacteriol., 128(1), 382-389 (1976)* En;en.

3212-U2 Purification and crystallization of $NADP^+$-specific isocitrate dehydrogenase from *Escherichia coli* using polyethylene glycol. Hackert,M.L.; Harris,B.A.; Poulsen,L.L. (Dep. Chem., Clayton Found. Biochem. Inst., Univ. Texas, Austin, TX 78712, USA) *Biochem. Biophys. Acta, 481(2), 340-347 (1977)* En;en.

3213-U2 The purification of restriction endonuclease EcoR I by precipitation involving polyethyleneimine. Bingham,A.H.A.; Sharman,A.F.; Atkinson,T. (Microbiol. Res. Establ., Porton Down, Salisbury, Wilts., UK) *FEBS Lett., 76(2), 250-256 (1977)* En.

3214-U2 A simple and rapid purification method for *Escherichia coli* DNA polymerase I. Rhodes,G.; Jentsch,K.D.; *Jovin,T.M. (Abt. Mol. Biol., Max-Planck-Inst. Biophys. Chem., D-3400 Gottingen-Nikolausberg, GFR) *J. Biol. Chem., 254(16), 7465-7467 (1979)* En;en.

3215-U2 Purification and properties of an L-glutaminase-L-asparaginase from *Pseudomonas acidovorans*. Davidson,L.; Brear,D.R.; Wingard,P.; Hawkins,J.; *Kitto,G.B. (Dep. Chem., Clayton Found. Biochem. Inst., Univ. Texas at Austin, TX 78712, USA) *J. Bacteriol., 129(3), 1379-1386 (1977)* En;en.

3216-U2 Large-scale purification and characterization of the exotoxin of *Pseudomonas aeruginosa*. Leppla,S.H. (Pathol. Div., US Army Med. Res. Inst. Infect. Dis., Fort Detrick, Frederick, MD 21701, USA) *Infect. Immun., 14(4), 1077-1086 (1976)* En;en.

3217-U2 A mild procedure to isolate the 34K, 35K and 36K porins of the outer membrane of *Salmonella typhimurium*. Nurminen,M. (Cent. Public Health Lab., Mannerheimintie 166, SF-00280 Helsinki 28, Finland) *FEMS Microbiol. Lett., 3(6), 331-334 (1978)* En.

3218-U2 A simple procedure for the purification of staphylococcal α-toxin. Dalen,A.B. (Mikrobiol. avd., MFH-bygget, N-5016 Haukeland sykehus, Norway) *Acta Pathol. Microbiol. Scand., Ser. B, 84(6), 326-332 (1976)* En;en.

3219-U2 A study on purification of staphylococcal enterotoxin A. Shibata,Y.; Morita,M.; Amano,Y.; Ishida,N. (Akita Prefectural Inst. Public Health, Akita, Japan) *Jap. J. Bacteriol., 31(2), 317-322 (1976)* Ja;en,ja.

3220-U2 A rapid procedure for purifying a restriction endonuclease from *Thermus thermophilus* (Tth I). Venegas,A.; Vicuna,R.; Alonso,A.; Valdes,F.; Yudelevich,A. (Lab. Bioquim., Dep. Biologia Cel., Univ. Catolica de Chile, Casilla 114-D, Santiago, Chile) *FEBS Lett., 109(1), 156-158 (1980)* En.

3221-U2 Ribosomal protein S1/S1A in bacteria. Visentin,L.P.; Hasnain,S.; Galin,W.; Johnson,K.G.; Griffith,D.W.; Wahba,A.J. (Cell Biochem. Group, Natl. Res. Counc., Canada, Ottawa, Ont. K1A 0R6, Canada) *FEBS Lett., 79(2), 258-263 (1977)* En.

3222-U2 A rapid method for the preparation of the three enzymes of bacitracin synthetase essentially free from other proteins. Roland,I.; Froyshov,O.; Laland,S.G. (Dep. Biochem., Univ. Oslo, Blindern, Norway) *FEBS Lett., 84(1), 22-24 (1977)* En.

3223-U2 Rapid, single-step purification of restriction endonucleases on cibacron blue F3GA-agarose. Baksi,K.; Rogerson,D.L.; *Rushizky,G.W. (Lab. Nutr. and Endocrinol., Natl. Inst. Arthritis, Metab., and Dig. Dis., NIH, Bethesda, MD 20014, USA) *Biochemistry (Wash.), 17(2), 4136-4139 (1978)* En;en.

3224-U2 A rapid method for the preparation of pure heavy enzyme of gramicidin S synthetase. Christiansen,C.; Aarstad,K.; Zimmer,T.L.; Laland,S.G. (Dep. Biochem., Univ. Oslo, Blindern, Oslo 3, Norway) *FEBS Lett., 81(1), 121-124 (1977)* En.

3225-U2 A simplified purification procedure for galactose oxidase. Tressel,P.; *Kosman,D.J. (Dep. Biochem., State Univ. New York, Buffalo, NY 14214, USA) *Anal. Biochem., 105(1), 150-153 (1980)* En;en.

3226-U2 A rapid procedure for isolating the photosystem I reaction center in a highly enriched form. Alberte,R.S.; Thornber,J.P. (Barnes Lab., Dep. Biol., Univ. Chicago, 5630 S. Ingleside Ave., Chicago, IL 60637, USA) *FEBS Lett., 91(1), 126-130 (1978)* En.

3227-U2 Immunopathogenicity of the purified cardiac protein antigens. Natu,S.M.; Chaturvedi,U.C. (Dep. Pathol. and Bacteriol., K.G. Med. Coll., Lucknow, India) *Indian J. Med. Res., 65(4), 554-562 (1977)* En;en.

3228-U2 A simplified procedure for the purification of Cl-inactivator from human plasma. Interaction with complement subcomponents Clr and Cls. Reboul,A.; Arlaud,G.J.; Sim,R.B.; Colomb,M.G. (DRF/Biochim., Cent. d'Etudes Nucleaires, 85 x, 38041 Grenoble-Cedex, France) *FEBS Lett., 79(1), 45-50 (1977)* En.

3229-U2 A simple method of lymphocyte purification from human peripheral blood. Alderson,E.; Birchall,J.P.; Owen,J.J.T. (Dep. Anat., Med. Sch., Univ. Newcastle upon Tyne, Newcastle upon Tyne, UK) *J. Immunol. Methods, 11(3,4), 297-301 (1976)* En;en.

3230-U2 Large scale purification of α_2-macroglobulin from human plasma. Song,M.K.; Adham,N.F.; Rinderknecht,H. (Dep. Med., Veterans Adm. Hosp., Sepulveda, CA 91343, USA) *Biochem. Med., 14(2), 162-169 (1975)* En;en.

3231-U2 A simple two-step method for purification of secretory IgA from human colostrum. Romero Piffiguer,M.D.; Riera,C.M. (Dep. Immunol. and Serol., Fac. Chem. Sci., Cordoba Natl. Univ., Cordoba, Argentina) *J. Immunol. Methods, 30(2), 153-159 (1979)* En;en.

3232-U2 Purification of peroxidase by isoelectric focusing. Use of ultrastructural localization of immunoglobulins. Rule,A.H.; Schaumburg-Lever,G.; Patel,R.P.; Schmidt-Ullrich,B.; Okun,M.R. (Dep. Dermatol., Tufts Univ. Sch. Med., Boston, MA 02111, USA) *Immunochemistry, 13(10), 819-821 (1976)* En;en.

3233-U2 Purification of human T and B cells by a discontinuous density gradient of percoll. Gutierrez,C.; Bernabe,R.R.; Vega,J.; Kreisler,M. (Dep. Immunol., Clin. Puerta de Hierro, Univ. Autonoma Madrid, Madrid, Spain) *J. Immunol. Methods, 29(1), 57-63 (1979)* En;en.

3234-U2 Rapid purification of peanut agglutinin by sialic acid-less fetuin-Sepharose column. Irle,C. (Dep. Pathol., Univ. Geneva, Fac. Med., Geneva, Switzerland) *J. Immunol. Methods, 17(1-2), 117-121 (1977)* En.

3235-U2 [Polyethylene glycol for concentration and purification of duck hepatitis virus]. Shalaby,M.A.; Reda,I.M.; Tantawi,H.H.; El Guindi,M.M. (Inst. Mikrobiol., Veterinarmed. Fak., Univ. Cairo, Cairo, Egypt) *Monatsch. Veterinarmed., 31(14), 534-536 (1976)* De;de,en,ru.

3236-U2 Optimal conditions for elution of hepatitis B antigen after absorption onto colloidal silica. Pilot,J.; Goueffon,S.; Keros,R.G. (Groupe INSERM, U131, Hop. Antoine Beclere, 92141 Clamart, France) *J. Clin. Microbiol., 4(3), 205-207 (1976)* En;en.

3237-U2 Concentration and purification of mycobacteriophages with polyethylene glycol 6000. Vajda,B.P.; Haber,K.J. (Microbiol. Res. Group Hungarian Acad. Sci., 1529 Budapest, Piheno ut 1, Hungary) *Am. Rev. Respir. Dis., 114(1), 245-248 (1976)* En,en.

3238-U2 Concentration and purification of tick-borne encephalitis virus grown in suspension of chick embryo cells. Heinz,F.; Kunz,C. (Inst. Virol., Univ. Vienna, 1095 Vienna, Austria) *Acta Virol., 21(4), 301-307 (1977)* En;en.

3239-U2 Preparation of chlorophyll a by a nonchromatographic method. Kis,P. (Inst. Phys. Chem., University, Wahringerstr. 42, Vienna, Austria) *Anal. Biochem., 96(1), 126-129 (1979)* En;en.

3240-U2 Rapid magnetic purification of rosette-forming lymphocytes. Owen,C.S.; Winger,L.A.; Symington,F.W.; Nowell,P.C. (Dep. Pathol., Univ. Pennsylvania Sch. Med., Philadelphia, PA 19104, USA) *J. Immunol., 123(4), 1778-1780 (1979)* En;en.

See: 969, 975, 1014, 1024, 1177, 1178, 1245, 1253, 1254, 1255, 2364, 2971, 3031, 1187

Comparisons of techniques

3241-U2 Evaluation of some methods for the laboratory examination of sputum. Tebbutt,G.M.; Coleman,D.J. (Reg. Public Health Lab., E. Birmingham Hosp., Birmingham B9 5ST, UK) *J. Clin. Pathol., 31(8), 724-729 (1978)* En;en.

3242-U2 European group for rapid laboratory viral diagnosis. Amsterdam symposium on rapid diagnosis. Gardner,P.S. (Dep. Virol., R. Victoria Infirm., Newcastle-upon-Tyne, NE1 4LP, UK) *J. Gen. Virol., 39(1), 201-203 (1978)* En.

3243-U2 Rapid diagnosis of viral infections. Smith,T. (Mayo Clin. and Found., Rochester, MN 55901, USA) *Ann. Intern. Med., 88(5), 708-709 (1978)* En.

3244-U2 Immune adherence hemagglutination: alternative to complement-fixation serology. Lennette,E.T.; Lennette,D.A. (Div. Virol., Joseph Stokes, Jr., Res. Inst., Child. Hosp. Philadelphia, Philadelphia, PA 19104, USA) *J. Clin. Microbiol., 7(3), 282-285 (1978)* En;en.

3245-U2 Comparative studies on the demonstration and formation of serum antibodies against Borna virus. Danner,K.; Luthgen,K.; Herlyn,M.; Mayr,A. (Inst. Med. Mikrobiol., Infekt.- und Seuchenmed., Fachber. Tiermed. Ludwig Maximilians Univ., Veterinarstr. 13, 8000 Munchen 22, GFR) *Zentralbl. Veterinarmed., B, 25(5), 345-355 (1978)* De;de,en,es,fr.

3246-U2 Comparative activity of immunofluorescent antibody and complement-fixing antibody in cytomegalovirus infection. Betts,R.F.; George,S.D.; Rundell,B.B.; Freeman,R.B.; Douglas,R.G.,Jr (Infect. Dis. Unit, Univ. Rochester Sch. Med., Rochester, NY 14642, USA) *J. Clin. Microbiol., 4(2), 151-156 (1976)* En;en.

3247-U2 Immunological tests for the detection of the hepatitis B surface antigen (HBsAg). Kimura,R.T.; Tachibana,C.F.; Cury,V.L.; Takeda,A.K. (Inst. Adolfo Lutz, Ave. Dr. Arnaldo 355, Caixa Postal 7027, 01000 Sao Paulo, Brazil) *Rev. Inst. Adolfo Lutz, 38(2), 83-86 (1978)* Pt;en,pt.

3248-U2 [Comparative evaluation of different immunological techniques for detection of HBsAg.] Castagnari,L.; Delia,S.; Russo,V.; Sebastiani,A. (Univ. degli Studi, Roma, Rome, Italy) *Ann. Sclavo, 18(1), 31-37 (1976)* It;en.

3249-U2 Comparative study of an enzyme immunoassay and of a radioimmunoassay for the detection of hepatitis B surface antigen. Crovari,P.; Rivano,R.; Bennicelli,C.; Zanacchi,P.; De Flora,S. (Ist. Ig., Univ. Genova, Via Pastore, I-16132 Genova, Italy) *Boll. Ist. Sieroter. Milan, 57(4), 434-443 (1978)* En;en,it.

3250-U2 [Immuno-enzymatic detection of HB antigens: a comparison of its sensitivity]. Pouliquen,A.; Sylvestre,R.; Ayed,K. (Cent. Secteur Transfusion Broussais-Cochin, 33 rue du Fg St.-Jacques, F75014 Paris, France) *Nouve. Presse Med., 7(44), 4056 (1978)* Fr.

3251-U2 [Comparative study of the different methods of detection of the antigen HBs. Results concerning 3006 sera]. Catelle,A. (Lab. Microbiol. CHU Brabois, Route de Neufchateau, 54500 Vandoeuvre-Les-Nancy, France) *Ann. Med. Nancy, 16(Nov.), 1051-1056 (1977)* Fr;en,fr.

3252-U2 Hepatitis B antigen card test III: sensitivity and specificity. Stevens,R.W.; McQuillan,G.; Dence,D.; Kelly,J. (Div. Lab. and Res., New York State Dep. Health, New Scotland Ave., Albany, NY 12201, USA) *Am. J. Clin. Pathol., 66(1), 59-64 (1976)* En;en.

3253-U2 Detection, by three techniques, of hepatitis B surface antigen (HBsAg) and determination of HBsAg and anti-HBs titres in patients with chronic liver disease. Chiaramonte,M.; Heathcote,J.; Crees,M.; *Sherlock,S. (Dep. Med., R. Free Hosp., Pond St., London NW3 2QG, UK) *Gut, 18(1), 1-6 (1977)* En;en.

3254-U2 The present state of HBsAg detection. Seidl,S. (Dep. Immunohaematol., Univ. Frankfurt am Main, Sandhofstr. 1, Postfach 730367, 6000 Frankfurt/Main, GFR) *Rev. Fr. Transfus. Immuno-Hematol., 21(2), 503-508 (1978)*

En;fr.

3255-U2 Hepatitis B detection systems: sensitivity and performance evaluation. Schultz,W.W. (Dep. Microbiol., Naval Med. Res. Inst., Bethesda, MD 20014, USA) *Mil. Med., 143(7), 471-474 (1978)* En;en.

3256-U2 Immunodiagnosis of hepatitis B. Critical review of tests involving antigen HBs for the detection of infected blood. Pillot,J. (Groupa INSERM U 131, Hop. Antoine Beclere, 157, rue de la Porte des Trivaux, 92140 Clamart, France) *Ann. Biol. Clin., 35(1), 3-17 (1977)* Fr;en,fr.

3257-U2 [A virological and epidemiological study of the infections due to *Herpesvirus hominis* (type 1 and 2)]. Gerna,G.; Torsellini Gerna,M.; Cattaneo,E.; Achilli,G.; Cereda,P.; Fiori,G.P. (Ist. Mal. Infettive, Univ. Pavia, Pavia, Italy) *G. Mal. Infett. Parassit., 31(2), 109-112 (1979)* Pt;en,pt.

3258-U2 Comparison of immunological methods in the diagnosis of influenza. Aymard,M. (Lab. Virol., Cent. Natl. de la Grippe, Fac. Med., 8, ave. Rockefeller, 69373 Lyon Cedex 2, France) *Bull. Inst. Pasteur, 75(4), 309-321 (1977)* En;en.

3259-U2 Comparative measurement of influenza virus antibodies in horse sera by the single radial hemolysis test and the hemagglutination inhibition test. Bockmann,J. (Inst. Virol. der Vet.-Med. Univ., Linke bahngasse 11, A-1030 Wien, Austria) *Zentralbl. Bakteriol. Hyg., I, Abt. A, 238(1), 1-8 (1977)* De;de,en.

3260-U2 Hemolysis-in-gel and neutralization tests for determination of antibodies to mumps virus. Grillner,L.; Blomberg,J. (Dep. Virol., Inst. Med. Microbiol., Univ. Goteborg, Goteborg, Sweden) *J. Clin. Microbiol., 4(1), 11-15 (1976)* En;en.

3261-U2 A comparison between the haemagglutination inhibition and complement fixation tests for Newcastle disease. Allan,W.H.; Gough,R.E. (Cent. Vet. Lab., New Haw, Weybridge, Surrey, UK) *Res. Vet. Sci., 20(1), 101-103 (1976)* En;en.

3262-U2 Comparison of the 'enzyme-linked immunosorbent assay' (ELISA) with the conventional bioassay on *Solanum demissum* A6 for detection of potato virus Y. Walter,C.; Sander,E. (Inst. Biol. II, Univ. Tubingen, Auf der Morgenstelle 28, D-7400 Tubingen, GFR) *Z. Pflanzenkr. Pflanzenschutz., 86(11), 662-666 (1979)* De;de,en.

3263-U2 Bovine rotavirus diagnosis: comparison of various antigens and serological tests. Mohammed,K.A.; Babiuk,L.A.; Saunders,J.R.; Acres,S.D. (Dep. Vet. Microbiol. and Vet. Infect. Dis. Organization, Univ. Saskatchewan, Saskatoon, Sask. S7N 0W0, Canada) *Vet. Microbiol., 3(2), 115-127 (1978)* En;en.

3264-U2 Comparison of electron microscopy, enzyme-linked immunosorbent assay, solid-phase radioimmunoassay, and indirect immunofluorescence for detection of human rotavirus antigen in faeces. Birch,C.J.; Lehmann,N.I.; Hawker,A.J.; Marshall,J.A.; Gust,I.D. (Virol. Dep., Fairfield Hosp. Communic. Dis., Yarra Bend Road, Fairfield 3078, Vic., Australia) *J. Clin. Pathol., 32(7), 700-705 (1979)* En;en.

3265-U2 Routine diagnosis of human rotaviruses in stools. Zissis,G.; Lambert,J.P.; De Kegel,D. (Dep. Microbiol., St. Pierre Hosp., Free Univ. Brussels, Brussels, Belgium) *J. Clin. Pathol., 31(2), 175-178 (1978)* En;en.

3266-U2 Comparison of five diagnostic methods for the detection of rotavirus antigens in calf faeces. Ellens,D.J. de Leeuw,P.W.; Straver,P.J.; van Balken,J.A.M. (Cent. Vet. Inst., Virol. Dep., 39 Houtribweg. Lelystad, Netherlands) *Med. Microbiol. Immunol., 166(1-4), 157-163 (1978)* En;en.

3267-U2 Rubella immunity: immunofluorescence test as a rapid and reliable alternative to the haemagglutination inhibition test. Schilt,U. (Inst. Hyg. und Med. Mikrobiol., Univ. Bern, CH-3010 Bern, Friedbuhlstr. 51, Switzerland) *Dtsch. Med. Wochenschr., 104(1), 22-24 (1979)* De;de,en.

3268-U2 A comparison of controlled pore glass chromatography and ultra centrifugation for detecting IgM antibody in congenital rubella infection. Robertson,P.W.; Giannikos,J.; Bishop,G.A. (Div. Microbiol., Prince of Wales Hosp., High St., Randwick, NSW 2031, Australia) *Med. J. Aust., 2(7), 293-295 (1978)* En;en.

3269-U2 Rubella antibody assay by the immunoperoxidase technique: comparison with the hemagglutination inhibition test for determination of immune status. Gerna,G.; Chambers,R.W. (Viral Diagn. Serv., Dep. Pathol., Georgetown Univ., Washington, DC 20007, USA) *J. Infect. Dis., 133(4), 469-472 (1976)* En;en.

3270-U2 [Rubella diagnosis with the single radial haemolysis test]. Heide,K.G.; Steinmann,J. (Abt. Hyg. Microbiol., Klin. Univ., Brunswiker Str. 2-6, D-2300 Kiel 1, GFR) *Dtsch. Med. Wochenschr., 103(44), 1726-1727 (1978)* De.

3271-U2 Comparison of radial haemolysis with haemagglutination inhibition in estimating rubella antibody. Appleton,P.N.; Macrae,A.D. (Public Health Lab., City and Sherwood Hosp., Nottingham NG5 1PH, UK) *J. Clin. Pathol., 31(5), 479-482 (1978)* En;en.

3272-U2 Evaluation and comparison of two assays for detection of immunity to rubella infection. Brody,J.P.; Binkley,J.H.; Harding,S.A. (Dep. Pathol., Univ. Virginia Sch. Med., Charlottesville, VA 22908, USA) *J. Clin. Microbiol., 10(5), 708-711 (1979)* En;en.

3273-U2 Comparison of methods for detecting specific IgM antibody in infants with congenital rubella. Pattison,J.R.; Jackson,C.M.; Hiscock,J.A.; Cradock-Watson,J.E.; Ridehalgh,M.K.S. (Dep. Med. Microbiol., King's Coll. Hosp. Med. Sch., Denmark Hill, London SE5 8RX, UK) *J. Med. Microbiol., 11(4), 411-418 (1978)* En;en.

3274-U2 Antibody assays for varicella-zoster virus: comparison of enzyme immunoassay with neutralization,

immune adherence hemagglutination, and complement fixation. Forghani,B.; Schmidt,N.J.; Dennis,J. (Viral and Rickettsial Dis. Lab., State Of California Dep. Health, Berkeley, CA 94704, USA) *J. Clin. Microbiol., 8(5), 545-552 (1978)* En;en.

3275-U2 Comparison of the efficiency of two methods for virus concentration from river water environment in a model experiment. Wallnerova,Z.; Simkova,A. (Res. Inst. Epidemiol. and Microbiol., 872 27 Bratislava, Sasinkova 9, Czechoslovakia) *J. Hyg. Epidemiol. Microbiol. Immunol., 22(2), 152-161 (1978)* En;de,en,es,fr.

3276-U2 Laboratory techniques for rapid diagnosis of viral infections: a memorandum. Anon. (Virus Dis. Div. Commun. Dis., WHO, 1211 Geneva 27, Switzerland) *Bull. WHO, 55(1), 33-37 (1977)* En.

3277-U2 Ultrastructural visualization of virus-induced surface antigens. Comparative studies of three immunohistochemical techniques. Morris,R.E.; Fritz,R.B. (Dep. Microbiol., Univ. Michigan, Sch. Med., Ann Arbor, MI, USA) *J. Histochem. Cytochem., 23(11), 855-862 (1975)* En;en.

3278-U2 Evaluation of the sensitivity of immunoprecipitation for serodiagnosis of influenza-bacterial pneumonias. Soloviev,V.D.; Neklyudova,I.; Fedorova,Yu.B.; Gumennik,A.E.; Kitsak,V.Ya.; Orlova,N.G. (Cent. Adv. Training Inst. Doctors, Moscow, USSR) *Vopr. Virusol., No.1, 29-32 (1977)* Ru;en,ru.

3279-U2 Manual for rapid laboratory viral diagnosis. /[WHO Offset Publication No. 47]. Almeida,J.D.; Atanasiu,P.; Bradley,D.W.; Gardner,P.S.; Maynard,J.; Schuurs,S.W.; Voller,A.; Yolken,R.H. (Wellcome Res. Lab. Beckenham, Kent, UK). *Publ. by:* WHO, Geneva, Switzerland, 1979, 48 pp. ISBN 92-4-170047-5 at Sw.Fr.6.En.Sw;Fr.En.

3280-U2 Comparison of selected diagnostic tests for detection of motile *Aeromonas* septicemia in fish. Eurell,T.E.; Lewis,D.H.; Grumbles,L.C. (Dep. Vet. and Microbiol., Coll. Vet. Med., Texas A and M Univ., College Station, TX 77483, USA) *Am. J. Vet. Res., 39(8), 1384-1386 (1978)* En;en.

3281-U2 Comparison of counter-immunoelectrophoresis with other serological tests in the diagnosis of human brucellosis. Diaz,R.; Maravi-Poma,E.; Rivero,A. (Dep. Bacteriol., Fac. Med., Univ. Navarra, Pamplona, Spain) *Bull. WHO, 53(4), 417-424 (1976)* En;en,fr.

3282-U2 A comparison of methods for the serological diagnosis of *Brucella ovis* infection. Cox,J.C.; Gorrie,C.J.R.; Nairn,R.C.; Ward,H.A. (Dep. Immunol., Commonw. Serum Lab., Poplar Road, Parkville, Vic. 3052, Australia) *Br. Vet. J., 133(5), 442-445 (1977)* En;en.

3283-U2 Rapid diagnosis of *Haemophilus influenzae* type b infections by latex particle agglutination and counterimmunoelectrophoresis. Ward,J.I.; Siber,G.R.; Scheifele,D.W.; Smith,D.H. (Cent. Dis. Control, Bact. Dis. Div., Bur. Epidemiol., Atlanta, GA 30333, USA) *J. Pediatr., 93(1), 37-42 (1978)* En;en.

3284-U2 A comparison of three rapid methods for the detection of β-lactamase activity in *Haemophilus influenzae*. Skinner,A.; Wise,R. (Dep. Med. Microbiol., Dudley Road Hosp., Birmingham B18 7QH, UK) *J. Clin. Pathol., 30(11), 1030-1032 (1977)* En;en.

3285-U2 Comparison of six methods of detecting *Salmonella typhimurium* infection of chickens. Williams,J.E.; Whittemore,A.D. (Southeast Poult. Res. Lab., ARS, USDA, PO Box 1072, Athens, GA 30601, USA) *Avian Dis., 20(4), 728-734 (1976)* En;en.

3286-U2 Rapid identification of group A, B, C and G beta-haemolytic streptococci by a modification of the coagglutination technique. Comparison of results obtained by co-agglutination, fluorescent antibody test, counterimmunoelectrophoresis, and precipitin technique. Arvilommi,H.; Uurasmaa,O.; Nurkkala,A. (Public Health Lab., SF-40620 Jyvaskyla 62, Finland) *Acta Pathol. Microbiol. Scand., Ser. B, 86(2). 107-111 (1978)* En;en.

3287-U2 [Determination of group B streptococci *(S. agalactiae)* using serological methods (precipitation, fluorescent antibody technique) and the CAMP test]. Remmers,J. (Inst. Hyg. und Technol. Fleisches TiHo 3 Hannover 1, Bischofsholer Damm 15, GFR) *Arch. Lebensmittelhyg., 26(3), 104-106 (1975)* De;de,en.

3288-U2 Comparison of RPR 'teardrop' card test, VDRL, and FTA-ABS test results on sera from persons with suspected yaws in Colombia. Hopkins,D.R.; Florez,D. (665 Huntington Ave., Boston, MA 02115, USA) *Br. J. Vener. Dis., 53(4), 218-220 (1977)* En;en.

3289-U2 Evaluation of the quantitative Rappaport rapid plate test in the diagnosis and treatment of syphilis. Ghinsberg,R.; Blumstein,G.; Meir,E.; Kafman,R. (Dr. Ephraim Rappaport Natl. Ref. Cent. for Treponematoses, Public Health Lab., Minist. Health, Abu Kabir, Tel Aviv, Israel) *Isr. Med. Sci., 14(10), 1079-1082 (1978)* En.

3290-U2 Interpretation of serological tests for syphilis. Kelly,D. (Dep. Public Health Supervision, Health and Welfare Build., 63-79 George St., Brisbane, Queensl. 4000, Australia) *Autralas. J. Dermatol., 19(3), 91-93 (1978)* En.

3291-U2 A new look at the serology of treponemal disease. O'Neill,P. (Dep. Clin. Microbiol., St. Thomas' Hosp., London SE1 7EH, UK) *Br. J. Vener. Dis., 52(5), 296-299 (1976)* En;en.

3292-U2 Qualitative evaluation of the Reagin screen test. Black,D.A.; Ray,P.E.; *Therrell,B.L. (Clin. Chem. Branch, Bur. Lab., Texas Dep. Health Resour., Austin, TX 78756, USA) *J. Clin. Microbiol., 4(1), 16-18 (1976)* En;en.

3293-U2 Comparison of the microhemagglutination assay for antibodies to *Treponema pallidum* and the automated fluorescent treponemal antibody-absorption test. Dowden,S.J.; Millian,S.J. (Dep. Virol. and Immunol., Serol. Sect. Bur. Lab., City of New York Dep. Health 455 First Av.

New York, NY 10016, USA) *Health Lab. Sci., 12(1), 20-27 (1975)* En;en.

3294-U2 Evaluation of methods to detect antibodies against *Aspergillus fumigatus*. Kurup,V.P.; Fink,J.N. (Res. Serv./151B, Veterans Adm. Cent., Wood, WI 53193, USA) *Am. J. Clin. Pathol., 69(4), 414-417 (1978)* En;en.

3295-U2 Comparison of methods for the determination of cell viability in stored baker's yeast. Park-kinen,E.; Oura,E.; Suomalainen,H. (Res. Lab. State Alcohol Monopoly (Alko), Box 350, SF-00101 Helsinki 10, Finland) *J. Inst. Brew., 82(5), 283-285 (1976)* En;en.
[Plating and staining techniques].

3296-U2 Computer-aided baker's yeast fermentations. Wang,H.Y.; Cooney,C.L.; Wang,D.I.C. (Dep. Nutr. and Food Sci., Masachusetts Inst. Technol., Cambridge, MA 02139, USA) *Biotechnol. Bioeng., 19(1), 69-86 (1977)* En;en.

3297-U2 The comparison of counterimmunoelectrophoresis with indirect haemagglutination test for detection of antibodies in experimentally infected guinea pigs with *Toxocara canis*. Enayat,M.; Pezeshki,M. (Pathobiol. Dep., Sch. Public Health, POB 1310, Univ. Tehran, Iran) *J. Helminthol., 51(2), 143-148 (1977)* En;en.

3298-U2 Simultaneous determination of albumin and IgG in serum and CSF: comparison of electroimmunodiffusion and immunonephelometry. Schuller,E.; Tompe,L.; Delasnerie,N. (Lab. Biochim. Systeme Nerveux, Hop. Salpetriere, 75634 Paris Cedex 13, France) *Biomed. Express, 23(5), 189-192 (1975)* En;en,fr.

3299-U2 Technical remarks on the detection of ABH polymorphism in human body fluids. Flori,A. (Ist. Med. Legale, Univ. Cattolica del Sacro Cuore, Via Pineta Sacchetti, 644, 00168 Rome, Italy) *Forensic Sci., 7(1), 91-94 (1976)* En.

3300-U2 Comparison of a simplified radioimmunoassay with a gas-liquid chromatographic method for urinary aldosterone. Etienne,J.; Pardaens,Y. (Serv. Biol. Clin., Hop. Univ. Brugmann, Univ. Libre Bruxelles, 1, avenue J.J. Croca, 1020 Bruxelles, Belgium) *Clin. Chim. Acta, 72(2), 201-204 (1976)* En;en.

3301-U2 Radioimmunoassay and automated enzyme immunoassay for the determination of thyroxine. Riesen,W.F.; Muacevic,B.; Jaggi,M. (Inst. Klin. Exp. Tumorforsch., Univ. Bern, Tiefenauspital, CH 3004 Bern, Switzerland) *J. Clin. Chem. Clin. Biochem., 16(7), 387-389 (1978)* En;de,en.

3302-U2 Determination of digoxin in serum. Comparison of radioimmunoassay and a heterogeneous enzyme immunoassay. Borner,K.; Rietbrock,N. (Insst. Klin. Chem. und Klin. Biochem., Freien Univ. Berlin, Klin. Steglitz, Hindenburgdamm 30, D-1000 Berlin 45, GFR) *J. Clin. Chem. Clin. Biochem., 16(6), 335-342 (1978)* De;de,en.

See: 151, 624, 1774, 1868, 2317, 1916, 2018, 1815, 2642

Author Index

Subject Index